Recent Progress in Astronomy

Volume II

Recent Progress in Astronomy
Volume II

Edited by **Audria Baldwin**

New York

Published by Callisto Reference,
106 Park Avenue, Suite 200,
New York, NY 10016, USA
www.callistoreference.com

Recent Progress in Astronomy: Volume II
Edited by Audria Baldwin

International Standard Book Number: 978-1-63239-540-5 (Hardback)

Printed in the United States of America.

Contents

Preface

We spend 365 nights every year staring at the beauty of the sky. Glaring at every possible light and wondering if there was a world beyond the one we live in. There are ample unanswered questions concerning our universe, galaxy and all the matter it is made up of. For centuries, scientists have studied all celestial bodies known to man such as moon, planets, stars, galaxies and more. The physics, chemistry, mathematics, origin and behavior of these subjects have been largely debated and there are countless theories explaining the evolution of the earth; The Big Bang Theory being the most popular and scientific of them all. But the truth is that no one knows the absolute truth. Endless researchers are trying to solve these mysteries as you read this.

The study of the universe or astronomy in scientific terms; is a vast field of knowledge. Astronomy is one of the most native sciences known to mankind and it has been a practice across the globe. Some of the most ancient civilizations such as Babylonians, Greeks, Chinese, Indians, Iranians and Maya had a tradition of observing the night sky through well-crafted methods and procedures. The Egyptian and Nubian monuments are widely considered as astronomical artifacts from the ancient cultures. As the science evolved, it fragmented into observational and theoretical branches during the 20th century.

This book comprises some of the most sought-after topics in the advancement of astronomy and shall prove to be helpful for everyone trying to get a detailed account of the discipline. We would like to thank our researchers and writers from all the parts of the world for their efforts and contributions.

Editor

A Principle Component Analysis of Galaxy Properties from a Large, Gas-Selected Sample

Yu-Yen Chang,[1,2,3] **Rikon Chao,**[2,4] **Wei-Hao Wang,**[3] **and Pisin Chen**[1,2,5]

[1] *Department of Physics and Graduate Institute of Astrophysics, National Taiwan University,*
  Taipei 10617, Taiwan
[2] *Leung Center for Cosmology and Particle Astrophysics, National Taiwan University,*
  Taipei 10617, Taiwan
[3] *Institute of Astronomy and Astrophysics, Academia Sinica, Taipei 10617, Taiwan*
[4] *Department of Electrical Engineering, National Taiwan University, Taipei 10617, Taiwan*
[5] *Kavli Institute for Particle Astrophysics and Cosmology, SLAC National Accelerator Laboratory,*
  Stanford University, Stanford, CA 94305, USA

Correspondence should be addressed to Yu-Yen Chang, rainswl@gmail.com

Academic Editor: Gary Wegner

Disney et al. (2008) have found a striking correlation among global parameters of H_I-selected galaxies and concluded that this is in conflict with the CDM model. Considering the importance of the issue, we reinvestigate the problem using the principal component analysis on a fivefold larger sample and additional near-infrared data. We use databases from the Arecibo Legacy Fast Arecibo *L*-band Feed Array Survey for the gas properties, the Sloan Digital Sky Survey for the optical properties, and the Two Micron All Sky Survey for the near-infrared properties. We confirm that the parameters are indeed correlated where a single physical parameter can explain 83% of the variations. When color (g-i) is included, the first component still dominates but it develops a second principal component. In addition, the near-infrared color (i-J) shows an obvious second principal component that might provide evidence of the complex old star formation. Based on our data, we suggest that it is premature to pronounce the failure of the CDM model and it motivates more theoretical work.

1. Introduction

One way to understand our universe is to gain insights into the structure of galaxies. For one thing, it helps to reveal the role of dark matter in their formation and dynamics. The cosmological model that consists of a cosmological constant and the cold dark matter in addition to the ordinary baryon matter and radiation (ΛCDM) has been able to successfully explain the evolution of the cosmic structures especially at large scales. By measuring CMB fluctuations COBE [4, 5], WMAP [6–9], Type Ia supernovae [10], and gravitational lensing [11], this model of cosmology has by now been established as the standard model of cosmology. Among its various implications, it suggests a hierarchical, bottom-up history of structure formation that evolves from small fluctuations to galaxies, clusters, and eventually superclusters. On the other side, the success of

ΛCDM has not been as clearcut at small scales. There still exist several inconsistencies between ΛCDM and the observations at small scales. For example, the simulations based on ΛCDM have revealed more number of galactic satellites [12–15] and less number of disk galaxies [16] than what have been observed. Besides, the degree of emptiness in the voids is also inconsistent between theory and observation [17]. While the ΛCDM can explain the galaxy rotational curves at large radii [18], the relatively higher density at the galactic core than that predicted by the ΛCDM, that is, the so-called cusp-core problem, is still unresolved [19].

The hierarchical galaxy formation scenario has been actively investigated previously (e.g., [20–22]), and the tight correlation between selected galactic parameters has been studied in the past [23–30]. Moreover, the overall correlation among all major galactic variables has also been thoroughly investigated recently by Disney et al. ([31]; hereafter D08) and Garcia-Appadoo et al. ([32]; hereafter G09). Based on 195 galaxies, the two studies found remarkably strong correlations between six galactic variables: the 90%-light radius (R_{90}), the 50%-light radius (R_{50}), the HI mass (M_{HI}), the dynamical mass (M_d), the luminosity (L), and the color. G09 found strong correlations among the galactic properties, including a common dynamical mass-to-light ratio within the optical radii, a correlation between surface brightness and luminosity, and a common HI surface density. D08 further showed that all the parameters can be included in one correlation and suggested that this is in conflict with the notion of the hierarchical structure formation scenario and the ΛCDM model.

Considering the importance of this issue, we set out to reinvestigate the analysis made by D08 and G09. We note that their claim actually involves two separate subissues. First, there is the issue of whether the galactic dynamics is indeed controlled by a single parameter. Second, even if this is true, there still remains the issue of whether such a fact necessarily concludes the failure of the hierarchical structure formation scenario and the cold dark matter model. In our view the positive conclusion of the former does not necessarily lead to the affirmative conclusion of the latter. In this paper we demonstrate tentative evidence that there may exist more than one principal component among the global parameters of galaxies with regard to the first issue. As for the second issue, we argue about the importance of the nongravitational baryon physics in galactic structure formation, which renders the naive extrapolation of the hierarchical structure formation scenario from cosmic to galactic scales questionable.

In order to scrutinize the correction issue, we performed a similar analysis on global parameters of galaxies with a significantly larger database and two additional parameters, L_J and M_J, based on the infrared J band. We include a total of 1022 galaxies from the Arecibo Legacy Fast Arecibo L-band Feed Array Survey (ALFALFA; [33–36]) for the atomic gas properties, and the Sloan Digital Sky Survey (SDSS; [37]) for the optical properties. Recently, Toribio et al. published two papers ([38, 39]; hereafter T11) with these two surveys. In T11, they assembled three samples to analyse the data. They argued that HI emission provides the most reliable way to determine the morphologies and conclude that color and HImass of gas-rich galaxies cannot be very closed. In addition, they also found the correlation between mass, radius, luminosity and (g-r) by principal component analysis (PCA; [40]) for one of their subsamples. Among our 1022 galaxies, 479 of them have also been detected by the Two Micron All Sky Survey (2MASS; [41]). We use these galaxies to study their near-infrared properties, which contain more stellar information.

Near-infrared studies of HI-selected galaxies had been attempted by various groups (e.g., G09; [42, 43]). Our motivation of adding the near-infrared data in our analysis is the following. The optical emission is sensitive to young stars. The near-infrared emission, on the other hand, is less affected by the young stars and is therefore a better tracer of the total stellar mass, which dominates the baryonic matter at galactic scales. We believe that the inclusion of the infrared data would provide us additional and independent information on the baryonic mass assembly history of the galaxies.

Through the PCA, we confirm that except the color, all other observables, from HI, optical to near-infrared bands, are highly correlated and dominated by a single parameter. This is true both in the optical and in the near-infrared bands and this confirms the results in D08. In addition, we also see a second component from color, especially in our near-infrared data. Because near-infrared color provides the information of integrated star formation history, it may be an evidence for complex formation history, whereas a valid structure formation theory needs to explain this observation.

The organization of this paper is as follows. We describe the data and sources in Section 2. Then several variables are adopted and applied to statistical analysis in Section 3. In Section 4, we summarize and discuss our results.

2. Data

Our samples are the blind 21 cm survey from ALFALFA. This selects gas rich galaxies which also contain low luminosity and low surface brightness galaxies in higher proportion than those in an optical selection. The optical data for this study come from the SDSS DR7, which covers 12,000 deg^2 for imaging and provides spectra of 930,000 galaxies. Here we briefly describe how we select SDSS counterparts to the ALFALFA sources, and we refer to G09 for more detailed discussion about identification.

The Arecibo Telescope has a beam size of $3\rlap.{''}5$ at 21 cm. Since the majority of the HI detections have S/N > 10 [34–36], we adopted a conservative searching radius of $10''$. We found 1233 SDSS galaxies that appear to be detected by ALFALFA. We then excluded Virgo galaxies, because their neutral hydrogen is known to undergo strong environmental impact (e.g., [44]). We also excluded galaxies whose half-light radii are too small ($<1''$) comparing to the SDSS resolution. These small half-light radii either arise from misidentifications (from stars) or would result in large uncertainties. We are left with a large sample of 1022 SDSS nearby galaxies, which have distances smaller than 254 Mpc. Among these galaxies, 889 have g magnitudes that are <18 and 120 have g = 18–20. The cumulative number counts of SDSS galaxies (e.g., [45]) are $\sim$60 deg^{-2} at g < 18 and $\sim$450 deg^{-2} at g = 18–20. Given our $10''$ search radius, we therefore expect at most 1.3 misidentifications in our 889 g < 18 galaxies and additional 1.3 misidentifications in our 120 g = 18 − −20 galaxies. These number are sufficiently small and misidentified galaxies should not impact our analyses.

The total stellar masses of galaxies are more directly reflected by the near-infrared observations. We added the data from 2MASS to our samples. 2MASS provides J, H, and K_s-band observations of the entire sky as well as a point source catalog and an extended source catalog. We use the 2MASS All-Sky Extended Source Catalog (XSC) to find the galaxies in the HI samples. To understand the quality of the identification, we first compared the 2MASS and SDSS coordinates of the 2MASS-detected galaxies. We found that more than 90% of them have offsets between 2MASS and SDSS that are well within their half-light radii. We visually inspected all galaxies with large offsets of $> 2''$ and found small number of cases that are likely misidentifications as well as ongoing mergers. We excluded these galaxies from our samples. Because 2MASS is shallower than SDSS, we are left with 481 reliably identified galaxies in the near-infrared. To match the same aperture of the color for SDSS and 2MASS catalog, 479 galaxies are left in our near-infrared subsamples.

From the ALFALFA-released catalog 1, 2, and 3, we obtained 1796 HI data, out of which 1265 galaxies could be found in the SDSS DR7 database. There are 32 galaxies within this set that are too faint in the optical to have reliable magnitudes and luminosities. Hence we finally used the remaining 1022 galaxies in our analyses. We have also analyzed the 195 galaxies of D08 and G09 from HIPASS [46–49] and have obtained similar results. However, since the definitions of the observational variables, such as those of the rotational velocity, are not entirely consistent between the two data sets, we only report on the results from ALFALFA. These 1022 galaxies can be regarded as a blind HI-selected sample. We deduced from the data six variables, which are R_{50} (half-light radius in units of pc), R_{90} (90%-light radius in units of pc), L_r (luminosity in r band in solar units), M_{HI} (HI mass in solar units), M_d (dynamical mass in solar units), and color (g-i).

The variables R_{50} and R_{90} represent the radii in the Petrosian system [45, 50, 51]. In SDSS, the parameters are `petroR50` and `petroR90`, respectively. Because the Petrosian system is based on circular objects, we corrected the radii with the major-to-minor axis ratios, which are the parameters `deVAB_r` or `expAB_r` in SDSS. To do this, we follow the result in [52] and [48, 49]. The authors fitted the corrections from circular to elliptical apertures as functions of major-to-minor axis ratios. We directly adopted their formulas for our corrections. By comparing the likelihoods of the de Vaucouleur and the exponential models, we chose the one with the larger likelihood between `deVAB_r` and `expAB_r`. L_r was derived from the Petrosian system and calculated from the Petrosian magnitude, `petroMag_r`. In order to have the same aperture for 2MASS, we use Petrosian color, which is from `petroMag_g` and `petroMag_i`. The variable M_d is calculated by $(\Delta V)^2 R_{90}/G$, in which ΔV is the rotational velocity from the HI spectra and corrected for the inclination with the major-to-minor axis ratio as what we did for R_{50} and R_{90}. M_{HI} is acquired directly from the ALFALFA database and it is derived from the HI flux. This estimation is based on the assumption that the masses from optical radius are proportional to dark matter halos.

To make sure that the masses from the HI measurement can describe the dynamical mass, we compare them with the masses based on the stellar velocity dispersion, which is $M_{dyn} = K_v \rho_0^2 R_e/G$, as in [1–3]. Here, ρ_0 is the velocity dispersion, R_e is the effective radius, $K_v = 73.32/[10.465 + (n - 0.95)^2 + 0.954]$, and n is the Sérsic index. In SDSS, the velocity dispersion can be calculated by $\rho = \rho_{ap}(8R_{ap}/R_e)^{0.066}$ [53], where $R_{ap} = 1''5$, R_e is from the best fitting circularized Sérsic profile, and ρ_{ap} is the SDSS measurement within the $3''$ fibers. The Sérsic data are from NYU Value-Added

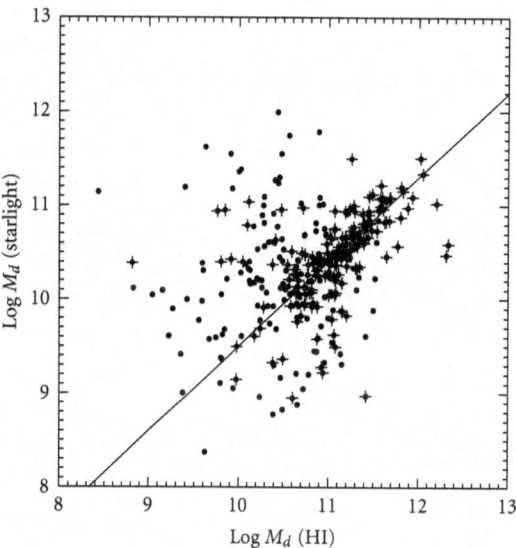

FIGURE 1: Comparison between dynamical masses from HI as D08/G09 and dynamical masses from starlight [1–3]. The dots are the 320 galaxies in ALFALFA/SDSS samples and cross symbols are the 164 galaxies from ALFALFA/SDSS/2MASS subsamples. The solid line indicates that the HI dynamical masses are on-average $\sim 4\times$ larger than the stellar dynamical masses.

Galaxy catalog (VAGC [54]). After matching, we have 320 of our 1022 main galaxy sample and 164 of our 479 2MASS subsample.

Figure 1 is the comparison between the HI and the stellar dynamical masses. There is an apparent sequence, indicating that the two masses trace similar dynamics for most of the galaxies. On the other hand, the HI dynamical masses are on-average $\sim 4\times$ larger than the stellar dynamical masses. This ratio is indicated by the solid line in Figure 1. This is not too surprising, since neutral gas (and R_{90}) can trace the dynamical mass to a larger distance in a dark matter halo. In our subsequent analyses, we adopted HI dynamical masses. This is consistent with the work of D08/G09 and provides a larger sample here.

Because there is high degeneracy between J, H, and K_s bands, we only chose J band to represent the near-infrared data. Therefore, we acquired R_J (half-light radius in J band in units of pc) and L_J (luminosity in J band in solar units) of the galaxies that overlap in the ALFALFA, SDSS, and 2MASS catalogs to gain a better insight into the stars of the galaxies. More specifically, L_J is derived from the magnitude in J band, which is the parameter `j_m_ext` in the 2MASS database. This magnitude is based on the extrapolation of the fit to the surface brightness profile. And R_J is the integrated half-flux radius of J band, which is the parameter `j_r_eff` in the 2MASS database. We adopted i-J for the color of the near-infrared subsample. The aperture of `petroMag_i` in SDSS is twice of the Petrosian radius, $2\times$`petroRad_i`. Thus we interpolate the J-band magnitude in 2MASS at the same aperture as SDSS. In our 481 2MASS subsamples, 479 of them have sufficient data at different apertures to interpolate

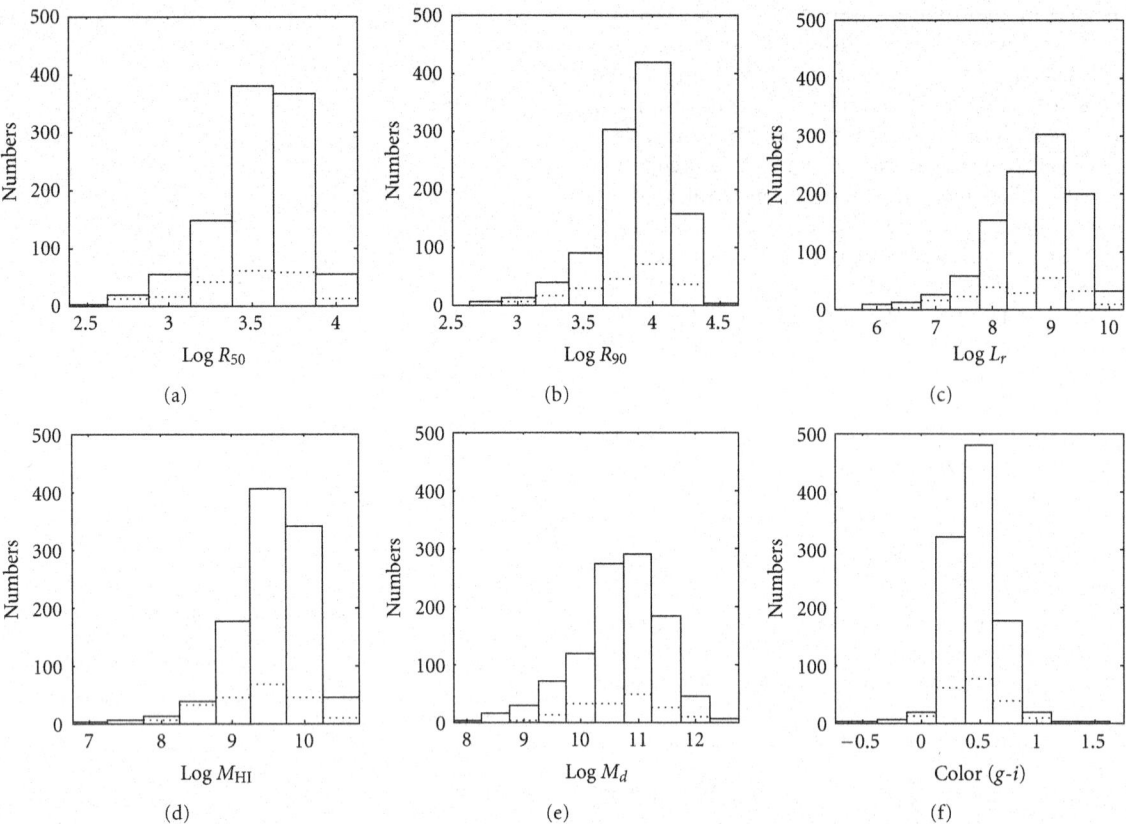

FIGURE 2: Distribution of the variables. The solid histograms are our samples and the dashed histograms are the samples of D08 and G09. We could only find 157 galaxies with rotational velocities for M_d in D08 and G09.

3. Methods and Results

Our sample of 1022 galaxies is not only larger than that in D08 and G09 but also covers broader ranges of size, luminosity, and mass (Figure 2). Despite that our sample is still dominated by L_* galaxies, the minimum value of L_r in our sample is much smaller than that of D08 and G09. As for the HI and the dynamical masses, although the median values of the two samples are similar, our sample contains a substantial amount of lower mass galaxies. Our sample also covers a broader range of the g-r color. Because of the larger sampling and the wider range of the galactic properties, our sample is in general more representative. However since 2MASS is shallower, our 2MASS subsample of 479 galaxies is relatively speaking less representative than that from SDSS. Even so, our 2MASS subsample is still substantially larger than that of D08 and G09.

It is important to investigate whether our HI-selected sample is biased against early type galaxies, since these galaxies are usually gas poor. To do so, we identify spheroidal and disk galaxies in our sample based on the morphology with a method similar to that in [55]. In the SDSS database, there are de Vaucouleur and exponential models for each galaxy. By comparing the likelihood and the fractions of the two models for our 1022 galaxies, we found that 804 galaxies are disk like and 218 galaxies are spheroidal like.

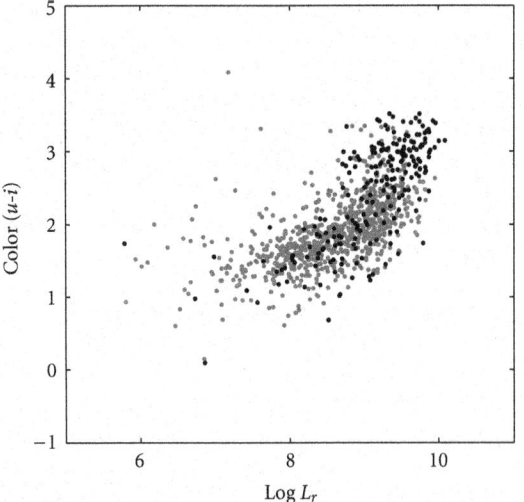

FIGURE 3: The u-i versus L_r diagram. Gray dots are disk galaxies that are better described by exponential models. Black dots are spheroidal galaxies that are better described by de Vaucouleur models.

the 2MASS magnitude. Therefore our final near-infrared subsample contains 479 galaxies.

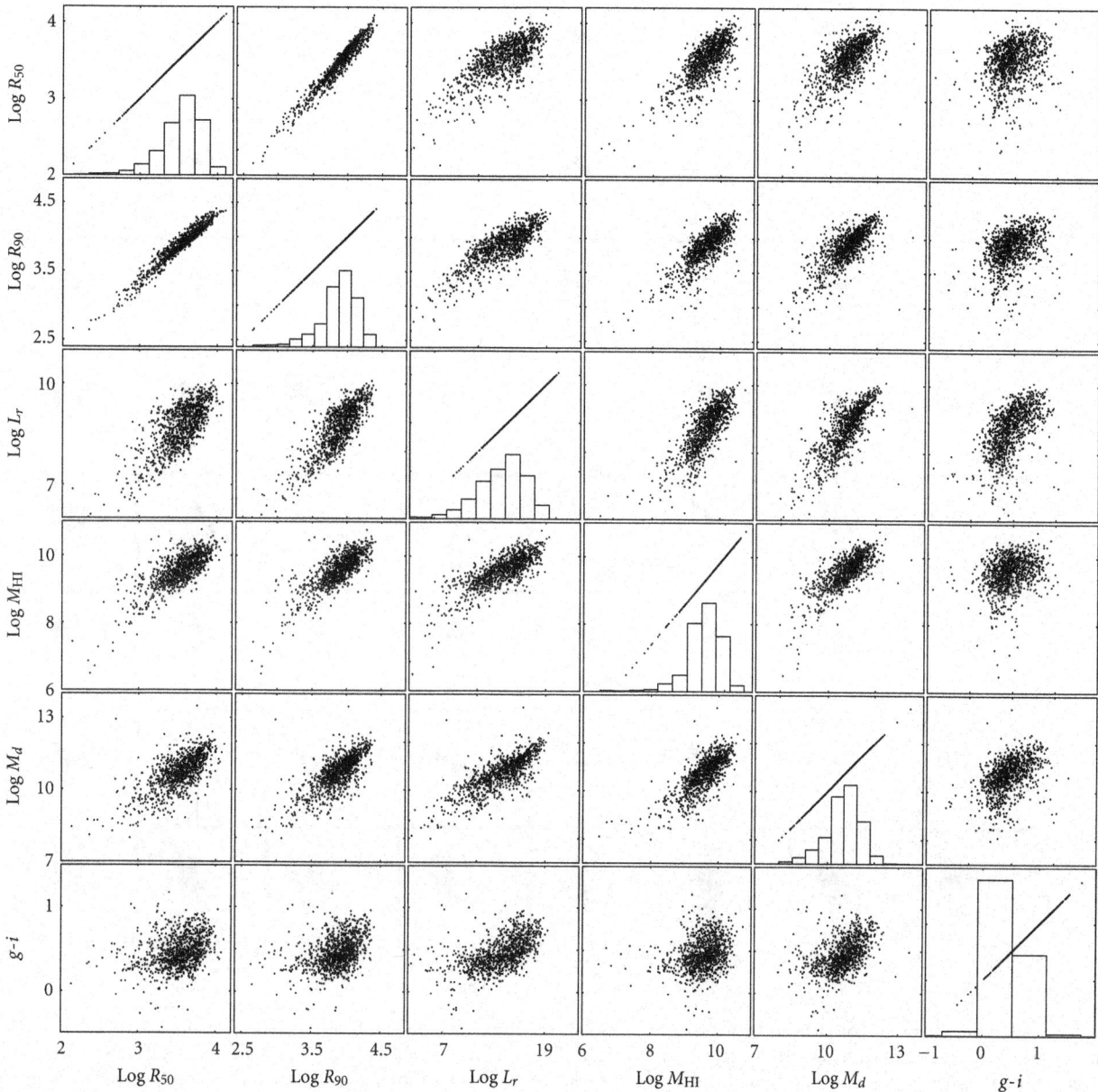

FIGURE 4: Scatter plots showing correlations between six measured variables. All the variables are in solar units and with logarithmic representation. The diagonal line is the histograms, which have vertical scales from 0 to 700.

In Figure 3, we show a color-luminosity diagram for our 1022 galaxies. The spheroidal galaxies are in general more luminous and redder than the disk galaxies. This is consistent with what we expect for elliptical and spiral galaxies. Most importantly, in the color-luminosity space, the spheroidal galaxies are redder than the blue cloud although they do not yet form a complete red sequence. Our sample thus appears to include a fair number of red and elliptical galaxies. Although the bias against extremely gas-poor galaxies can be hardly avoided here, fortunately, we found no major difference between these two types in our subsequent studies. We therefore believe that the omission of extremely gas-poor galaxies should not have caused major systematic bias in our analysis.

Relations among the key parameters can be inferred from the 1022 galaxies in both ALFALFA and SDSS. For instance, it is found that the half-light radius is proportional to the 90%-light radius ($R_{50} \propto R_{90}$; [56]); the r-band luminosity is proportional to the cubic power of the half-light radius ($L_r \propto R_{50}^3$; [57]); the HI mass is proportional to the square of the half-light radius ($M_{HI} \propto R_{50}^2$; [43, 58]); finally, the dynamical mass is proportional to the r-band luminosity ($M_d \propto L_r$; [25]). We found that all the correlations are evident even after including the near-infrared variables, except the color (Figures 4 and 5).

As a whole, the correlations between color and other variables are much weaker than other correlations. We tested various combinations of colors and found that g-i gives a

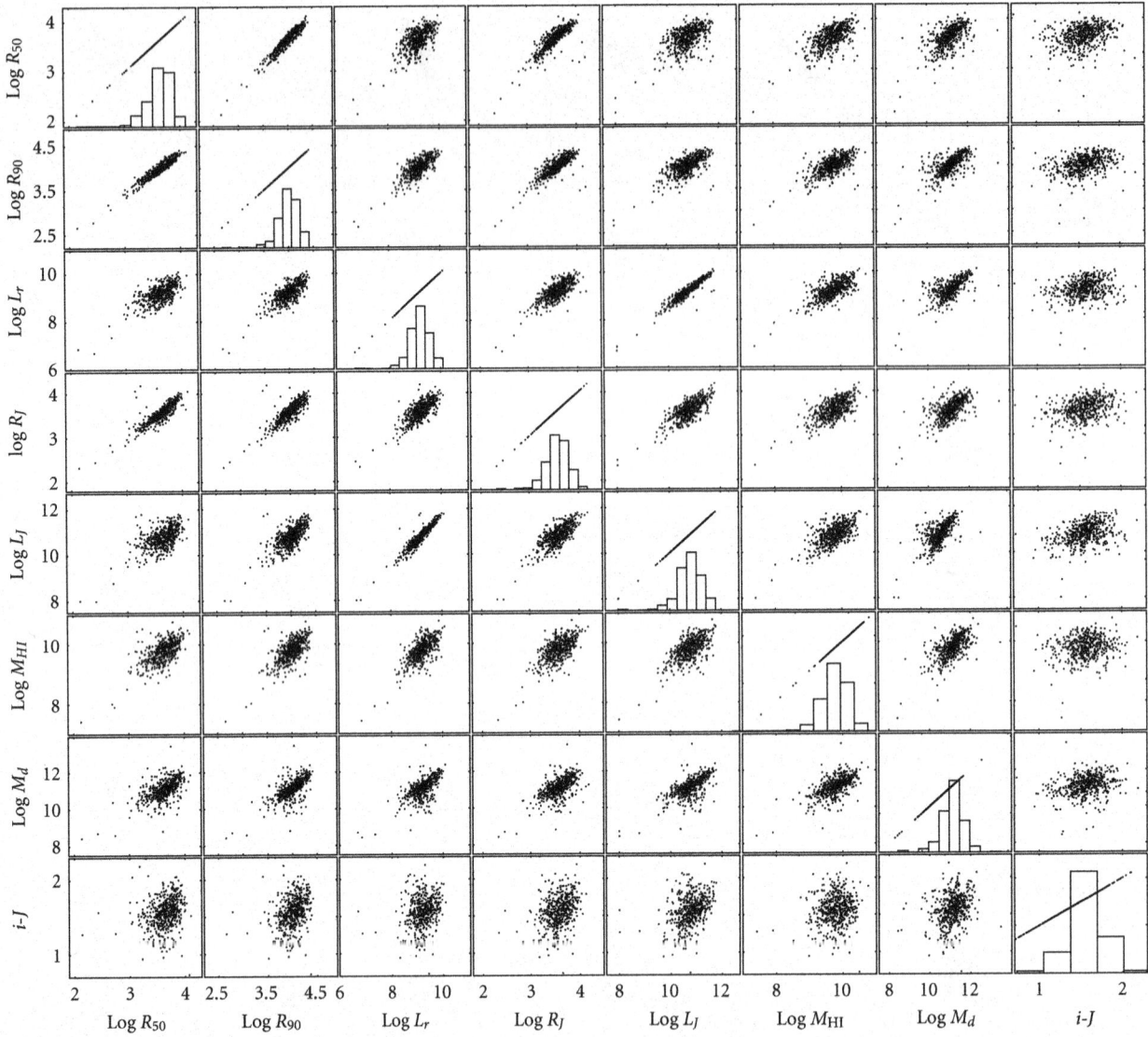

FIGURE 5: Scatter plots showing correlations between eight measured variables, including 2MASS data and reducing to 479 galaxies. All the variables are in solar units and with logarithmic representation. The diagonal line is the histograms, which have vertical scales from 0 to 350. There are small numbers of outliers in many of the plots. They are likely misidentifications or bad photometry, and they do not impact our analyses.

larger PCA correlation coefficient than other colors (e.g., g-r, adopted by D08). The reason could be the larger wavelength difference between g and i. Among the three 2MASS bands, the result based on J is somewhat better in the PCA, possibly because of the better signal-to-noise ratio. Hence, we adopted i-J for the color when we analyzed the 479 galaxies in 2MASS. Nevertheless, our analyses show that all the correlation coefficients are smaller than 0.7, indicating that they are not so highly correlated with other parameters. Intuitively, more luminous galaxies tend to be redder because their colors are dominated by older stars. In fact, the color is more complex than any other variable because of the bias introduced by the very luminous young stars.

We conducted PCA to find correlations among the variables. PCA typically produces a series of new variables

called the principal components, namely, PC1, PC2, and so forth. The correlations between these principal components and the original variables then reveal the general correlations between the particular variable and the others. In our case we found that the first principal component, PC1, is highly correlated with the six observational variables. We notice that while the color is less correlated to PC1, possibly because of the recent star formation, it is much more correlated with the second principal component, PC2. In addition to the investigation into the diagram of correlations, the eigenvalues of the correlation matrices of the original variables give quantitative information of the degree of correlations. For the 1022 galaxies based on SDSS, the eigenvalues of PC1 through PC6 are 4.29, 0.92, 0.39, 0.20, 0.18, and 0.02 (Figure 6), respectively, where the maximum possible value

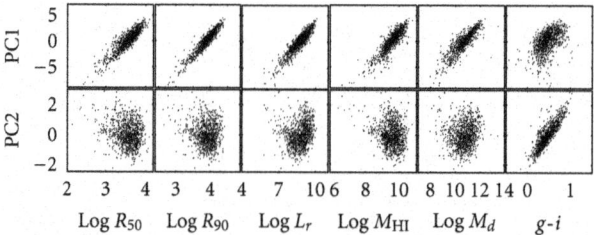

FIGURE 6: The PCA results for 1022 galaxies from ALFALFA and SDSS with colors. Here we only show the strongest ones, PC1 and PC2, because other principal components are not significant by PCA criterions. PC1 is well correlated with all the variables. In the first row, the color is still correlated with the other five variables and PC1. In the second row, the rightmost plot shows that the color is even more strongly correlated with a new principal component, PC2, which is not correlated with other variables. In this plot, $PC1 = 0.44 \log R_{50} + 0.46 \log R_{90} + 0.44 \log L_r + 0.43 \log M_{HI} + 0.42 \log M_d + 0.20(g\text{-}i)$ and $PC2 = -0.21 \log R_{50} - 0.13 \log R_{90} + 0.10 \log L_r - 0.22 \log M_{HI} + 0.04 \log M_d + 0.94(g\text{-}i)$.

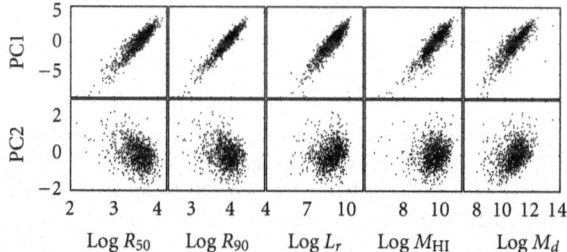

FIGURE 7: The PCA results for 1022 galaxies from ALFALFA and SDSS without colors. Here we only show the strongest ones, PC1 and PC2, because other principal components are not significant by PCA criterions. PC1 is well correlated with all the variables. In this plot, $PC1 = 0.45 \log R_{50} + 0.47 \log R_{90} + 0.45 \log L_r + 0.44 \log M_{HI} + 0.43 \log M_d$ and $PC2 = -0.59 \log R_{50} - 0.43 \log R_{90} + 0.28 \log L_r + 0.18 \log M_{HI} + 0.60 \log M_d$.

is 6. Based on common PCA criteria, eigenvalues larger than 1 are considered significant. We therefore only plot PC1 and PC2. Next we conducted PCA without color, and we found the eigenvalues of PC1 through PC5 to be 4.15, 0.41, 0.22, 0.20, and 0.02 (Figure 7), respectively, where the maximum possible value is 5.

The aforementioned observations confirm the finding of D08. All the observed parameters are tightly correlated with PC1. The eigenvalue of PC1 indicates that it can explain 83% of the variance in the data (when color is not included). Color itself forms a second principal component. This might be explained by the fact that the optical color tends to be strongly affected by recent star formation activities and thus carries extra information that is unrelated to the global formation history of the galaxies. We will come back to the issue of color in Section 4. D08 claimed that the strong dominance of PC1 implies a single physical parameter to govern the structure of galaxies. However, this may be an oversimplified view of galaxies due to limited observational

parameters. To test this, we extend the PCA to the near-infrared.

We included the 2MASS J-band radius and the luminosity to investigate the role of stars, which dominate the baryonic matter in galaxies. For the 479 galaxies detected by 2MASS, we found the eigenvalues of PC1 through PC8 to be 5.40, 1.01, 0.61, 0.45, 0.32, 0.16, 0.03, and 0.02 (Figure 8), respectively, where the maximum possible value is 8. Conducting PCA without color again, we found that the eigenvalues of PC1 through PC7 are 5.35, 0.62, 0.45, 0.31, 0.19, 0.05, and 0.02 (Figure 9), respectively, where the maximum possible value is 7.

The overall trends in the previous near-infrared PCA are similar to those in the optical PCA. When color is not included, PC1 dominates and can explain 76% of the variance in the data. The importance of PC2 slightly increases from 8% in the optical case to 9% in the near-infrared case here. When color is included, it forms another principal component by itself. These again confirm the observations of D08 that only one physical parameter governs the dynamics of galaxies.

A subtle but surprising difference between the optical and near-infrared PCAs is the behavior of color. In the optical PCA in Figure 6, although the $g\text{-}i$ color forms a second principal component, it still weakly correlates with other parameters and it is part of the first principal component. On the other hand, in the near-infrared PCA in Figure 8, the $i\text{-}J$ color almost does not involve in the first principal component and itself forms a second component that is almost independent of other parameters. A potential issue here is the combination of SDSS i and 2MASS J. It is the reason we choose consistent apertures in SDSS and 2MASS to calculate the color. To test whether other minor differences between the two photometric system hamper the correlation, we replaced the $i\text{-}J$ color with the pure 2MASS $J\text{-}K$ color, and we found consistent results. In addition, the median errors in g and i for the 1022 SDSS galaxies are 0.015 and 0.018, respectively, and the median errors in i and J for the 479 2MASS galaxies are 0.011 and 0.075, respectively. These translate to typical color errors of 0.023 in $g\text{-}i$ and 0.076 in $i\text{-}J$. Both values are significantly smaller than the color dynamical ranges shown in Figures 4 and 5, meaning that the distribution of the data is not dominated by measurement errors. We therefore believe that the lack of correlation between the near-infrared color and other parameters is real. We will discuss more on this result in Section 4.

4. Discussion and Conclusion

For the 195 galaxies analyzed in D08 and G09, the correlations among the parameters R_{50}, R_{90}, L_g, M_{HI}, and M_d are obvious. By selecting significant 5 times larger, 1022 overlapping samples from SDSS and ALFALFA, we also performed the PCA and confirmed through the high eigenvalues of the correlation matrices that the correlations are similarly strong. It follows that the radius, luminosity, HI mass, and dynamical mass of those chosen galaxies are tightly correlated. D08 shows that the basic parameters of galaxies

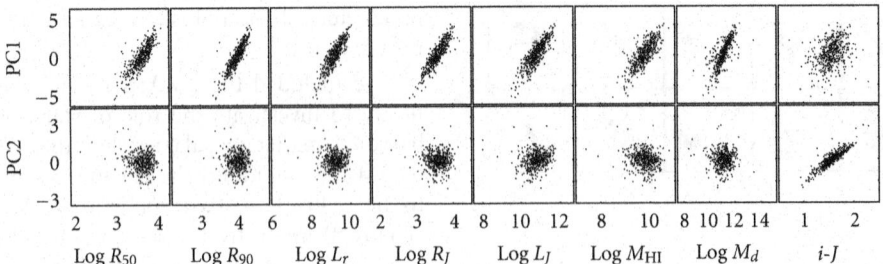

FIGURE 8: The PCA results for 479 galaxies from ALFALFA, SDSS, and 2MASS with colors. Here we only show the strongest ones, PC1 and PC2, because other principal components are not significant by PCA criterions. PC1 is well correlated with all the variables. The color is correlated with other variables and PC1 in the first row as well as strongly correlated with PC2 as in Figure 6. In this plot, PC1 $= 0.37 \log R_{50} + 0.39 \log R_{90} + 0.39 \log L_r + 0.39 \log R_J + 0.38 \log L_J + 0.36 \log M_{\mathrm{HI}} + 0.34 \log M_d + 0.12(i-J)$ and PC2 $= -0.22 \log R_{50} - 0.18 \log R_{90} - 0.05 \log L_r + 0.08 \log R_J + 0.18 \log L_J - 0.08 \log M_{\mathrm{HI}} - 0.04 \log M_d + 0.93(i-J)$.

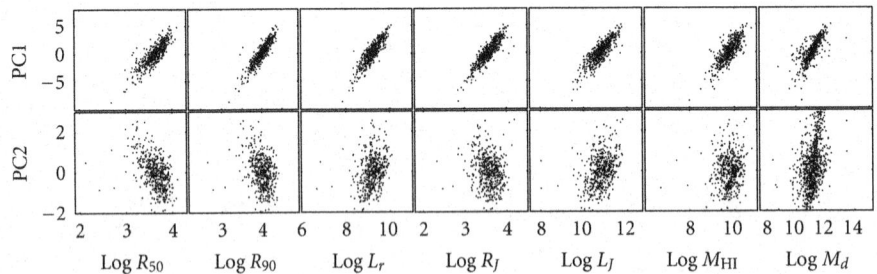

FIGURE 9: The PCA results for 479 galaxies from ALFALFA, SDSS, and 2MASS without colors. Here we only show the strongest ones, PC1 and PC2, because other principal components are not significant by PCA criterions. PC1 is well correlated with all the variables, as in Figure 7. In this plot, PC1 $= 0.38 \log R_{50} + 0.40 \log R_{90} + 0.49 \log L_r + 0.49 \log R_J + 0.38 \log L_J + 0.36 \log M_{\mathrm{HI}} + 0.34 \log M_d$ and PC2 $= -0.59 \log R_{50} - 0.39 \log R_{90} + 0.36 \log L_r - 0.22 \log R_J + 0.46 \log L_J + 0.14 \log M_{\mathrm{HI}} + 0.29 \log M_d$.

are highly correlated and that there exists only one dominant principal component. From our studies, we know that this is true in both optical and near-infrared bands. Based on the hierarchical galaxy formation assumption, the dark matter halo formation has been well studied through the merger tree process (e.g., [20, 21, 59–64]). D08 believes that this scenario may not be consistent with the simple relation between all basic galactic parameters because the processes of merger would break the original galactic structures. Although our analysis has confirmed D08's finding, we do not find it compelling or sufficient reason to reject ΛCDM. On the other hand, the dynamical mass-to-light ratio enclosed by the optical radius that strongly anticorrelates with surface brightness has been studied [65–67]. It is not shown in our data. It may be caused by the selection effects from L_* galaxies and insufficient dynamic range in surface brightness with SDSS and 2MASS data. Nevertheless, the strong correlations between parameter in D09 and our gas-selected samples are obvious in both optical and near-infrared data.

However, the color appears to be much less correlated, as T11 mentioned. This indeed complicates the situation and may be explained by a more sophisticated theory that would include, for example, the influence of recent star formation and very luminous young stars. Indeed these studies have already been pursued by many authors (e.g., [55, 59, 68–71]. D08 suggests that the optical color (g-r in their case) consists of two components: the systematic component that

correlates with other parameters (and therefore involves in the first principal component), and the rogue component that is more or less random (and forms the second principal component). It is tempting to assume that the systematic component comes from the established stellar populations and is related to the global formation history of the galaxies, while the random component is related to the ongoing star formation activities and can be short-lived events in the formation history. Surprisingly, our near-infrared analyses suggest a different story. In our near-infrared PCA, the color is even more uncorrelated with other parameters, comparing to the optical color.

One may believe that optical colors like g-i can be more strongly affected by recent star formation, whereas the i-J color may be a better tracer of integrated star formation history because it is less affected by ongoing star formation and dust extinction. If this is correct, the result that g-i is more related to other dynamical properties suggests that the recent star formation history is more controlled by the current dynamical structure galaxies. On the other hand, our result that i-J is unrelated to other dynamical properties suggests that the formation history of the old stellar components is a more chaotic process. In other words, the stronger second component may be an indication of complex formation and evolution history of galaxies. If this is the correct interpretation of our data, then the evidence of complex merger history predicted by the CDM hierarchical formation model had been hidden in the near-infrared colors and was not revealed by D08. However, we believe that this has to be tested by more observations and simulations, and it is premature to suggest the failure of the CDM model.

To sum up, we believe that the 5 times larger sample size strengthens the claim in D08 that there is one dominant parameter in galactic structure. In addition, our near-infrared analyses also provide additional insight into the color of galaxies. The tight correlations and the uncorrelated color part between the parameters in the optical and near-infrared data provide potentially powerful observational constraints on the hierarchical structure formation theory and any other cosmology models.

Acknowledgments

This work is supported by Taiwan National Science Council under Project no. NSC97-2112-M-002-026-MY3 (P. Chen and Y.-Y. Chang), NSC98-2112-M-001-003-MY2 and NSC99-2112-M-001-012-MY3 (W.- H. Wang and Y.-Y. Chang), and by US Department of Energy under Contract no. DE-AC03- 76SF00515. The authors also thank the support of the National Center for Theoretical Sciences of Taiwan. They are grateful to L. Lin, H. Hirashita, M. Disney, P. Kroupa, and H.-W. Rix for useful comments and discussion. The Arecibo Observatory is part of the National Astronomy and Ionosphere Center, which is operated by Cornell University under a cooperative agreement with the National Science Foundation. Funding for the SDSS and SDSS-II has been provided by the Alfred P. Sloan Foundation, the Participating Institutions, the National Science Foundation, the US Department of Energy, the National Aeronautics and Space Administration, the Japanese Monbukagakusho, the Max Planck Society, and the Higher Education Funding Council for England. The SDSS website is http://www.sdss.org/. The SDSS is managed by the Astrophysical Research Consortium for the participating institutions. The participating institutions are the American Museum of Natural History, Astrophysical Institute Potsdam, University of Basel, University of Cambridge, Case Western Reserve University, University of Chicago, Drexel University, Fermilab, the Institute for Advanced Study, the Japan Participation Group, Johns Hopkins University, the Joint Institute for Nuclear Astrophysics, the Kavli Institute for Particle Astrophysics and Cosmology, the Korean Scientist Group, the Chinese Academy of Sciences (LAMOST), Los Alamos National Laboratory, the Max-Planck-Institute for Astronomy (MPIA), the Max-Planck-Institute for Astrophysics (MPA), New Mexico State University, Ohio State University, University of Pittsburgh, University of Portsmouth, Princeton University, the United States Naval Observatory, and the University of Washington. This paper makes use of data products from the Two Micron All Sky Survey, which is a joint project of the University of Massachusetts and the Infrared Processing and Analysis Center/California Institute of Technology, funded by the National Aeronautics and Space Administration and the National Science Foundation.

References

[1] E. N. Taylor, M. Franx, J. Brinchmann, A. Van Der Wel, and P. G. Van Dokkum, "On the masses of galaxies in the local universe," *Astrophysical Journal Letters*, vol. 722, no. 1, pp. 1–19, 2010.

[2] D. A. Wake, M. Franx, and P. G. van Dokkum, "Which galaxy property is the best indicator of its host dark matter halo properties?" http://arxiv.org/abs/1201.1913.

[3] D. A. Wake, P. G. van Dokkum, and M. Franx, "Revealing velocity dispersion as the best indicator of a galaxy's color, compared to stellar mass, surface mass density or morphology," http://arxiv.org/abs/1201.4998.

[4] G. F. Smoot, C. L. Bennett, A. Kogut et al., "Structure in the COBE differential microwave radiometer first-year maps," *Astrophysical Journal Letters*, vol. 396, no. 1, pp. L1–L5, 1992.

[5] C. L. Bennett, A. Kogut, G. Hinshaw et al., "Cosmic temperature fluctuations from two years of COBE differential microwave radiometers observations," *Astrophysical Journal Letters*, vol. 436, no. 2, pp. 423–442, 1994.

[6] C. L. Bennett, M. Halpern, G. Hinshaw et al., "First-year Wilkinson Microwave Anisotropy Probe (WMAP) observations: preliminary maps and basic results," *Astrophysical Journal*, vol. 148, no. 1, pp. 1–27, 2003.

[7] C. L. Bennett, R. S. Hill, G. Hinshaw et al., "First-year Wilkinson Microwave Anisotropy Probe (WMAP) observations: foreground emission," *Astrophysical Journal*, vol. 148, no. 1, pp. 97–117, 2003.

[8] E. Komatsu, J. Dunkley, M. R. Nolta et al., "Five-year wilkinson microwave anisotropy probe observations: cosmological interpretation," *Astrophysical Journal*, vol. 180, no. 2, pp. 330–376, 2009.

[9] J. Dunkley, D. N. Spergel, E. Komatsu et al., "Five-year wilkinson microwave anisotropy probe (WMAP) observations: bayesian estimation of cosmic microwave background polarization maps," *Astrophysical Journal Letters*, vol. 701, no. 2, pp. 1804–1813, 2009.

[10] A. G. Riess, A. V. Filippenko, P. Challis et al., "Observational evidence from supernovae for an accelerating universe and a cosmological constant," *Astronomical Journal*, vol. 116, no. 3, pp. 1009–1038, 1998.

[11] F. Zwicky, "On the masses of nebulae and of clusters of nebulae," *Astrophysical Journal*, vol. 86, p. 217, 1937.

[12] B. Moore, S. Ghigna, F. Governato et al., "Dark matter substructure within galactic halos," *Astrophysical Journal Letters*, vol. 524, no. 1, pp. L19–L22, 1999.

[13] A. Klypin, A. V. Kravtsov, O. Valenzuela, and F. Prada, "Where are the missing galactic satellites?" *Astrophysical Journal Letters*, vol. 522, no. 1, pp. 82–92, 1999.

[14] J. R. Primack, "Cosmology: small-scale issues," *New Journal of Physics*, vol. 11, no. 10, Article ID 105029, 2009.

[15] A. Kravtsov, "Dark matter substructure and dwarf galactic satellites," *Advances in Astronomy*, vol. 2010, Article ID 281913, 2010.

[16] J. Sommer-Larsen and A. Dolgov, "Formation of disk galaxies: warm dark matter and the angular momentum problem," *Astrophysical Journal Letters*, vol. 551, no. 2, pp. 608–623, 2001.

[17] P. J. E. Peebles, "The void phenomenon," *Astrophysical Journal Letters*, vol. 557, no. 2, pp. 495–504, 2001.

[18] W. J. G. De Blok and S. S. McGaugh, "The dark and visible matter content of low surface brightness disc galaxies," *Monthly Notices of the Royal Astronomical Society*, vol. 290, no. 3, pp. 533–552, 1997.

[19] J. J. Dalcanton and C. J. Hogan, "Halo cores and phase-space densities: observational constraints on dark matter physics and structure formation," *Astrophysical Journal Letters*, vol. 561, no. 1, pp. 35–45, 2001.

[20] S. D. M. White and C. S. Frenk, "Galaxy formation through hierarchical clustering," *Astrophysical Journal Letters*, vol. 379, no. 1, pp. 52–79, 1991.

[21] S. Cole, C. G. Lacey, C. M. Baugh, and C. S. Frenk, "Hierarchical galaxy formation," *Monthly Notices of the Royal Astronomical Society*, vol. 319, no. 1, pp. 168–204, 2000.

[22] C. M. Baugh, "A primer on hierarchical galaxy formation: the semi-analytical approach," *Reports on Progress in Physics*, vol. 69, no. 12, article R02, pp. 3101–3156, 2006.

[23] S. M. Faber and R. E. Jackson, "Velocity dispersions and mass-to-light ratios for elliptical galaxies," *Astrophysical Journal*, vol. 204, pp. 668–683, 1976.

[24] R. B. Tully and J. R. Fisher, "A new method of determining distances to galaxies," *Astronomy and Astrophysics*, vol. 54, no. 3, pp. 661–673, 1977.

[25] G. Gavazzi, D. Pierini, and A. Boselli, "The phenomenology of disk galaxies," *Astronomy and Astrophysics*, vol. 312, no. 2, pp. 397–408, 1996.

[26] M. R. Blanton, D. W. Hogg, N. A. Bahcall et al., "The broadband optical properties of galaxies with redshifts 0.02 < Z < 0.22," *Astrophysical Journal Letters*, vol. 594, no. 1 I, pp. 186–207, 2003.

[27] M. Bernardi, R. K. Sheth, J. Annis et al., "Early-type galaxies in the sloan digital sky survey. II. Correlations between observables," *Astronomical Journal*, vol. 125, no. 4, pp. 1849–1865, 2003.

[28] G. Kauffmann, T. M. Heckman, S. D. M. White et al., "Stellar masses and star formation histories for 105 galaxies from the Sloan Digital Sky Survey," *Monthly Notices of the Royal Astronomical Society*, vol. 341, no. 1, pp. 33–53, 2003.

[29] G. Kauffmann, S. D. M. White, T. M. Heckman et al., "The environmental dependence of the relations between stellar mass, structure, star formation and nuclear activity in galaxies," *Monthly Notices of the Royal Astronomical Society*, vol. 353, no. 3, pp. 713–731, 2004.

[30] M. R. Blanton, M. Geha, and A. A. West, "Testing cold dark matter with the low-mass tully-fisher relation," *Astrophysical Journal Letters*, vol. 682, no. 2, pp. 861–873, 2008.

[31] M. J. Disney, J. D. Romano, D. A. Garcia-Appadoo, A. A. West, J. J. Dalcanton, and L. Cortese, "Galaxies appear simpler than expected," *Nature*, vol. 455, no. 7216, pp. 1082–1084, 2008.

[32] D. A. Garcia-Appadoo, A. A. West, J. J. Dalcanton, L. Cortese, and M. J. Disney, "Correlations among the properties of galaxies found in a blind H i survey, which also have SDSS optical data," *Monthly Notices of the Royal Astronomical Society*, vol. 394, no. 1, pp. 340–356, 2009.

[33] L. Staveley-Smith, W. E. Wilson, T. S. Bird et al., "The Parkes 21 cm multibeam receiver," *Publications of the Astronomical Society of Australia*, vol. 13, no. 3, pp. 243–248, 1996.

[34] R. Giovanelli, M. P. Haynes, B. R. Kent et al., "The arecibo legacy fast alfa survey. III. H I source catalog of the northern virgo cluster region," *Astronomical Journal*, vol. 133, no. 6, pp. 2569–2583, 2007.

[35] A. Saintonge, R. Giovanelli, M. P. Haynes et al., "The arecibo legacy fast alfa survey. V. the H i source catalog of the anti-virgo region at $\delta = +27°$," *Astronomical Journal*, vol. 135, no. 2, pp. 588–604, 2008.

[36] B. R. Kent, R. Giovanelli, M. P. Haynes et al., "The Arecibo legacy fast alfa survey. VI. Second HI source catalog of the Virgo cluster region," *Astronomical Journal*, vol. 136, no. 2, pp. 713–724, 2008.

[37] K. N. Abazajian, J. K. Adelman-Mccarthy, M. A. Agüeros et al., "The seventh data release of the sloan digital sky survey," *Astrophysical Journal*, vol. 182, no. 2, pp. 543–558, 2009.

[38] M. C. Toribio, J. M. Solanes, R. Giovanelli, M. P. Haynes, and K. L. Masters, "H I content and optical properties of field galaxies from the ALFALFA survey. I. Selection of a control sample," *Astrophysical Journal Letters*, vol. 732, no. 2, article 92, 2011.

[39] M. C. Toribio, J. M. Solanes, R. Giovanelli, M. P. Haynes, and A. M. Martin, "H i content and optical properties of field galaxies from the ALFALFA survey. II. Multivariate analysis of a galaxy sample in low-density environments," *Astrophysical Journal Letters*, vol. 732, no. 2, article 93, 2011.

[40] I. T. Jolliffe, *Principal Component Analysis*, vol. 1, Springer, 1986.

[41] M. F. Skrutskie, R. M. Cutri, R. Stiening et al., "The two Micron All Sky Survey (2MASS)," *Astronomical Journal*, vol. 131, no. 2, pp. 1163–1183, 2006.

[42] J. Brinchmann, S. Charlot, S. D. M. White et al., "The physical properties of star-forming galaxies in the low-redshift Universe," *Monthly Notices of the Royal Astronomical Society*, vol. 351, no. 4, pp. 1151–1179, 2004.

[43] J. L. Rosenberg, S. E. Schneider, and J. Posson-Brown, "Gas and stars in an H I-selected galaxy sample," *Astronomical Journal*, vol. 129, no. 3, pp. 1311–1330, 2005.

[44] A. Chung, J. H. van Gorkom, J. D. Kenney, H. Crowl, and B. Vollmer, "VLA imaging of virgo spirals in atomic gas (VIVA). I. The atlas and the H i properties," *Astronomical Journal*, vol. 138, no. 6, pp. 1741–816, 2009.

[45] N. Yasuda, M. Fukugita, V. K. Narayanan et al., "Galaxy number counts from the Sloan Digital Sky Survey commissioning data," *Astronomical Journal*, vol. 122, no. 3, pp. 1104–1124, 2001.

[46] M. J. Meyer, M. A. Zwaan, R. L. Webster et al., "The HIPASS catalogue—I. Data presentation," *Monthly Notices of the Royal Astronomical Society*, vol. 350, no. 4, pp. 1195–1209, 2004.

[47] K. Abazajian, J. K. Adelman-McCarthy, M. A. Agüeros et al., "The second data release of the sloan digital sky survey," *Astronomical Journal*, vol. 128, no. 1, pp. 502–512, 2004.

[48] A. A. West, D. A. Garcia-Appadoo, J. J. Dalcanton et al., "H i-Selected Galaxies in the Sloan Digital Sky Survey. II. The colors of Gas-Rich galaxies," *Astronomical Journal*, vol. 138, no. 3, pp. 796–807, 2009.

[49] A. A. West, D. A. Garcia-Appadoo, J. J. Dalcanton et al., "H I-selected galaxies in the sloan digital sky survey. I. Optical data," *Astronomical Journal*, vol. 139, no. 2, pp. 315–328, 2010.

[50] V. Petrosian, "Surface brightness and evolution of galaxies," *Astrophysical Journal*, vol. 209, pp. L1–L5, 1976.

[51] M. R. Blanton, J. Dalcanton, D. Eisenstein et al., "The luminosity function of galaxies in SDSS commissioning data," *Astronomical Journal*, vol. 121, no. 5, pp. 2358–2380, 2001.

[52] A. A. West, *H i selected galaxies in the sloan digital sky survey*, Ph.D. thesis, University of Washington, 2005.

[53] M. Cappellari, R. Bacon, M. Bureau et al., "The SAURON project—IV. The mass-to-light ratio, the virial mass estimator and the Fundamental Plane of elliptical and lenticular galaxies," *Monthly Notices of the Royal Astronomical Society*, vol. 366, no. 4, pp. 1126–1150, 2006.

[54] M. R. Blanton, D. J. Schlegel, M. A. Strauss et al., "New York University Value-Added Galaxy Catalog: a galaxy catalog based

on new public surveys," *Astronomical Journal*, vol. 129, no. 6, pp. 2562–2578, 2005.

[55] I. Strateva, Ž. Ivezić, G. R. Knapp et al., "Color separation of galaxy types in the Sloan Digital Sky Survey imaging data," *Astronomical Journal*, vol. 122, no. 4, pp. 1861–1874, 2001.

[56] P. C. van der Kruit, "The three-dimensional distribution of light and mass in disks of spiral galaxies," *Astronomy and Astrophysics*, vol. 192, no. 1-2, pp. 117–127, 1988.

[57] M. R. Blanton, D. W. Hogg, N. A. Bahcall et al., "The galaxy luminosity function and luminosity density at redshift $z = 0.1$," *Astrophysical Journal Letters*, vol. 592, no. 2 I, pp. 819–838, 2003.

[58] M. P. Haynes and R. Giovanelli, "Neutral hydrogen in isolated galaxies. IV—results for the Arecibo ample," *Astronomical Journal*, vol. 89, pp. 758–800, 1984.

[59] J. Kormendy, "Integrated colors of bright galaxies in the u, b, V system," *Astrophysical Journal*, vol. 218, p. 333, 1977.

[60] M. S. Longair, *Galaxy Formation*, vol. 1, Springer, 2000.

[61] J. Kormendy and R. C. Kennicutt, "Secular evolution and the formation of pseudobulges in disk galaxies," *Annual Review of Astronomy and Astrophysics*, vol. 42, pp. 603–683, 2004.

[62] R. G. Bower, A. J. Benson, R. Malbon et al., "Breaking the hierarchy of galaxy formation," *Monthly Notices of the Royal Astronomical Society*, vol. 370, no. 2, pp. 645–655, 2006.

[63] J. Kormendy, D. B. Fisher, M. E. Cornell, and R. Bender, "Structure and formation of elliptical and spheroidal galaxies," *Astrophysical Journal*, vol. 182, no. 1, pp. 216–309, 2009.

[64] F. Fontanot, G. De Lucia, P. Monaco, R. S. Somerville, and P. Santini, "The many manifestations of downsizing: hierarchical galaxy formation models confront observations," *Monthly Notices of the Royal Astronomical Society*, vol. 397, no. 4, pp. 1776–1790, 2009.

[65] M. A. Zwaan, J. M. van der Hulst, W. J. G. de Blok, and S. S. McGaugh, "The Tully-Fisher relation for low surface brightness galaxies: implications for galaxy evolution," *Monthly Notices of the Royal Astronomical Society*, vol. 273, pp. L35–L38, 1998.

[66] S. S. McGauon and W. J. G. De Blok, "Testing the hypothesis of modified dynamics with low surface brightness galaxies and other evidence," *Astrophysical Journal Letters*, vol. 499, no. 1, pp. 66–81, 1998.

[67] R. Reyes, R. Mandelbaum, J. E. Gunn, J. Pizagno, and C. N. Lackner, "Calibrated Tully-Fisher relations for improved estimates of disk rotation velocities," http://arxiv.org/abs/1106.1650.

[68] G. de Vaucouleurs, "Integrated colors of bright galaxies in the u, b, V system," *Astrophysical Journal Supplement Series*, vol. 5, p. 233, 1961.

[69] R. B. Tully, J. R. Mould, and M. Aaronson, "A colormagnitude relation for spiral galaxies," *Astrophysical Journal*, vol. 257, pp. 527–537, 1982.

[70] D. W. Hogg, M. R. Blanton, J. Brinchmann et al., "The dependence on environment of the color-magnitude relation of galaxies," *Astrophysical Journal*, vol. 601, no. 1, pp. L29–L32, 2004.

[71] Y. Chen, J. D. Lowenthal, and M. S. Yun, "Color-magnitude relation and morphology of low-redshift ulirgs in sloan digital sky survey," *Astrophysical Journal Letters*, vol. 712, no. 2, pp. 1385–1402, 2010.

Multimessengers from Core-Collapse Supernovae: Multidimensionality as a Key to Bridge Theory and Observation

Kei Kotake,[1,2] Tomoya Takiwaki,[2] Yudai Suwa,[3] Wakana Iwakami Nakano,[4] Shio Kawagoe,[5] Youhei Masada,[6] and Shin-ichiro Fujimoto[7]

[1] *Division of Theoretical Astronomy, National Astronomical Observatory of Japan, 2-21-1, Osawa, Mitaka, Tokyo 181-8588, Japan*
[2] *Center for Computational Astrophysics, National Astronomical Observatory of Japan, Mitaka, Tokyo 181-8588, Japan*
[3] *Yukawa Institute for Theoretical Physics, Kyoto University, Oiwake-cho, Kitashirakawa, Sakyo-ku, Kyoto 606-8502, Japan*
[4] *Department of Aerospace Engineering, Tohoku University, 6-6-01 Aramaki-Aza-Aoba, Aoba-ku, Sendai 980-8579, Japan*
[5] *Knowledge Dissemination Unit, Oshima Lab, Institute of Industrial Science, The University of Tokyo, 4-6-1 Komaba, Meguro-ku, Tokyo 153-8505, Japan*
[6] *Department of Computational Science, Graduate School of System Informatics, Kobe University, Nada, Kobe 657-8501, Japan*
[7] *Kumamoto National College of Technology, 2659-2 Suya, Goshi, Kumamoto 861-1102, Japan*

Correspondence should be addressed to Kei Kotake, kkotake@th.nao.ac.jp

Academic Editor: Andrew Fruchter

Core-collapse supernovae are dramatic explosions marking the catastrophic end of massive stars. The only means to get direct information about the supernova engine is from observations of neutrinos emitted by the forming neutron star, and through gravitational waves which are produced when the hydrodynamic flow or the neutrino flux is not perfectly spherically symmetric. The multidimensionality of the supernova engine, which breaks the sphericity of the central core such as convection, rotation, magnetic fields, and hydrodynamic instabilities of the supernova shock, is attracting great attention as the most important ingredient to understand the long-veiled explosion mechanism. Based on our recent work, we summarize properties of gravitational waves, neutrinos, and explosive nucleosynthesis obtained in a series of our multidimensional hydrodynamic simulations and discuss how the mystery of the central engines can be unraveled by deciphering these multimessengers produced under the thick veils of massive stars.

1. Introduction

The majority of stars more massive than $\sim 8M_\odot$ end their lives as core-collapse supernovae. They have long attracted the attention of astrophysicists because they have many facets playing important roles in astrophysics. They are the mother of neutron stars as well as black holes; they play an important role for acceleration of cosmic rays; they influence galactic dynamics triggering further star formation; they are gigantic emitters of neutrinos and gravitational waves. They are also a major site for nucleosynthesis, so, naturally, any attempt to address human origins may need to begin with an understanding of core-collapse supernovae (CCSNe).

Current estimates of CCSN rates in our Galaxy predict one event every $\sim 40 \pm 10$ years [1]. When a CCSN event occurs in our galactic center, copious numbers of neutrinos are produced, some of which may be detected on the earth. Such "supernova neutrinos" will carry valuable information from deep inside the core. In fact, the detection of neutrinos from SN1987A (albeit in the Large Magellanic Cloud) opened up the *Neutrino Astronomy*, which is an alternative to conventional astronomy by electromagnetic waves [2, 3]. Even though there were just two dozen neutrino events from SN1987A, these events have been studied extensively (yielding ~ 500 papers) and have allowed us to have a confidence that our basic picture of the supernova physics is correct (e.g., [4], see [5, 6] for a recent review). Recently significant progress has been made in the large water Cherenkov detectors such as Super-Kamiokande [7] and IceCube [8], and also in the liquid scintillator detector as KamLAND [9]. If a supernova occurs in our Galactic center (~10 kpc), about $10^5 \bar{\nu}_e$ events are estimated to be detectable by IceCube (e.g., [10] and references therein). Those successful neutrino detections are important not only to study the supernova physics but also to unveil the nature of neutrinos itself such as the neutrino oscillation parameters and the mass hierarchy (e.g., [5] for a recent review).

CCSNe are now about to start even another astronomy, *Gravitational-Wave Astronomy*. Currently long-baseline laser interferometers such as LIGO (USA, e.g., [11]), VIRGO (Italy) (http://www.ego-gw.it/) GEO600 (Germany)(http://geo600.aei.mpg.de/), and TAMA300 (Japan, e.g., [12]) are currently operational and preparing for the first observation (see, e.g., [13] for a recent review), by which the prediction by Einstein's theory of General Relativity (GR) can be confirmed. These instruments are being updated to their *Advanced* status and may start taking data, possibly detecting GWs (gravitational waves) for the first time, as soon as 2015 (see [14] for a recent review). In fact, *Advanced* LIGO/VIRGO, which is an upgrade of the initial LIGO and VIGRO, are expected to be completed by 2015 and will increase the observable detection volume by a factor of ~1000 [15]. The Large-scale Cryogenic Gravitational-wave Telescope (LCGT [16], renamed recently as KAGRA) was funded in late 2010, which is being built under the Kamioka mine and is expected to take its first data in 2016. At such a high level of precision, those GW detectors are sensitive to many different sources, including chirp, ring-down, and merger phases of black-hole and neutron star binaries (e.g., [17–19]), neutron star normal mode oscillations (e.g., [20]), rotating neutron star mountains (e.g., [21]), and CCSN explosions (e.g., [22–24] for recent reviews), on the final of which we focus in this paper.

According to the Einstein's theory of GR (e.g., [25]), no GWs can be emitted if gravitational collapse of the supernova core proceeds perfectly spherically symmetric. To produce GWs, the gravitational collapse should proceed aspherically and dynamically. Observational evidence gathered over the last few decades has pointed towards CCSNe indeed being generally aspherical (see [26] for a recent review). The most unequivocal example is SN1987A. To explain the light-curve, a large amount of mixing of Ni outward to the H-He interface and of H inward into the He-core was required (e.g., Woosley [27]; Blinnikov et al. [28]; Utrobin [29], and Woosley & Heger [30] for a review). Such mixing processes coming from the Rayleigh-Taylor instability at the interface with different compositions after shock passage have been examined extensively so far by 2D (e.g., [31–33]) and 3D simulations [34]. The asymmetry of iron and nickel lines in SN1987A was proposed to be explained, if the explosion occurs in a jet-like [35, 36] or in a unipolar manner [37]. More directly, the HST images of SN1987A are showing that the expanding envelope is elliptical with the long axis aligned with the rotation axis inferred from the ring ([38], see however [39] for a recent counterargument). The aspect ratio and position angle of the symmetry axis are consistent with those predicted earlier from the observations of speckle and linear polarization. What is more, the linear polarization became greater as time passed (e.g., [40–42]), a fact which has been used to

argue that the central engine of the explosion is responsible for the nonsphericity (e.g., [33, 43–46]). By performing a series of time-dependent, non-LTE (local thermodynamic equilibrium), radiation-transport simulations (e.g., [47–49] , see [50, 51] for an alternative approach for the light-curve modeling), Dessart and Hillier [52] recently pointed out that asymmetry in the ejecta can explain the increase in the continuum polarization observed at the nebular phases [53, 54]. Dense knots, indications of ejecta clumpiness, and filaments seen in supernova remnants by HST in the visual [55, 56] and by ROSAT, Chandra, and XMM-Newton [57–59] in X-rays also provide evidence that small- and large-scale inhomogeneities (and perhaps even fragmentation) are a common feature in supernova explosions.

Advancing ability of the HST has enabled the direct observation of the progenitors of nearby CCSNe from pre-explosion images (see [60] for a review). Although observational measurements of the progenitor masses are currently not many and still highly uncertain (see [61] for collective references therein), evidence has accumulated that SN type II plateau (II-P) comes predominantly from stars in the range about of $8–16 M_\odot$ [60]. A generic explosion energy of the SN II-P in the mass range is roughly on the order of 10^{51} erg [37], however, a large diversity of the explosion signatures (i.e., explosion energy and the synthesized Ni mass) have been so far observed from quite similar progenitors [60]. For example, the inferred ^{56}Ni mass and the kinetic energies differ by a factor of five between SNe 2005cs and 2003gd, both of them are among the *golden events* in which enough information was obtained to give an accurate estimate on a color or spectral type of the progenitor and the initial mass. More massive stars are expected to lose much of their mass and explode as hydrogen-stripped SNe (Ib/c and IIb). Among them, the type Ic-BL SNe, which are associated with long gamma-ray bursts [62], all show much broader lines than SNe Ic. Due to the large kinetic energies of 2–5×10^{52} erg, they have been referred to as "hypernovae" (e.g., [63] for a recent review). These events are likely to come from interacting binaries in which the primary exploding star has a mass lower than what is usually associated with evolution to the massive WR phase (e.g., [64–67], see however [68, 69]). In addition to the two branches mentioned above, Nomoto et al. [70] predicted yet another branch, in which the SNe II-P with higher progenitor mass result in fainter explosions. By connecting to candidate SN explosion mechanisms or to progenitor structures, it is indeed best if one could obtain a unified picture to understand the mentioned wide diversity which is not only dependent on the progenitor masses but also on the evolution scenarios (a single versus binary evolution). But, this is an area for future study firstly because it is too computationally expensive to perform a long-term simulation that follows multidimensional (multi-D) dynamics consistently from the onset of core-collapse, through explosion, up to the nebular phase (for example see [71] for the most up-to-date 1D modeling of spectra and light curves for binary models.), secondly because we are still inaccessible to multi-D stellar evolution models, in which multi-D modeling has been a major undertaking (see [72, 73] for recent developments).

From a theoretical point of view, clarifying what makes the dynamics of the supernova engine deviate from spherical symmetry is essential in understanding the GW emission mechanism. Here it is worth mentioning that GWs are primary observables, which imprint a live information of the central engine, because they carry the information directly to us without being affected in propagating from the stellar center to the earth. On the other hand, SN neutrinos are exposed to a number of (external) environmental effects, including self-interaction that induces collective neutrino flavor oscillations predominantly in the vicinity of the neutrino sphere (see [74–77] for reviews of the rapidly growing research field and collective references therein) and the Mikheyev-Smirnov-Wolfenstein (MSW) effect [78] both in the stellar envelope and in the earth (see [5] for a review). Although the time profiles of neutrino signals can be potentially used like a tomography to monitor the envelope profile (e.g., [79] for a recent review), SN neutrinos generally could provide a rather indirect information about the central engine compared to GWs.

The breaking of the sphericity in the supernova engine has been considered as the most important ingredient to understand the explosion mechanism, for which supernova theorists have been continuously keeping their efforts for the past ~40 years. Currently multi-D simulations based on refined numerical models have shown several promising scenarios. Among the candidates are the neutrino heating mechanism aided by convection and hydrodynamic instabilities of the supernova shock (e.g., [80] for a review), the acoustic mechanism [81], or the magnetohydrodynamic (MHD) mechanism (e.g., [22] see references therein). To pin down the true answer among the candidate mechanisms, GW signatures, albeit being a primary observable, will not be solely enough and a careful analysis including neutrinos and photons should be indispensable. The current neutrino detectors is ready to broadcast the alert to astronomers to let them know the arrival of neutrinos (e.g., Supernova Early Warning Systems (SNEWS) [82]). In addition to optical observations using largest 8–10 m telescopes such as VLT and Subaru telescope (e.g., [83, 84]), the planned thirty-meter telescope (TMT) (http://www.tmt.org/) with a refined spectropolarimetric technique is expected to detect more than 20 events of the SN polarization per year (Tanaka in private communication). This should provide valuable information to understand the SN asymmetry with increasing statistics. Not only for understanding the origin of nonspherical ejecta morphology (especially for nearby event) but more importantly for understanding the origin of heavy elements, it is of crucial importance to accurately determine nucleosynthesis in the SN ejecta, which naturally requires a multi-D numerical modeling (see, [30, 85] for as recent reviews, it is worth mentioning that radioactive decay can affect the shape of the light curve for some peculiar SN 1987A-like events [37], in which explosive nucleosynthesis plays an important role for the light-curve modeling).

Putting things together, the multidimensionality determines the explosion mechanism, in turn we may extract the information that traces the multidimensionality by the SN multimessengers, which would be only possible by a careful analysis on GWs, neutrinos, and photons. In this paper, we hope to bring together various of our published and unpublished findings from our recent multi-D supernova simulations and the obtained predictions of the SN multimessengers so far (for other high-energy astrophysical sources such as magnetars, gamma-ray bursts, and coalescing binaries, see [86–89] for recent reviews). Before we go into details from the next sections, we first have to draw a caution that the current generation of simulation results that we report in this article should depend on the next generation calculations by which more sophistication can be made not only in determining the efficiency of neutrino-matter coupling (the so-called neutrino transport calculation), but also in the treatment of general relativity. Therefore we provide here only a snapshot of the moving (long-run) documentary film whose headline we (boldly) chose to entitle as "multimessengers from CCSNe to bridge theory and observation."

Among the mentioned candidate mechanisms, we focus on the neutrino-heating mechanism in Section 2 and the MHD mechanism in Section 3, respectively. In each section, we first briefly summarize the properties of the explosion dynamics and then move on to discuss possible properties of the SN multimessengers paying particular attention to their detectability. It may be best if we can cover these SN messengers once for all in this paper, but unfortunately not. What we have studied so far is limited to GWs and explosive nucleosynthesis in the neutrino-heating mechanism, and to GWs and neutrino signals in the MHD mechanism. To compensate the uncovered fields, the related references will be given. Although a number of excellent reviews already exist on various topics in this paper, this one might go beyond such reviews by its new perspectives on the multimessenger astronomy.

2. Neutrino-Heating Mechanism

CCSN simulations have been counted as one of the most challenging subjects in computational astrophysics. The four fundamental forces of nature are all at play; the collapsing iron core bounces due to strong interactions; weak interactions determine the energy and lepton number loss in the core via the transport of neutrinos; electromagnetic interactions determine the properties of the stellar gas; GR plays an important role due to the compactness of the protoneutron star and also due to high velocities of the collapsing material outside. Naturally, such physical richness ranging from a microphysical scale (i.e., femto-meter scale) of strong/weak interactions to a macrophysical scale of stellar explosions has long attracted the interest of researchers, necessitating a worldwide multidisciplinary collaboration to clarify the theory of massive stellar core-collapse and the formation mechanisms of compact objects.

Ever since the first numerical simulation of such events [90], the neutrino-heating mechanism [91–93], in which a stalled bounce shock could be revived via neutrino absorption on a timescale of several hundred milliseconds after bounce, has been the working hypothesis of supernova theorists for these ~45 years. However, the simplest spherically-symmetric (1D) form of this mechanism fails to blow up canonical massive stars [94–97]. Pushed by mounting observations of the blast morphology (e.g., [26]) mentioned above, it is now almost certain that the breaking of the spherical symmetry is the key to solve the supernova problem. So far a number of multi-D hydrodynamic simulations have been reported, which *demonstrated* that hydrodynamic motions associated with convective overturn (e.g., [98–102]) as well as the Standing-Accretion-Shock-Instability (SASI, e.g., [44, 103–113] see references therein) can help the onset of the neutrino-driven explosion.

To test the neutrino-heating mechanism in the multi-D context, it is of crucial importance to solve accurately the neutrino-matter coupling in spatially nonuniform hydrodynamic environments. For the purpose, one ultimately needs to solve the six-dimensional (6D) neutrino radiation transport problem (three in space, three in the momentum space of neutrinos), which is a main reason why the supernova simulations stand out from other astrophysical simulations due to their complexity. In the final sentence of the last paragraph, we wrote "demonstrated" because the neutrino heating was given by hand as an input parameter in most of the simulations cited above (see, however [101, 102]). The neutrino heating proceeds dominantly via the charged current interactions ($\nu_e + n \rightleftarrows e^- + p$, $\bar{\nu}_e + p \rightleftarrows e^+ + n$) in the gain region. The neutrino heating rate in the gain region can be roughly expressed as $Q^+ \propto L_\nu \langle \mu_\nu \rangle^{-1}/r^2$, where L_ν is the neutrino luminosity emitted from the surface of neutrino sphere and it determines the amplitude of the neutrino heating as well as cooling, and r and $\langle \mu_\nu \rangle$ are the distance from the stellar center and the flux factor (this quantity represents the degree of anisotropy in neutrino emission; $\langle \mu_\nu \rangle \sim 0.25$ near at the neutrino sphere, $\langle \mu_\nu \rangle = 1$ in the free-streaming limit ($r \rightarrow \infty$)), respectively (e.g., [114]). For example, L_ν is treated as an input parameter in the so-called "light-bulb" approach (e.g., [100]). This is one of the most prevailing approximations in recent 3D simulations [111, 112, 115, 116] because it is handy to study multi-D effects on the neutrino heating mechanism (albeit on the qualitative grounds). To go beyond the light-bulb scheme, L_ν should be determined in a self-consistent manner. For the purpose, one needs to tackle with neutrino transport problem, only by which energy as well as angle dependence of the neutrino distribution function can be determined without any assumptions. Since the focus of this review is on the SN multimessengers, a detailed discussion of various approximations and numerical techniques taken in the recent radiation-hydrodynamic SN simulations cannot be provided. Table 1 is not intended as a comprehensive compilation, but we just want to summarize milestones that have recently reported the neutrino-driven *exploding* models so far.

In Table 1, the first column ("Progenitor") shows the progenitor model employed in each simulation. The abbreviations of "NH," "WHW," and "WW" mean [117–119]. The second column shows SN groups with the published or submitted year of the corresponding work. "MPA" stands for the CCSN group in the Max Planck Institute for Astrophysics led by Janka and Müller. "Princeton+" stands for the group chiefly consisting of the staffs in the Princeton University (Burrows, Murphy), Caltech (Ott), Hebrew University (Livne), Université de Provence (Dessart), and their collaborators. The SN group in the Basel university is led by Liebendörfer and Thielemann. "OakRidge+" stands for the SN group mainly consisting of the Florida Atlantic University (Bruenn) and the Oak Ridge National Laboratory (Mezzacappa, Messer) and their collaborators. Tokyo+ is the SN group chiefly consisting of the staffs in the National Astronomical Observatory of Japan (myself, Takiwaki), Kyoto university (Suwa), Numazu College (Sumiyoshi), Waseda University (Yamada), and their collaborators. The third column represents the mechanism of explosions which are basically categorized into two (to date), namely by the neutrino-heating mechanism (indicated by "ν-driven") or by the acoustic mechanism ("Acoustic"). "Dim." in the fourth column is the fluid space dimensions which is one-, two-, or three-dimension (1-,2-,3D). The abbreviation "N" stands for "Newtonian," while "PN"—for "Post-Newtonian"—stands for some attempt at inclusion of general relativistic effects, and "GR" denotes full relativity. t_{exp} in the fifth column indicates an approximate typical timescale when the explosion initiates and E_{exp} represents the explosion energy normalized by Bethe (=10^{51} erg) given at the postbounce time of t_{pb}, both of which are attempted to be sought in literatures (but if we cannot find them, we remain them as blank "—"). In the final column of "ν transport," "Dim" represents ν momentum dimensions and the treatment of the velocity-dependent term in the transport equations is symbolized by $\mathcal{O}(v/c)$. The definition of the "RBR," "IDSA," and "MGFLD" will be given soon in the following.

Due to the page limit of this paper, we have to start the story only after 2006 (see, e.g., [80] for a complete review, and also [120] for a similar table before 2006). The first news of the exploding model was reported by the MPA group. By performing radiation-neutrino-hydrodynamic simulations which includes one of the best available neutrino transfer approximations, they reported 1D and 2D explosions for the $8.8M_\odot$ star [121, 122] whose progenitor has a very tenuous outer envelope with steep density gradient (a characteristic property of AGB stars). Also in 1D, the Basel+ group reported explosions for 10 and $15M_\odot$ progenitors of WHW02 triggered by the hypothesised first-order QCD phase transition in the protoneutron star (PNS) [123]. To date, these two are the only modern numerical results where the neutrino-driven mechanism succeeded in 1D. In the 2D MPA simulations, they obtained explosions for a nonrotating $11.2M_\odot$ progenitor of WHW02 [124], and then for a $15M_\odot$ progenitor [125] of WW95 with a relatively rapid rotation imposed (by comparing the precollapse angular velocity to the one predicted in a recent stellar evolution calculation [126, 127]). They newly brought in the so-called "ray-by-ray" approach (indicated by "RBR" in the table), in which the neutrino transport is solved along a given radial direction assuming that the hydrodynamic medium for the direction is spherically symmetric. This method, which reduces the 2D problem partly to 1D, fits

TABLE 1: Selected lists of recent neutrino-radiation hydrodynamic milestones reported by many SN groups around the world ("Group"), which obtained explosions by the neutrino-heating mechanism (indicated by "ν-driven") or the acoustic mechanism ("Acoustic") (See text for more details).

Progenitor	Group (Year)	Mechanism	Dim. (Hydro)	t_{exp} (ms)	E_{exp} (B) @t_{pb} (ms)	ν transport (Dim, $\mathcal{O}(v/c)$)
8.8$M_\odot$ (NH88)	MPA (2006, 2011)	ν-driven	1D(2D) (PN)	~200	0.1 (~800)	Boltzmann 2, $\mathcal{O}(v/c)$
	Princeton+ (2006)	ν-driven	2D (N)	≲125	0.1 —	MGFLD 1, (N)
10$M_\odot$ (WHW02)	Basel (2009)	ν+ (QCD transition)	1D (GR)	255	0.44 (350)	Boltzmann 2, (GR)
11$M_\odot$ (WW95)	Princeton+ (2006)	Acoustic	2D (N)	≳550	~0.1* (1000)	MGFLD 1, (N)
11.2$M_\odot$ (WHW02)	MPA (2006, 2012)	ν-driven	2D (PN, *C-GR*)	~100 *~200*	~0.005, *0.025* ~200, *900*	"RBR" Boltzmann, 2, $\mathcal{O}(v/c)$
	Princeton+ (2007)	Acoustic	2D (N)	≳1100	~0.1* (1000)	MGFLD 1, (N)
	Tokyo+ (2011)	ν-driven	3D (N)	~100	0.01 (300)	IDSA 1, (N)
12$M_\odot$ (WHW02)	Oak Ridge+ (2009)	ν-driven	2D (PN)	~300	0.3 (1000)	"RBR" MGFLD 1, $\mathcal{O}(v/c)$
13$M_\odot$ (WHW02) (NH88)	Princeton+ (2007)	Acoustic	2D (N)	≳1100	~0.3* (1400)	MGFLD 1, (N)
	Tokyo+ (2010)	ν-driven	2D (N)	~200	0.1 (500)	IDSA 1, (N)
15$M_\odot$ (WW95) (WHW02)	MPA (2009,2012)	ν-driven	2D (PN, *C-GR*)	~600 *~400*	0.025, *0.125* (~700, *800*)	Boltzmann 2, $\mathcal{O}(v/c)$
	Princeton+	Acoustic	2D (N)	—	— (—)	MGFLD 1, (N)
	Oak Ridge+ (2009)	ν-driven	2D (PN)	~300	~0.3 (600)	"RBR" MGFLD 1, $\mathcal{O}(v/c)$
20$M_\odot$ (WHW02)	Princeton+ (2007)	Acoustic	2D (N)	≳1200	~0.7* (1400)	MGFLD 1, (N)
25$M_\odot$ (WHW02)	Princeton+ (2007)	Acoustic	2D (N)	≳1200	— (—)	MGFLD 1, (N)
	Oak Ridge+ (2009)	ν-driven	2D (PN)	~300	~0.7 (1200)	"RBR" MGFLD 1, $\mathcal{O}(v/c)$

well with their original 1D Boltzmann solver [94] (note that the ray-by-ray approach has an advantage compared to other approximation schemes, such that it can fully take into account the available neutrino reactions (e.g., [128] for references therein) and also give us the most accurate solution for a given angular direction). For 2D hydrodynamic simulations with the ray-by-ray transport, one needs to solve the 4D radiation transport problem (two in space and two in the neutrino momentum space). Regarding the explosion energies obtained in the MPA simulations, their values at their final simulation time are typically underpowered by one or two orders of magnitudes to explain the canonical supernova kinetic energy ($\sim 10^{51}$ erg). But the explosion energies presented in their figures are still growing with time,

and they could be as high as 1 B if they were able to follow a much longer evolution as discussed in [124]. Very recently, Mueller et al. [129] reported that more energetic explosions are obtained for the 11.2 and 15$M_\odot$ stars in their GR 2D simulations compared to the corresponding post-Newtonian models (e.g., [125]). In the table, the data in italic character represent their most up-to-date results based on the 2D GR simulations using conformally-flatness approximation (e.g., [130, 131], indicated by "*C-GR*" in the table).

It is rather only recently that fully 2D multiangle Boltzmann transport simulations become practicable by the Princeton+ group [132, 133]. In this case, one needs to handle the 5D problem for 2D simulations (two in space, and three in the neutrino momentum space). However this scheme is very computationally expensive currently to perform long-term supernova simulations. In fact, the most recent 2D work by Brandt et al. [133] succeeded in following the dynamics until ~400 ms after bounce for a nonrotating and a rapidly rotating $20M_\odot$ model of WHW02, but explosions seemingly have not been obtained in such an earlier phase either by the neutrino-heating or the acoustic mechanism.

In the table, "MGFLD" stands for the multigroup Flux-limited diffusion scheme which eliminates the angular dependence of the neutrino distribution function (see, e.g., [134] for more details). For 2D simulations, one needs to solve the 3D problem, namely two in space, and one in the neutrino momentum space. By implementing the MGFLD algorithm to the CHIMERA code in a ray-by-ray fashion (e.g., [135]), Bruenn et al. (2009) obtained neutrino-driven explosions for nonrotating progenitors in a relatively wide range in 12, 15, 20, $25M_\odot$ of WHW02 (see Table 1). These models tend to start exploding at around 300 m after bounce, and the explosion energy for the longest running model of the $25M_\odot$ progenitor is reaching to 1 B at 1.2 s after bounce [135].

On the other hand, the 2D MGFLD simulations implemented in the VULCAN code [136] obtained explosions for a variety of progenitors of 11, 11.2, 13, 15, 20, and $25M_\odot$ not by the neutrino-heating mechanism but by the acoustic mechanism (for the $8.8M_\odot$ progenitor, they obtained neutrino-driven explosions (see Table 1)). The acoustic mechanism relies on the revival of the stalled bounce shock by the energy deposition via the acoustic waves that the oscillating protoneutron stars (PNSs) would emit in a much delayed phase (~1 second) compared to the conventional neutrino-heating mechanism (~300–600 milliseconds). If the core pulsation energy given in Burrows et al. [45] could be used to measure the explosion energy in the acoustic mechanism, they reach to 1 B after 1000 ms after bounce.

By performing 2D simulations in which the spectral neutrino transport was solved by the isotropic diffusion source approximation (IDSA) scheme [137], the Tokyo+ group reported explosions for a nonrotating and rapidly rotating $13M_\odot$ progenitor of NH88. They pointed out that a stronger explosion is obtained for the rotating model comparing to the corresponding nonrotating model. The IDSA scheme splits the neutrino distribution into two components (namely the streaming and trapped neutrinos), both of which are solved using separate numerical techniques (see [137] for more details). The approximation level of the IDSA scheme is basically the same as the one of the MGFLD. The main advantage of the IDSA scheme is that the fluxes in the transparent region can be determined by the nonlocal distribution of sources rather than the gradient of the local intensity like in MGFLD. A drawback in the current version of the IDSA scheme is that heavy lepton neutrinos (ν_x, i.e., ν_μ, ν_τ and their antiparticles) as well as the energy-coupling weak interactions have yet to be implemented. Extending the 2D modules in [138] to 3D, they recently reported explosions in the 3D models for an $11.2M_\odot$ progenitor of WHW02 [139]. By comparing the convective motions as well as neutrino luminosities and

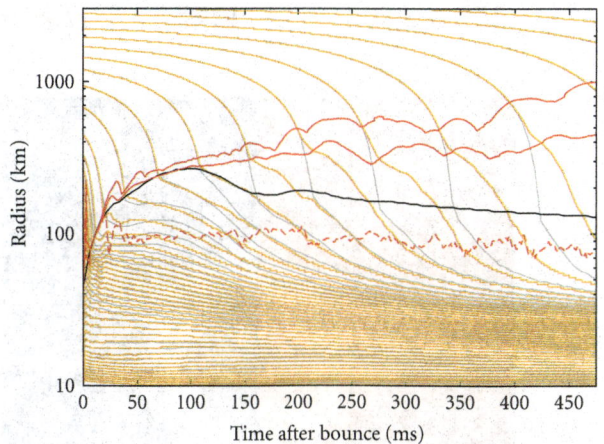

FIGURE 1: Time evolution of 1D (thin gray lines) or 2D (thin orange lines) hydrodynamic simulation [138] of a $13M_\odot$ progenitor [117]. Thick lines in red (for 2D) and black (for 1D) show the position of shock waves, noting for 2D that the maximum (top) and average (bottom) shock position are shown. The red-dashed line represents the position of the gain radius, which is similar to the 1D case (not shown).

energies between their 2D and 3D models, they pointed out whether 3D effects would help explosions or not is sensitive to the employed numerical resolutions. They argued that next-generation supercomputers are at least needed to draw a robust conclusion of the 3D effects.

Having summarized a status of the current supernova simulations, one might easily see a number of issues that remain to be clarified. First of all, the employed progenitors usually rather scatter (e.g., Table 1). Different SN groups seem to have a tendency to employ different progenitors, providing different results. By climbing over a wall which may have rather separated exchanges among the groups, a detailed comparison for a given progenitor needs to be done seriously in the multi-D results (as have been conducted in the Boltzmann 1D simulations between the MPA, Basel+, and Oak Ridge+ groups [140]).

In the next section, we briefly summarize the findings obtained in our 2D [138] and 3D [139] simulations, paying particular attention to how multidimensionality such as SASI, convection, and rotation could affect the neutrino-driven explosions.

2.1. Multidimensionality in Multi-D Radiation Hydrodynamic Simulations. Figure 1 depicts the difference between the time evolutions of 1D (thin gray lines) or 2D (thin orange lines) simulation of the $13M_\odot$ progenitor model. Until ~100 ms after bounce, the shock position of the 2D model (thick red line) is similar to the 1D model (thick black line). Later on, however, the shock for 2D does not recede as for 1D, but gradually expands and reaches 1000 km at about 470 ms after bounce. Comparing the position of the gain radius (red-dashed line) to the shock position for 1D (thick black line) and 2D (thick red line), one can see that the advection time of the accreting material in the gain region can be longer in 2D

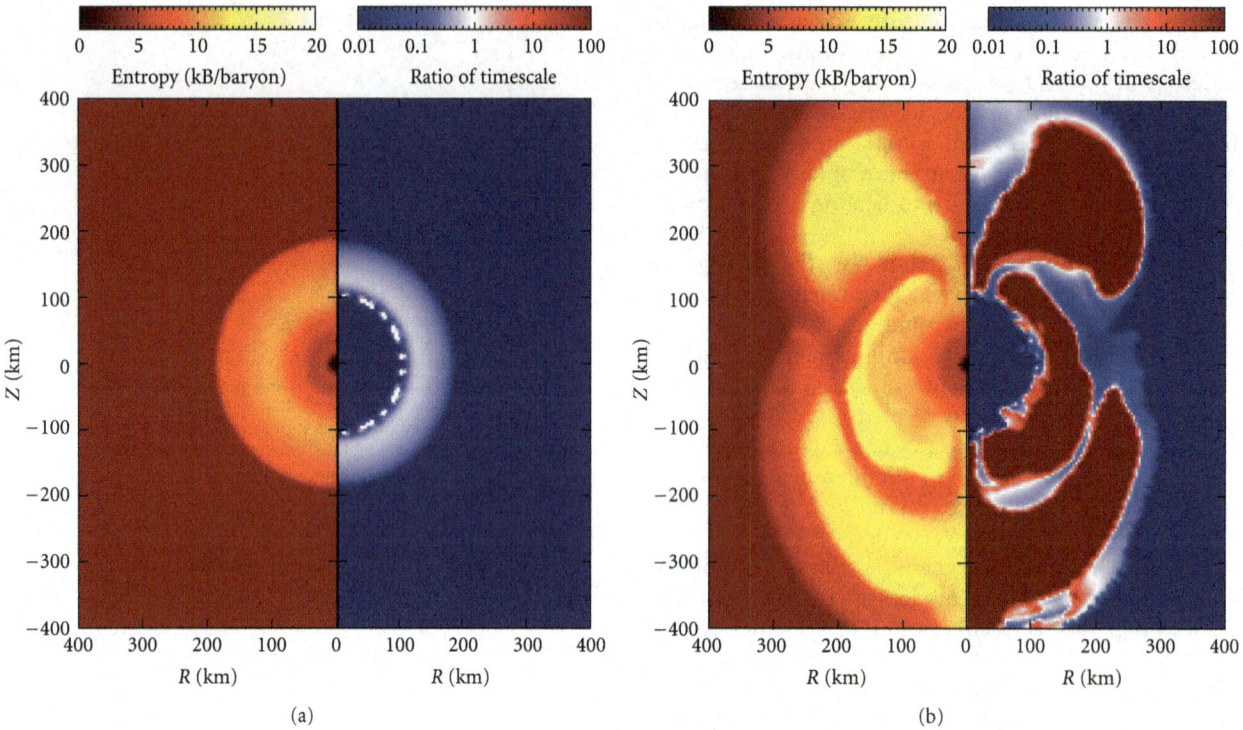

FIGURE 2: Snapshot of the distribution of entropy (left half) and the ratio of the advection to the heating timescale (right half) for models of the 1D (left) and 2D (right) models at 200 ms after bounce. These figures are taken from [138].

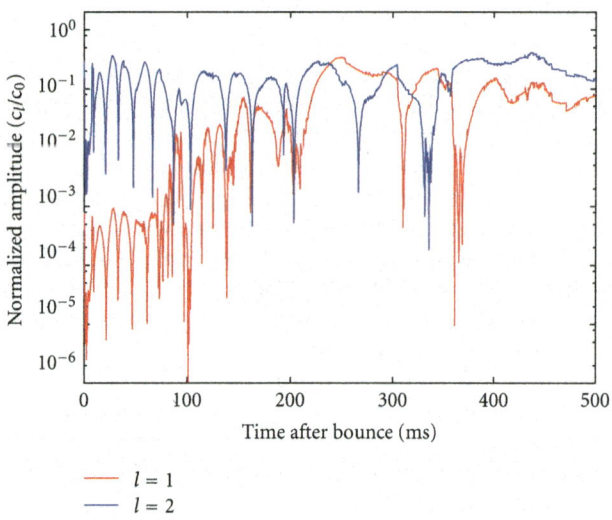

FIGURE 3: SASI activity for the rotating model versus postbounce time. Shown are the coefficients of the dipole ($\ell = 1$) and quadrupole ($\ell = 2$) modes of the spherical harmonics of the aspherical shock position, normalized to the amplitude of the $\ell = 0$ mode (taken from [138]).

than in 1D. This longer exposure of cool matter in the heating region to the irradiation of hot outstreaming neutrinos from the PNS is essential for the increased efficiency of the neutrino heating in multi-D models. This can be also depicted in Figure 2. The right half shows $\tau_{\mathrm{adv}}/\tau_{\mathrm{heat}}$, which is

the ratio of the advection to the neutrino heating timescale. This quantity is known as a useful quantity to diagnose the success ($\tau_{\mathrm{adv}}/\tau_{\mathrm{heat}} \gtrsim 1$, i.e., the neutrino-heating timescale is shorter than the advection timescale of material in the gain region) or failure ($\tau_{\mathrm{adv}}/\tau_{\mathrm{heat}} \lesssim 1$) of the neutrino-driven explosion (e.g., [141]). For the 2D model (right panel), it can be shown that the condition of $\tau_{\mathrm{adv}}/\tau_{\mathrm{heat}} \gtrsim 1$ is satisfied behind the aspherical shock, which is deformed predominantly by the SASI ($\ell = 2$ mode at the snapshot), while the ratio is shown to be smaller than unity in the whole region behind the spherical standing accretion shock (left panel: 1D). Here SASI that develops due to the advection-acoustic cycle in the supernova core [142, 143], is a uni- and bipolar sloshing of the stalled bounce shock with pulsational strong expansion and contraction (seen as oscillations in the red curves in Figure 1). Comparing the left half of each panel, the entropy for the 2D model is shown to be larger than for the 1D model. This is also the evidence that the neutrino heating works more efficiently in 2D.

To see clearly the effects of stellar rotation on the postbounce evolution, a rapidly rotating model was taken in [138], in which a constant angular frequency of $\Omega_0 = 2$ rad/s is imposed inside the iron core with a dipolar cut off ($\propto r^{-2}$) outside. This corresponds to $\beta \sim 0.18\%$ with β being the ratio of the rotational to the gravitational energy when the simulation starts. For the rotating model, the dominant mode of the shock deformation after bounce is almost always the $\ell = 2$ mode for the rotating model (e.g., Figure 3, although the $\ell = 1$ mode can be as large as the $\ell = 2$ mode when the SASI enters the nonlinear regime ($\gtrsim 200$ ms after bounce). In contrast to this rotation-induced $\ell = 2$ deformation, the $\ell = 1$ mode tends to be larger than the $\ell = 2$ mode for the 2D models without rotation in the saturation phase.

The top panel of Figure 4 shows the comparison of the explosion energies for the 2D models with and without rotation (models with "-rot" or "-2D," resp.). Although the explosion energies depend on the numerical resolutions quantitatively, they show a continuous increase for the rotating models. The explosion energies for the models without rotation, on the other hand, peak around 180 ms when the neutrino-driven explosion sets in (see also Figure 1), and show a decrease later on. The reason for the greater explosion energy for models with rotation is due to the bigger mass of the exploding material. This is because the northsouth symmetric ($\ell = 2$) explosion can expel more material than for the unipolar explosion (compare the left and right of the bottom panel in Figure 4). The explosion energies when we terminated the simulation are less than $\lesssim 10^{50}$ erg for all the computed models. For the rotating models, we are tempted to speculate that the explosion energies could increase later on by a linear extrapolation. However, in order to identify the robust feature of an explosion, a longer-term simulation with improved input physics is needed. In combination with the assumed rapid rotation, magnetic fields should be also taken into account as in [46], which we are going to study as a followup of our rotating 2D models (Suwa et al. in preparation).

Extending our 2D modules mentioned above, we are currently running 3D simulations for an $11.2 M_\odot$ star of Woosley et al. [118] with spectral neutrino transport that is solved by the IDSA scheme in a ray-by-ray manner [139]. We briefly summarize the results in the following.

The left panel of Figure 5 shows mass-shell trajectories for the 3D (red lines) and 1D model (green line), respectively. At around 300 ms after bounce, the average shock radius for the 3D model exceeds 1000 km in radius. On the other hand, an explosion is not obtained for the 1D model, which is in agreement with Buras et al. [124]. The right panel of Figure 5 shows a comparison of the average shock radius versus postbounce time. In the 2D model, the shock expands rather continuously after bounce. This trend is qualitatively consistent with the 2D result by Buras et al. [124] (see their Figure 15 for model s112_128_f). The reason that the average shock of our 2D model expands much faster than theirs would come from the neglected effects in this work including general relativistic effects, inelastic neutrino-electron scattering, and cooling by heavy-lepton neutrinos. All of them could give a more optimistic condition to produce explosions. Apparently these ingredients should be appropriately implemented, which we hope to be practicable in the next-generation 3D simulations.

Comparing the shock evolution between our 2D (green line in the right panel of Figure 5) and 3D models (red line), the shock is shown to expand much faster for 2D. The pink line labeled by "3D low" is for the low resolution 3D model, in which the mesh numbers are taken to be half of the standard model. Note that the 3D computational grid consists of 300 logarithmically spaced, radial zones to cover

from the center up to 5000 km and 64 polar (θ) and 32 azimuthal (ϕ) uniform mesh points, which are used to cover the whole solid angle. The low resolution 3D model has one-half of the mesh numbers in the ϕ direction ($n_\phi = 16$), while fixing the mesh numbers in other directions. Comparing with our standard 3D model (red line), the shock expansion becomes less energetic for the low resolution model (later than ~ 150 ms). Above results indicate that explosions are easiest to obtain in 2D, followed in order by 3D, and 3D (low). At first sight, this may look contradicted with the finding by Nordhaus et al. [116] who pointed out that explosions could be more easily obtained in 3D than in 2D. The reason of the discrepancy is summarized shortly as it follows.

Figure 6 compares the blast morphology for our 3D (left panel) and 2D (right) model (note that the polar axis is tilted (about $\pi/4$) both in the left and middle panel). In the 3D model (left panel), nonaxisymmetric structures are clearly seen. By performing a tracer-particle analysis, the maximum residency time of material in the gain region is shown to be longer for 3D than 2D due to the nonaxisymmetric flow motions (see Figure 7). This is one of advantageous aspects of 3D models to obtain the neutrino-driven explosions. On the other hand, our detailed analysis showed that convective matter motions below the gain radius become much more violent in 3D than in 2D, making the neutrino luminosity larger for 3D (see [139] for more details). Nevertheless the emitted neutrino energies are made smaller due to the enhanced cooling. Due to these competing ingredients, the neutrino-heating timescale becomes shorter for the 3D model, leading to a smaller net-heating rate compared to the corresponding 2D model (Figure 8). Note here that the spectral IDSA scheme, by which the feedback between the mass accretion and the neutrino luminosity can be treated in a self-consistent manner (not like the light-bulb scheme assuming a constant luminosity), sounds quite efficient in the first-generation 3D simulations.

As seen from Figure 5, an encouraging finding in [139] was that the shock expansion tends to become more energetic for models with finer resolutions. These results would indicate whether these advantages for driving 3D explosions can or cannot overwhelm the disadvantages is sensitive to the employed numerical resolutions (It is of crucial importance to conduct a convergence test in which a numerical gridding is changed in a systematic way (e.g., [144]).) To draw a robust conclusion, 3D simulations with much more higher numerical resolutions and also with more advanced treatment of neutrino transport as well as of gravity are needed, which could be hopefully practicable by utilizing forthcoming petaflops-class supercomputers.

In addition to the 3D effects, impacts of GR on the neutrino-driven mechanism stand out among the biggest open questions in the supernova theory. It should be remembered that using newly derived Einstein equations [145], the consideration of GR was standard in the pioneering era of supernova simulations (e.g., [146]). One year after Colgate and White [90], Schwartz [147] reported the first fully GR simulation of stellar collapse to study the supernova mechanism, who implemented a gray transport of neutrino

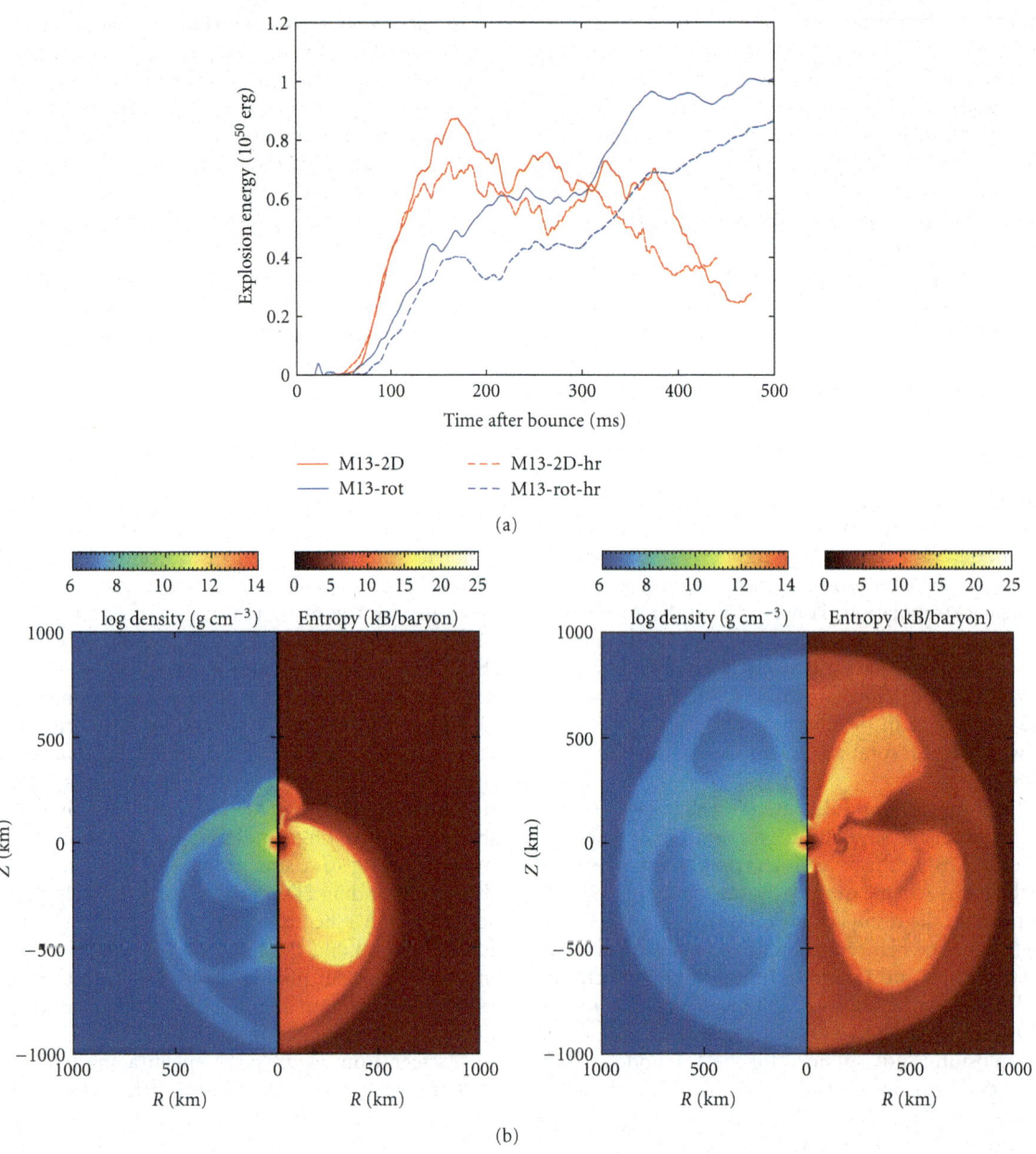

FIGURE 4: Time evolution of the explosion energy versus postbounce time for 2D models with and without rotation (a). The explosion energy is defined as the total energy (internal plus kinetic plus gravitational), integrated overall matter where the sum of the corresponding specific energies is positive. Models with "-hr" indicate the ones with higher numerical resolution, in which the mesh numbers in the lateral direction are doubled. (b) are snapshots of the density (left half) and the entropy (right half) for 1D (left) and 2D rotating (right) models at the epoch when the shock reaches 1000 km. These figures are taken from [138].

diffusion in the 1D GR hydrodynamics (citing from his paper, "*In this calculation, the neutrino luminosity of the core is found to be* 10^{54} *erg/s, or 1/2 a solar rest mass per second !! ... This is the mechanism which the supernova explodes.*" The neutrino luminosity rarely becomes so high in the modern simulations, but it is surprising that the potential impact of GR on the neutrino-heating mechanism was already indicated in the very first GR simulation). Using GR Boltzmann equations derived by Lindquist [148], Wilson [149] developed a 1D GR-radiation-hydrodynamic code including a more realistic (at the time) description of the

collisional term than the one in [147]. By performing 1D GR hydrodynamic simulations that included a leakage scheme for neutrino cooling, hydrodynamical properties up to the prompt shock stagnation were studied in detail [150–152]. These pioneering studies, albeit using a much simplified neutrino physics than today, did provide a bottom-line of our current understanding of the supernova mechanism (see [153] for a complete list of references for the early GR studies). In the middle of the 1980s, Bruenn [134] developed a code that coupled 1D GR hydrodynamics to the MGFLD transport of order (v/c) including the so-called standard set

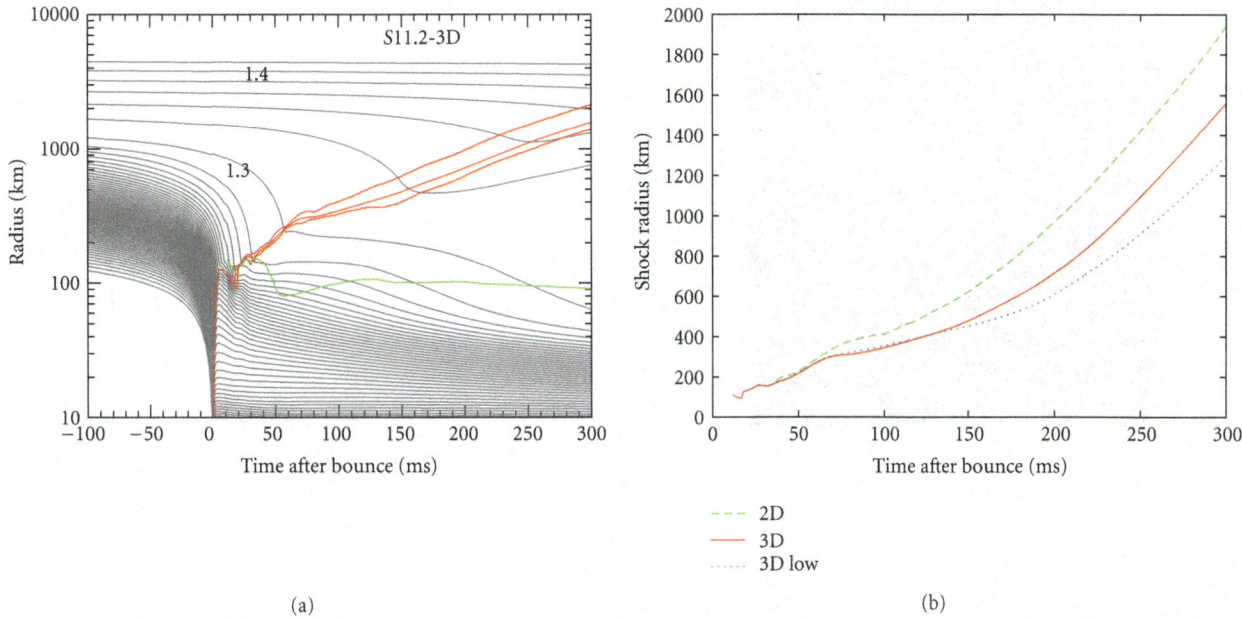

(a)   (b)

FIGURE 5: Time evolution of our 3D model [139] visualized by mass shell trajectories in thin gray lines (left panel). Thick red lines show the position of shock waves, noting that the maximum (top), average (middle), and the minimum (bottom) shock position are shown, respectively. The green line represents the shock position of the 1D model. "1.30" and "1.40" indicate the mass in unit of $M_\odot$ enclosed inside the mass-shell. Right panel shows the evolution of average shock radii for the 2D (green line) and 3D (red line) models. The "3D low" (pink line) corresponds to the low resolution 3D model, in which the mesh numbers are taken to be half of the standard model (see text). These figures are taken from [139].

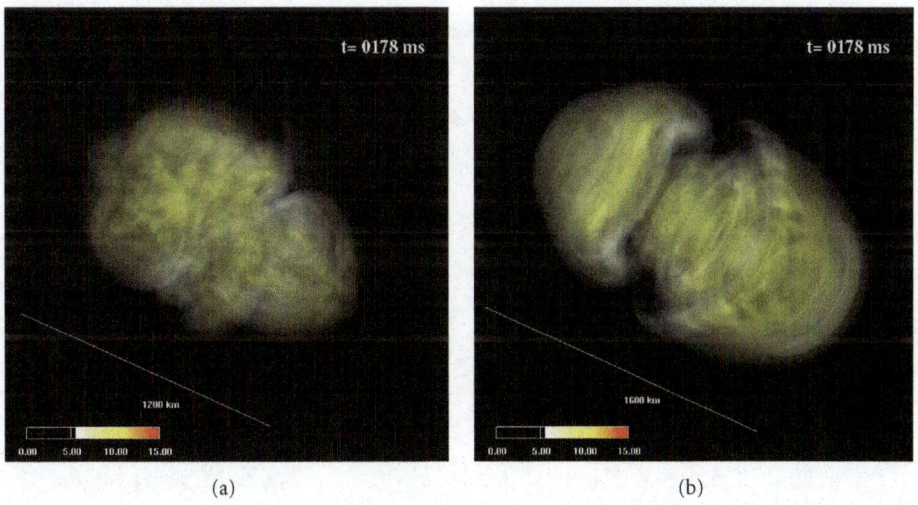

(a)   (b)

FIGURE 6: Volume rendering of entropy showing the blast morphology in our 3D (left) and 2D (right) model for the $11.2M_\odot$ progenitor of Woosley et al. [118] (at $t = 178$ ms after bounce), respectively. The linear scale is indicated in each panel. These figures are taken from [139].

of neutrino interactions. Since the late 1990s, the ultimate 1D simulations, in which the GR Boltzmann transport is coupled to 1D GR hydrodynamics, have been made feasible by Yamada et al. [154–157] (very recently, they reported their success to develop the first multiangle, multienergy neutrino transport code in 3D [158]) and by Mezzacappa et al. [153, 159–161] (and by their collaborators).

Among them, Bruenn et al. [153] firstly showed that the neutrino luminosity and the average neutrino energy of any neutrino flavor during the shock reheating phase increase when switching from Newtonian to GR hydrodynamics. They also pointed out that the increase is larger in magnitude compared to the decrease due to redshift effects and gravitational time dilation. By employing the

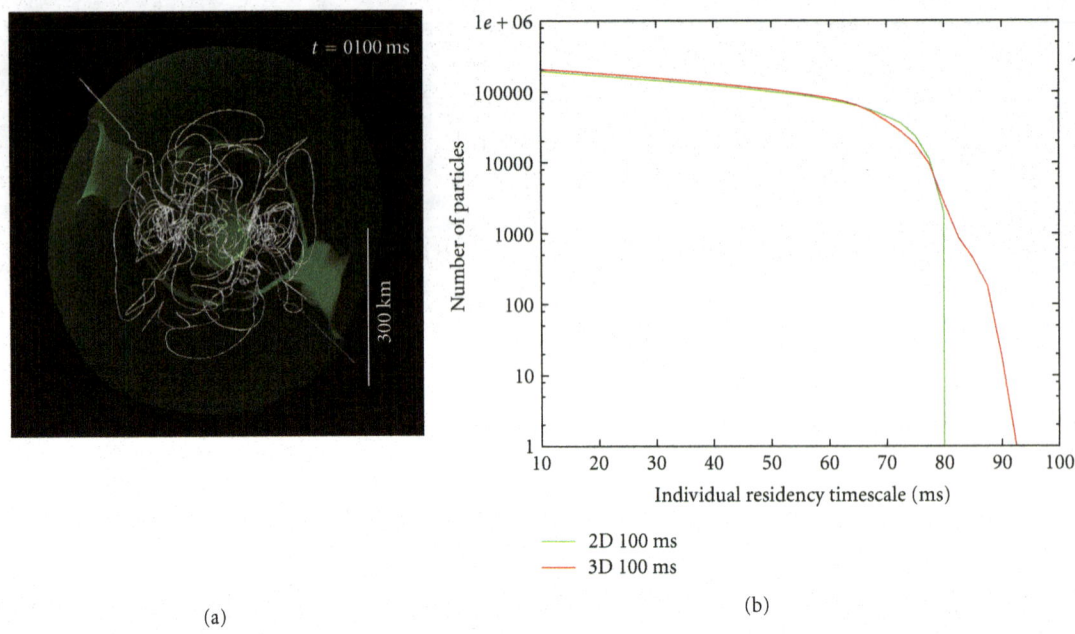

(a)   (b)

FIGURE 7: The left panel shows streamlines of selected tracer particles advecting through the shock wave to the PNS seen from the polar direction in our 3D model [139]. Several surfaces of constant entropy marking the position of the shock wave (greenish outside) and the PNS (indicated by the central sphere) are shown with the linear scale given in the right-bottom edge. The right panel shows the number of tracer-particles travelling in the gain region as a function of their individual residency time between the 2D and 3D model, at 100 ms after bounce. If the left panel were for 2D models, the streamlines would be seen as a superposition of circles with different diameters. The maximum residency timescale for the 3D model is shown to be longer than that in the corresponding 2D model. These figures are taken from [139].

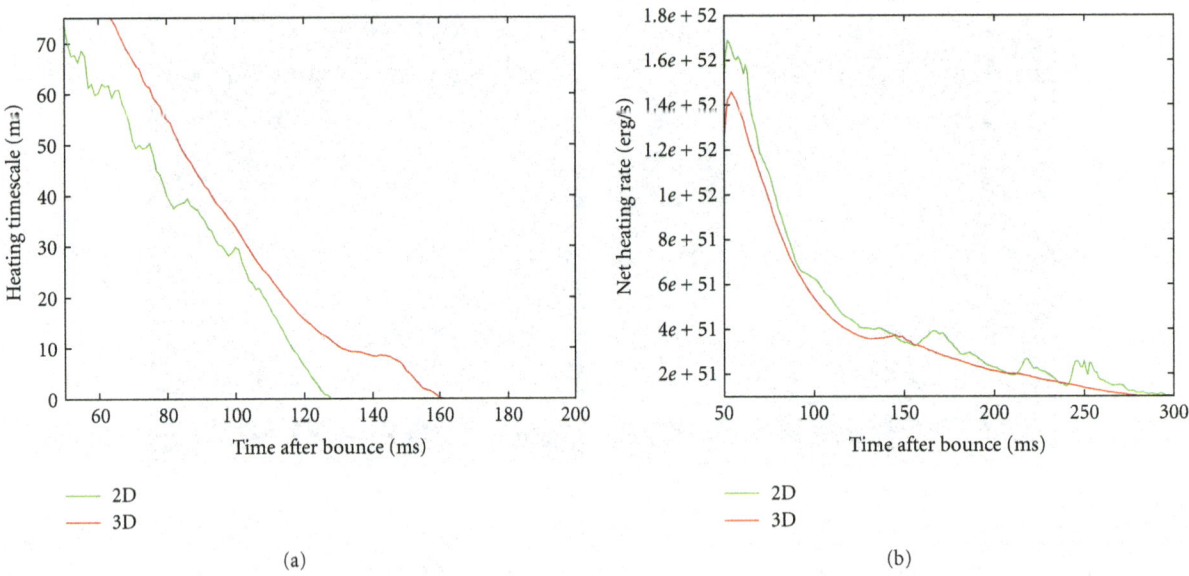

(a)   (b)

FIGURE 8: Time evolution of neutrino-heating timescale (left) and total net rate of neutrino heating (right) in our 2D (green line) and 3D models (red line). These figures are taken from [139].

current best available weak interactions, Lentz et al. [162] reported the update of Bruenn et al. [153] very recently. They showed that the omission of observer corrections in the transport equation particularly does harm to drive the neutrino-driven explosions. A disadvantageous trend of

GR to produce the neutrino-driven explosions has been commonly observed in these full-fledged 1D simulations; the residency time of material in the gain region becomes shorter due to the stronger gravitational pull. Due to this competition between the gain and loss effects in the end,

GR works disadvantageously to facilitate the neutrino-driven explosions in 1D. In fact, the maximum shock extent in the postbounce phase is shown to be 20% smaller when switching from Newtonian to GR hydrodynamics (e.g., Figure 2 in [162]).

Among the most up-to-date multi-D models with spectral neutrino transport mentioned earlier (Table 1), the GR effects are best attempted to be modeled by using a modified gravitational potential that takes into account a 1D, post-Newtonian correction [124, 125, 135, 163]. A possible drawback of this prescription is that a conservation law for the total energy cannot be guaranteed by adding an artificial term in the Poisson equation. Since the energy reservoir of the supernova engines is the gravitational binding energy, any potential inaccuracies in the argument of gravity should be avoided. There are a number of relativistic simulations of massive stellar collapse in full GR (e.g., 2D [164] or 3D [165, 166], and references therein) or using the conformally-flatness approximation (CFC) (e.g., [130, 131]). Although extensive attempts have been made to include microphysics such as by the Y_e formula [167] or by the neutrino leakage scheme [168], the effects of neutrino heating have yet to be included in them, which has been a main hindrance to study the GR effects on the multi-D neutrino-driven mechanism (see, however, [129, 169]).

Putting things together, a complete "realistic" supernova model should naturally be done in full GR (MHD) with multi-D GR Boltzmann neutrino transport, in which a microphysical treatment of equation of state (EOS) and nuclear-neutrino interactions are implemented as realistically as possible. Unfortunately none of the currently published SN simulations satisfy the "ultimate" requirement (e.g., Table 1). In this sense, all the mentioned studies employ some approximations (with different levels of sophistication) towards the final goal.

In the same way, theoretical predictions of the SN multimessengers that one can obtain by analyzing the currently available numerical results, cannot unambiguously give us the final answer yet. Again we hereby note the feature of this paper which shows only a snapshot of the moving theoretical terrain. Keeping this caveat in mind, it is also true that a number of surprising GW features of CCSNe have been reported recently both by the first-principle simulations (e.g., in Table 1) and also by idealized simulations in which explosions are parametrically initiated mostly by the light-bulb scheme. As will be mentioned in the next section, the latter approach is also useful to get a better physical understanding of the GW signatures obtained in the first-principle simulations (in this sense, these two approaches are complimentary in understanding the GW signatures). Having shortly summarized a current status of CCSN simulations, we are now ready to move on to focus on the GW signatures from the next section.

2.2. Gravitational Waves. The paper by Müller [170] entitled as *"Gravitational Radiation from Collapsing Rotating Stellar Cores"* unquestionably opened our eyes to the importance of making the GW prediction based on realistic SN numerical modeling (see e.g., Section 2 in [23] for a summary of more earlier work which had mainly focused on the GW emission in very idealized systems such as in homogeneous spheroids and ellipsoids) (needless to say, this kind of approach is still very important to extract the physics of the GW emission mechanism). As one may expect from the title of his paper, rapid rotation, if it would exist in the precollapse iron core, leads to significant rotational flattening of the collapsing and bouncing core, which produces a time-dependent quadrupole (or higher) GW emission. Following the first study by Müller, most studies of the past thirty have focused on the so-called bounce signals (e.g., [170–184]), and references therein).

As summarized by Ott [23], a number of important progresses have been recently made to understand features of the bounce signals by extensive 2D GR studies using the CFC approximation [180–182] and also by fully GR 3D simulations [177] both including realistic EOSs and a deleptonization effect based on 1D-Boltzmann simulations [185].

For the bounce signals having a strong and characteristic signature, the iron core must rotate enough rapidly. However recent stellar evolution calculations suggest that rapid rotation assumed in most of the previous studies is not canonical for progenitors with neutron star formations [126, 127, 186]. To explain the observed rotation periods of radio pulsars, the precollapse rotation periods are estimated to be larger than ~100 sec [186]. In such a slowly rotating case, the detection of the bounce signals becomes very hard even by next-generation laser interferometers for a Galactic supernova (e.g., [175]).

Besides the rapid rotation, convective matter motions and anisotropic neutrino emission in the postbounce phase are expected to be primary GW sources with comparable amplitudes to the bounce signals (e.g., [187] for a review). Thus far, various physical ingredients for producing asphericities and the resulting GWs in the postbounce phase have been studied, such as the roles of precollapse density inhomogeneities [102, 188, 189], moderate rotation of the iron core [190], nonaxisymmetric rotational instabilities [191, 192], g-modes [193] and r-modes pulsations [20] of PNSs, and the SASI [194–198] (see also [199] for the GW signals at the black hole formation). Among them, the most promising GW sources may be convection and SASI, because the degree of the initial inhomogeneities [200] and the growth of rotational instabilities as well as the r-modes and g-modes pulsations are rather uncertain.

Based on 2D simulations that parametrize the neutrino heating and cooling by the light-bulb scheme (which we shortly call as *parametric SASI simulations* hereafter), we pointed out that the GW amplitudes from anisotropic neutrino emission (as anisotropic matter motions generate GWs, anisotropic neutrino emission also gives rise to GWs, which has been originally pointed out in late 1970's by Epstein [201] and Turner and Wagoner [202] (see recent progress in [203]). It is expected as a primary GW source also in gamma-ray bursts [204–206] and Pop III stars [207]) increase almost monotonically with time, and that such signals may be visible to next-generation detectors for a

Galactic source [195, 196]. By performing such a parametric simulation but without the excision inside the PNS, Murphy et al. [198] showed that the GW signals from matter motions can be a good indicator of the explosion geometry. These features qualitatively agree with the ones obtained by Yakunin et al. [208] who reported exploding 2D simulations in which the ray-by-ray MGFLD neutrino transport is solved with the hydrodynamics (e.g., Oak-Ridge+ simulations in Table 1). Marek et al. [197] analyzed the GW emission based on their long-term 2D ray-by-ray Boltzmann simulations, which seem very close to produce explosions [125] (e.g., MPA simulations in Table 1). They also confirmed that the GWs from neutrinos with continuously growing amplitudes (but with the different sign of the amplitudes in [195, 196, 208]), are dominant over the ones from matter motions. They proposed that the third-generation class detectors such as the Einstein Telescope are required for detecting the GW signals with a good signal-to-noise ratio.

Regarding the GW predictions in 3D models, Müller and Janka [189] coined the first study to analyze the GW signature of 3D nonradial matter motion and anisotropic neutrino emission from prompt convection in the outer layers of a PNS during the first 30 ms after bounce. Their first 3D calculations using the light-bulb recipe were forced to be performed in a wedge of opening angle of 60°. Albeit with this limitation (probably coming from the computer power at that time), they obtained important findings that because of smaller convective activities inside the cone with slower overturn velocities, the GW amplitudes of their 3D models are more than a factor of 10 smaller than those of the corresponding 2D models, and the wave amplitudes from neutrinos are a factor of 10 larger than those due to nonradial matter motions. With another pioneering (2D) study by Burrows and Hayes [188] it is worth mentioning that those early studies had brought new blood into the conventional GW predictions, which illuminated the importance of the theoretical prediction of the neutrino GWs.

A series of findings obtained by Fryer et al. in early 2000s [101, 209] have illuminated also the importance of the 3D modeling. By running their 3D Newtonian Smoothed-Particle-Hydrodynamic (SPH) code coupled to a gray flux-limited neutrino transport scheme, they studied the GW emission due to the inhomogeneous core-collapse, core rotation, low-modes convection, and anisotropic neutrino emission. Although the early shock-revival and the subsequent powerful explosions obtained in these SPH simulations have yet to be confirmed by other groups, their approach paying particular attention to the multiple interplay between the explosion dynamics, the GW signatures, the kick and spins of pulsars, and also the nonspherical explosive nucleosynthesis, blazed a new path on which current supernova studies are progressing.

We also studied the GW signals from 3D models that mimic neutrino-driven explosions aided by the SASI [194, 196, 210]. In the series of our 3D experimental simulations, the light-bulb scheme was used to obtain explosions and the initial conditions were derived from a steady-state approximation of the postshock structure and the dynamics only outside an inner boundary at 50 km was solved. Based on the results, we show in the following that features of the gravitational waveforms obtained in 3D models are significantly different than those in 2D, which tells us a necessity of 3D modeling for a reliable prediction of the GW signals.

2.2.1. Stochastic Nature of Gravitational Waves in 3D Simulations. Figure 9 shows the evolution of 3D hydrodynamic features from the onset of the nonlinear regime of SASI (top left) until the shock breakout (this corresponds to the shock emergence at the outer boundary of the computational domain (~2000 km in radius)) (bottom right) with the gravitational waveform from neutrinos inserted in each panel. After about 100 ms, the deformation of the standing shock becomes remarkable marking the epoch when the SASI enters the nonlinear regime (top left of Figure 9). At the same time, the gravitational amplitudes begin to deviate from zero. Comparing the top two panels in Figure 10, which shows the total amplitudes (top panel, neutrino + matter) and the neutrino contribution only (bottom), it can be seen that the overall structures of the waveforms are predominantly determined by the neutrino-originated GWs with the slower temporal variations ($\gtrsim$30–50 ms), to which the GWs from matter motions with rapid temporal variations ($\lesssim$10 ms) are superimposed.

As seen from the top right through bottom left to right panels of Figure 9, the major axis of the growth of SASI is shown to be not aligned with the symmetric axis (z-axis in the figure) and the flow inside the standing shock wave is not symmetric with respect to this major axis (see the first and third quadrants in Figure 9). This is a generic feature in the computed 3D models, which is in contrast to the axisymmetric case. The GW amplitudes from SASI in 2D showed an increasing trend with time due to the symmetry axis, along which SASI can develop preferentially [195, 196]. Free from such a restriction, a variety of the waveforms is shown to appear (see waveforms inserted in Figure 5). Furthermore, the 3D standing shock can also oscillate in all directions, which leads to the smaller explosion anisotropy than 2D. With these two factors, the maximum amplitudes seen either from the equator or the pole become smaller than 2D. On the other hand, their sum in terms of the total radiated energy are found to be almost comparable between 2D and 3D models, which is likely to imply the energy equipartition with respect to the spatial dimensions.

The (two) pair panels of Figure 10 show the gravitational waveforms for models with different neutrino luminosity. The input luminosity for the pair panels (models A (top two) and B (bottom two)) differs only 0.5% (although any seed perturbation would demonstrate the chaotic behavior, we simply chose the value of 0.5% as a reference). Despite the slight difference in the luminosities, the waveforms of each polarization are shown to exhibit no systematic similarity when seen from the pole or equator. This also reflects a chaotic nature of the SASI influenced by small differences (see also [211]).

It should be noted that the approximations taken in the simulation, such as the excision inside the PNS with its

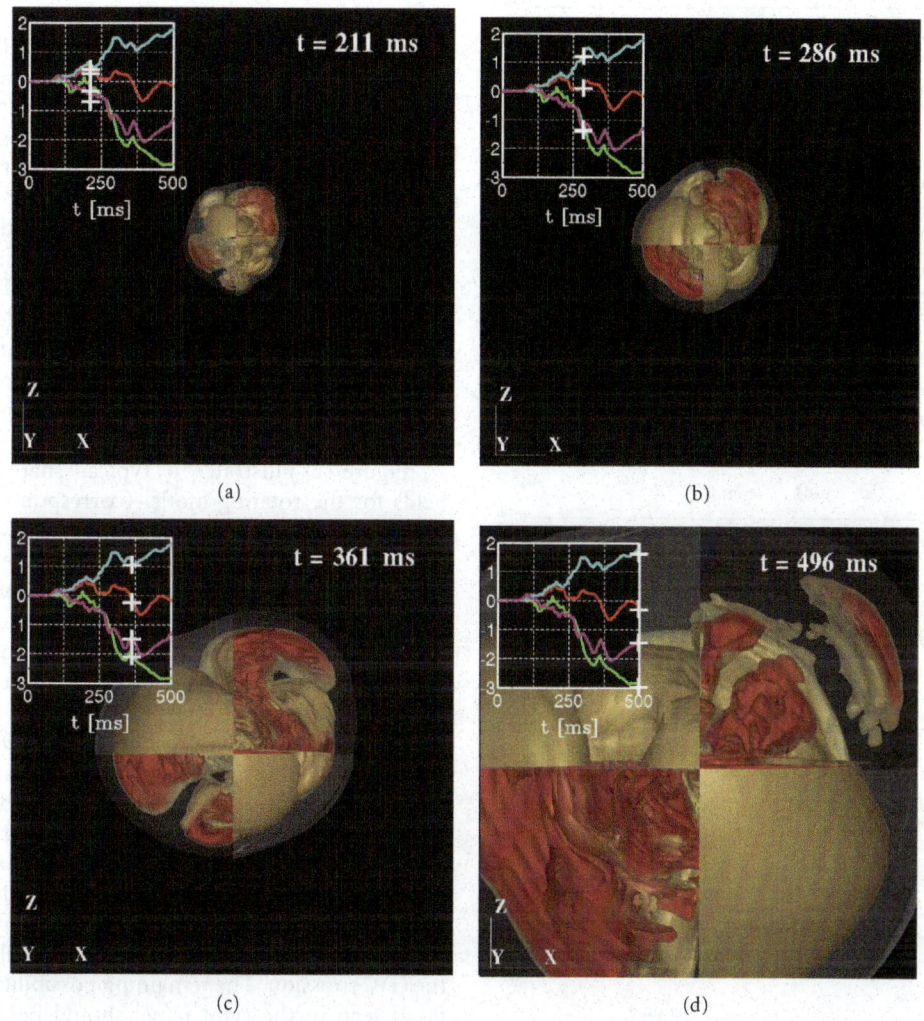

FIGURE 9: Four snapshots of the entropy distributions of a representative 3D supernova explosion model (corresponding to model A in [194]). The second and fourth quadrants of each panel show the surface of the standing shock wave. In the first and third quadrants, the profiles of the high entropy bubbles (colored by red) inside the section cut by the ZX plane are shown. The side length of each plot is 1000 km. The insets show the gravitational waveforms from anisotropic neutrino emissions, with "+" on each curve representing the time of the snapshot. Note that the colors of the curves are taken to be the same as the top panel of Figure 10. This figure is taken from [194].

fixed inner boundary and the light bulb approach with the isotropic luminosity constant with time, are the very first step to model the dynamics of the neutrino-heating explosion aided by SASI and study the resulting GWs. As already mentioned, the excision of the central regions inside PNSs may hinder the efficient gravitational emission of the oscillating neutron star [193] and the nonaxisymmetric instabilities [184, 192] of the PNSs, and the enhanced neutrino emissions inside the PNSs [197]. Bearing these caveats in mind, a piece of encouraging news is that the gravitational waveforms obtained in the 2D radiation-hydrodynamic simulations [208] are similar to the ones obtained in our 2D study using the light-bulb scheme [196]. More recently, Müller et al. [211] confirmed the stochastic nature by analyzing the GW features obtained in their 3D models [115]. The obtained GW amplitudes as well as the degree of anisotropic neutrino

emission are almost comparable to our results, which are considerably smaller than 2D models [190, 197, 198, 208].

2.2.2. Breaking of the Stochasticity due to Stellar Rotation.
More recently, we studied the effects of stellar rotation on the stochastic nature of the GWs mentioned above [210]. In 3D, the modes of SASI are divided into *sloshing* modes and *spiral* modes (e.g., [111]). Asymmetric $m = 0$ modes so far studied in 2D models and axisymmetric $m \neq 0$ modes are classified into the *sloshing* modes, where m stands for the azimuthal index of the spherical harmonics Y_l^m. In the latter situation, the $\pm m$ modes degenerate so that the $+m$ modes has the same amplitudes as the $-m$ modes. If random perturbations or uniformly rotating flows are imposed on these axisymmetric flows in the postbounce phase, the degeneracy is broken and

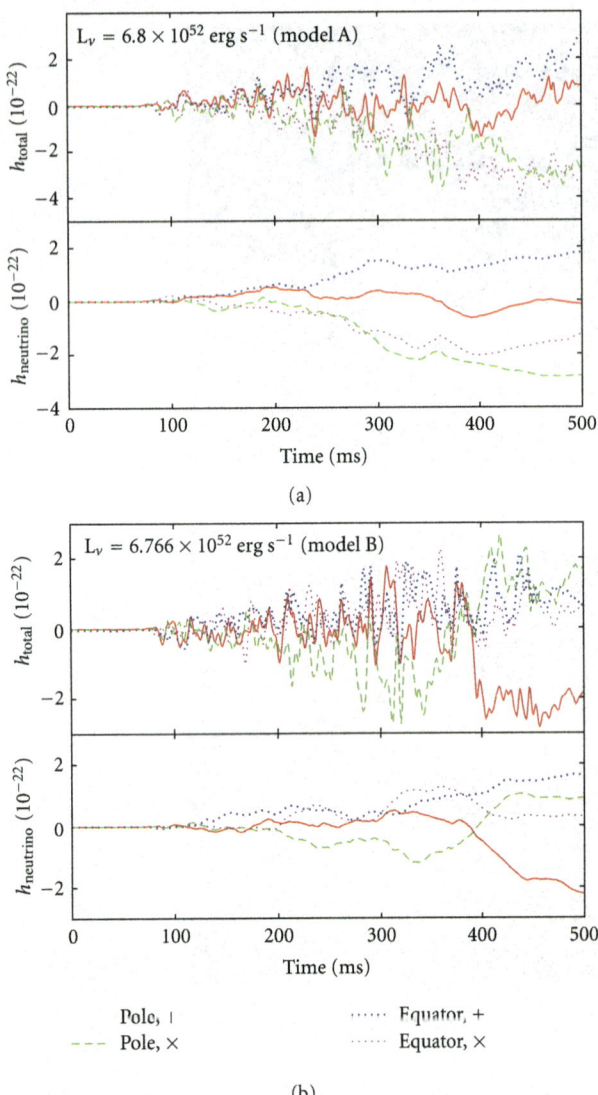

FIGURE 10: Gravitational waveforms from neutrinos (bottom) and from the sum of neutrinos and matter motions (top), seen from the polar axis and along the equator (indicated by "Pole" and "Equator") with polarization (+ or × modes) for two representative 3D models of A and B (see [194] for details). The distance to the SN is assumed to be 10 kpc.

the rotational modes emerge [112, 212]. In this situation, the $+m$ modes has the different amplitudes from $-m$ modes. Such rotating nonaxisymmetric $m \neq 0$ modes are called as *spiral* modes. These nonaxisymmetric modes are expected to bring about a breakthrough in our supernova theory, because they can help to produce explosions more easily compared to 2D due to its extra degree of freedom [116], and also because they may have a potential to generate pulsar spins ([212, 213], see also [115, 214]). These findings illuminating the importance of stellar rotation motivated us to clarify how rotation that gives a special direction to the system (i.e., the spin axis), could affect the stochastic GW features which we observed in the absence of rotation (Section 2.2.1).

Figure 11 shows the gravitational waveform for a typical 3D model with rotation (left: total amplitudes, right: neutrino only). To construct a model with rotation, we give a uniform rotation on the flow advecting from the outer boundary of the iron core as in [112], whose specific angular momentum is assumed to agree with recent stellar evolution models [126, 127]. Comparing to Figure 10 (for models without rotation), one can clearly see a sudden rise in the GW amplitude after around 500 ms for the rotating model (blue line in Figure 11), which is the plus mode of the neutrino GW seen from the equator. By systematically changing the initial angular momentum and the input neutrino luminosities from the PNS, we computed fifteen 3D models in [210] and found that the GW features mentioned above were common among the models.

Figure 12 illustrates a typical snapshot of the flow fields for the rotating model (corresponding to the one in Figure 11) when the spiral SASI modes have already entered the nonlinear regime, seen from the pole (right panel) or from the equator (left panel), respectively. From the left panel, one may guess the presence of the sloshing modes that happen to develop along the rotational axis (Z-axis) at this epoch. It should be emphasized that although the dominance of $h_{\nu,+}^{\text{equ}}$ observed in the current 3D simulations is similar to the one obtained in previous 2D studies [196], its origin has nothing to do with the coordinate symmetry axis. The preferred direction here is determined by the spin axis. Free from the 2D axis effects, the major axis of the sloshing SASI mode changes stochastically with time, and the flow patterns behind the standing shock simultaneously change in every direction like the nonrotating models. As a result, the sloshing modes can make only a small contribution to the GW emission. The remaining possibility is that the spiral flows seen in the right panel should be a key importance to understand the GW feature mentioned above. In fact, by analyzing the matter distribution on the equatorial plane, we find that the compression of matter is more enhanced in the vicinity of the equatorial plane due to the growth of the spiral SASI modes, leading to the formation of the spiral flows circulating around the spin axis with higher temperatures. As a result, the neutrino emission seen parallel to the spin axis becomes higher than the ones seen from the other direction. Remembering that the lateral-angle (θ) dependent function of the GW formulae (e.g., in (9) in [196]) is positive near the north and south polar caps, the dominance of the polar neutrino luminosities leads to make the positively growing feature of $h_{\nu,+}^{\text{equ}}$ in Figure 11 (blue line). From the spectral analysis of the gravitational waveform (Figure 13), it can be readily seen that it is not easy to detect these neutrino-originated GW signatures with slower temporal evolution ($\gtrsim O(10)$ ms) by ground-based detectors whose sensitivity is limited mainly by the seismic noises at such lower frequencies. However these signals may be detectable by the recently proposed future space interferometers like Fabry-Perot type DECIGO ([217], black line in Figure 13). Contributed by the neutrino GWs in the lower frequency domains, the total GW spectrum tends to become rather flat over a broad frequency range below ~100 Hz. These GW features obtained in the context of the SASI-aided neutrino-driven mechanism

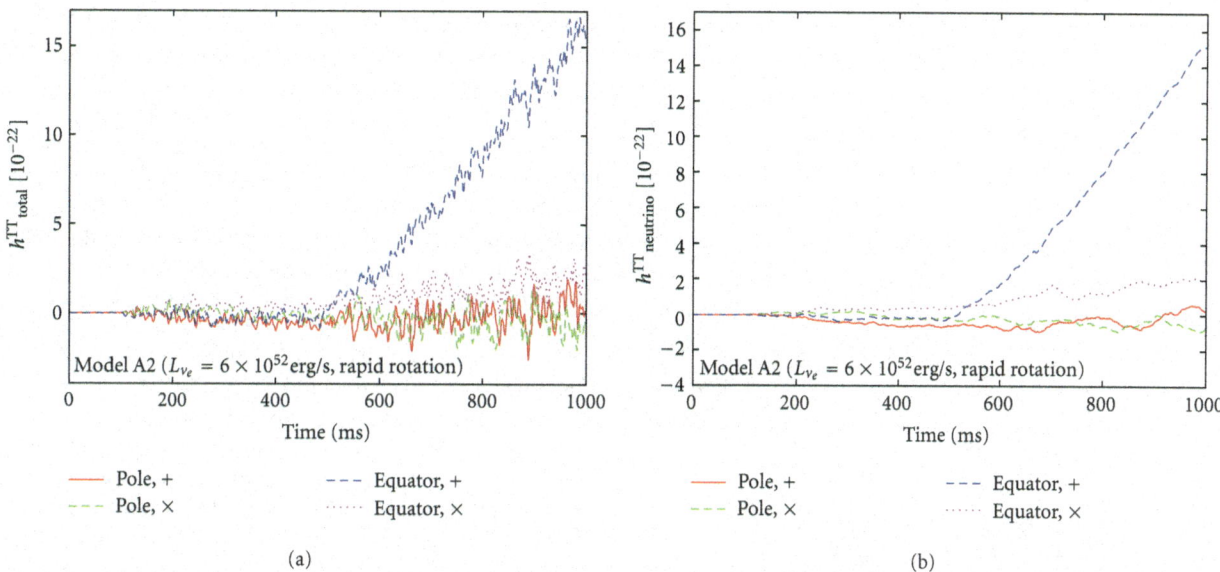

(a) (b)

FIGURE 11: Gravitational waveforms from the sum of neutrinos and matter motions (left) and only from neutrinos (right) for a 3D model with rotation (from [210]). The time is measured from the epoch when the neutrino luminosity is injected from the surface of the neutrino sphere. For this 3D model with rotation, the rotational flow is imposed to advect to the PNS surface at around $t = 400$ ms. The supernova is assumed to be located at the distance of 10 kpc.

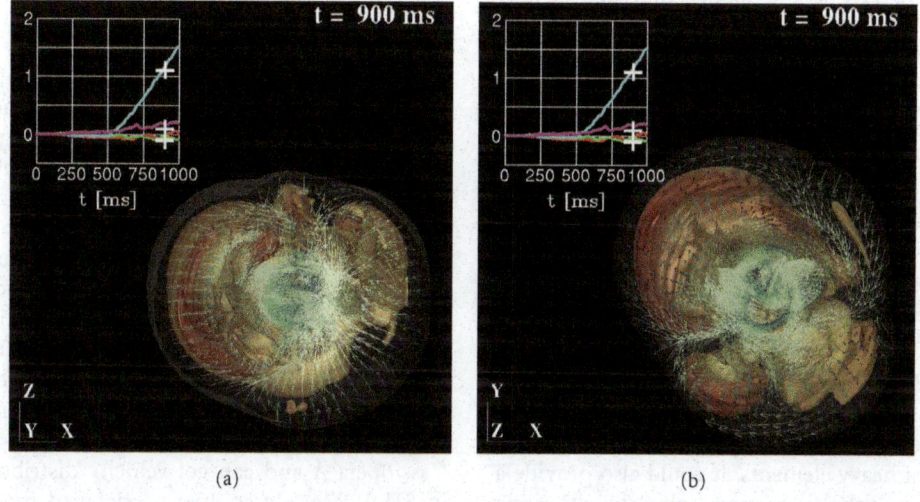

(a) (b)

FIGURE 12: Partial cutaway of the entropy isosurfaces and the velocity vectors on the cutting plane for a 3D model that includes rotation. Left and right panels are for the equatorial and polar observer, respectively. The insets show the gravitational waveforms with "+" on each curves representing the time of the snapshot. Note that the colors of the curves are taken to be the same as the top panel of Figure 10. This figure is taken from [210].

are different from the ones expected in the other candidate mechanisms, such as the MHD mechanism (e.g., [219]) and the acoustic mechanism [193]. Therefore the detection of such signals could be expected to provide an important probe into the explosion mechanism (e.g., [23, 220]).

We like to draw a caution that most of the 3D models cut out the PNS and the neutrino transport is approximated by a simple light-bulb scheme [210] or by the gray transport scheme [211]. Needless to say, these exploratory approaches are but the very first step to model the neutrino-heating

explosion and to study the resulting GWs. As already mentioned, the excision of the central regions inside PNSs truncates the feedback between the mass accretion to PNS and the resulting neutrino luminosity, which should affect the features of the neutrino GWs. By the cutout, efficient GW emission of the oscillating neutron star [193] and nonaxisymmetric instabilities [183, 184] of the PNSs, and the enhanced neutrino emissions inside the PNSs [197] cannot be treated in principle. To elucidate the GW signatures in a more quantitative manner, full 3D simulations with spectral

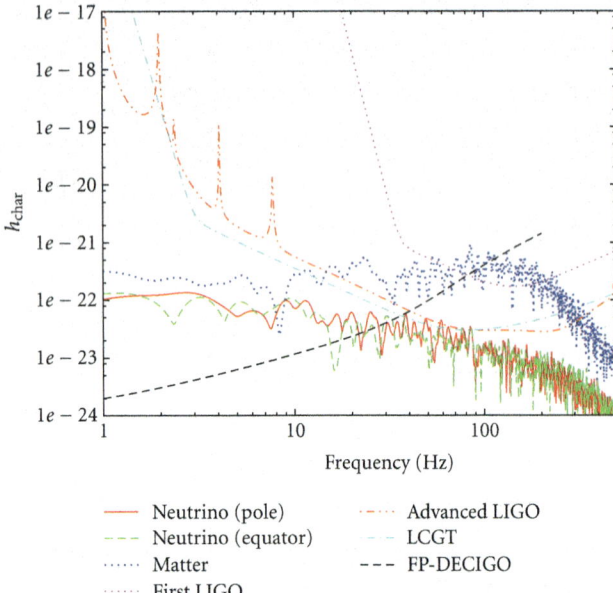

— Neutrino (pole)   ·-·-· Advanced LIGO
- - - Neutrino (equator)   ····· LCGT
····· Matter   - - - FP-DECIGO
····· First LIGO

FIGURE 13: Spectral distributions of GWs from matter motions ("Matter") and neutrino emission ("Neutrino") seen from the pole or the equator for a representative 3D rotating model (e.g., [210]) with the expected detection limits of TAMA300 [12], first LIGO and advanced LIGO [215], Large-scale Cryogenic Gravitational wave Telescope (LCGT) [216] and Fabry-Perot type DECIGO [217]. It is noted that h_{char} is the characteristic gravitational wave strain defined in [218]. The distance to the supernova is assumed to be 10 kpc. Note that for the matter signal, the + mode seen from the polar direction is plotted (from [210]).

neutrino transport are apparently needed (e.g., Section 2.1). This is unquestionably a vast virgin territory awaited to be explored for the future.

2.3. Explosive Nucleosynthesis. In this section, we proceed to discuss possible signatures of supernova nucleosynthesis. The study of nucleosynthesis is of primary importance to unveil the origins of heavy elements. It could also provide a valuable information of the ejecta morphology by observing the aspherical distributions of the synthesized elements especially for a nearby CCSN event (note that nucleosynthesis is not critical for the modeling of the light-curve and spectra for the most frequent types of SNe II-P). In the following, we first present a short overview paying particular attention to explosive nucleosynthesis, and then discuss possible observational signatures that would imprint information of multidimensionalities of the supernova engine.

When in a successful explosion the shock passes through the outer shells, its high temperature induces an explosive nucleosynthesis on short timescales (e.g., [118, 119, 221], and collective references in [222]). The observational determination of the masses of the three main radioactive isotopes ^{56}Ni, ^{57}Ni, and ^{44}Ti sets one of the main constraints on the explosion dynamics, because the production of these elements is sensitive to the track of density and temperature that the expanding material traces (e.g., [223]). During the

shock propagation, iron group elements such as ^{56}Ni and its daughter nucleus ^{56}Co are predominantly produced, which are radioactive with a lifetime of 8.8 days and 111.5 days, respectively. Most CCSNe enter the so-called nebular phase after the first few months when the expanding ejecta becomes optically thin in the continuum. In the early nebular phase, ^{56}Co is the major nuclear power source. As long as the decay particles are trapped by the ejecta, the radiation energy supplied by radioactivity is emitted instantaneously, so that the light curve can be described by an exponential decay with time, simply tracing the decay of the ^{56}Co nuclide. To explain the bolometric light curve of SN1987A in such a phase, the ^{56}Ni mass was determined to be $0.07 M_\odot$ [224].

After several years of explosion, the radioactive output from the ejecta no longer balances with the instantaneous input by radioactivity, because the reprocessing timescale is going to be longer [225]. The bolometric light curve is affected by the delayed release of the ionization energy. After that, a self-consistent modeling is needed, in which one should include a detailed calculation of the gamma-ray/positron thermalization and a determination of the time-dependent temperature, ionization, and excitation (e.g., [225–228] and references therein). Such a time-dependent modeling by Fransson and Kozma [225] revealed the ^{57}Ni mass of $\sim 3.3 \times 10^{-3} M_\odot$ of SN1987A, which agrees well with observations (e.g., [229, 230] and collective references in [231]).

By the similar reason to ^{57}Ni just mentioned above, the determination of the ^{44}Ti is also complicated. The most recent study by Jerkstrand et al. [228] gives an estimate of the ^{44}Ti to be $1.5^{+0.5}_{-0.5} \times 10^{-4} M_\odot$, which is in good agreement with the eight-year spectrum analysis of SN1987A (e.g., [232], see also [233]). As shown above, the amount of ^{44}Ti is typically one order-of-magnitude smaller than that of ^{57}Ni, however it is crucially important for young supernova remnants due to its long lifetime ($\sim$86 years). It is worth mentioning that NASA will launch the satellite NuSTAR (Nuclear Spectroscopic Telescope Array) to study ^{44}Ti production in CCSNe. The detector will be able to map out the ^{44}Ti distribution of the supernova remnant Cassiopeia A and can get velocity distributions of the ^{44}Ti in SN 1987A. By comparing detailed modeling of the SN nucleosynthesis in the context of 2D and 3D models (e.g., [234, 235]), these are expected to provide both direct probes of the explosion asymmetry.

Ever since SN1987A, challenges to the classical spherical modeling [119, 221, 236, 237] have been built also in the SN nucleosynthesis (likewise in the explosion theory and GWs mentioned so far). For many years it has been customary to simulate explosions and the effects of the shock wave on the explosive nucleosynthesis by igniting a thermal bomb in the star's interior or by initiating the explosion by a strong push with a piston. 2D simulations with manually imparted asymmetries showed that bipolar explosion scenarios could account for enhanced ^{44}Ti synthesis along the poles as indicated in SN1987A (e.g., [35]). More recently, 3D effects have been more elaborately studied ([238], see also [239]) as well as the impacts of different explosions by employing a number of progenitors [240] or by assuming a jet-like explosion [241, 242], which is one of the possible candidates of hypernovae (e.g., [243]).

In addition to the above-mentioned work, nucleosynthesis in a more realistic simulation that model multi-D neutrino-driven explosions has been also extensively studied (e.g., [33, 43, 244] and references therein). Although a small network has ever been included in the computations, these 2D simulations employing a light-bulb scheme [33] or a more accurate gray transport scheme [43, 44] have made it possible to elucidate nucleosynthesis inside from the iron core after the shock-revival up to explosion in a more consistent manner. Kifonidis et al. [43] demonstrated that the SASI-aided low-mode explosions can naturally explain masses and distributions of the synthesized elements observed in SN1987A. Their recent 3D results by Hammer et al. [34] show that the 3D effects change the velocity profiles as well as the growth of the Rayleigh-Taylor instability, which affects properties of the SN ejecta. In simulations with spectral neutrino transport, a new nucleosynthesis process, the so-called νp process, is reported to successfully explain some light proton-rich (p-)nuclei including ^{92,94}Mo and ^{96,98}Ru (e.g., [245–247] and references therein). It should be noted that self-consistent simulations are currently too computationally expensive to follow the dynamics of the expanding shock far outside the central iron core ($\gg 1$ s after bounce) where explosive nucleosynthesis takes place. On the other hand, the light-bulb scheme is not accurate to determine a sensitive balance between neutrino captures proceeding via ν_e or $\bar{\nu}_e$, which is decisive for quantifying the νp processes. Therefore the two approaches, namely, light-bulb scheme versus self-consistent neutrino transport, may be regarded as playing a complimentary role at present.

In the next section, we briefly summarize our findings on explosive nucleosynthesis in our 2D models that utilize the light-bulb scheme to trigger explosions (e.g., [248] for more details). By changing neutrino luminosities from PNSs systematically (L_ν), we discuss how the multidimensionality formed by the SASI and convection could impact on the explosive nucleosynthesis. We employ a nonrotating $15 M_\odot$ star with solar metalicity, which has been a best-studied supernova progenitor. The abundance pattern of the synthesized elements is estimated by a post-processing procedure, in which we have adopted a large nuclear network continuously maintained by the Joint Institute for Nuclear Astrophysics (JINA) REACLIB project [249]. Our readers might wonder why we are returning to 2D results here again after referring to 3D studies in Section 2.2. This is simply because our 3D project for the nucleosynthesis is currently in progress, and we like to note again the feature of this paper which shows only a snapshot of the moving theoretical terrain.

2.3.1. Symmetry Breaking in Explosive Nucleosynthesis.
Figure 14 shows how explosion energies (left panel), explosion timescales, as well as the PNS masses (right panel) change with the input neutrino luminosity (L_{ν_e}) for the 2D parametric explosion models (e.g., [248] for more details). As seen, higher L_{ν_e} makes explosion energy greater, the onset

of explosion (t_{exp}) earlier, the PNS masses (M_{PNS}) smaller. The PNS masses range in $(1.54–1.62)M_\odot$ for models with higher luminosity ($L_{\nu_e} \gtrsim 4.25 \times 10^{52}$ erg s^{-1}, panel (b), blue line), which are comparable to the baryonic mass (around $1.5M_\odot$) of neutron stars observed in binaries [250]. The explosion energies for these high luminosity models (panel (a), red line) are also close to the canonical supernova kinetic energy of 10^{51} erg. To discuss explosive nucleosynthesis in the following, we thus choose to focus on the model with $L_{\nu_e} = 4.5 \times 10^{52}$ erg s^{-1} as a reference.

For the model, Figure 15 shows a snapshot of internal energy (left panel) and abundances of ejecta (O, Si, Fe) when the SASI-aided low-mode shock is propagating in the O-rich layers at around ~ 20000 km along the pole. Ejecta with higher internal energies (left panel, colored by red) concentrate on the shock front particularly in polar regions ($\theta < \pi/4, \theta > 3\pi/4$), where both Si and O burning proceed to produce abundantly ^{56}Ni and ^{28}Si. These aspherical element distributions might be responsible for anisotropies of the SN ejecta as observed in SN1987A. More recently, Kimura et al. [251, 252] have analyzed the metal distribution of the Cygnus loop and pointed out that the progenitor of the Cygnus loop is a CCSN explosion whose progenitor mass ranges in ~ 12–$15 M_\odot$. Since the material in this middle-aged supernova remnant has not been completely mixed yet, the observed asymmetry is considered to still remain a trace of inhomogeneity produced at the moment of explosion. The asymmetries of the SN ejecta obtained in our 2D explosion model of a $15 M_\odot$ progenitor have a close correlation with the ones in the Cygnus loop (see [248] for more detailed comparison). This might allow us to speculate that globally asymmetric explosions induced by the SASI-aided neutrino-heating mechanism could account for the origin.

Concerning the synthesized amount of ^{56}Ni and ^{44}Ti, they are $\sim 0.06 M_\odot$ and $4 \times 10^{-5} M_\odot$, respectively, for the model (bottom panel in Figure 16). For SN1987A, masses of ^{56}Ni and ^{44}Ti are deduced to be $\sim 0.07 M_\odot$ [27, 253] and 1–$2 \times 10^{-4} M_\odot$ ([36], and references therein). The obtained mass of ^{44}Ti is, therefore, not enough for SN1987A as well as for Cas A ($1.6^{+0.6}_{-0.3} \times 10^{-4} M_\odot$) [254]. The shortage of ^{44}Ti is a long-standing problem, which might be solved by 3D effects [34]. For the abundance patterns, our reference model (top and middle panels in Figure 16) is shown to reproduce a similar trend to the solar system (top panel, Figure 16), which is in sharp contrast to the model with lower neutrino luminosity (middle panel, Figure 16).

Finally, we note that there should be at least two barriers to get over in the current-generation studies of the CCSN nucleosynthesis. As mentioned already, the first one is the need of radiation hydrodynamic simulations that follow the dynamics from gravitational collapse, shock-revival, shock propagation in the stellar mantle, to stellar explosions in a self-consistent manner. The second one is the need of computing light curves, spectra, and the degree of polarizations in 2D or 3D hydrodynamic models, for which one needs to solve the multi-D photon transport coupled with multi-D hydrodynamics. Unfortunately however, the marriage of the two items, albeit ultimately needed for clarifying the photon messengers, may not be done immediately due to

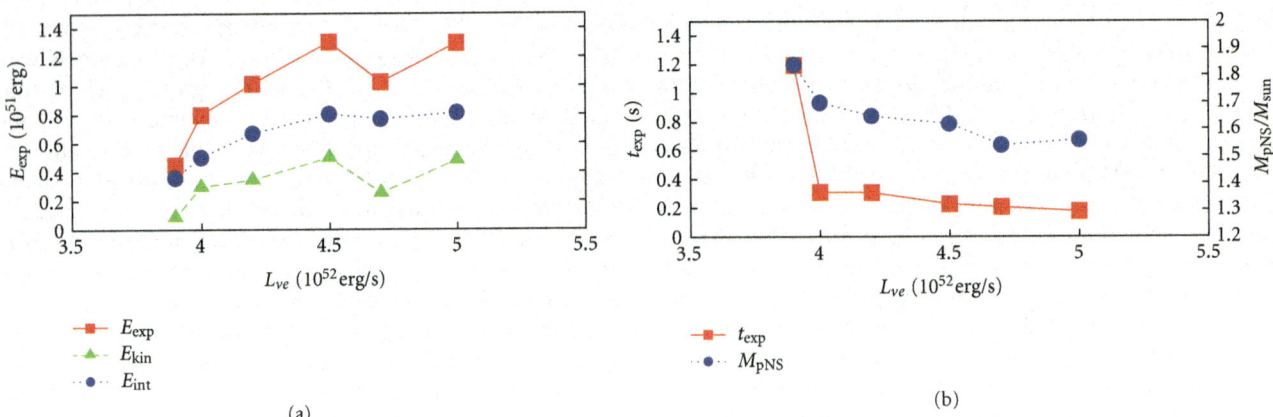

FIGURE 14: Panel (a) shows explosion energies (red line) versus input neutrino luminosities (L_{ν_e}), where green and blue lines show contributions from kinetic and thermal energy, respectively. Panel (b) shows the mass of protoneutron star (M_{PNS}) and the explosion timescales (t_{exp}) as a function of L_{ν_e}. M_{PNS} is estimated as the enclosed mass exceeding 10^{12} g cm^{-3} and t_{exp} as the time scale when the mass ejection rate at 100 km drops down to $0.1 M_\odot$/s in our 2D simulations (from [248]).

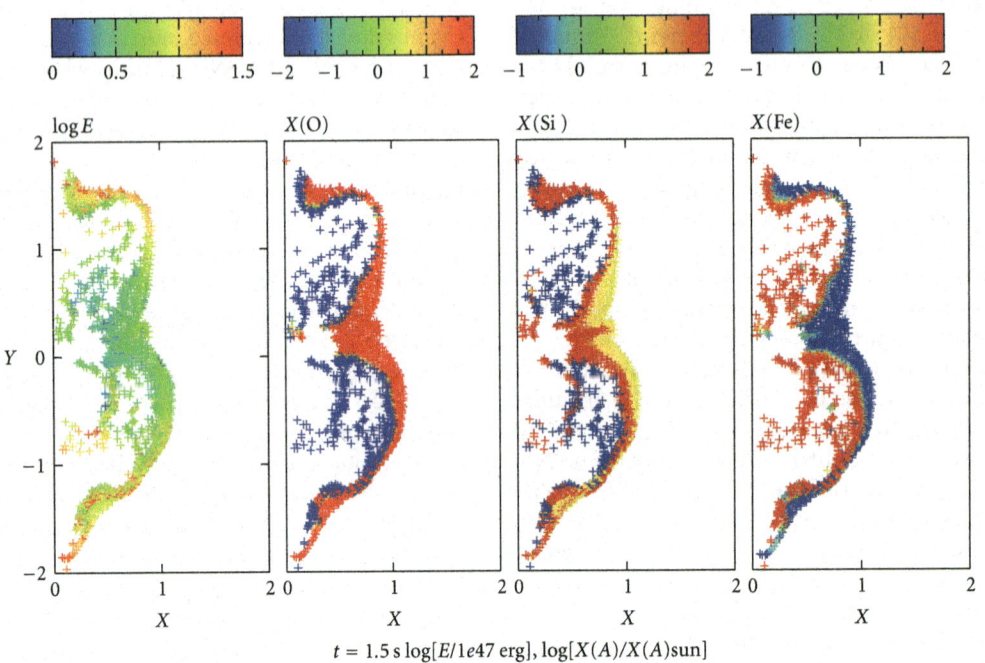

FIGURE 15: Distribution of internal energy (left panel) and abundances of the ejecta (O, Si, Fe from left to right, normalized to solar) when the nonspherical shock is propagating in the O-rich layers at around ~20000 km along the pole (~1.5 s after bounce). For this model, the input neutrino luminosity is set as $L_{\nu_e} = 4.5 \times 10^{52}$ erg s^{-1}. Note that Fe in the right panel is the sum of ^{56}Fe, ^{56}Ni, and ^{54}Fe (from [248]).

their computational expensiveness. For example, one needs to follow the dynamics more than ~1 day after the onset of gravitational core-collapse for computing the explosive nucleosynthesis, because the envelop of a typical supernova progenitor extends up to a radius of ~10^{12} cm. It takes furthermore more than ~1 week before the shock propagation enters to the so-called homologous phase. At present, the best available numerical simulation to this end is limited to 2D (e.g., [244]). For discussing anisotropies in emission-line profiles, one needs to follow the dynamics later than ~1 year

after explosions when the remnant becomes transparent. It is a very challenging (but very important) task for the first principle simulations to overcome the big gaps regarding the very different timescales. An encouraging news is that several groups are pursuing it with the use of advanced numerical techniques such as the adaptive-mesh-refinement approach [116] or Ying-Yang gridding [115]. With an accelerating power (like peta- or exa-scale) of supercomputers, it would be unsurprising that the next (or the next after next!) generation supernova modellers can execute the ultimate

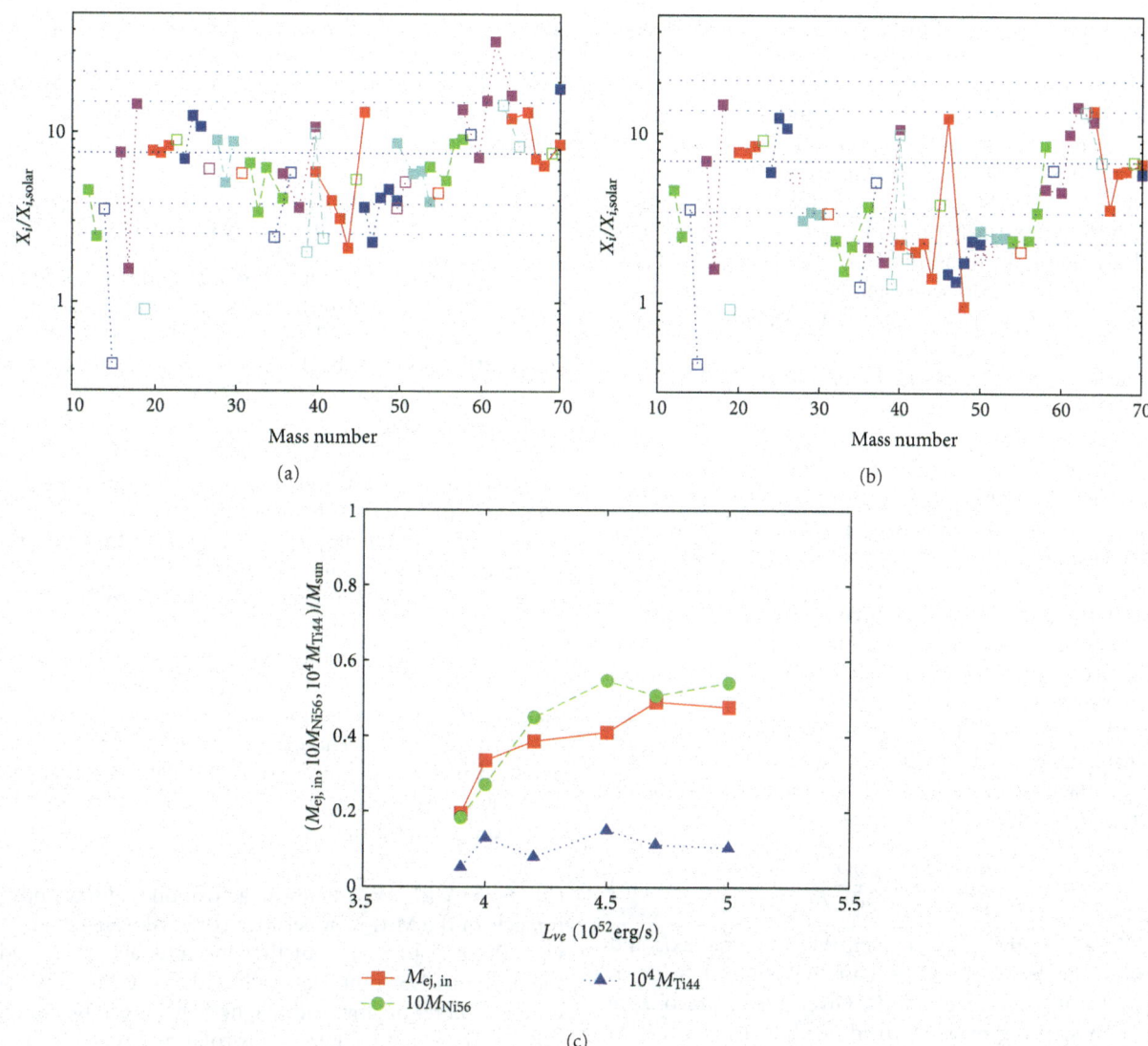

FIGURE 16: Bottom panel shows masses of ejecta ($M_{ej,in}$) and masses of radioactives (^{56}Ni and ^{44}Ti) as a function of the input neutrino luminosity. Top and middle panels show element abundance yields ($X_i/X_{\odot,i}$, normalized to solar) versus mass number for our reference model of $L_{\nu_e} = 4.5 \times 10^{52}$ erg s^{-1} and for the middle panel $L_{\nu_e} = 3.9 \times 10^{52}$ erg s^{-1}, respectively. Thick horizontal-dashed lines represent a factor equals to that of ^{16}O, while two normal and two thin lines denote a factor equals to that of ^{16}O times 2, 1/2, 3, and 1/3, respectively. These figures are taken from [248].

ab initio simulations, whose findings would be immediately used to make a more precise prediction of the photon messengers.

3. MHD Mechanism

In this section, we move on to focus on the MHD mechanism. After we briefly summarize the current status of this topic below, we mention the explosion dynamics that proceeds by the field-wrapping mechanism in Section 3(a) and discuss properties of the so-called magnetorotational

instability (MRI) in Sections 3(b) and 3(c) based on our preliminary results. Possible signatures of GWs and neutrinos are discussed in Sections 3.1 and 3.2, respectively.

Numerical simulations of MHD stellar explosions have started already in early 1970s shortly after the discovery of pulsars [255–258]. However, it is rather only recently that the MHD studies come back to the front-end topics in the supernova research followed by a number of extensive MHD simulations (e.g., [46, 175, 207, 259–270] for references therein). Main reasons for this activity are observations indicating very asymmetric explosions [38, 40], and the interpretation of magnetars [271, 272], collapsars and their relevance to gamma-ray bursts [273–275] as a possible outcome of the magnetorotational core-collapse of massive stars.

The MHD mechanism of stellar explosions relies on the extraction of rotational free energy of collapsing progenitor core via magnetic fields. Hence a high angular momentum of the core is preconditioned for facilitating the mechanism [276]. Given (a rapid) rotation of the precollapse core, there are at least two ways to amplify the initial magnetic fields to a dynamically important strength, namely, by the field wrapping by means of differential rotation that naturally develops in the collapsing core, and by the MRI (MRI, see [16]). Each of them we are going to review briefly in the following.

(a) *Field-Wrapping Mechanism.* Three dimensional plots of Figure 17 are useful to see how the field wrapping occurs. For the 2D model taken from our 2D special relativistic (SR) MHD simulation [268], the precollapse magnetic field is set to be as high as 10^{12} G that is uniform and parallel to the rotational axis and the initial β parameter (ratio of initial rotational energy to the absolute value of the initial gravitational energy) is taken to be 0.1% with a uniform rotation imposed in the iron core. From the left panel, it can be seen that the field lines are strongly twisted around the rotational axis. From the induction equation of ideal MHD equations, the time evolution of the toroidal fields (B_ϕ) can be expressed as (e.g., [276]),

$$\frac{\partial B_\phi}{\partial t} \approx B_X \left(X \frac{\partial \Omega}{\partial X} \right), \qquad (1)$$

where X denotes the distance from the rotational axis and B_X represents the X component of the poloidal fields in the cylindrical coordinates (X, Z). Given that a precollapse iron core has strong magnetic field (such as $B_{X,0} \sim 10^{12}$ G with $B_{X,0}$ representing the initial poloidal fields) and rapid rotation (such as $P_{\text{init}} \lesssim 4$ s with P_{init} being the precollapse rotation period), the typical amplification of B_ϕ near core bounce may be estimated as

$$\Delta B_\phi \approx \frac{t}{P_{\text{rot}}} B_X \sim 10^{15} \, \text{G} \left(\frac{t}{100 \, \text{ms}} \right) \left(\frac{P_{\text{rot}}}{1 \, \text{ms}} \right) \left(\frac{B_{X,0}}{10^{12} \, \text{G}} \right), \qquad (2)$$

where t represents the typical timescale when the field amplification via the wrapping processes saturates, P_{rot} denotes the rotational period near bounce. Equation (2) shows that the toroidal fields grow linearly with time by wrapping the poloidal fields (B_X), which is the so-called Ω dynamo (e.g., [265]). The 10^{15} G class magnetic fields are already dynamically important strength to affect the postbounce hydrodynamics as will be mentioned below.

The white lines in the right panel in Figure 17 show the streamlines of matter. A fallback of material just outside of the jet head advecting downwards to the equator (like a cocoon) is seen. In this jet with a cocoon-like structure, the magnetic pressure is always dominant over the matter pressure. As estimated above, the typical field strength behind the shock is $\sim 10^{15}$ G. The critical strength of the toroidal magnetic field to induce the *magnetic* shock revival

(i.e., the revival of the stalled bounce shock due to the MHD mechanism) is estimated as follows. The matter in the regions behind the stalled shock is pushed inwards by the ram pressure of the accreting matter. This ram pressure is estimated as,

$$P = 4 \times 10^{28} \left(\frac{\rho}{10^{10} \, \text{g/cm}^3} \right) \left(\frac{\Delta v}{2 \times 10^9 \, \text{cm/s}} \right)^2 \, \text{erg/cm}^3, \qquad (3)$$

where ρ and Δv denote typical density and radial velocity in the vicinity of the stalled shock. When the toroidal magnetic fields are amplified as large as $\sim 10^{15}$ G due to the field wrapping behind the shock, the resulting magnetic pressure, $B^2/8\pi \sim (10^{15})^2/8\pi \sim 10^{29}$ erg/cm^3, can overwhelm the ram pressure, leading to the magnetic shock revival (e.g., Figure 18). The onset timescale of the magnetic shock revival is sharply dependent on the precollapse magnetic fields and rotation (e.g., [46]). As the initial field strength is larger with more rapid rotation imposed, the interval from the shock stall to the MHD-driven shock revival becomes shorter. The speed of the jet head is typically mildly relativistic (at most $\sim 0.4c$), with c being the speed of light (see also the right panel of Figure 18). At the shock breakout of the iron core, the explosion energies generally exceed 10^{51} erg for models that produce MHD explosions in a shorter delay after the stall of the bounce shock. These features are in accord with those obtained in more detailed MHD simulations with spectral neutrino transport (e.g., [46, 273]) (see, e.g., [219] for a more detailed comparison).

(b) *On the MRI.* We are now in a position to discuss possible impacts of the MRI in supernova cores. Akiyama et al. [277] were the first to point out that the interfaces surrounding the nascent PNSs quite generally satisfy the MRI instability criteria. Therefore any seed magnetic fields can be amplified exponentially in the differentially rotating layers, much faster than the linear amplification due to the field wrapping (e.g., (2)). After the MRI enters to the saturated state, the field strength might reach $\sim 10^{15\text{-}16}$ G, which is high enough to affect the supernova dynamics. Not only in the field amplification, the MRI plays a crucial role also in operating the MHD turbulence (see [16, 278, 279]). The turbulent viscosity sustained by the MRI can convert a fraction of the shear rotational energy to the thermal energy of the system. Thompson et al. [141] suggested that the additional energy input by the viscous heating can help the neutrino-driven supernova explosion. Followed by the exponential field amplification and additional heating, a natural outcome of the magnetorotational core-collapse is expected to be the formation of energetic bipolar explosions, which might be observed as hypernovae (e.g., Section 2.3).

Note again that bipolar explosions obtained in the previous section with the assumption of a strong precollapse magnetic field ($\gtrsim 10^{12}$ G), are predominantly driven by the field wrapping processes, not by the MRI. Shibata et al. [280] pointed out in their fully 2D GRMHD core-collapse simulations that more than 10 grids are at least needed to capture the fastest growing mode of the MRI (see also [281]). As well-known, the growth rate of MRI-unstable

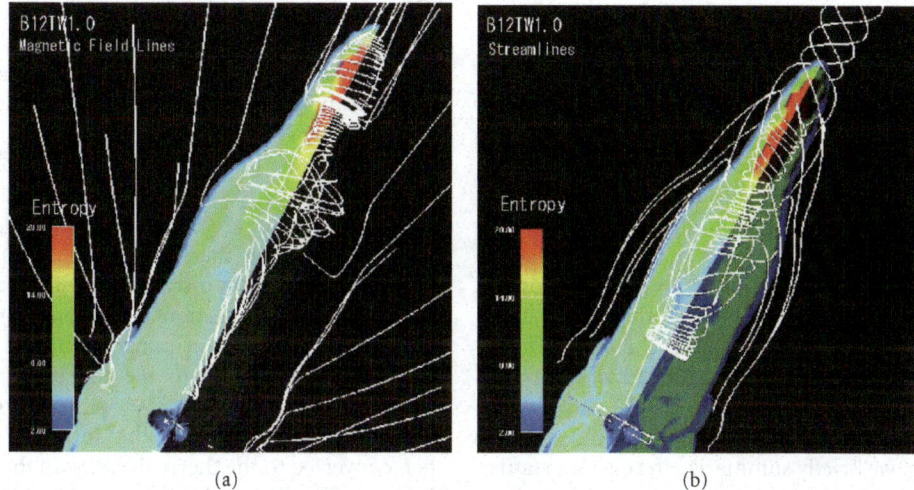

FIGURE 17: Three-dimensional plots of entropy with the magnetic field lines (a) and the streamlines of the matter (b) during the jet propagation at 20 ms and 94 ms after bounce, respectively (for model B12TW1 taken from [268]). The outer edge of the sphere colored by blue represents the radius of 7.5×10^7 cm. These panels highlight not only the wound-up magnetic field around the rotational axis (a), but also the fallback of the matter from the jet head of the downwards to the equator, making a cocoon-like structure behind the jet (b).

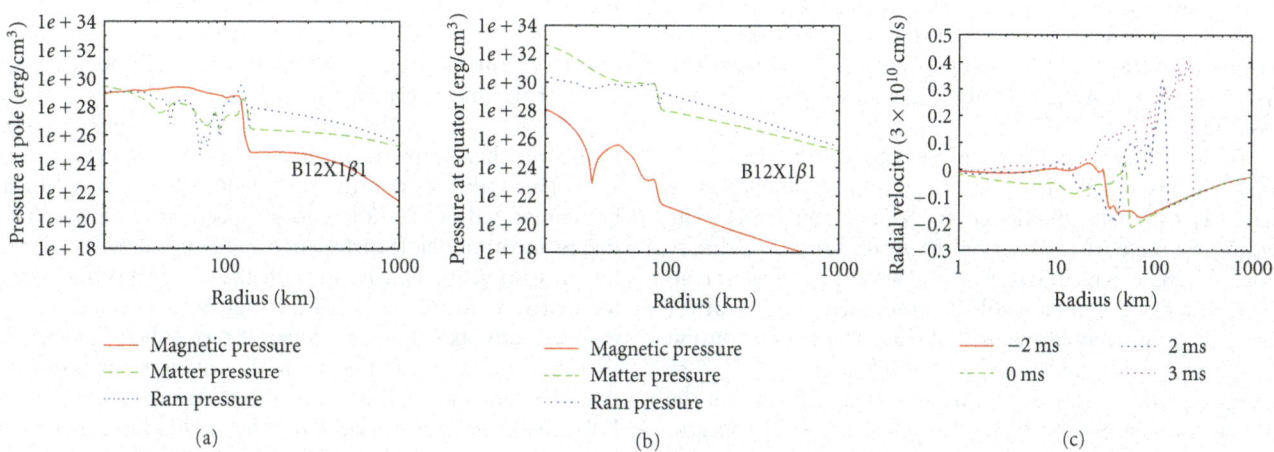

FIGURE 18: Left and middle panels show magnetic pressure (red line) versus ram pressure (blue line) for a typical MHD model (same as Figure 17) along the polar axis (a) or the equatorial plane (b) at 2 ms after the revival of the stalled bounce shock. The matter pressure is shown by green line as a reference. The right panel shows a velocity profile along the pole measured from the shock-stall (0.0 ms in the panel). For the equator, the magnetic pressure is much less than the ram pressure (middle panel), while the magnetic pressure amplified by the field wrapping along the pole becomes as high as the ram pressure of the infalling material at the shock front ((a), note that the shock position can be inferred by the discontinuity at around 150 km), leading to the MHD-driven shock formation (c). These figures are taken from [219].

modes depend on the product of the field strength and the wave number of the mode [16]. When a rapidly precollapse core is strongly magnetized as 10^{11} G (like in the case of collapsar progenitor), the wavelength is on the order of km in the postbounce core. In this case, the exponential field growth of the MRI was successfully captured by high-resolution simulations [280]. On the other hand, the fastest growing modes drop below several meters [282] for canonical supernova progenitors, which rotates much more slowly with weaker initial fields ($\sim 10^9$ G) as predicted by recent stellar evolution models [126, 127]. At present, it is computationally too expensive to resolve those small scales in the global MHD simulations, typically more than two or three orders-of-magnitudes smaller than their typical finest grid size. To reveal the nature of the MRI, local simulations focusing on a small part of the MRI-unstable regions are expected to be quite useful as traditionally studied in the context of accretion disks (see [16]).

Obergaulinger et al. [282] were the first to study the growth and saturation level of the MRI in the supernova environment (see, [283] for extensive simulations that focus on the MRI in relativistic outflows in the context of gamma-ray bursts and active galactic nuclei). To ease a drawback of the local shearing box simulation, they employed the shearing disk boundary conditions by which global radial density stratification can be taken into account. By performing such a *semiglobal* simulation systematically in 2D and 3D, they derived scaling laws for the termination of the MRI. As estimated in [277], the MRI was shown to amplify the seed fields exceeding 10^{15} G. These important findings may open several questions that motivate us to join in this effort, such as how the nonlinear properties as well as the scaling laws could be in the subsequent nonlinear state and whether the viscous and resistive heating driven by the MRI turbulence could or could not affect the supernova mechanism.

In the following, we briefly summarize the current status of our MRI project, in which we perform a local shearing box simulation to study the MRI-driven turbulence and their nonlinear properties bearing in mind the application to the supernova core (Masada, Takiwaki, and Kotake in preparation). The final goal of this project is to determine how much the conversion from the rotational energy (that the differential rotation taps) to the thermal energy occurs via the MRI turbulence, which was treated parametrically in [141]. As will be shown later, our preliminary results suggest that the additional energy supply would affect the neutrino-driven explosion in the case of rapidly rotating cores at late postbounce phase.

To study the nonlinear properties of the MRI, we solve viscous MHD equations by a finite-difference code that was originally developed by Sano et al. [284]. The hydrodynamic part is based on the second-order Godunov scheme [285], which consists of Lagrangian and remap steps. The exact Riemann solver is modified to account for the effect of tangential magnetic fields. The field evolution is calculated with Consistent MoC-CT method [286]. The energy equation is solved in the conservative form and the viscous terms are consistently calculated in the Lagrangian step. The advantages of our scheme are its robustness for strong shocks and the satisfaction of the divergence-free constraint of magnetic fields [287, 288].

Rigorously speaking, 3D simulations are absolutely required for the study of the MRI because only by them the disruption of channel structures, mode couplings, and the growth of parasitic instabilities can be accurately determined (e.g., [289, 290]). But in the following, we are only able to show our 2D results assuming axisymmetry, which are currently being updated to 3D. Bearing these caveats in mind, we briefly illustrate fundamental properties of the MRI (which is not always familiar with supernova modellers,) and discuss their possible impacts on the explosion dynamics.

(c) *Local Simulations in Axisymmetry.* As it is well known, there are three typical evolutionary stages observed in our local simulations. They are (i) exponential growth stage, (ii) transition stage, and (iii) nonlinear turbulent stage [278, 291]. Each stage is denoted by white, dark gray, and light gray shaded regions in Figure 19.

In the stage (i), the channel structure of the magnetic field exponentially evolves and inversely cascades to the larger spatial scale with small structures merging as the magnetic field is amplified. This is because the channel modes of the MRI are an exact solution for the nonlinear MHD equations [292]. The temporal evolution of channel structures for the toroidal magnetic field and their inversely cascading nature are shown in panels (a) and (b) of Figure 20. Then in the stage (ii), the channel structure of the magnetic field is disrupted via the parasitic instability and magnetic reconnection at the transition stage [282, 292]. The channel disruption induces a drastic phase-shift from the coherent structure to the turbulent tangled structure of the field as is illustrated by panel (c) of Figure 20. The magnetic energy stored in the amplified magnetic field is then converted to the thermal energy of the system. Finally in the stage (iii), the nonlinear stage emerges after the channel structures are disrupted to be a turbulence state driven by the MRI (see panel (d) in Figure 20). Figure 19(a) shows that, at this stage, the turbulent Maxwell stress ($M_{r\phi} = -B_r B_\phi/4\pi$) dominates over the Reynolds stress ($R_{r\phi} = \rho v_r \delta v_\phi$). Therefore the turbulent Maxwell stress is the main player to convert the free energy stored in the differential rotation into the thermal energy of the system.

Figure 19(b) shows the power spectra of the magnetic energy in the nonlinear turbulent stage. The logarithmic slope of the spectrum roughly coincides with $-11/3$ in the regime $10 \lesssim \bar{k} \lesssim 100$ while it steepens at higher wave-numbers where numerical effects become important (like k^{-6}). This indicates that the turbulent state in the MRI can be represented by the Kolmogorov spectrum for a homogeneous, incompressible turbulence that scales as $k^{-11/3}$. These features would be consistent with the 3D properties of the MRI-driven turbulence [278]. It has been pointed out in the previous axisymmetric local shearing box simulation that the channel structures do neither disrupt nor transit to the turbulent state without a relatively strong resistive effect [282, 293]. We however adopted the simulation box with a large aspect ratio ($L_x/L_z = 4$) and quasi-isothermal equation of state ($\gamma \simeq 1$), both of which could facilitate to disrupt the channel structures [294] as in 3D simulations.

Figure 21 shows the time- and volume-averaged Maxwell (green squares) and Reynolds stresses (orange diamonds, panel (a)), and magnetic energy (blue circles, panel (b)) in the turbulent stage as a function of the initial magnetic energy. It is shown that the Maxwell stress depends on the initial magnetic flux and it provides the following scaling relation:

$$\langle\langle M_{r\phi} \rangle\rangle \propto E_{\mathrm{mag,ini}}, \tag{4}$$

with $E_{\mathrm{mag,ini}} \equiv B_{\mathrm{ini}}^2/8\pi$. This suggests that the magnetic flux initially penetrating the system controls the nonlinear transport properties of the MRI-driven turbulence. This is qualitatively consistent with the previous results in the accretion disk [278, 284, 295, 296] while the power-law index obtained here is slightly larger. Note that the magnetic energy at the nonlinear turbulent stage has a linear dependence on the initial magnetic flux (because $\langle\langle E_{\mathrm{mag}} \rangle\rangle \propto E_{\mathrm{mag,ini}}^{1/2} \propto B_{\mathrm{ini}}$

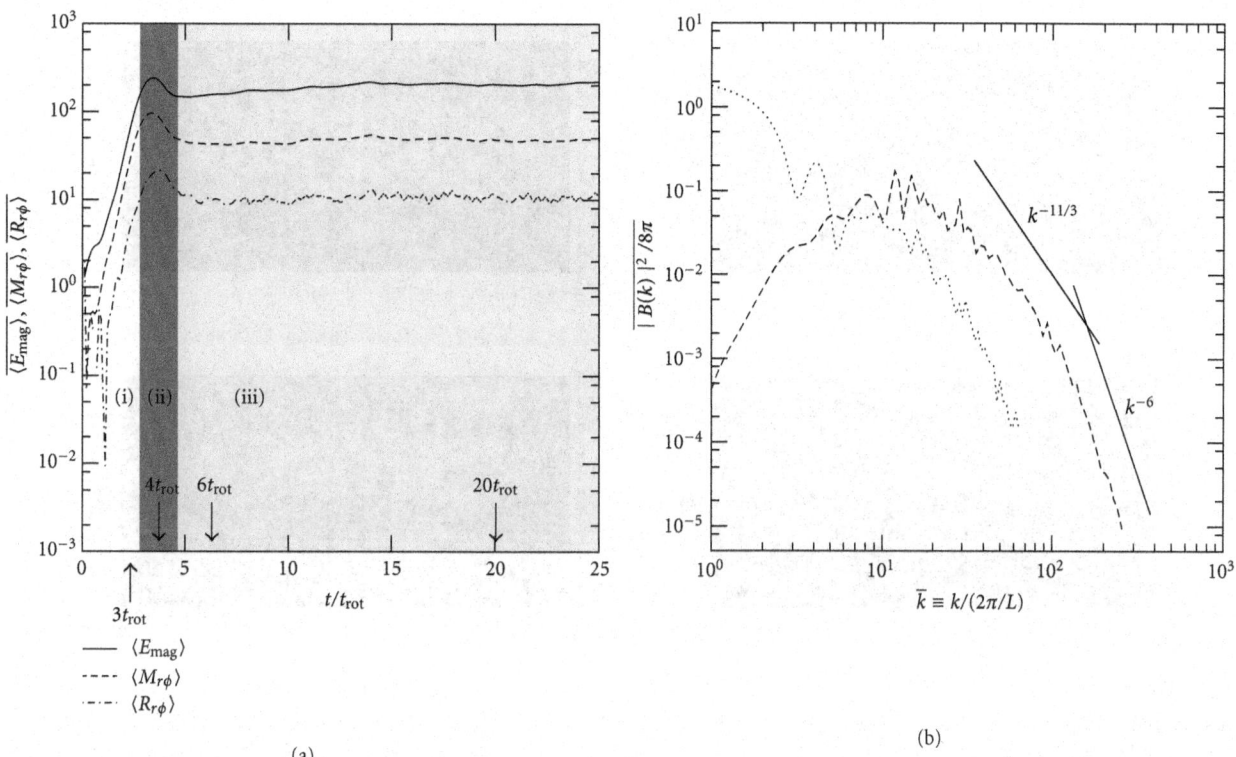

(a)

(b)

FIGURE 19: (a) Temporal evolution of the volume averaged magnetic energy, Maxwell and Reynolds stresses in the fiducial run are represented by thick, dashed, and dash-dotted curves, respectively. The vertical axis is normalized by the initial magnetic energy $E_{mag,0} \equiv B_0^2/8\pi$, that is $\overline{\langle E_{mag} \rangle} \equiv \langle E_{mag} \rangle / E_{mag,0}$, $\overline{\langle M_{r\phi} \rangle} \equiv \langle M_{r\phi} \rangle / E_{mag,0}$, and $\overline{\langle R_{r\phi} \rangle} \equiv \langle R_{r\phi} \rangle / E_{mag,0}$ where single bracket denotes the volume average of the physical variables. Horizontal axis is normalized by the rotational period $t_{rot} \equiv 2\pi/\Omega$. Three typical evolutionary stages, (i) exponential growth stage, (ii) transition stage, and (iii) nonlinear turbulent stage, are denoted by white, dark gray, and light gray shaded regions. (b) The amplitude of power spectra of the magnetic energy $|B(k)|^2$ along the k_x and k_z axes. The dashed and dotted curves describes the k_x and k_z components respectively. The vertical axis is normalized by the initial magnetic energy $E_{mag,0}$. The wave-numbers shown in the horizontal axis are normalized by the $2\pi/L$. The upper thick curve demonstrates the Kolmogorov slope for isotropic incompressible turbulence $\propto k^{-11/3}$. The lower slope is proportional to k^{-6} just for the reference. In our local simulations, we prepare a 2D shearing box which is threaded by the initial vertical field $B_0 = 1.4 \times 10^{12}$ G with imposing initial gas pressure $P_0 = 4 \times 10^{28}$ dyn cm^{-2}, angular velocity $\Omega_0 = 2000$ sec^{-1} and shear parameter $q_0 = 1.25$, respectively. Here the shear parameter represents the degree of differential rotation as $q \equiv -d \ln \Omega / d \ln r$ with r being the radial coordinate. The other parameters are fixed to mimic the properties in the vicinity of the PNS's surface in the postbounce phase (i.e., the density is taken to be $\rho_0 = 10^{12}$ g cm^{-3} and the box size is taken to be $L_x = 4$ km (horizontal) $\times L_z = 1$ km (vertical)).

(Figure 21(b))). This is also qualitatively consistent with Hawley et al. [278]. In this way, we investigated parameter dependence of the rotational shear (q parameter) as well as the initial pressure (P) on the saturated value of the Maxwell stress and deduced a scaling relation that yields $M_{r\phi} \propto B^2 P^{1/4} [q/(2-q)]^{1/4} \Omega^{1/2}$. Here the shear parameter represents the degree of differential rotation as $q \equiv -d \ln \Omega / d \ln r$ with r being the radial coordinate.

Using the scaling relation obtained in our local simulations, the heating rate maintained by the MRI turbulence can be estimated as:

$$\epsilon_{MRI} = M_{r\phi} q\Omega$$
$$= \left[f \left(\frac{B_{pb}}{B_0} \right)^2 \left(\frac{P_{pb}}{P_0} \right)^{1/4} s_{pb}^{1/4} \left(\frac{\Omega_{pb}}{\Omega_0} \right)^{1/2} \right] q\Omega$$
$$\simeq 10^{30} \text{ erg cm}^{-3} \text{ sec}^{-1}$$

$$\times f_{30} q_{pb,1} s_{pb,1}^{1/4} \left(\frac{B_{pb}}{10B_0} \right)^2 \left(\frac{P_{pb}}{100P_0} \right)^{1/4} \left(\frac{\Omega_{pb}}{\Omega_0} \right)^{3/2}, \tag{5}$$

where f_{30} represents a typical amplification factor of the stress normalized by 30 (e.g., Figure 21), $q_{pb,1}$ is the shear rate normalized by unity, and $s_{pb,1}$ is the ratio of the vorticity to the shear normalized by unity. Since we are interested in the growth of the MRI in the postbounce phase, the original magnetic flux, gas pressure angular velocity, and the shear rate $B_0 = 1.4 \times 10^{12}$ G, $P_0 = 4 \times 10^{28}$ erg cm^{-3}, $\Omega_0 = 2000$ rad s^{-1}, and $q = 1.25$ are replaced by $B_{pb}, P_{pb}, \Omega_{pb}$ and q_{pb} with some amplification factors.

A typical volume in which the MRI-driven turbulence is active, may be estimated as $V_{MRI} = 4\pi R^2 h \simeq 10^{21}cm^3 R_7^2 h_6$, where $R_7 = R/10^7$ cm is typical radius of the PNS normalized by 10^7 cm, and $h_6 = h/10^6$ cm is the typical radial width of

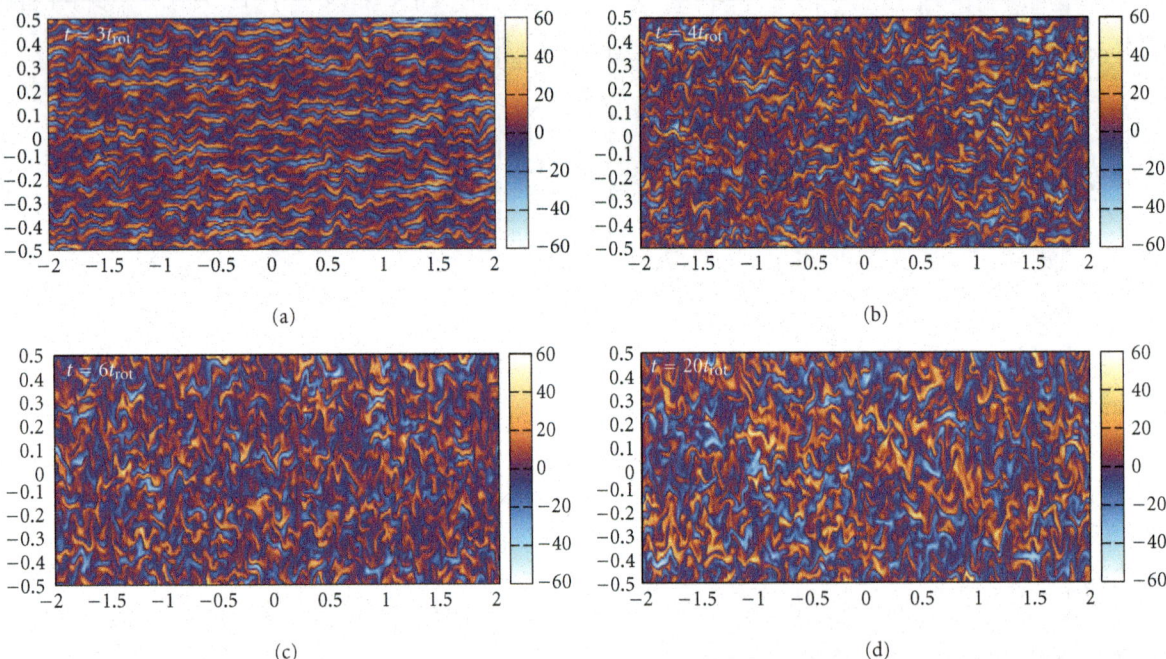

FIGURE 20: Four snapshots representing the temporal evolution of MRI from the linear to the nonlinear phase. Panels (a), (b), (c) and (d) correspond to the time slice $t = 3t_{\text{rot}}$, $4t_{\text{rot}}$, $6t_{\text{rot}}$, and $20t_{\text{rot}}$ respectively. The color bar indicates the size of the toroidal magnetic field normalized by the strength of the initial vertical magnetic field ($B_0 = 1.4 \times 10^{12}$ G). Note that the vertical and horizontal are normalized by the typical spatial scale of the local portion in the protoneutron stars which is selected as $L = 1$ km in the fiducial run. Three typical evolutionary stages are observed, (i) exponential growth stage, (ii) transition stage, and (iii) nonlinear turbulent stage (compare with Figure 19).

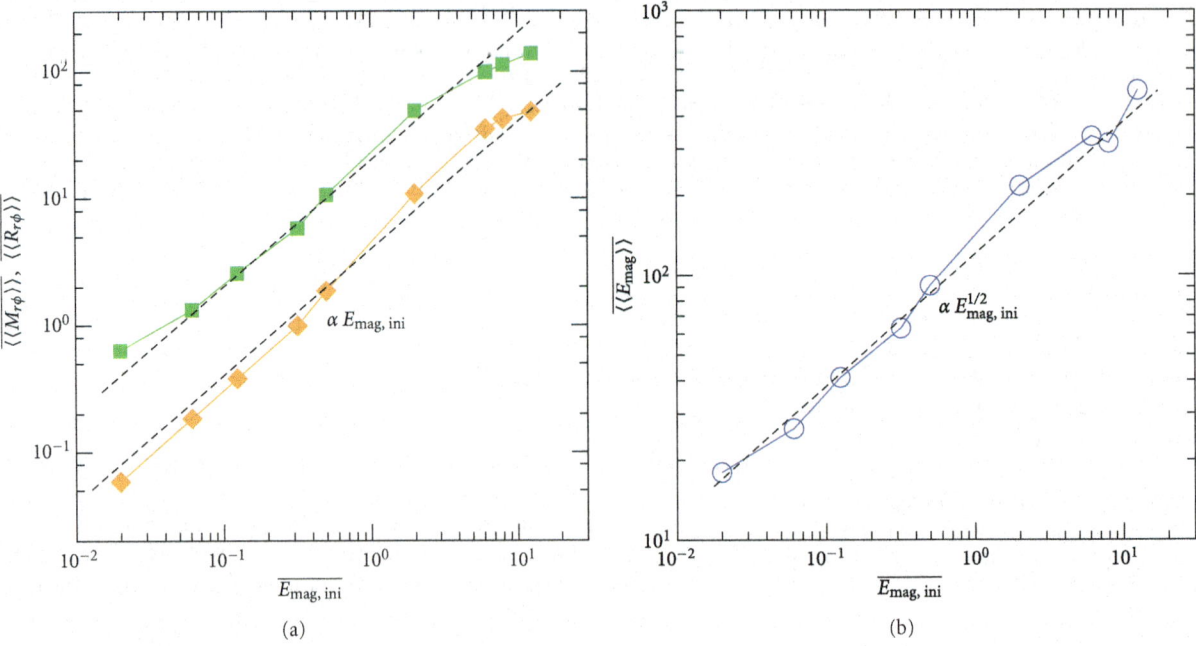

FIGURE 21: The dependence of the time- and volume-averaged (a) Maxwell (green squares) and Reynolds stresses (orange diamonds) and (b) magnetic energy (blue circles) and in the turbulent state on the strength of the initial magnetic energy E_{ini}. Both the vertical and horizontal axes are normalized by the initial magnetic energy of the fiducial model $E_{\text{mag},0}$. The initial field strength is varied from $\sim 10^{11}$ to $\sim 6 \times 10^{12}$ G. The dashed lines denote reference slopes proportional to (a) $E_{\text{mag}}^{1/2}$ and (b) E_{mag}, respectively. We take the time-average of volume-averaged quantities during $90t_{\text{rot}} < t < 100t_{\text{rot}}$.

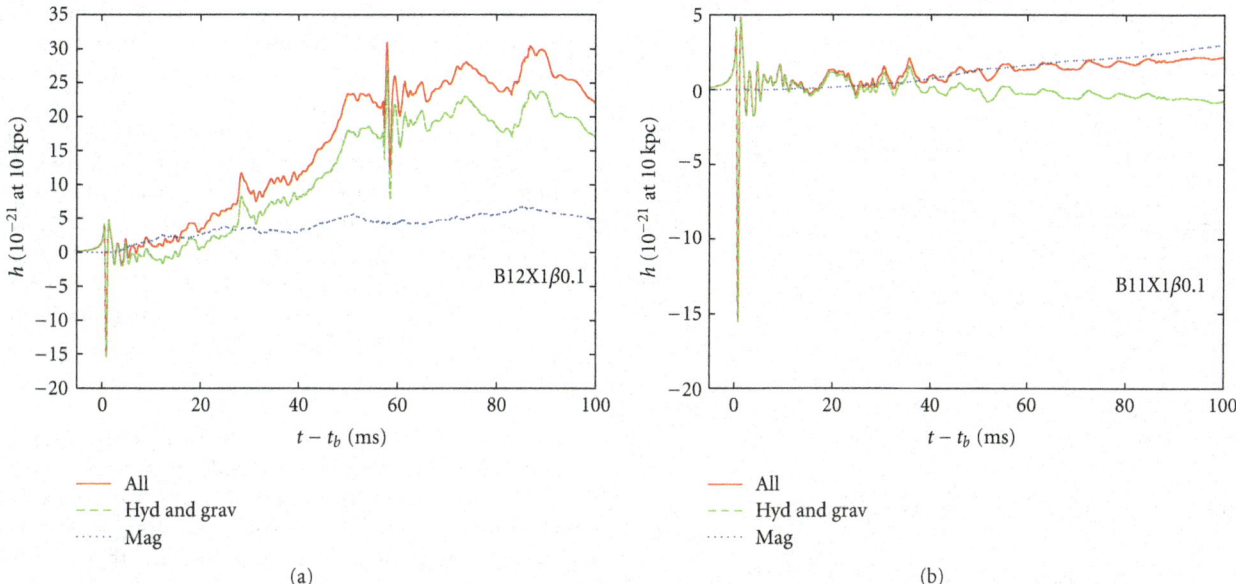

FIGURE 22: Gravitational waveforms with an increasing trend (left) or not (right panel). Initial rotation parameter is set to be $\beta = 0.1\%$, while the precollapse magnetic field is taken as 10^{12} G (left panel) and 10^{11} G (right panel), respectively (from [219]). The total wave amplitudes are shown by the red line, while the contribution from the magnetic fields and from the sum of hydrodynamic and gravitational parts are shown by blue and green lines, respectively (e.g., (14) and (11), (13)). Note that the bounce GW signals (($t - t_b \lesssim 20$ ms) are not affected by the magnetic fields significantly, and they are categorized into the so-called type I or II waveforms.

the MRI-active layer normalized by 10^6 cm. Then the energy releasing rate $L_{\mathrm{MRI}} \equiv \epsilon_{\mathrm{MRI}} V_{\mathrm{MRI}}$ becomes:

$$L_{\mathrm{MRI}} \sim 10^{51} R_7^2 h_6 \, \mathrm{erg\,sec^{-1}}$$
$$\times f_{30} q_{\mathrm{pb},1} s_{\mathrm{pb},1}^{1/4} \left(\frac{B_{\mathrm{pb}}}{10 B_0} \right)^2 \left(\frac{P_{\mathrm{pb}}}{100 P_0} \right)^{1/4} \left(\frac{\Omega_{\mathrm{pb}}}{\Omega_0} \right)^{3/2}. \quad (6)$$

Typical timescales of neutrino-driven explosions observed in recent supernova simulations are $\gtrsim 400$ ms after bounce (e.g., [125, 138]). So the energy deposition could be as high as 10^{50} erg if the core rotates rather rapidly as taken in the above estimation. If this is the case, the turbulent heating due to the MRI is expected to play an important role for assisting the neutrino-driven explosion.

It should be noted that nonaxisymmetric modes of the parasitic instability could disrupt the channel structures more effectively than in 2D [289, 290]. To obtain more accurate estimate of the amplification factor of f (6), 3D simulations are unquestionably required, which we are currently undertaking [297].

3.1. *Gravitational Waves.* As already mentioned in Section 2.2, rapid rotation, necessary for producing strong bounce GW signals, is likely to obtain ~1% of massive star population (e.g., [62]). This can be really the case, albeit minor, for progenitors of rapidly rotating metal-poor stars, which experience the so-called chemically homogeneous evolution [274, 275]. In such a case, the MHD mechanism could work to produce energetic explosions, which are receiving great attention recently as a possible relevance to magnetars and collapsars (e.g., [298, 299], see [273, 300, 301] for collective references), which are presumably linked to the

formation of long-duration gamma-ray bursts (GRBs) (e.g., [302] for review).

Among the previous studies focusing on the bounce signals (e.g., references in Section 2.2), only a small portion of papers has been spent on determining the GW signals in the MHD mechanism [175, 260, 264, 265, 269, 280]. This may be because the MHD effects on the dynamics as well as their influence over the GW signals can be visible only for cores with precollapse magnetic fields over $B_0 \gtrsim 10^{12}$ G [175, 264]. Considering that the typical magnetic-field strength of GRB progenitors is at most ~ $10^{11\text{-}12}$ G [274], this is already an extreme situation. In a more extremely case of $B_0 \sim 10^{13}$ G, a secularly growing feature in the waveforms was observed [263, 269, 280]. In the following, we summarize the GW signatures based on our 2D SRMHD simulations [219] and discuss how a peculiarity of the MHD mechanism could be imprinted in the GW signals.

The left panel of Figure 22 shows the gravitational waveform with the increasing trend, which is obtained for a model with strong precollapse magnetic field ($B_0 = 10^{12}$ G) also with rapid rotation initially imposed (β parameter = 0.1%). Such a feature cannot be observed for a weakly magnetized model ($B_0 = 10^{11}$ G, right panel). To understand the origin of the increasing trend, it is straightforward to look into the quadrupole GW formula, which can be expressed as

$$h_{ij}^{TT}(\mathbf{X}, t) \stackrel{\ell=2,m=0}{=} \frac{1}{R} A_{20}^{E2} \left(t - \frac{R}{c} \right) T_{ij}^{E2,20}(\theta, \phi), \quad (7)$$

where $T_{ij}^{E2,20}(\theta, \phi)$ is

$$T_{ij}^{E2,20}(\theta, \phi) = \frac{1}{8} \sqrt{\frac{15}{\pi}} \sin^2 \theta, \quad (8)$$

(e.g., [303]). A_{20}^{E2} can be expressed as

$$A_{20}^{E2} = A_{20\,(hyd)}^{E2} + A_{20\,(mag)}^{E2} + A_{20\,(grav)}^{E2}. \tag{9}$$

On the right hand side, the first term is related to anisotropic kinetic energies which we call as hydrodynamic part,

$$A_{20\,(hyd)}^{E2} = \frac{G}{c^4} \frac{32\pi^{3/2}}{\sqrt{15}} \int_0^1 d\mu \int_0^\infty r^2 dr f_{20\,(hyd)}^{E2}, \tag{10}$$

$$f_{20\,(hyd)}^{E2} = \rho_* W^2 \left(v_r^2 (3\mu^2 - 1) + v_\theta^2 (2 - 3\mu^2) - v_\phi^2 \right. \\ \left. - 6v_r v_\theta \mu \sqrt{1 - \mu^2} \right), \tag{11}$$

the second term is related to anisotropy in gravitational potentials which we call as the gravitational part,

$$A_{20\,(grav)}^{E2} = \frac{G}{c^4} \frac{32\pi^{3/2}}{\sqrt{15}} \int_0^1 d\mu \int_0^\infty r^2 dr f_{20\,(grav)}^{E2}, \tag{12}$$

$$f_{20\,(grav)}^{E2} = \left[\rho h \left(W^2 + \left(\frac{v_k}{c} \right)^2 \right) + \frac{2}{c^2} \left(p + \frac{|b|^2}{2} \right) \right. \\ \left. - \frac{1}{c^2} \left((b^0)^2 + (b_k)^2 \right) \right] \\ \times \left[-r\partial_r \Phi (3\mu^2 - 1) + 3\partial_\theta \Phi \mu \sqrt{1 - \mu^2} \right], \tag{13}$$

and finally the third term is related to anisotropy in magnetic energies that we refer to as magnetic part,

$$A_{20\,(mag)}^{E2} = -\frac{G}{c^4} \frac{32\pi^{3/2}}{\sqrt{15}} \int_0^1 d\mu \int_0^\infty r^2 dr f_{20\,(mag)}^{E2}, \\ f_{20\,(mag)}^{E2} = \left[b_r^2 (3\mu^2 - 1) + b_\theta^2 (2 - 3\mu^2) - b_\phi^2 \right. \\ \left. - 6b_r b_\theta \mu \sqrt{1 - \mu^2} \right]. \tag{14}$$

The right panel of Figure 23 shows contributions to the total GW amplitudes (6) for the strongly magnetized model (left panel of Figure 22), in which the left-hand-side panels are for the sum of the hydrodynamic and gravitational part, namely $\log(\pm[f_{20\,(hyd)}^{E2} + f_{20\,(grav)}^{E2}])$ (left top(+)/bottom(−)) (11), (13), and the right-hand-side panels are for the magnetic part, namely $\log(\pm f_{20\,(mag)}^{E2})$ (right top(+)/bottom(−)) (e.g., (14)). By comparing the top two panels in Figure 19, it can be seen that the positive contribution is overlapped with the regions where the MHD outflows exist. The major positive contribution is from the kinetic term of the MHD outflows with large radial velocities (e.g., $+\rho_* W^2 v_r^2$ in (11)). The magnetic part also contributes to the positive trend (see top right-half in the right panel (labeled by mag(+))). This comes from the toroidal magnetic fields (e.g., $+b_\phi^2$ in (14)), which dominantly contribute to drive MHD explosions.

Figure 24 shows the GW spectra for a pair models of B12X5β1 and B11X5β1 like in Figure 22 that with or without the increasing trend. Regardless of the presence (left panel) or absence (right panel) of the increasing trend, the peak amplitudes in the spectra are around 1 kHz. This is because they come from the GWs near at bounce when the MHD effects are minor. On the other hand, the spectra for lower frequency domains (below ∼100 Hz) are much larger for the

model with the increasing trend (left panel) than without (right panel). This reflects a slower temporal variation of the secular drift inherent to the increase-type waveforms (e.g., Figure 22). It is true that the GWs in the low frequency domains mentioned above are relatively difficult to detect due to seismic noises, but a recently proposed future space interferometers like Fabry-Perot type DECIGO is designed to be sensitive in the frequency regimes [217, 305] (e.g., the black line in Figure 24). Our results suggest that these low-frequency signals, if observed, could be one important messenger of the increase-type waveforms that are likely to be associated with MHD explosions exceeding 10^{51} erg.

3.2. *Neutrino Signals.* As of now, SN1987A remains the only astrophysical neutrino source outside of our solar system. Even though the detected events were just two dozen, these events have been studied extensively (yielding ∼500 papers) and have allowed us to have a confidence that our basic picture of the supernova physics is correct (e.g., [4], see [5, 6] for a recent review). For a next nearby event, it is almost certain that the flagship detectors like Super-Kamiokande and IceCube are able to detect a SN "neutrino light curve" with high statistics. For the detectors, the horizon to the sources now extends out to about 100 kpc and thus covers our galaxy and its satellites. A future megaton-class detector reaches as far as Andromeda galaxy at a distance of 780 kpc (e.g., Hyper-Kamiokande, Memphys, and LBNE) and large-scale scintillator (e.g., HALO [307] that is an upgrade of SNO) or liquid-Argon detectors (GLACIER [308] that is an upgrade of ICARUS) will also play an important role (see [309]) for a complete list and details).

Before core bounce, neutrinos of different flavors are initially tapped in their relative neutrino spheres with increasing their Fermi energies as the central density becomes higher. After bounce, the shock dissociates nuclei into free nucleon, which drastically increases the number of protons, leading to a sharp increase in electron capture and production of the prompt ν_e burst. At this *neutronization* epoch (in the first 10–20 ms postbounce), the largest difference in the light-curves among ν_e and the rest of the neutrino species can be seen (e.g., Figure 1 in [310]). Since neutrino oscillations take place only when there is a difference in the neutrino fluxes among different species (see Section 3 [22] for more details on the neutrino oscillations), the oscillation effects in the neutronization epoch would be strongest. Unfortunately, however, they are very hard to measure because the currently-running detectors are primarily sensitive to $\bar{\nu}_e$ signals that are produced by the inverse beta process $\bar{\nu}_e + p \rightarrow e^+ + n$. It should be noted that the detection the ν_e bursts is important for studying not only neutrino oscillations but also for determining the direction and distance to the source [310]. In a megaton-class Čherenkov detector in which gadolinium is added to catch neutrons [311], and large liquid-Argon detectors (like ICARUS and GLACIER cited above) are expected to be powerful ν_e detectors.

During the subsequent accretion phase (approximately before 1 s postbounce), the bounce shock stagnates and matter falls in through the stalled shock to the center,

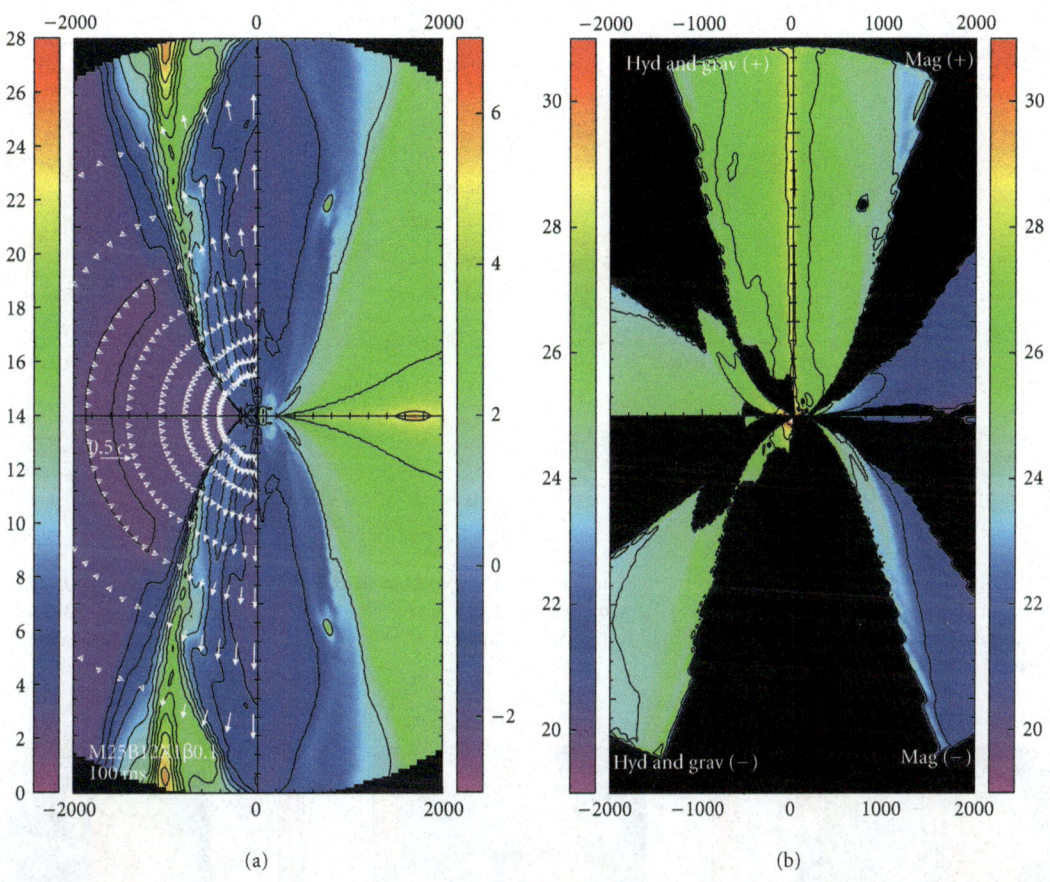

(a)   (b)

FIGURE 23: Left panel shows distributions of entropy (k_B/baryon) (left) and logarithm of plasma β (right) for model B12X1β0.1 at 100 ms after bounce. The white arrows in the left-hand side show the velocity fields, which are normalized by the scale in the middle left edge ($0.5c$). Right panel shows the sum of the hydrodynamic and gravitational parts (indicated by "hyd and grav" in the left-hand side) and the magnetic part (indicated by "mag" in the right-hand side), respectively. The top and bottom panels represent the positive and negative contribution (indicated by ($+$) or ($-$)) to A_{20}^{E2}, respectively (see text for more detail). The side length of each plot is 4000 (km) × 8000 (km). This figure is taken from [219].

releasing gravitational energy that powers neutrino emission. For relatively well-studied progenitors in the mass range of 10.8 to 25$M_\odot$ (e.g., [96, 121, 310, 312, 313] and references therein), one generally finds $L_{\nu_e} \sim L_{\bar{\nu}_e} > L_{\nu_x}$ and $\langle E_{\nu_x} \rangle > \langle E_{\bar{\nu}_e} \rangle \gtrsim \langle E_{\nu_e} \rangle$ with L_{ν_i} and $\langle E_{\nu_i} \rangle$ representing luminosity and average neutrino energy for the neutrino species of $i = \nu_e, \bar{\nu}_e, \nu_x (; \nu_\mu, \nu_\tau$ and their antiparticles). The accretion phase is followed by the PNS (protoneutron star) cooling in which the accretion stops and the PNS settles to become a neutron star after a successful revival of the stalled bounce shock into explosions. Since the explosion mechanism is still under debate (see discussion in Section 2), it is at present difficult to tell the accurate turn-over time. But it should be less than 1-2 s postbounce, otherwise the central PNS would collapse into a black hole [314]. In the PNS cooling phase, the luminosities of all neutrino species become similar $L_{\nu_e} \sim L_{\bar{\nu}_e} \sim L_{\nu_x}$ (with the energy hierarchy of $\langle E_{\nu_x} \rangle > \langle E_{\bar{\nu}_e} \rangle > \langle E_{\nu_e} \rangle$) and decrease monotonically with time. Due to the similarity in the luminosities and spectra among $\bar{\nu}_e$ and ν_x and also due

to the darkening with time, it is much harder to see flavor oscillation effects by the currently running detectors.

Neutrinos streaming out of the iron core interact with matter via the Mikheyev-Smirnov-Wolfenstein (MSW) effect (e.g., [78]) firstly in propagating through progenitor envelope and then through the Earth before reaching detectors. Such effects have been extensively investigated so far from various points of view, with a focus such as on the progenitor dependence of the early neutrino burst (e.g., [310, 315]) and on the Earth matter effects (e.g., [316, 317]). The conversion efficiency via the MSW effect depends sharply on density, equivalently electron fraction gradients, thus sensitive to discontinuities produced by the passage of supernova shocks. Such shock effects have been also extensively studied (e.g., [318–322], see [79] for a recent review). The shock passage to the so-called high-resonance regions may be observed as a sudden decrease in ν_e events in the case of normal mass hierarchy (or $\bar{\nu}_e$ in the case of inverted mass hierarchy). Such features monitoring time evolution of the density profile like

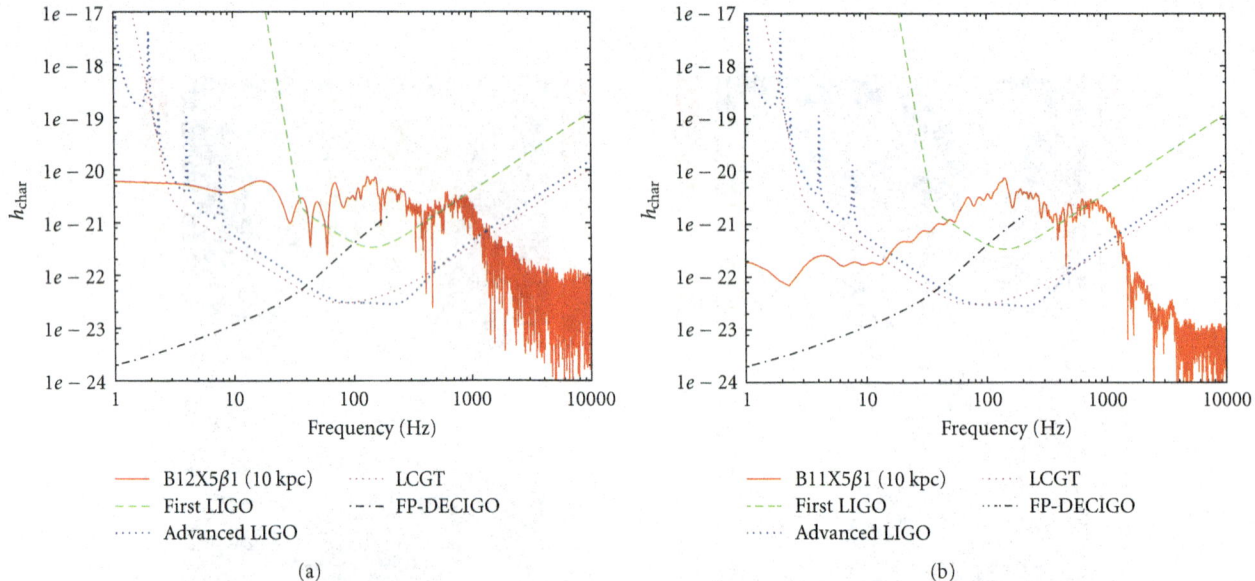

FIGURE 24: Gravitational-wave spectrum for representative models with the expected detection limits of the first LIGO [11], the advanced LIGO [304], Large-scale Cryogenic Gravitational wave Telescope (LCGT) [216], and Fabry-Perot type DECIGO [217, 305]. It is noted that h_{char} is the characteristic gravitational wave strain defined in [218]. The supernova is assumed to be located at the distance of 10 kpc. This figure is taken from [219].

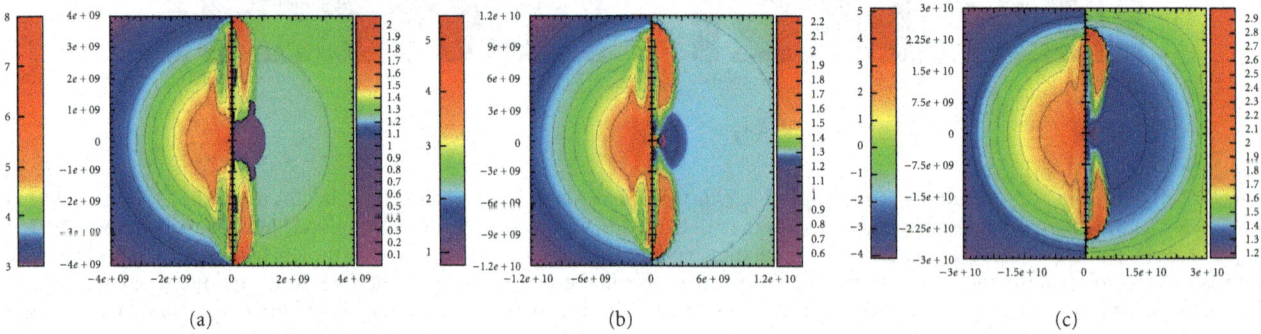

FIGURE 25: Snapshots showing MHD explosions of CCSNe at 0.8, 2.0, and 4.3 sec after core bounce from left to right (e.g., [268]). In each figure, contour of the logarithmic density (g/cm^3) (left) and entropy per baryon (k_B) (right) are shown. The unit of the horizontal and the vertical axis is in (cm). Note that the difference of the length scale for each panel and that the outermost radius of the progenitor is $\sim 3 \times 10^{10}$ (cm).

a tomography, thus could provide a powerful test of the mixing angle and the mass hierarchy (e.g., [320, 323–325]).

It is rather recently that the importance of collective flavor oscillations was widely recognized. Neutrinos streaming out of the neutrino spheres are so dense that they provide a large matter effect for each other. The collective effect usually takes place between the neutrino sphere and the mentioned MSW region. Regardless of the inherent nonlinearity and the presence of multiangle effects, the final outcome for the emergent neutrino flux seems to be converging after a series of extensive study over the past years at least for the 1D models. Although the collective effects will not be significant to assist the neutrino-driven explosions (e.g., [326, 327], see however [328, 329]), they are predicted to emerge as a

distinct observable feature in their energy spectra (see [74–77] for reviews of the rapidly growing research field and collective references therein).

An important lesson from SN1987A is that for explaining the duration of the events, there was no other energy-loss channel but for the ordinary neutrinos in the context of the Standard Model of particle physics (e.g., [6, 330]). The next SN event could provide an opportunity to study also a nontrivial property of neutrinos, such as the magnetic dipole moments. The resonant spin-flavor conversion has been also studied both analytically (e.g., [331–333] and references therein) and numerically (e.g., [334]), which can transform some of the prompt ν_e burst into $\bar{\nu}_e$ in highly magnetized supernova envelopes, leading to a huge $\bar{\nu}_e$ burst.

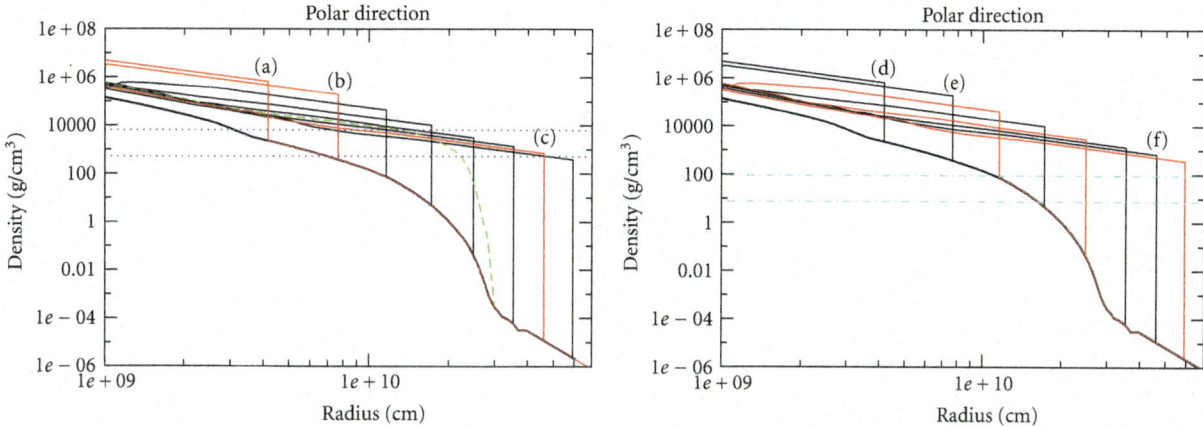

FIGURE 26: The density profile in the polar direction every 0.4 s as a function of radius. (a), (b) and (c) correspond to 0.4, 0.8 and 2.8 s, respectively. The horizontal blue lines show approximately density of the H-resonance for different neutrino energies which are 5 (above) and 60 MeV (below), respectively. The green line is an original density profile for 2.0 s. Same as the left panel, but for 1.2 s (d), 2.0 s (e), and 3.2 s (f), respectively. Note that a sharpness of the shock along the polar direction is modified from the one in the MHD simulations (green line showing an original density profile for 2.0 s), which is made to be inevitably blunt due to the employed numerical scheme using an artificial viscosity to capture shocks (taken from [306]).

The mentioned important ingredients related to the flavor conversions due to the MSW effects, collective effects, and the electromagnetic effects in the supernova environment have been studied often one by one in each study without putting all the effects together, possibly in order to highlight the new ingredient. In this sense, all the studies mentioned above should be regarded as complimentary towards the precise predictions of SN neutrinos.

Here it should be mentioned that most of those rich phenomenology of SN neutrinos have been based on 1D spherically symmetric simulations (e.g., [75, 310, 312, 315, 319, 320, 335–338] and references therein). Apart from a simple parametrization to mimic anisotropy and stochasticity of the shock (e.g., [339–341]), there have been not so many studies focusing how global asymmetry of the explosion dynamics obtained in recent supernova simulations (e.g., Sections 2 and 3) would affect the neutrino oscillations (see, however, [79, 342]). This is mainly due to the lack of multi-D supernova models, which are very computationally expensive to continue the simulations till the shock waves propagate outward until they affect neutrino transformations. It is only recently that several studies along this line have been reported [133, 343], in which the collective or MSW effects are rendered to be treated in an approximate manner. By analyzing the 2D results of a $15M_\odot$ model by Marek and Janka [125] who included one of the best available neutrino transfer approximations (e.g., Table 1), Lund et al. [343] pointed out that fast time variations caused by convection and SASI lead to significant modulations around a few hundred Hz, which can be visible in IceCube or future megaton-class detectors for the galactic SN source. Based on the 2D results of $20M_\odot$ models by Ott et al. [132] who reported the first 2D multiangle transport simulations, Brandt et al. [133] also obtained the similar

results. From their rapidly rotating model, they also pointed out that a rapid rotation of precollapse SN cores imprints strong asymmetries in the neutrino flux [344] as well as in its light curves [133, 345].

In the following, we briefly summarize our findings [306] in which we studied exploratory the neutrino oscillations in the context of the MHD mechanism. As mentioned in Section 3, it is recently possible for special relativistic simulations to follow the dynamics of MHD explosions continuously, starting from the onset of gravitational collapse, through core-bounce, the magnetic shock-revival, till the shock-propagation to the stellar surface. Based on our models [268], we calculated numerically the neutrino flavor conversion in the highly nonspherical envelope through the MSW effect. The neutrino transport was simply treated by a leakage scheme (e.g., [268]). The emergent neutrino spectra was assumed to take a Fermi-Dirac type distribution function and the neutrino temperature was estimated to take the average matter temperature on the neutrino sphere. As explained in Section 3(a), the density profile along the polar and equatorial direction is very different due to the strong explosion anisotropy inherent to the MHD explosions. In the following, we pay attention to the anisotropic shock effect on the MSW effect. As will be discussed, we could observe a sharp dip in the neutrino event only seen from a polar direction, albeit depending on the mass hierarchy and the mixing angle of θ_{13}. An advantage of the MHD models is that the shock revival can occur much faster after bounce compared to the other proposed mechanism. Therefore, the neutrino luminosity could remain higher than those in the other mechanisms, which could potentially enhance the detectability due to the early shock arrival to the resonance region. For simplicity, the effects of neutrino self-interactions are treated very phenomenologically and

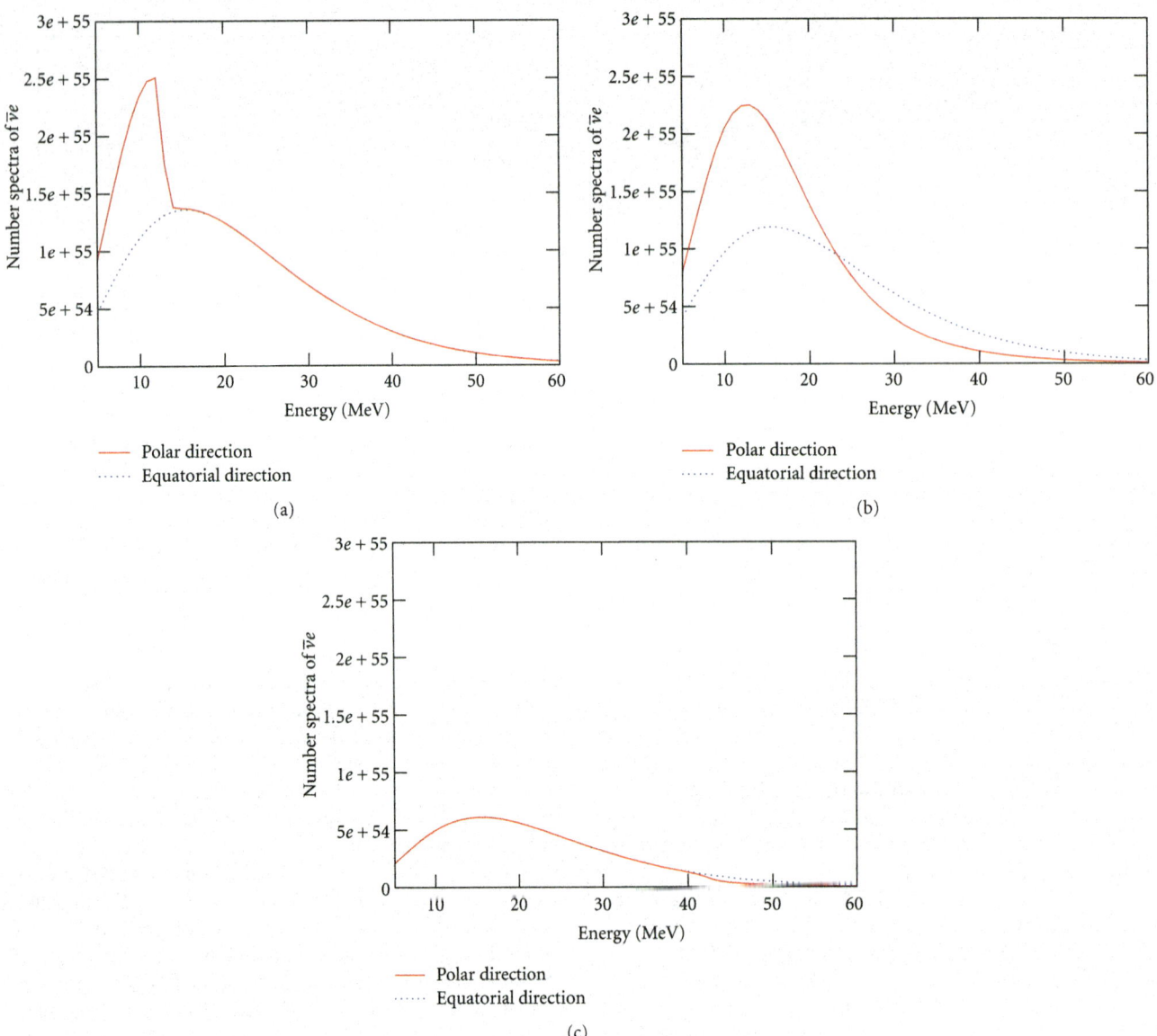

FIGURE 27: $\bar{\nu}_e$ spectra at the surface of the star in the case of (a), (b), and (c). We assume that the inverted mass hierarchy and $\sin^2 2\theta_{13} = 10^{-3}$. The solid lines (red lines) and the dotted lines (blue lines) are the spectra of polar direction and equatorial direction, respectively. Note in this figure that the neutrino luminosity and spectra are modeled to be the same as those obtained in a 1D full-scale numerical simulation by the Lawrence Livermore group [91]. These figures are taken from [306].

the resonant spin-flavor is not considered. Even though far from comprehensive in this respect, we presented the first discussion how the magneto-driven explosion anisotropy has impacts on the emergent neutrino spectra and the resulting event number observed by the SK for a future Galactic supernova (e.g., [306]).

3.2.1. Possible Neutrino Signatures from MHD Explosions.
One prominent feature of the MHD models is a high degree of the explosion asphericity. Figure 25 shows several snapshots featuring typical hydrodynamics of the model, from near core-bounce (0.8 s, left), during the shock-propagation

(2.0 s, middle), till near the shock breakout from the star (4.3 s, right), in which time is measured from the epoch of core bounce. It can be seen that the strong shock propagates outwards with time along the rotational axis. On the other hand, the density profile hardly changes in the equatorial direction, which is a generic feature of the MHD explosions.

As well known, the flavor conversion through the pure-matter MSW effect occurs in the resonance layer, where the density is

$$\rho_{\text{res}} \sim 1.4 \times 10^3 \text{ g/cm}^3 \left(\frac{\Delta m^2}{10^{-3} \text{ eV}^2}\right)\left(\frac{10 \text{ MeV}}{E_\nu}\right)\left(\frac{0.5}{Y_e}\right) \cos 2\theta,$$

$$(15)$$

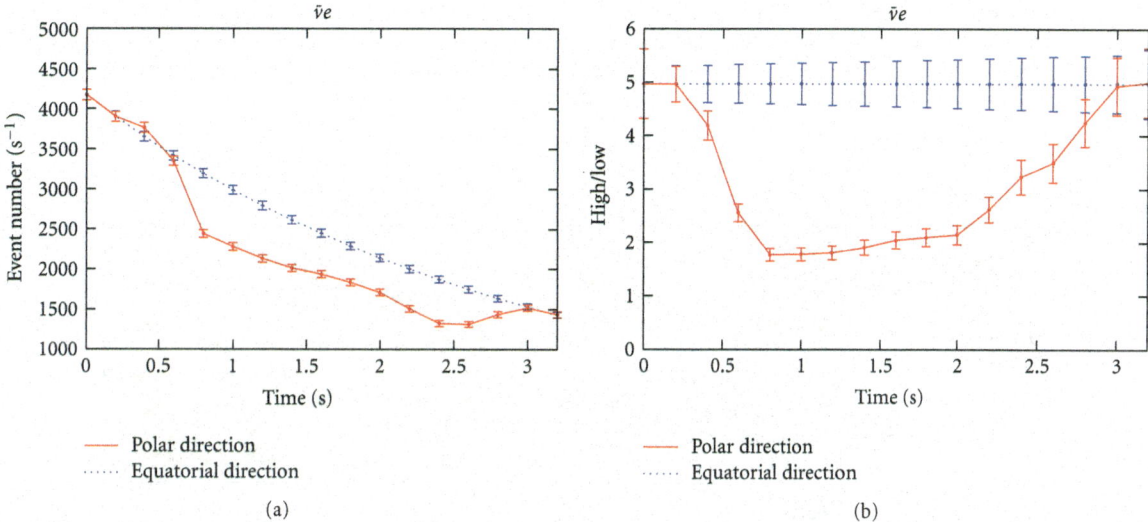

FIGURE 28: Same as Figure 27 but for the expected event number of $\bar{\nu}_e$ (left panel) in the SK as a function of the time. The solid line (red line) and the the dotted line (blue line) are for the polar and the equatorial directions, respectively. The supernova is assumed to be located at the distance of 10 kpc. The error bars represent the 1σ statistical errors only. Right panel is for the ratio of high-to-low-energy events which can be a useful quantity to characterize the change (dip in the right panel) of neutrino signals by the shock effect (see [306] for details).

where Δm^2 is the mass squared difference, E_ν is the neutrino energy, Y_e is the number of electrons per baryon, and θ is the mixing angle. Since the inner supernova core is too dense to allow MSW resonance conversion, we focus on two resonance points in the outer supernova envelope. One that occurs at higher density is called the H-resonance, and the other, which occurs at lower density, is called the L-resonance. Δm^2 and θ correspond to Δm_{13}^2 and θ_{13} at the H-resonance and to Δm_{12}^2 and θ_{12} at the L-resonance.

Figure 26 are evolutions of the density profiles in the polar direction every 0.4 s as a function of radius. Above and below horizontal lines show approximately density of the resonance for different neutrino energies which are 5 and 60 MeV, respectively. Blue lines (left panel) show the range of the density of the H-resonance, and sky-blue lines (right panel) show that of the L-resonance. Along the polar axis, the shock wave reaches to the H-resonance, $\sim O\,(10^3)\,\mathrm{g/cm^3}$ at $\sim$ 0.5 s, and the L-resonance, $\sim O\,(1)\,\mathrm{g/cm^3}$ at ~1.2 s. It should be noted that those timescales are very early in comparison with the ones predicted in the neutrino-driven explosion models, typically ~5 s and ~15 s for the H- and L-resonances, respectively (e.g., [320, 325]). This arises from the fact that the MHD explosion is triggered promptly after core bounce, which is in sharp contrast to the neutrino-driven *delayed* explosion models [320, 325]. The progenitor of the MHD models, possibly linked to long-duration gamma-ray bursts, is more compact due to a chemically homogeneous evolution [274, 275], which is also the reason for the early shock-arrival to the resonance regions.

Figure 27 is the energy spectra of $\bar{\nu}_e$ for (a), (b), and (c) in the case of inverted mass hierarchy. The solid lines (red lines) and the dotted lines (blue lines) are the spectra of the polar direction and the equatorial direction, respectively. At the panel (a), an enhancement of the low-energy side

of the polar direction is seen, which is as a result of the supernova shock reaching to the H-resonance region. The survival probability of $\bar{\nu}_e$ can remain non-zero when the steep decline of the density at the shock front changes the resonance into non-adiabatic (see [306] for more details). As the shock propagates from the high density region to low energy region, the shock effect transits from low-energy side to high-energy side in spectra, because the density at the resonance point, ρ_{res}, is proportional to E_ν^{-1}. In fact, the energy spectrum of the polar direction becomes softer in the high-energy side than for the equatorial direction as shown in Figure 27(b).

Figure 28 shows the time evolution of the event number in SK, where the solid line (red line) and the dotted line (blue line) are for the polar and the equatorial direction, respectively. The shock effect is clearly seen. The event number of the polar direction shows a steep decrease, marking the shock passage to the H-resonance layer. The change of the event number by the shock passage is about 36% of the event number without the shock. Since the expected events are ~2500 at the sudden decrease (Figure 28), it seems to be quite possible to identify such a feature by the SK class detectors. Such a large number imprinting the shock effect, possibly up to two-orders-of magnitudes larger than the ones predicted in the neutrino-driven explosions (e.g., Figure 11 in [320]), is thanks to the mentioned early shock-arrival to the resonance layer, peculiar for the MHD explosions.

Finally we exploratory discuss possible impacts of the spectral swap between $\bar{\nu}_e$ and ν_x, which is one of a probable outcome of the collective neutrino oscillations [346]. In our model, the original neutrino flux of $\bar{\nu}_e$ is larger than that of ν_x for the low-energy side, which is vice versa for the high-energy side. Due to the spectral swap, the above inequality sign reverses (see [306] for more detail). As a result, the

TABLE 2: A summary illustrating the candidate supernova mechanisms (vertical direction in the Table) and their expected features of the emergent multimessengers (horizontal direction). Regarding the photon messengers, possible signatures associated with nucleosynthesis are only partly presented (see Section 1 for a number of important findings regarding the light-curve and spectra modeling).

Mechanism \ Messenger	Gravitational waves	Neutrinos	Photons (nucleosynthesis)
Neutrino-heating mechanism — Canonical rotation	Stochastic (convection and SASI)	Stochastic (convection and SASI)	νp process Anisotropic explosive nucleosynthesis
Rapid rotation	Stochastic (excess for equator) (spiral SASI modes)	Polar excess	νp process (?)
Fails: black-hole forming	Burst signals (bounce and BH formation)	Disappearing signals	No photon (?)
MHD mechanism	Burst and tail (rapid rotation, non axisymmetric instabilities, and magnetic fields)	(i) Polar excess (?) (ii) Anisotropy in SK events (MSW effect) (iii) $\bar{\nu}_e$ bursts (RSF)	(i) νp process (?) (ii) r-process cites (?) (iii) Path to hypernovae (?)

event number in the polar direction is expected to increase by the influence of the shock. To draw a robust conclusion to the self-interaction effects, one needs to perform a more sophisticated analysis such as the multiangle approach (e.g., [76]), which is a major undertaking.

4. Summary and Discussion

The aim of writing this paper was to provide an overview of what we currently know about the explosion mechanism, neutrinos, gravitational waves, and explosive nucleosynthesis in CCSNe to bridge theory and observation through multimessenger astronomy. Not to inflate the volume, we had to limit ourselves to focus primarily on the findings based on our numerical studies, which are thus limited to GWs (Section 2.2) and explosive nucleosynthesis (Section 2.3) in the neutrino-heating mechanism (Section 2), and to GWs (Section 3.1) and neutrino signals (Section 3.2) in the MHD mechanism (Section 3). Apparently our current presentation is weak on the photon side mainly because there is a big gap to connect the outcomes obtained in the multi-D neutrino-radiation-hydrodynamic simulations covering only the very centre of the star, to the light-curve and spectra modeling that requires multi-D photon-radiation-hydrodynamic modeling (far) after the shock-breakout of the star. The gap is not only the matter of physical scale, but also the matter of numerical difficulty. Note that Table 2 is not intended to show an overview covering everything concerning the SN multimessengers, which is far beyond the scope of this review. We boldly present the incomplete figure

here, hoping that the shortage might give momentum to theorists for fixing the problems and for gaining much more comprehensive multimessenger perspectives for the future.

In the neutrino-heating mechanism, stellar rotation holds a key importance to characterize the SN multimessengers. The whole story may be summarized as follows. If a precollapse iron core has a "canonical" rotation rate as predicted by recent stellar evolution calculations [126, 127], collapse-dynamics before bounce proceeds spherically and the postbounce structures interior to the PNS are essentially spherical. Typically later than ~100 ms after bounce, the bounce shock turns to a standing accretion shock, and the activities of convective overturns and SASI become vigorous with time behind the standing shock. Since anisotropies of the neutrino flux and matter motions in the postbounce phase are governed by the nonlinear hydrodynamics, GWs emitted in this epoch also change stochastically with time (indicated by "Stochastic" in the table, e.g., Section 3.1 for more details). For detecting these signals for a Galactic CCSN event with a good signal-to-noise ratio, we need next generation detectors such as the advanced LIGO, LCGT, and the Fabry-Perot (FP-) type DECIGO. Neutrino signals at the nonlinear epoch also change stochastically with time, which is expected to be detected by currently running detector such as by IceCube and SK [133, 197, 343] for the galactic event. It should be also mentioned that detailed neutrino transport including the most up-to-date interaction rates should be accurately implemented in the multi-D simulations to make a reliable prediction of the neutrino signals (e.g., [312, 313]). When material behind the stalled shock successfully absorbs enough neutrino energy to be gravitationally unbound from the iron core, the νp process is expected to successfully explain some light proton-rich nuclei in the neutrino-driven wind phase [245–247]. When the revived supernova shock (successfully) attains the kinetic energy as energetic as 10^{51} erg, explosive nucleosynthesis that proceeds in the SASI-aided low-mode explosions is expected to partly explain the solar abundance yields as well as the observed nonuniform morphology of the synthesized elements (e.g., Section 2.3.1).

One important notice here is that explosion energies obtained in the state-of-the-art multi-D simulations are typically underpowered by one or two orders-of-magnitudes to explain the canonical supernova kinetic energy ($\sim 10^{51}$ erg, e.g., Table 1). Moreover, the softer EOS, such as of Lattimer and Swesty [347] (LS) EOS with an incompressibility at nuclear densities, K, of 180 MeV, is employed in those simulations (e.g., in multi-D simulations of the MPA, Oak Ridge+, and Tokyo+ groups in Table 1). On top of a striking evidence that favors a stiffer EOS based on the nuclear experimental data ($K = 240 \pm 20$ MeV, [348]), the soft EOS may not account for the recently observed massive neutron star of $\sim 2M_\odot$ [349] (see the maximum mass for the LS180 EOS in [314, 350]). With a stiffer EOS, the explosion energy may be even lower as inferred from Marek and Janka [125] who did not obtain the neutrino-driven explosion for their model with $K = 263$ MeV (on the other hand, they obtained 2D explosions for Shen EOS ($K = 281$ MeV) (Janka, private communication)). What is then missing furthermore? We may get the answer by going to 3D simulations (Section 2.1) or by taking into account new ingredients, such as exotic physics in the core of the protoneutron star [123, 351], viscous heating by the magnetorotational instability (Section 3(b)), or energy dissipation via Alfvén waves [352]. In addition to the 3D effects, GR is expected to help the onset of multi-D neutrino-driven explosions [169, 353] (e.g., [129] in 2D simulations but with detailed neutrino transport or [169] in 3D simulations but with a very approximate neutrino transport). At present, the most up-to-date neutrino transport code can treat the multienergy and multiangle transport in 2D [132] and even in 3D simulations [158] but only limited to the Newtonian simulations. Extrapolating a current growth rate of supercomputing power, the dream simulation (i.e., 6D simulations with full Boltzmann neutrino transport in full GR simulations) is likely to be practicable not in the distant future (in our lifetime!) presumably by using the exa-scale platforms.

In addition to these numerical advancements, the physical understanding of the supernova theory via analytical studies is also in a steady progress. What determines the saturation levels of SASI? A careful analysis on the parasitic instabilities has been reported to answer this question [354]. What determines mode couplings between small-scale convection eddies and large-scale SASI modes? To apply the theory of turbulence [355] should put a milestone to address this question. Besides the two representative EOSs of Lattimer-Swesty and Shen, new sets of EOSs have been recently reported [356–358]. Some of them will enable SN modellers to go beyond a single nucleus approximation, which is an important improvement for an accurate description of neutrino-nucleus interaction.

Rotation, albeit depending on its strength, can give a special direction (i.e., spin axis) in the supernova cores. As discussed in Section 2.2.2, stochastic nature of the GWs becomes weak in the presence of rotation as a result of the spiral SASI (It should be noted that the feature can be seen even in a slowly rotating progenitor that agrees with recent stellar evolution calculation [126, 127] (see Section 2.2.2 for more details)) (indicated by "excess for equator" in Table 2). Although it is not easy to detect these low-frequency GWs from anisotropic neutrino emission by currently running laser interferometers, a recently proposed space-based interferometers (like FP-type DECIGO) would permit detection for a galactic event. Contributed by the neutrino GWs in the lower frequency domain, the total GW spectrum is expected to become rather flat over a broad frequency range below ~ 1 kHz. This GW feature obtained in the context of the SASI-aided neutrino-driven mechanism is different from the one in the other candidate supernova mechanisms, such as the MHD mechanism (Section 3.1) and the acoustic mechanism [193], thus could provide an important probe into the explosion mechanism. Rapid rotation induces a polar excess of neutrino emission with a harder spectrum than on the equator [133, 344, 345] (indicated by "polar excess" in the table). This should be also the case of the MHD mechanism, which deserves further investigation for a quantitative discussion (symbolized by Polar excess (?) in the table). These (rotation-induced) neutrino signatures, if observed, will carry an important information about the angular momentum profiles hidden deep inside massive stellar cores. Concerning the νp process and explosive synthesis in rapidly rotating cores (as well as in the MHD explosions), there still remains a vast virgin territory for further investigation (symbolized by "νp process (?)" in the table).

If the neutrino-heating mechanism (or some other mechanisms) fails to blow up massive stars, central PNSs collapse to BHs. Recent GR simulations by Ott et al. [199] show that the significant GW emission is associated at the moment of the BH formation, which can be a promising target of the advanced LIGO for a galactic source. As pointed out by Sumiyoshi et al. [359], the disappearing neutrino signals that mark the epoch of BH formations also can be a target of SK and IceCube [10]. When quarks and hyperons appear in the postbounce core, a neutrino burst produced by the sudden EOS softening and by the subsequent rebounce [123] is likely to be detected by IceCube for a galactic event [10]. The interval between core-bounce and the BH formation depends on the details of exotic physics in the super-dense PNS cores [360, 361]. All of these observational signatures should provide an important probe into the so-called dense QCD region in the QCD phase diagram (e.g., [362] for recent review) to which lattice calculations are hardly accessible at present. Concerning photons, no optical outbursts are expected in the BH-forming SNe (indicated by "no photon (?)" in Table 2), if not for rapid rotation and strong magnetic fields prior to core-collapse.

If a precollapse core rotates enough rapidly (typically initial rotation period less than 4 s) with strong magnetic fields (higher than $\sim 10^{11}$ G), the MHD mechanism can produce bipolar explosions along the rotational axis predominantly driven by the field wrapping processes (Section 3(a)). If the MRI can be sufficiently resolved in global simulations, the MRI would exponentially amplify the initial magnetic fields to a dynamically important strength within several rotational periods (Sections 3(b) and 3(c)). The next-generation supercomputing resources are needed again to see the outcome. If such simulations would be executable, the MHD outflows will be produced even for more weakly magnetized cores than currently predicted. The GW signals in the MHD explosions are characterized by a burst-like bounce signal plus a secularly growing tail (Section 3.1, indicated by "burst and tail" in the table). Similar to the neutrino GWs in the neutrino-driven mechanism, a future detector (like FP-type DECIGO) is again necessary for detecting the low-frequency tail component. As discussed in Section 3.2.1, the pole to equator anisotropy of the shock propagation in the MHD explosions affect the neutrino signals through the MSW effects. The anisotropic shock passage to the H-resonance regions leads to a sudden decrease in the SK events (~ 2500 events for a galactic source), which is quite possible to identify by the SK-class detectors (indicated by "anisotropy in SK events" in Table 2). Since the MHD mechanism has such distinct signatures, a planned joint analysis of neutrino and GW data [363] would be potentially very powerful to tell the MHD mechanism from the other candidate mechanisms. If the resonant spin-flavor conversion (indicated as RSF in Table 2) occurs in the highly magnetized core, the neutronization burst of ν_e converts to that of $\bar{\nu}_e$, which is thus expected to be a probe into the magnetic moment of neutrinos (e.g., [334]). Finally the MHD explosions are considered to be a possible r-process cite [364, 365] (see collective references in [247] for other plausible r-process cites). This is because the mass ejection from the iron core occurs in much shorter timescales compared to the delayed neutrino-heating mechanism, so that the ejecta can stay neutron rich before it becomes proton-rich via neutrino capture reactions. Most of these possibilities regarding the MHD mechanism have been proposed so far in simulations with a crude treatment of neutrino transport, which is now rather easily reexamined by the state-of-the-art multi-D simulations (if the code is MHD).

Finally, what is about the story if the MHD mechanism fails? (indicated by "path to hypernovae in Table 2). The rapidly rotating PNS collapses into a BH, probably leading to the formation of accretion disk around the BH. Neutrinos emitted from the accretion disk heat matter in the polar funnel region to launch outflows or strong magnetic fields in the cores of the order of 10^{15} G also play an active role both in driving the magneto-driven jets and in extracting a significant amount of energy from the BH (e.g., [366–368] and see references therein). This picture, often referred as collapsar (e.g., [298, 299]), has been the working hypothesis as a central engine of long-duration GRBs for these 10 years (see references in [367–371] for other candidate mechanism including magnetar models). However, it is still controversial whether the generation of the relativistic outflows predominantly proceeds via the neutrino-heating mechanism or the MHD mechanism. In contrast to a number of findings illustrated in Table 2, much little things are known about the BH-forming supernovae and also about the collapsar. This is mainly because the requirement for making a *realistic* numerical modeling is very computationally expensive, which (at least) necessitates the multi-D MHD simulations not only with GR for handling the BH formation (thus full GR), but also with the multiangle neutrino transfer for treating highly anisotropic neutrino radiation from the accretion disk [372]. To get a unified picture of massive stellar death, we need to draw a schematic picture (like Table 2) also in the case of the BH-forming supernovae. We anticipate that our long-lasting focus (and experience) on the explosion dynamics of canonical CCSNe will be readily applicable to clarifying the origin of the gigantic explosion energy of hypernovae and also unraveling (ultimately) the central engine of long-duration GRBs. This should provide yet another grand challenge in computational astrophysics in the next decades.

As repeatedly mentioned so far, all the numerical results in this paper should be tested by the next-generation calculations by which more sophistication is made not only in the treatment of multi-D radiation transport (both of neutrino and photon), but also in multi-D hydrodynamics including stellar rotation and magnetic fields in full GR. From an optimistic point of view, our understanding on every issue raised in this paper can progress in a step-by-step manner at the same pace as our available computational resources will be growing bigger and bigger from now on. Since 2009, several neutrino detectors form the Supernova Early Warning Systems (SNEWS) to broadcast the alert to astronomers to let them know the arrival of neutrinos [82]. Currently, Super-Kamiokande, LVD, Borexino, and IceCube contributes to the SNEWS, with a number of other neutrino and GW detectors planning to join in the near future. This is a very encouraging news towards the high-precision multimessenger astronomy. The interplay between the detailed numerical modeling, the advancing supercomputing resources, and the multimessenger astronomy, will remain a central issue for advancing our understanding of the theory of massive stellar core-collapse for the future. A documentary film recording our endeavors to make practicable the "multimessenger astronomy of CCSNe" seems not to show "fin" immediately and is becoming even longer as far as the three components evolve with time. To raise the edifice, it is becoming increasingly important to bring forward a worldwide, multidisciplinary collaboration among different research groups. We believe that such an approach would provide the shortest cut to get a deeper understanding of a number of unsettled and exciting issues that we were only able to touch in this paper.

Acknowledgments

K. Kotake is thankful to K. Sato for continuing encouragements. He also wishes to thank his collaborators, S. Yamada, M. Liebendörfer, N. Ohnishi, K. Sumiyoshi, K. Nakamura, T. Kuroda, M. Hashimoto, S. Harikae, N. Yasutake, N. Nishimura, K. Shaku, and H. Suzuki. The authors would like to acknowledge helpful exchanges and stimulating discussions with E. Müller, T. Foglizzo, H.-T. Janka, C. D. Ott, C. Fryer, J. Novak, S. Ando, P. Cerdá-Durán, M. Obergaulinger, J. Murphy, R. Fernández, E. O'Connor, M. Shibata, T. Font, and J. M. Ibañez. They are also grateful to our anonymous referees who gave us a number of valuable comments to enhance the quality of this paper. Numerical computations were carried out in part on XT4 and general common use computer system at the center for Computational Astrophysics, CfCA, the National Astronomical Observatory of Japan. This study was supported in part by the Grants-in-Aid for the Scientific Research from the Ministry of Education, Science and Culture of Japan (nos. 19540309, 20740150, 23540323, 23340069, and 24244036) and by HPCI Strategic Program of Japanese MEXT.

References

[1] S. Ando, J. F. Beacom, and H. Yüksel, "Detection of neutrinos from supernovae in nearby galaxies," *Physical Review Letters*, vol. 95, no. 17, 4171101 pages, 2005.

[2] K. Hirata, T. Kajita, M. Koshiba, M. Nakahata, and Y. Oyama, "Observation of a neutrino burst from the supernova SN1987A," *Physical Review Letters*, vol. 58, no. 14, pp. 1490–1493, 1987.

[3] R. M. Bionta, G. Blewitt, C. B. Bratton et al., "Observation of a neutrino burst in coincidence with supernova 1987A in the large magellanic cloud," *Physical Review Letters*, vol. 58, no. 14, pp. 1494–1496, 1987.

[4] K. Sato and H. Suzuki, "Analysis of neutrino burst from the supernova 1987A in the large magellanic cloud," *Physical Review Letters*, vol. 58, no. 25, pp. 2722–2725, 1987.

[5] G. Raffelt, "Neutrinos and the stars," in press, http://arxiv.org/abs/1201.1637.

[6] G. G. Raffelt, "Physics with supernovae," *Nuclear Physics B*, vol. 110, pp. 254–267, 2002.

[7] Y. Totsuka, "Neutrino astronomy," *Reports on Progress in Physics*, vol. 55, no. 3, article 377, 1992.

[8] K. Hultqvist and IceCube collaboration, "IceCube: physics, status, and future," *Nuclear Instruments and Methods in Physics Research A*, vol. 626-627, supplement, pp. S6–S12, 2011.

[9] A. Suzuki, "Present status of KamLand," *Nuclear Physics B*, vol. 77, no. 1–3, pp. 171–176, 1999.

[10] IceCube Collaboration, R. Abbasi, Y. Abdou et al., "IceCube sensitivity for low-energy neutrinos from nearby supernovae," in press, http://arxiv.org/abs/1108.0171.

[11] B. Abbott, R. Abbott, R. Adhikari et al., "Upper limits from the LIGO and TAMA detectors on the rate of gravitational-wave bursts," *Physical Review D*, vol. 72, no. 12, Article ID 122004, 16 pages, 2005.

[12] M. Ando and The TAMA Collaboration, "Current status of the TAMA300 gravitational-wave detector," *Classical and Quantum Gravity*, vol. 22, no. 18, pp. S881–S889, 2005.

[13] J. Hough, S. Rowan, and B. S. Sathyaprakash, "The search for gravitational waves," *Journal of Physics B*, vol. 38, no. 9, article S497, 2005.

[14] M. van der Sluys, "Gravitational waves from compact binaries," . In press, http://arxiv.org/abs/1108.1307.

[15] G. M. Harry and The LIGO Scientific Collaboration, "Advanced LIGO: the next generation of gravitational wave detectors," *Classical and Quantum Gravity*, vol. 27, no. 8, Article ID 084006, 2010.

[16] S. A. Balbus and J. F. Hawley, "Instability, turbulence, and enhanced transport in accretion disks," *Reviews of Modern Physics*, vol. 70, no. 1, pp. 1–53, 1998.

[17] B. S. Sathyaprakash and B. F. Schutz, "Physics, astrophysics and cosmology with gravitational waves," *Living Reviews in Relativity*, vol. 12, article 2, 2009.

[18] J. Faber, "Status of neutron star-black hole and binary neutron star simulations," *Classical and Quantum Gravity*, vol. 26, no. 11, Article ID 114004, 2009.

[19] M. D. Duez, "Numerical relativity confronts compact neutron star binaries: a review and status report," *Classical and Quantum Gravity*, vol. 27, no. 11, Article ID 114002, 2010.

[20] N. Andersson, V. Ferrari, D. I. Jones et al., "Gravitational waves from neutron stars: promises and challenges," *General Relativity and Gravitation*, vol. 43, no. 2, pp. 409–436, 2011.

[21] C. J. Horowitz, "Multi-messenger observations of neutron rich matter," *International Journal of Modern Physics E*, vol. 20, pp. 2077–2100, 2011.

[22] K. Kotake, K. Sato, and K. Takahashi, "Explosion mechanism, neutrino burst and gravitational wave in core-collapse supernovae," *Reports on Progress in Physics*, vol. 69, no. 4, article 971, 2006.

[23] C. D. Ott, "The gravitational-wave signature of core-collapse supernovae," *Classical and Quantum Gravity*, vol. 26, no. 6, Article ID 063001, 2009.

[24] C. L. Fryer and K. C. B. New, "Gravitational waves from gravitational collapse," *Living Reviews in Relativity*, vol. 14, no. 1, 2011.

[25] S. L. Shapiro and S. A. Teukolsky, *Black Holes, White Dwarfs and Neutron Stars: The Physics of Compact Objects*, The National Science Foundation, Wiley-Interscience, New York, NY, USA, 1983.

[26] L. Wang and J. C. Wheeler, "Spectropolarimetry of supernovae," *Annual Review of Astronomy and Astrophysics*, vol. 46, no. 1, pp. 433–474, 2008.

[27] S. E. Woosley, "SN 1987A—after the peak," *The Astrophysical Journal*, vol. 330, pp. 218–253, 1988.

[28] S. Blinnikov, P. Lundqvist, O. Bartunov, K. Nomoto, and K. Iwamoto, "Radiation hydrodynamics of SN 1987A. I. Global analysis of the light curve for the first 4 months," *The Astrophysical Journal*, vol. 532, no. 2, pp. 1132–1149, 2000.

[29] V. P. Utrobin, "The light curve of supernova 1987A: the structure of the presupernova and radioactive nickel mixing," *Astronomy Letters*, vol. 30, no. 5, pp. 293–308, 2004.

[30] S. E. Woosley and A. Heger, "Nucleosynthesis and remnants in massive stars of solar metallicity," *Physics Reports*, vol. 442, no. 1—, pp. 269–283, 2007.

[31] D. Arnett, B. Fryxell, and E. Mueller, "Instabilities and nonradial motion in SN 1987A," *The Astrophysical Journal*, vol. 341, pp. L63–L66, 1989.

[32] E. Müller, B. Fryxell, and D. Arnett, "Instability and clumping in SN 1987A," *Astronomy and Astrophysics*, vol. 251, no. 2, pp. 505–514, 1991.

[33] K. Kifonidis, T. Plewa, H. T. Janka, and E. Müller, "Non-spherical core collapse supernovae I. Neutrino-driven convection, Rayleigh-Taylor instabilities, and the formation and propagation of metal clumps," *Astronomy and Astrophysics*, vol. 408, no. 2, pp. 621–649, 2003.

[34] N. J. Hammer, H. T. Janka, and E. Müller, "Three-dimensional simulations of mixing instabilities in supernova explosions," *The Astrophysical Journal*, vol. 714, no. 2, pp. 1371–1385, 2010.

[35] S. Nagataki, M. A. Hashimoto, K. Sato, and S. Yamada, "Explosive nucleosynthesis in axisymmetrically deformed type II supernovae," *The Astrophysical Journal*, vol. 486, no. 2, pp. 1026–1035, 1997.

[36] S. Nagataki, "Effects of jetlike explosion in SN 1987A," *Astrophysical Journal Supplement Series*, vol. 127, no. 1, pp. 141–157, 2000.

[37] V. P. Utrobin and N. N. Chugai, "Supernova 2000cb: high-energy version of SN 1987A," *Astronomy and Astrophysics*, vol. 532, article A100, 6 pages, 2011.

[38] L. Wang, J. C. Wheeler, P. Hoflich et al., "The axisymmetric ejecta of supernova 1987A," *The Astrophysical Journal*, vol. 579, no. 2, article 671, 2002.

[39] K. Kjær, B. Leibundgut, C. Fransson, A. Jerkstrand, and J. Spyromilio, "The 3-D structure of SN 1987A's inner ejecta," *Astronomy and Astrophysics*, vol. 517, article A51, 10 pages, 2010.

[40] L. Wang, D. A. Howell, P. Hoflich, and J. C. Wheeler, "Bipolar Supernova Explosions," *The Astrophysical Journal*, vol. 550, article 1030, 2001.

[41] D. C. Leonard, A. V. Filippenko, M. Ganeshalingam et al., "A non-spherical core in the explosion of supernova SN 2004dj," *Nature*, vol. 440, no. 7083, pp. 505–507, 2006.

[42] D. C. Leonard, A. V. Filippenko, M. Ganeshalingam et al., "A non-spherical core in the explosion of supernova SN 2004dj," *Nature*, vol. 440, no. 7083, pp. 505–507, 2006.

[43] K. Kifonidis, T. Plewa, L. Scheck, H. T. Janka, and E. Müller, "Non-spherical core collapse supernovae," *Astronomy and Astrophysics*, vol. 453, no. 2, pp. 661–678, 2006.

[44] L. Scheck, K. Kifonidis, H. T. Janka, and E. Müller, "Multidimensional supernova simulations with approximative neutrino transport," *Astronomy and Astrophysics*, vol. 457, no. 3, pp. 963–986, 2006.

[45] A. Burrows, E. Livne, L. Dessart, C. D. Ott, and J. Murphy, "Features of the acoustic mechanism of core-collapse supernova explosions," *The Astrophysical Journal*, vol. 655, no. 1 I, pp. 416–433, 2007.

[46] A. Burrows, L. Dessart, E. Livne, C. D. Ott, and J. Murphy, "Simulations of magnetically driven supernova and hypernova explosions in the context of rapid rotation," *The Astrophysical Journal*, vol. 664, no. 1, article 416, 2007.

[47] L. Dessart and D. J. Hillier, "Supernova radiative-transfer modelling: a new approach using non-local thermodynamic equilibrium and full time dependence," *Monthly Notices of the Royal Astronomical Society*, vol. 405, no. 4, pp. 2141–2160, 2010.

[48] L. Dessart and D. J. Hillier, "Time-dependent effects in photospheric-phase type II supernova spectra," *Monthly Notices of the Royal Astronomical Society*, vol. 383, no. 1, pp. 57–74, 2008.

[49] L. Dessart and D. J. Hillier, "Quantitative spectroscopy of photospheric-phase type II supernovae," *Astronomy and Astrophysics*, vol. 437, no. 2, pp. 667–685, 2005.

[50] D. Kasen, R. C. Thomas, and P. Nugent, "Time-dependent Monte Carlo radiative transfer calculations for three-dimensional supernova spectra, light curves, and polarization," *The Astrophysical Journal*, vol. 651, no. 1, article 366, 2006.

[51] D. Kasen and S. E. Woosley, "Type II supernovae: model light curves and standard candle relationships," *The Astrophysical Journal*, vol. 703, no. 2, pp. 2205–2216, 2009.

[52] L. Dessart and D. J. Hillier, "Synthetic line and continuum linear-polarization signatures of axisymmetric type II supernova ejecta," *Monthly Notices of the Royal Astronomical Society*, vol. 415, no. 4, pp. 3497–3519, 2011.

[53] D. C. Leonard, A. V. Filippenko, M. Ganeshalingam et al., "A non-spherical core in the explosion of supernova SN 2004dj," *Nature*, vol. 440, no. 7083, pp. 505–507, 2006.

[54] R. Chornock, A. V. Filippenko, W. Li, and J. M. Silverman, "LArge late-time asphericities in three type IIP supernovae," *The Astrophysical Journal Letters*, vol. 713, no. 2, pp. 1363–1375, 2010.

[55] R. A. Fesen, M. C. Hammell, J. Morse et al., "Discovery of outlying high-velocity oxygen-rich ejecta in Cassiopeia A," *The Astrophysical Journal*, vol. 636, pp. 859–872, 2006.

[56] R. A. Fesen, J. A. Zastrow, M. C. Hammell, J. M. Shull, and D. W. Silvia, "Ejecta knot flickering, mass ablation, and fragmentation in Cassiopeia A," *The Astrophysical Journal*, vol. 736, article 109, 2011.

[57] B. Aschenbach, R. Egger, and J. Trümper, "Discovery of explosion fragments outside the Vela supernova remnant shock-wave boundary," *Nature*, vol. 373, no. 6515, pp. 587–590, 1995.

[58] J. P. Hughes, C. E. Rakowski, D. N. Burrows, and P. O. Slane, "Nucleosynthesis and mixing in Cassiopeia A," *The Astrophysical Journal*, vol. 528, no. 2, pp. L109–L113, 2000.

[59] M. Miceli, F. Bocchino, and F. Reale, "Physical and chemical inhomogeneities inside the Vela SNR shell. Indications of ejecta shrapnels," *The Astrophysical Journal*, vol. 676, no. 2, article 1064, 2008.

[60] S. J. Smartt, "Progenitors of core-collapse supernovae," *Annual Review of Astronomy and Astrophysics*, vol. 47, pp. 63–106, 2009.

[61] J. W. Murphy, Z. G. Jennings, B. Williams, J. J. Dalcanton, and A. E. Dolphin, "The progenitor mass of SN 2011dh from stellar population analysis," *The Astrophysical Journal*, vol. 742, article L4, 2011.

[62] S. E. Woosley and J. S. Bloom, "The supernova gamma-ray burst connection," *Annual Review of Astronomy and Astrophysics*, vol. 44, no. 1, pp. 507–556, 2006.

[63] M. Modjaz, "Stellar forensics with the supernova-GRB connection—Ludwig Biermann award lecture 2010," *Astronomische Nachrichten*, vol. 332, no. 5, pp. 434–447, 2011.

[64] P. Podsiadlowski, N. Langer, A. J. T. Poelarends, S. Rappaport, A. Heger, and E. Pfahl, "The effects of binary evolution on the dynamics of core collapse and neutron star kicks," *The Astrophysical Journal*, vol. 612, no. 2, pp. 1044–1051, 2004.

[65] P. Podsiadlowski, N. Ivanova, S. Justham, and S. Rappaport, "Explosive common-envelope ejection: implications for gamma-ray bursts and low-mass black-hole binaries," *Monthly Notices of the Royal Astronomical Society*, vol. 406, article 840, 2010.

[66] C. L. Fryer, P. A. Mazzali, J. Prochaska et al., "Constraints on type Ib/c supernovae and gamma-ray burst progenitors,"

Publications of the Astronomical Society of the Pacific, vol. 119, no. 861, article 1211, 2007.

[67] N. Smith, W. Li, A. V. Filippenko, and R. Chornock, "Observed fractions of core-collapse supernova types and initial masses of their single and binary progenitor stars," *Monthly Notices of the Royal Astronomical Society*, vol. 412, pp. 1522–1538, 2011.

[68] S. C. Yoon and N. Langer, "Evolution of rapidly rotating metal-poor massive stars towards gamma-ray bursts," *Astronomy and Astrophysics*, vol. 443, no. 2, pp. 643–648, 2005.

[69] S. C. Yoon, S. E. Woosley, and N. Langer, "Type Ib/c supernovae in binary systems. I. Evolution and properties of the progenitor stars," *The Astrophysical Journal*, vol. 725, no. 1, pp. 940–954, 2010.

[70] K. Nomoto, N. Tominaga, H. Umeda, C. Kobayashi, and K. Maeda, "Nucleosynthesis yields of core-collapse supernovae and hypernovae, and galactic chemical evolution," *Nuclear Physics A*, vol. 777, pp. 424–458, 2006.

[71] L. Dessart, D. J. Hillier, E. Livne et al., "Core-collapse explosions of Wolf-Rayet stars and the connection to type IIb/Ib/Ic supernovae," *Monthly Notices of the Royal Astronomical Society*, vol. 414, no. 4, pp. 2985–3005, 2011.

[72] W. D. Arnett and C. Meakin, "Towards realistic progenitors of core-collapse supernovae," *The Astrophysical Journal Letters*, vol. 733, no. 2, article 78, 2011.

[73] C. A. Meakin and D. Arnett, "Turbulent convection in stellar interiors. I. Hydrodynamic simulation," *The Astrophysical Journal*, vol. 667, pp. 448–475, 2007.

[74] G. G. Raffelt and A. Y. Smirnov, "Self-induced spectral splits in supernova neutrino fluxes," *Physical Review D*, vol. 76, no. 8, Article ID 081301, 5 pages, 2007.

[75] H. Duan, G. M. Fuller, J. Carlson, and Y. Z. Qian, "Flavor evolution of the neutronization neutrino burst from an O–Ne–Mg core-collapse supernova," *Physical Review Letters*, vol. 100, no. 2, Article ID 021101, 4 pages, 2008.

[76] H. Duan, G. M. Fuller, and Y. Qian, "Collective neutrino oscillations," *Annual Review of Nuclear and Particle Science*, vol. 60, pp. 569–594, 2010.

[77] B. Dasgupta, "Physics and astrophysics opportunities with supernova neutrinos," ArXiv e-prints.

[78] S. P. Mikheev and A. I. Smirnov, "Resonant amplification of ν oscillations in matter and solar-neutrino spectroscopy," *Nuovo Cimento C*, vol. 9, no. 1, pp. 17–26, 1986.

[79] H. Duan and J. P. Kneller, "Neutrino flavour transformation in supernovae," *Journal of Physics G*, vol. 36, no. 11, Article ID 113201, 2009.

[80] H. T. Janka, K. Langanke, A. Marek, G. Martínez-Pinedo, and B. Müller, "Theory of core-collapse supernovae," *Physics Reports*, vol. 442, no. 1—6, pp. 38–74, 2007.

[81] A. Burrows, L. Dessart, C. D. Ott, and E. Livne, "Multidimensional explorations in supernova theory," *Physics Reports*, vol. 442, p. 23, 2007.

[82] P. Antonioli, R. T. Fienberg, F. Fleurot et al., "SNEWS: the supernova early warning system," *New Journal of Physics*, vol. 6, article 114, 2004.

[83] F. Patat, D. Baade, P. Hoflich, J. R. Maund, L. Wang, and J. C. Wheeler, "VLT spectropolarimetry of the fast expanding type Ia SN 2006X," *Astronomy and Astrophysics*, vol. 508, no. 1, pp. 229–246, 2009.

[84] M. Tanaka, K. S. Kawabata, K. Maeda et al., "Spectropolarimetry of the unique type Ib supernova 2005bf: larger

asymmetry revealed by later-phase data," *The Astrophysical Journal*, vol. 699, no. 2, article 1119, 2009.

[85] F. K. Thielemann, R. Hirschi, M. Liebendörfer, and R. Diehl, "Massive Stars and Their Supernovae," in *Lecture Notes in Physics*, R. Diehl, D. H. Hartmann, and N. Prantzos, Eds., vol. 182, pp. 153–232, Springer, Berlin, Germany, 2011.

[86] E. Chassande-Mottin, M. Hendry, P. J. Sutton, and S. Márka, "Multimessenger astronomy with the Einstein telescope," *General Relativity and Gravitation*, vol. 43, no. 2, pp. 437–464, 2011.

[87] S. Márka, The LIGO Scientific Collaboration, and The Virgo Collaboration, "The path to the enhanced and advanced LIGO gravitational-wave detectors," *Classical and Quantum Gravity*, vol. 26, no. 11, Article ID 114013, 2009.

[88] T. Pradier and Antares Collaboration, "The Antares neutrino telescope and multi-messenger astronomy," *Classical and Quantum Gravity*, vol. 27, no. 19, Article ID 194004, 2010.

[89] Y. Aso, Z. Márka, C. Finley, J. Dwyer, K. Kotake, and S. Márka, "Search method for coincident events from LIGO and IceCube detectors," *Classical and Quantum Gravity*, vol. 25, no. 11, Article ID 114039, 2008.

[90] S. A. Colgate and R. H. White, "The hydrodynamic behavior of supernovae explosions," *The Astrophysical Journal*, vol. 143, article 626, 1966.

[91] J. R. Wilson, *Numerical Astrophysics*, 1985.

[92] H. A. Bethe and J. R. Wilson, "Revival of a stalled supernova shock by neutrino heating," *The Astrophysical Journal*, vol. 295, pp. 14–23, 1985.

[93] H. A. Bethe, "Supernova mechanisms," *Reviews of Modern Physics*, vol. 62, no. 4, pp. 801–866, 1990.

[94] M. Rampp and H. T. Janka, "Spherically symmetric simulation with Boltzmann neutrino transport of core collapse and postbounce evolution of a 15 M star," *The Astrophysical Journal Letters*, vol. 539, no. 1, article L33, 2000.

[95] M. Liebendörfer, A. Mezzacappa, and F. K. Thielemann, "Conservative general relativistic radiation hydrodynamics in spherical symmetry and comoving coordinates," *Physical Review D*, vol. 63, no. 10, Article ID 104003, 14 pages, 2001.

[96] T. A. Thompson, A. Burrows, and P. A. Pinto, "Shock breakout in core-collapse supernovae and its neutrino signature," *The Astrophysical Journal*, vol. 592, no. 1, article 434, 2003.

[97] K. Sumiyoshi, S. Yamada, H. Suzuki, H. Shen, S. Chiba, and H. Toki, "Postbounce evolution of core-collapse supernovae: long-term effects of the equation of state," *The Astrophysical Journal*, vol. 629, no. 2, article 922, 2005.

[98] M. Herant, W. Benz, W. R. Hix, C. L. Fryer, and S. A. Colgate, "Inside the supernova: a powerful convective engine," *The Astrophysical Journal*, vol. 435, no. 1, pp. 339–361, 1994.

[99] A. Burrows, J. Hayes, and B. A. Fryxell, "On the nature of core-collapse supernova explosions," *The Astrophysical Journal Letters*, vol. 450, no. 2, pp. 830–850, 1995.

[100] H. T. Janka and E. Müller, "Neutrino heating, convection, and the mechanism of type-II supernova explosions," *Astronomy and Astrophysics*, vol. 306, article 167, 1996.

[101] C. L. Fryer, D. E. Holz, and S. A. Hughes, "Gravitational wave emission from core collapse of massive stars," *The Astrophysical Journal*, vol. 565, no. 1, pp. 430–446, 2002.

[102] C. L. Fryer, "Neutron star kicks from asymmetric collapse," *The Astrophysical Journal Letters*, vol. 601, no. 2, article L175, 2004.

[103] J. M. Blondin, A. Mezzacappa, and C. Demarino, "Stability of standing accretion shocks, with an eye toward core-collapse

supernovae," *The Astrophysical Journal Letters*, vol. 584, no. 2, pp. 971–980, 2003.

[104] L. Scheck, T. Plewa, H. T. Janka, K. Kifonidis, and E. Müller, "Pulsar recoil by large-scale anisotropies in supernova explosions," *Physical Review Letters*, vol. 92, no. 1, Article ID 011103, 4 pages, 2004.

[105] N. Ohnishi, K. Kotake, and S. Yamada, "Numerical analysis of standing accretion shock instability with neutrino heating in supernova cores," *The Astrophysical Journal*, vol. 641, pp. 1018–1028, 2006.

[106] N. Ohnishi, K. Kotake, and S. Yamada, "Inelastic neutrino-helium scatterings and standing accretion shock instability in core-collapse supernovae," *The Astrophysical Journal*, vol. 667, pp. 375–381, 2007.

[107] T. Foglizzo, L. Scheck, and H. T. Janka, "Neutrino-driven convection versus advection in core-collapse supernovae," *The Astrophysical Journal*, vol. 625, pp. 1436–1450, 2006.

[108] J. W. Murphy and A. Burrows, "Criteria for core-collapse supernova explosions by the neutrino mechanism," *The Astrophysical Journal*, vol. 688, no. 2, pp. 1159–1175, 2008.

[109] R. Fernández and C. Thompson, "Stability of a spherical accretion shock with nuclear dissociation," *The Astrophysical Journal*, vol. 697, no. 2, pp. 1827–1841, 2009.

[110] R. Fernández and C. Thompson, "Dynamics of a spherical accretion shock with neutrino heating and alpha-particle recombination," *The Astrophysical Journal*, vol. 703, no. 2, pp. 1464–1464, 2009.

[111] W. Iwakami, K. Kotake, N. Ohnishi, S. Yamada, and K. Sawada, "Three-dimensional simulations of standing accretion shock instability in core-collapse supernovae," *The Astrophysical Journal*, vol. 678, no. 2, article 1207, 2008.

[112] W. Iwakami, K. Kotake, N. Ohnishi, S. Yamada, and K. Sawada, "Effects of rotation on standing accretion shock instability in non-linear phase for core-collapse supernovae," *The Astrophysical Journal*, vol. 700, no. 1, article 232, 2009.

[113] R. Fernández, "The spiral modes of the standing accretion shock instability," *The Astrophysical Journal*, vol. 725, no. 2, pp. 1563–1580, 2010.

[114] H. T. Janka, "Conditions for shock revival by neutrino heating in core-collapse supernovae," *Astronomy and Astrophysics*, vol. 368, pp. 527–560, 2001.

[115] A. Wongwathanarat, N. J. Hammer, and E. Müller, "An axis-free overset grid in spherical polar coordinates for simulating 3D self-gravitating flows," *Astronomy and Astrophysics*, vol. 514, article A48, 14 pages, 2010.

[116] J. Nordhaus, A. Burrows, A. Almgren, and J. Bell, "Dimension as a key to the neutrino mechanism of core-collapse supernova explosions," *The Astrophysical Journal*, vol. 720, no. 1, article 694, 2010.

[117] K. Nomoto and M. Mashimoto, "Presupernova evolution of massive stars," *Physics Reports*, vol. 163, no. 1–3, pp. 13–36, 1988.

[118] S. E. Woosley, A. Heger, and T. A. Weaver, "The evolution and explosion of massive stars," *Reviews of Modern Physics*, vol. 74, no. 4, pp. 1015–1071, 2002.

[119] S. E. Woosley and T. A. Weaver, "The evolution and explosion of massive stars. II. Explosive hydrodynamics and nucleosynthesis," *Astrophysical Journal Supplement Series*, vol. 101, article 181, 1995.

[120] C. Y. Cardall, "Supernova neutrino challenges," *Nuclear Physics B*, vol. 145, no. 1–3, pp. 295–300, 2005.

[121] F. S. Kitaura, H. T. Janka, and W. Hillebrandt, "Explosions of O–Ne–Mg cores, the crab supernova, and subluminous type II-P supernovae," *Astronomy and Astrophysics*, vol. 450, no. 1, pp. 345–350, 2006.

[122] H. T. Janka, B. Müller, F. S. Kitaura, and R. Buras, "Dynamics of shock propagation and nucleosynthesis conditions in O–Ne–Mg core supernovae," *Astronomy and Astrophysics*, vol. 485, no. 1, pp. 199–208, 2008.

[123] I. Sagert, T. Fischer, M. Hempel et al., "Signals of the QCD phase transition in core-collapse supernovae," *Physical Review Letters*, vol. 102, no. 8, Article ID 081101, 4 pages, 2009.

[124] R. Buras, H. T. Janka, M. Rampp, and K. Kifonidis, "Two-dimensional hydrodynamic core-collapse supernova simulations with spectral neutrino transport II. Models for different progenitor stars," *Astronomy and Astrophysics*, vol. 457, no. 1, pp. 281–308, 2006.

[125] A. Marek and H. T. Janka, "Delayed neutrino-driven supernova explosions aided by the standing accretion-shock instability," *The Astrophysical Journal*, vol. 694, no. 1, article 664, 2009.

[126] A. Maeder and G. Meynet, "Stellar evolution with rotation. VI. The Eddington and Omega-limits, the rotational mass loss for OB and LBV stars," *Astronomy and Astrophysics*, vol. 361, pp. 159–166, 2000.

[127] A. Heger, S. E. Woosley, and H. C. Spruit, "Presupernova evolution of differentially rotating massive stars including magnetic fields," *The Astrophysical Journal*, vol. 626, no. 1, p. 350, 2005.

[128] R. Buras, M. Rampp, H. T. Janka, and K. Kifonidis, "Two-dimensional hydrodynamic core-collapse supernova simulations with spectral neutrino transport I," *Astronomy and Astrophysics*, vol. 447, no. 3, pp. 1049–1092, 2006.

[129] B. Mueller, H. T. Janka, and A. Marek, "General relativistic explosion models of core-collapse supernovae," ArXiv e-prints.

[130] H. Dimmelmeier, J. A. Font, and E. Müller, "Relativistic simulations of rotational core collapse I. Methods, initial models, and code tests," *Astronomy and Astrophysics*, vol. 388, no. 3, pp. 917–935, 2002.

[131] I. Cordero-Carrión, P. Cerdá-Durán, H. Dimmelmeier, J. L. Jaramillo, J. Novak, and E. Gourgoulhon, "Improved constrained scheme for the Einstein equations: an approach to the uniqueness issue," *Physical Review D*, vol. 79, no. 2, Article ID 024017, 17 pages, 2009.

[132] C. D. Ott, A. Burrows, L. Dessart, and E. Livne, "Two-dimensional multiangle, multigroup neutrino radiation-hydrodynamic simulations of postbounce supernova cores," *The Astrophysical Journal*, vol. 685, no. 2, article 1069, 2008.

[133] T. D. Brandt, A. Burrows, C. D. Ott, and E. Livne, "Results from core-collapse simulations with multi-dimensional, multi-angle neutrino transport," *The Astrophysical Journal Letters*, vol. 728, no. 1, article 8, 2011.

[134] S. W. Bruenn, "Stellar core collapse—numerical model and infall epoch," *Astrophysical Journal Supplement Series*, vol. 58, pp. 771–841, 1985.

[135] S. W. Bruenn, A. Mezzacappa, W. R. Hix et al., "2D and 3D core-collapse supernovae simulation results obtained with the CHIMERA code," ArXiv e-prints. In press.

[136] A. Burrows, E. Livne, L. Dessart, C. D. Ott, and J. Murphy, "A new mechanism for core-collapse supernova explosions," *The Astrophysical Journal*, vol. 640, no. 2, pp. 878–890, 2006.

[137] M. Liebendörfer, S.C. Whitehouse, and T. Fischer, "The isotropic diffusion source approximation for supernova neutrino transport," *The Astrophysical Journal*, vol. 698, no. 2, article 1174, 2009.

[138] Y. Suwa, K. Kotake, T. Takiwaki, S. C. Whitehouse, M. Liebendörfer, and K. Sato, "Explosion geometry of a rotating 13 M_{sun} star driven by the SASI-aided neutrino-heating supernova mechanism," *Publications of the Astronomical Society of Japan*, vol. 62, pp. L49–L53, 2010.

[139] T. Takiwaki, K. Kotake, and Y. Suwa, "Three-dimensional hydrodynamic core-collapse supernova simulations for an 11.2 Msun star with spectral neutrino transport," *The Astrophysical Journal*, vol. 749, no. 2, article 98, 2012.

[140] M. Liebendörfer, M. Rampp, H. T. Janka, and A. Mezzacappa, "Supernova simulations with Boltzmann neutrino transport: a comparison of methods," *The Astrophysical Journal*, vol. 620, no. 2, article 840, 2005.

[141] T. A. Thompson, E. Quataert, and A. Burrows, "Viscosity and rotation in core-collapse supernovae," *The Astrophysical Journal*, vol. 620, no. 2, article 861, 2005.

[142] T. Foglizzo and M. Tagger, "Entropic-acoustic instability in shocked accretion flows," *Astronomy and Astrophysics*, vol. 363, pp. 174–183, 2000.

[143] T. Foglizzo, "Entropic-acoustic instability of shocked Bondi accretion I. What does perturbed Bondi accretion sound like?" *Astronomy and Astrophysics*, vol. 368, pp. 311–324, 2001.

[144] F. Hanke, A. Marek, B. Mueller, and H. T. Janka, "Is strong SASI activity the key to successful neutrino-driven supernova explosions?" *The Astrophysical Journal*, vol. 755, no. 2, article 138, 2012.

[145] C. W. Misner and D. H. Sharp, "Relativistic equations for adiabatic, spherically symmetric gravitational collapse," *Physical Review*, vol. 136, no. 2, pp. B571–B576, 1964.

[146] M. M. May and R. H. White, "Hydrodynamic calculations of general-relativistic collapse," *Physical Review*, vol. 141, no. 4, pp. 1232–1241, 1966.

[147] R. A. Schwartz, "Gravitational collapse, neutrinos and supernovae," *Annals of Physics*, vol. 43, no. 1, pp. 42–73, 1967.

[148] R. W. Lindquist, "Relativistic transport theory," *Annals of Physics*, vol. 37, no. 3, pp. 487–518, 1966.

[149] J. R. Wilson, "A Numerical Study of Gravitational Stellar Collapse," *TheAstrophysical Journal*, vol. 163, article 209, 1971.

[150] K. A. van Riper, "General relativistic hydrodynamics and the adiabatic collapse of stellar cores," *The Astrophysical Journal*, vol. 232, pp. 558–571, 1979.

[151] K. A. van Riper and J. M. Lattimer, "Stellar core collapse. I—infall epoch," *The Astrophysical Journal*, vol. 249, pp. 270–289, 1981.

[152] K. A. van Riper, "Stellar core collapse. II—inner core bounce and shock propagation," *The Astrophysical Journal*, vol. 257, pp. 793–820, 1982.

[153] S. W. Bruenn, K. R. De Nisco, and A. Mezzacappa, "General relativistic effects in the core collapse supernova mechanism," *The Astrophysical Journal Letters*, vol. 560, no. 1, pp. 326–338, 2001.

[154] S. Yamada, "An implicit lagrangian code for spherically symmetric general relativistic hydrodynamics with an approximate riemann solver," *The Astrophysical Journal*, vol. 475, no. 2, article 720, 1997.

[155] S. Yamada, H. T. Janka, and H. Suzuki, "Neutrino transport in type II supernovae: Boltzmann solver vs. Monte Carlo method," *Astronomy and Astrophysics*, vol. 344, pp. 533–550, 1999.

[156] K. Sumiyoshi, S. Yamada, H. Suzuki, H. Shen, S. Chiba, and H. Toki, "Postbounce evolution of core-collapse supernovae: long-term effects of the equation of state," *The Astrophysical Journal*, vol. 629, no. 2, pp. 922–932, 2005.

[157] K. Sumiyoshi, S. Yamada, and H. Suzuki, "Dynamics and neutrino signal of black hole formation in nonrotating failed supernovae. I. Equation of state dependence," *The Astrophysical Journal*, vol. 667, pp. 382–394, 2007.

[158] K. Sumiyoshi and S. Yamada, "Neutrino transfer in three dimensions for core-collapse supernovae. I. Static configurations," *The Astrophysical Journal Supplement Series*, vol. 199, no. 1, article 17, 2012.

[159] A. Mezzacappa and R. A. Matzner, "Computer simulation of time-dependent, spherically symmetric spacetimes containing radiating fluids—formalism and code tests," *The Astrophysical Journal*, vol. 343, pp. 853–873, 1989.

[160] M. Liebendörfer, A. Mezzacappa, and F. K. Thielemann, "Conservative general relativistic radiation hydrodynamics in spherical symmetry and comoving coordinates," *Physical Review D*, vol. 63, no. 10, Article ID 104003, 14 pages, 2001.

[161] M. Liebendörfer, O. E. B. Messer, A. Mezzacappa, S. W. Bruenn, C. Y. Cardall, and F.-K. Thielemann, "A finite difference representation of neutrino radiation hydrodynamics in spherically symmetric general relativistic spacetime," *Astrophysical Journal Supplement Series*, vol. 150, no. 1, pp. 263–316, 2004.

[162] E. J. Lentz, A. Mezzacappa, O. E. Bronson Messer, M. Liebendörfer, W. R. Hix, and S. W. Bruenn, "On the requirements for realistic modeling of neutrino transport in simulations of core-collapse supernovae," *The Astrophysical Journal*, vol. 747, no. 1, article 73, 2012.

[163] R. Buras, M. Rampp, H. T. Janka, and K. Kifonidis, "Two-dimensional hydrodynamic core-collapse supernova simulations with spectral neutrino transport," *Astronomy and Astrophysics*, vol. 447, no. 3, pp. 1049–1092, 2006.

[164] M. Shibata and Y. I. Sekiguchi, "Magnetohydrodynamics in full general relativity: formulation and tests," *Physical Review D*, vol. 72, no. 4, Article ID 044014, 24 pages, 2005.

[165] M. Shibata and Y. I. Sekiguchi, "Three-dimensional simulations of stellar core collapse in full general relativity: nonaxisymmetric dynamical instabilities," *Physical Review D*, vol. 71, no. 2, Article ID 024014, 32 pages, 2005.

[166] C. D. Ott, H. Dimmelmeier, A. Marek et al., "3D collapse of rotating stellar iron cores in general relativity including deleptonization and a nuclear equation of state," *Physical Review Letters*, vol. 98, no. 26, Article ID 261101, 4 pages, 2007.

[167] M. Liebendörfer, "A simple parameterization of the consequences of deleptonization for simulations of stellar core collapse," *The Astrophysical Journal*, vol. 663, pp. 1042–1051, 2005.

[168] Y. Sekiguchi, "Stellar core collapse in full general relativity with microphysics," *Progress of Theoretical Physics*, vol. 124, no. 2, pp. 331–379, 2010.

[169] T. Kuroda, K. Kotake, and T. Takiwaki, "Fully general relativistic simulations of core-collapse supernovae with an approximate neutrino transport," *The Astrophysical Journal*, vol. 755, no. 1, article 11, 2012.

[170] E. Müller, "Gravitational radiation from collapsing rotating stellar cores," *Astronomy and Astrophysics*, vol. 114, no. 1, pp. 53–59, 1982.

[171] R. Mönchmeyer, G. Schaefer, E. Mueller, and R. E. Kates, "Gravitational waves from the collapse of rotating stellar cores," *Astronomy and Astrophysics*, vol. 246, no. 2, pp. 417–440, 1991.

[172] S. Yamada and K. Sato, "Gravitational radiation from rotational collapse of a supernova core," *The Astrophysical Journal*, vol. 450, article 245, 1995.

[173] T. Zwerger and E. Müller, "Dynamics and gravitational wave signature of axisymmetric rotational core collapse," *Astronomy and Astrophysics*, vol. 320, pp. 209–227, 1997.

[174] K. Kotake, S. Yamada, and K. Sato, "Gravitational radiation from axisymmetric rotational core collapse," *Physical Review D*, vol. 68, no. 4, Article ID 044023, 7 pages, 2003.

[175] K. Kotake, S. Yamada, K. Sato, K. Sumiyoshi, H. Ono, and H. Suzuki, "Gravitational radiation from rotational core collapse: effects of magnetic fields and realistic equations of state," *Physical Review D*, vol. 69, no. 12, Article ID 124004, 11 pages, 2004.

[176] M. Shibata and Y. I. Sekiguchi, "Gravitational waves from axisymmetric rotating stellar core collapse to a neutron star in full general relativity," *Physical Review D*, vol. 69, no. 8, Article ID 084024, 16 pages, 2004.

[177] C. D. Ott, A. Burrows, E. Livne, and R. Walder, "Gravitational waves from axisymmetric, rotating stellar core collapse," *The Astrophysical Journal*, vol. 600, no. 2, article 834, 2004.

[178] C. D. Ott, H. Dimmelmeier, A. Marek et al., "3D collapse of rotating stellar iron cores in general relativity including deleptonization and a nuclear equation of state," *Physical Review Letters*, vol. 98, no. 26, Article ID 261101, 4 pages, 2007.

[179] C. D. Ott, H. Dimmelmeier, A. Marek et al., "Rotating collapse of stellar iron cores in general relativity," *Classical and Quantum Gravity*, vol. 24, no. 12, pp. S139–S154, 2007.

[180] H. Dimmelmeier, J. A. Font, and E. Müller, "Relativistic simulations of rotational core collapse. II. Collapse dynamics and gravitational radiation," *Astronomy and Astrophysics*, vol. 393, pp. 523–542, 2002.

[181] H. Dimmelmeier, C. D. Ott, H. T. Janka, A. Marek, and E. Müller, "Generic gravitational-Wave signals from the collapse of rotating stellar cores," *Physical Review Letters*, vol. 98, no. 25, Article ID 251101, 4 pages, 2007.

[182] H. Dimmelmeier, C. D. Ott, A. Marek, and H. T. Janka, "Gravitational wave burst signal from core collapse of rotating stars," *Physical Review D*, vol. 78, no. 6, Article ID 064056, 28 pages, 2008.

[183] S. Scheidegger, T. Fischer, S. C. Whitehouse, and M. Liebendörfer, "Gravitational waves from 3D MHD core collapse simulations," *Astronomy and Astrophysics*, vol. 490, no. 1, pp. 231–241, 2008.

[184] S. Scheidegger, R. Kaeppeli, S. C. Whitehouse, T. Fischer, and M. Liebendöerfer, "The influence of model parameters on the prediction of gravitational wave signals from stellar core collapse," *Astronomy and Astrophysics*, vol. 514, article A51, 23 pages, 2010.

[185] M. Liebendörfer, "A simple parameterization of the consequences of deleptonization for simulations of stellar core collapse," *The Astrophysical Journal*, vol. 663, pp. 1042–1051, 2005.

[186] C. D. Ott, A. Burrows, T. A. Thompson, E. Livne, and R. Walder, "The spin periods and rotational profiles of neutron stars at birth," *The Astrophysical Journal*, vol. 164, no. 1, article 130, 2006.

[187] K. Kotake, "Multiple physical elements to determine the gravitational-wave signatures of core-collapse supernovae," *Comptes Rendus Physique*. In press.

[188] A. Burrows and J. Hayes, "Pulsar recoil and gravitational radiation due to asymmetrical stellar collapse and explosion," *Physical Review Letters*, vol. 76, pp. 352–355, 1996.

[189] E. Müller and H. T. Janka, "Gravitational radiation from convective instabilities in type II supernova explosions," *Astronomy and Astrophysics*, vol. 317, pp. 140–163, 1997.

[190] E. Müller, M. Rampp, R. Buras, H. T. Janka, and D. H. Shoemaker, "Toward gravitational wave signals from realistic core-collapse supernova models," *The Astrophysical Journal*, vol. 603, no. 1, pp. 221–230, 2004.

[191] M. Rampp, E. Mueller, and M. Ruffert, "Simulations of non-axisymmetric rotational core collapse," *Astronomy and Astrophysics*, vol. 332, pp. 969–983, 1998.

[192] C. D. Ott, H. Dimmelmeier, A. Marek et al., "3D collapse of rotating stellar iron cores in general relativity including deleptonization and a nuclear equation of state," *Physical Review Letters*, vol. 98, no. 26, Article ID 261101, 4 pages, 2007.

[193] C. D. Ott, A. Burrows, L. Dessart, and E. Livne, "A new mechanism for gravitational-wave emission in core-collapse supernovae," *Physical Review Letters*, vol. 96, no. 20, Article ID 201102, 4 pages, 2006.

[194] K. Kotake, W. Iwakami, N. Ohnishi, and S. Yamada, "Stochastic nature of gravitational waves from supernova explosions with standing accretion shock instability," *The Astrophysical Journal Letters*, vol. 697, no. 2, article L133, 2009.

[195] K. Kotake, N. Ohnishi, and S. Yamada, "Gravitational radiation from standing accretion shock instability in core-collapse supernovae," *The Astrophysical Journal*, vol. 655, no. 1, article 406, 2007.

[196] K. Kotake, W. Iwakami, N. Ohnishi, and S. Yamada, "Ray-tracing analysis of anisotropic neutrino radiation for estimating gravitational waves in core-collapse supernovae," *The Astrophysical Journal*, vol. 704, article 951, 2009.

[197] A. Marek, H. T. Janka, and E. Müller, "Equation-of-state dependent features in shock-oscillation modulated neutrino and gravitational-wave signals from supernovae," *Astronomy and Astrophysics*, vol. 496, no. 2, pp. 475–494, 2009.

[198] J. W. Murphy, C. D. Ott, and A. Burrows, "A model for gravitational wave emission from neutrino-driven core-collapse supernovae," ArXiv e-prints.

[199] C. D. Ott, C. Reisswig, E. Schnetter et al., "Dynamics and gravitational wave signature of collapsar formation," *Physical Review Letters*, vol. 106, no. 16, Article ID 161103, 4 pages, 2011.

[200] W. D. Arnett and C. Meakin, "Toward realistic progenitors of core-collapse supernovae," *The Astrophysical Journal Letters*, vol. 733, no. 2, article 78, 2011.

[201] R . Epstein, "The generation of gravitational radiation by escaping supernova neutrinos," *The Astrophysical Journal*, vol. 223, pp. 1037–1045, 1978.

[202] M. S. Turner and R. V. Wagoner, in *Sources of Gravitational Radiation*, L. L. Smarr, Ed., pp. 383–407, 1979.

[203] M. Favata, "The gravitational-wave memory effect," *Classical and Quantum Gravity*, vol. 27, no. 8, Article ID 084036, 2010.

[204] T. Hiramatsu, K. Kotake, H. Kudoh, and A. Taruya, "Gravitational wave background from neutrino-driven gamma-ray bursts," *Monthly Notices of the Royal Astronomical Society*, vol. 364, no. 3, pp. 1063–1068, 2005.

[205] Y. Suwa and K. Murase, "Probing the central engine of long gamma-ray bursts and hypernovae with gravitational waves and neutrinos," *Physical Review D*, vol. 80, no. 12, Article ID 123008, 11 pages, 2009.

[206] K. Kotake, T. Takiwaki, and S. Harikae, "Gravitational wave signatures of hyperaccreting collapsar disks," *The Astrophysical Journal*, vol. 755, no. 2, article 84, 2012.

[207] Y. Suwa, T. Takiwaki, K. Kotake, and K. Sato, "Gravitational wave background from population III stars," *The Astrophysical Journal Letters*, vol. 665, pp. L43–L46, 2007.

[208] K. N. Yakunin, P. Marronetti, A. Mezzacappa et al., "Gravitational waves from core collapse supernovae," *Classical and Quantum Gravity*, vol. 27, no. 19, Article ID 194005, 2010.

[209] C. L. Fryer, D. E. Holz, and S. A. Hughes, "Gravitational waves from stellar collapse: correlations to explosion asymmetries," *The Astrophysical Journal*, vol. 609, pp. 288–300, 2004.

[210] K. Kotake, W. Iwakami-Nakano, and N. Ohnishi, "Effects of rotation on stochasticity of gravitational waves in the nonlinear phase of core-collapse supernovae," *The Astrophysical Journal*, vol. 736, no. 2, article 124, 2011.

[211] E. Müller, H. T. Janka, and A. Wongwathanarat, "Parametrized 3D models of neutrino-driven supernova explosions: neutrino emission asymmetries and gravitatio-nalwave signals," *Astronomy & Astrophysics*, vol. 537, article A63, 20 pages, 2012.

[212] J. M. Blondin and A. Mezzacappa, "Pulsar spins from an instability in the accretion shock of supernovae," *Nature*, vol. 445, no. 7123, pp. 58–60, 2007.

[213] E. Rantsiou, A. Burrows, J. Nordhaus, and A. Almgren, "Induced rotation in three-dimensional simulations of core-collapse supernovae: implications for pulsar spins," *The Astrophysical Journal*, vol. 732, no. 1, article 57, 2011.

[214] T. Yamasaki and T. Foglizzo, "Effect of rotation on the stability of a stalled cylindrical shock and its consequences for core-collapse supernovae," *The Astrophysical Journal*, vol. 679, no. 1, pp. 607–615, 2008.

[215] K. S. Thorne, in *Particle and Nuclear Astrophysics and Cosmology in the Next Millenium*, E. W. Kolb and R. D. Peccei, Eds., p. 160, World Scientific, 1995.

[216] K. Kuroda and LCGT Collaboration, "Status of LCGT," *Classical and Quantum Gravity*, vol. 27, no. 8, Article ID 084004, 2010.

[217] S. E. A. Kawamura, "The Japanese space gravitational wave antenna—DECIGO," *Classical and Quantum Gravity*, vol. 23, no. 8, article S125, 2006.

[218] É. É. Flanagan and S. A. Hughes, "Measuring gravitational waves from binary black hole coalescences. II. The waves' information and its extraction, with and without templates," *Physical Review D*, vol. 57, no. 8, pp. 4566–4587, 1998.

[219] T. Takiwaki and K. Kotake, "Gravitational wave signatures of magnetohydrodynamically-driven core-collapse supernova explosions," *The Astrophysical Journal*, vol. 743, no. 1, article 30, 2011.

[220] C. D. Ott, "Probing the core-collapse supernova mechanism with gravitational waves," *Classical and Quantum Gravity*, vol. 26, no. 20, Article ID 204015, 2009.

[221] M. Hashimoto, "Supernova nucleosynthesis in massive stars," *Progress of Theoretical Physics*, vol. 94, no. 5, pp. 663–736, 1995.

[222] A. Jerkstrand, "Spectral modeling of nebular-phase supernovae," ArXiv e-prints.

[223] S. E. Woosley and R. D. Hoffman, "Co-57 and Ti-44 production in SN 1987A," *The Astrophysical Journal*, vol. 368, pp. L31–L34, 1991.

[224] P. Bouchet, M. M. Phillips, N. B. Suntzeff, C. Gouiffes, R. W. Hanuschik, and D. H. Wooden, "The bolometric lightcurve of supernova 1987A—part two—results from visible and infrared spectrophotometry," *Astronomy and Astrophysics*, vol. 245, no. 2, article 490, 1991.

[225] C. Fransson and C. Kozma, "The freeze-out phase of SN 1987A—implications for the light curve," *The Astrophysical Journal*, vol. 408, no. 1, pp. L25–L28, 1993.

[226] P. A. Mazzali, K. Nomoto, F. Patat, and K. Maeda, "The nebular spectra of the hypernova SN 1998bw and evidence for asymmetry," *The Astrophysical Journal*, vol. 559, no. 2, pp. 1047–1053, 2001.

[227] I. Maurer, A. Jerkstrand, P. A. Mazzali et al., "NERO—a postmaximum supernova radiation transport code," *Monthly Notices of the Royal Astronomical Society*, vol. 418, no. 3, pp. 1517–1525, 2011.

[228] A. Jerkstrand, C. Fransson, and C. Kozma, "The^{44}Ti-powered spectrum of SN 1987A," *Astronomy and Astrophysics*, vol. 530, article A45, 23 pages, 2011.

[229] G. F. Varani, W. P. S. Meikle, J. Spyromilio, and D. A. Allen, "Direct observation of radioactive cobalt decay in supernova 1987A," *Monthly Notices of the Royal Astronomical Society*, vol. 245, pp. 570–576, 1990.

[230] J. D. Kurfess, W. N. Johnson, R. L. Kinzer et al., "Oriented scintillation spectrometer experiment observations of Co-57 in SN 1987A," *The Astrophysical Journal*, vol. 399, no. 2, pp. L137–L140, 1992.

[231] A. V. Filippenko, "Optical spectra of supernovae," *Annual Review of Astronomy and Astrophysics*, vol. 35, pp. 309–355, 1997.

[232] N. N. Chugai, R. A. Chevalier, R. P. Kirshner, and P. M. Challis, "Hubble space telescope spectrum of SN 1987A at an age of 8 years: radioactive luminescence of cool gas," *The Astrophysical Journal Letters*, vol. 483, no. 2, pp. 925–940, 1997.

[233] P. Lundqvist, C. Kozma, J. Sollerman, and C. Fransson, "ISO/SWS observations of SN 1987A. II. A refined upper limit on the mass of ^{44}Ti in the ejecta of SN 1987A," *Astronomy and Astrophysics*, vol. 374, pp. 629–637, 2001.

[234] C.L. Fryer, G. Rockefeller, and M.S. Warren, "SNSPH: a parallel three-dimensional smoothed particle radiation hydrodynamics code," *The Astrophysical Journal*, vol. 643, no. 1, pp. 292–305, 2006.

[235] G. Magkotsios, F. X. Timmes, A. L. Hungerford, C. L. Fryer, P. A. Young, and M. Wiescher, "Trends in^{44}Ti and^{56}Ni from core-collapse supernovae," *Astrophysical Journal Supplement Series*, vol. 191, no. 1, article 66, 2010.

[236] F. K. Thielemann, K. Nomoto, and M. A. Hashimoto, "Core-collapse supernovae and their ejecta," *The Astrophysical Journal*, vol. 460, article 408, 1996.

[237] T. Rauscher, A. Heger, R. D. Hoffman, and S. E. Woosley, "Nucleosynthesis in massive stars with improved nuclear and stellar physics," *The Astrophysical Journal*, vol. 576, no. 1, pp. 323–348, 2002.

[238] A. L. Hungerford, C. L. Fryer, and G. Rockefeller, "Gamma rays from single-lobe supernova explosions," *The Astrophysical Journal*, vol. 635, no. 1, pp. 487–501, 2005.

[239] P. A. Young, C. L. Fryer, A. Hungerford et al., "Constraints on the progenitor of Cassiopeia A," *The Astrophysical Journal*, vol. 640, no. 2, article 891, 2006.

[240] C. C. Joggerst, A. Almgren, and S. E. Woosley, "Three-dimensional simulations of rayleigh-taylor mixing in core-collapse supernovae with castro," *The Astrophysical Journal*, vol. 723, no. 1, pp. 353–363, 2010.

[241] S. M. Couch, Wheeler J. C., and M. Milosavljević, "Aspherical core-collapse supernovae in red supergiants powered by nonrelativistic jets," *The Astrophysical Journal*, vol. 696, no. 1, pp. 953–970, 2009.

[242] N. Tominaga, "Aspherical properties of hydrodynamics and nucleosynthesis in jet-induced supernovae," *The Astrophysical Journal*, vol. 690, no. 1, pp. 526–536, 2009.

[243] K. Maeda and K. Nomoto, "Bipolar supernova explosions: nucleosynthesis and implications for abundances in extremely metal-poor stars," *The Astrophysical Journal*, vol. 598, no. 2, article 1163, 2003.

[244] A. Gawryszczak, J. Guzman, T. Plewa, and K. Kifonidis, "Non-spherical core collapse supernovae. III. Evolution towards homology and dependence on the numerical resolution," *Astronomy and Astrophysics*, vol. 521, article A38, 2010.

[245] J. Pruet, R. D. Hoffman, S. E. Woosley, H. T. Janka, and R. Buras, "Nucleosynthesis in early supernova winds. II. The role of neutrinos," *The Astrophysical Journal*, vol. 644, no. 2, pp. 1028–1039, 2006.

[246] C. Fröhlich, G. Martínez-Pinedo, M. Liebendörfer et al., "Neutrino-induced nucleosynthesis of $A > 64$ nuclei: the vp process," *Physical Review Letters*, vol. 96, no. 14, Article ID 142502, 4 pages, 2006.

[247] S. Wanajo and H. T. Janka, "The r-process in the neutrino-driven wind from a black-hole torus," *The Astrophysical Journal*, vol. 746, no. 2, article 180, 2012.

[248] S. I. Fujimoto, K. Kotake, M. A. Hashimoto, M. Ono, and N. Ohnishi, "Explosive nucleosynthesis in the neutrino-driven aspherical supernova explosion of a non-rotating $15\,M_{sun}$ star with solar metallicity," *The Astrophysical Journal*, vol. 738, no. 1, article 61, 2011.

[249] R. H. Cyburt, J. Ellis, B. D. Fields, F. Luo, K. A. Olive, and V. C. Spanos, "Nuclear reaction uncertainties, massive gravitino decays and the cosmological lithium problem," *Journal of Cosmology and Astroparticle Physics*, vol. 2010, article 32, 2010.

[250] J. Schwab, P. Podsiadlowski, and S. Rappaport, "Further evidence for the bimodal distribution of neutron star masses," *The Astrophysical Journal*, vol. 719, pp. 722–727, 2010.

[251] M. Kimura, H. Tsunemi, S. Katsuda, and H. Uchida, "Suzaku observations across the cygnus loop from the NorthEastern to the SouthWestern rim," *Publications of the Astronomical Society of Japan*, vol. 61, pp. S137–S145, 2009.

[252] H. Uchida, H. Tsunemi, S. Katsuda, M. Kimura, H. Kosugi, and H. Takahashi, "Abundance inhomogeneity in the Northern rim of the cygnus loop," *Publications of the Astronomical Society of Japan*, vol. 61, no. 3, pp. 503–510, 2009.

[253] T. Shigeyama, K. Nomoto, and M. Hashimoto, "Hydrodynamical models and the light curve of supernova 1987A in the large magellanic cloud," *Astronomy and Astrophysics*, vol. 196, no. 1-2, pp. 141–151, 1988.

[254] M. Renaud, J. Vink, A. Decourchelle et al., "The signature of^{44}Ti in Cassiopeia A revealed by IBIS/ISGRI on *INTEGRAL*," *The Astrophysical Journal Letters*, vol. 647, no. 1, article L41, 2006.

[255] J. M. LeBlanc and J. R. Wilson, "A numerical example of the collapse of a rotating magnetized star," *The Astrophysical Journal*, vol. 161, article 541, 1970.

[256] G. S. Bisnovatyi-Kogan, Y. P. Popov, and A. A. Samochin, "The magnetohydrodynamic rotational model of supernova explosion," *Astrophysics and Space Science*, vol. 41, no. 2, pp. 287–320, 1976.

[257] E. Müller and W. Hillebrandt, "A magnetohydrodynamical supernova model," *Astronomy and Astrophysics*, vol. 80, no. 2, pp. 147–154, 1979.

[258] E. M. D. Symbalisty, "Magnetorotational iron core collapse," *The Astrophysical Journal*, vol. 285, pp. 729–746, 1984.

[259] N. V. Ardeljan, G. S. Bisnovatyi-Kogan, and S. G. Moiseenko, "Nonstationary magnetorotational processes in a rotating magnetized cloud," *Astronomy and Astrophysics*, vol. 355, no. 3, pp. 1181–1190, 2000.

[260] S. Yamada and H. Sawai, "Numerical study on the rotational collapse of strongly magnetized cores of massive stars," *The Astrophysical Journal*, vol. 608, no. 2, article 907, 2004.

[261] K. Kotake, H. Sawai, S. Yamada, and K. Sato, "Magnetorotational effects on anisotropic neutrino emission and convection in core collapse supernovae," *The Astrophysical Journal*, vol. 608, no. 1, article 391, 2004.

[262] K. Kotake, S. Yamada, and K. Sato, "North-South neutrino heating asymmetry in strongly magnetized and rotating stellar cores," *The Astrophysical Journal*, vol. 618, pp. 474–484, 2005.

[263] M. Obergaulinger, M. A. Aloy, H. Dimmelmeier, and E. Müller, "Axisymmetric simulations of magnetorotational core collapse: approximate inclusion of general relativistic effects," *Astronomy and Astrophysics*, vol. 457, no. 1, pp. 209–222, 2006.

[264] M. Obergaulinger, M. A. Aloy, and E. Müller, "Axisymmetric simulations of magneto-rotational core collapse: dynamics and gravitational wave signal," *Astronomy and Astrophysics*, vol. 450, no. 3, pp. 1107–1134, 2006.

[265] P. Cerdá-Durán, J. A. Font, and H. Dimmelmeier, "General relativistic simulations of passive-magneto-rotational core collapse with microphysics," *Astronomy and Astrophysics*, vol. 474, no. 1, pp. 169–191, 2007.

[266] Y. Suwa, T. Takiwaki, K. Kotake, and K. Sato, "Magnetorotational collapse of population III stars," *Publications of the Astronomical Society of Japan*, vol. 59, pp. 771–785, 2007.

[267] T. Takiwaki, K. Kotake, S. Nagataki, and K. Sato, "Magneto-driven shock waves in core-collapse supernovae," *The Astrophysical Journal*, vol. 616, no. 2, article 1086, 2004.

[268] T. Takiwaki, K. Kotake, and K. Sato, "Special relativistic simulations of magnetically-dominated jets in collapsing massive stars," *The Astrophysical Journal*, vol. 691, no. 2, article 1360, 2009.

[269] S. Scheidegger, R. Kaeppeli, S. C. Whitehouse, T. Fischer, and M. Liebendoerfer, "The influence of model parameters on the prediction of gravitational wave signals from stellar core collapse,". In press, http://arxiv.org/abs/1001.1570.

[270] M. Obergaulinger and H. T. Janka, "Magnetic field amplification in collapsing, non-rotating stellar cores,". In press, http://arxiv.org/abs/1101.1198.

[271] R. C. Duncan and C. Thompson, "Formation of very strongly magnetized neutron stars—implications for gamma-ray bursts," *The Astrophysical Journal Letters*, vol. 392, no. 1, pp. L9–L13, 1992.

[272] J.M. Lattimer and M. Prakash, "Neutron star observations: prognosis for equation of state constraints," *Physics Reports*, vol. 442, no. 1–6, pp. 109–165, 2007.

[273] L. Dessart, A. Burrows, E. Livne, and C. D. Ott, "The proto-neutron star phase of the collapsar model and the route to long-soft gamma-ray bursts and hypernovae," *The Astrophysical Journal Letters*, vol. 673, no. 1, pp. L43–L46, 2008.

[274] S. E. Woosley and A. Heger, "The progenitor stars of gamma-ray bursts," *The Astrophysical Journal*, vol. 637, no. 2, article 914, 2006.

[275] S. C. Yoon and N. Langer, "Evolution of rapidly rotating metal-poor massive stars towards gamma-ray bursts," *Astronomy and Astrophysics*, vol. 443, no. 2, pp. 643–648, 2005.

[276] D. L. Meier, R. I. Epstein, W. D. Arnett, and D. N. Schramm, "Magnetohydrodynamic phenomena in collapsing stellar cores," *The Astrophysical Journal*, vol. 204, pp. 869–878, 1976.

[277] S. Akiyama, J. C. Wheeler, D. L. Meier, and I. Lichtenstadt, "The magnetorotational instability in core-collapse supernova explosions," *Astrophysical Journal Letters*, vol. 584, no. 2, pp. 954–970, 2003.

[278] J. F. Hawley, C. F. Gammie, and S. A. Balbus, "Local three-dimensional magnetohydrodynamic simulations of accretion disks," *The Astrophysical Journal*, vol. 440, article 742, 1995.

[279] Y. Masada, T. Sano, and H. Takabe, "Nonaxisymmetric magnetorotational instability in proto-neutron stars," *The Astrophysical Journal*, vol. 641, no. 1, pp. 447–457, 2006.

[280] M. Shibata, Y. T. Liu, S. L. Shapiro, and B. C. Stephens, "Magnetorotational collapse of massive stellar cores to neutron stars: simulations in full general relativity," *Physical Review D*, vol. 74, no. 10, Article ID 104026, 28 pages, 2006.

[281] Z. B. Etienne, Y. T. Liu, and S. L. Shapiro, "General relativistic simulations of slowly and differentially rotating magnetized neutron stars," *Physical Review D*, vol. 74, no. 4, Article ID 044030, 2006.

[282] M. Obergaulinger, P. Cerdá-Durán, E. Müller, and M. A. Aloy, "Semi-global simulations of the magneto-rotational instability in core collapse supernovae," *Astronomy and Astrophysics*, vol. 498, no. 1, pp. 241–271, 2009.

[283] W. Zhang, A. MacFadyen, and P. Wang, "Three-dimensional relativistic MHD simulations of the Kelvin-Helmholtz instability: magnetic field amplification by a turbulent dynamo," *The Astrophysical Journal Letters*, vol. 692, no. 1, article L40, 2009.

[284] T. Sano, S. Inutsuka, and S. M. Miyama, "A saturation mechanism of magnetorotational instability due to ohmic dissipation," *The Astrophysical Journal*, vol. 506, no. 1, pp. L57–L60, 1998.

[285] B. van Leer, "Towards the ultimate conservative difference scheme. V. A second-order sequel to Godunov's method," *Journal of Computational Physics*, vol. 32, no. 1, pp. 101–136, 1979.

[286] D. A. Clarke, "A consistent method of characteristics for multidimensional magnetohydrodynamics," *The Astrophysical Journal Letters*, vol. 457, no. 1, pp. 291–320, 1996.

[287] C. R. Evans and J. F. Hawley, "Simulation of magnetohydrodynamic flows—a constrained transport method," *The Astrophysical Journal*, vol. 332, pp. 659–677, 1988.

[288] J. M. Stone and M. L. Norman, "ZEUS-2D: a radiation magnetohydrodynamics code for astrophysical flows in two space dimensions. I—the hydrodynamic algorithms and tests," *Astrophysical Journal Supplement Series*, vol. 80, no. 2, pp. 753–790, 1992.

[289] M. E. Pessah and J. Goodman, "On the saturation of the magnetorotational instability via parasitic modes," *The Astrophysical Journal*, vol. 698, no. 1, pp. L72–L76, 2009.

[290] P. Longaretti and G. Lesur, "MRI-driven turbulent transport: the role of dissipation, channel modes and their parasites," *Astronomy and Astrophysics*, vol. 516, article A51, 2010.

[291] J. F. Hawley and S. A. Balbus, "A powerful local shear instability in weakly magnetized disks. III—long-term evolution in a shearing sheet," *The Astrophysical Journal*, vol. 400, no. 2, pp. 595–609, 1992.

[292] J. Goodman and G. Xu, "Parasitic instabilities in magnetized, differentially rotating disks," *The Astrophysical Journal*, vol. 432, no. 1, pp. 213–223, 1994.

[293] T. Sano and S. Inutsuka, "Saturation and thermalization of the magnetorotational instability: recurrent channel flows and reconnections," *The Astrophysical Journal*, vol. 561, no. 2, pp. L179–182, 2001.

[294] G. Bodo, A. Mignone, F. Cattaneo, P. Rossi, and A. Ferrari, "Aspect ratio dependence in magnetorotational instability shearing box simulations," *Astronomy and Astrophysics*, vol. 487, no. 1, pp. 1–5, 2008.

[295] T. Sano, S. Inutsuka, N. J. Turner, and J. M. Stone, "Angular momentum transport by magnetohydrodynamic turbulence in accretion disks: gas pressure dependence of the saturation level of the magnetorotational instability," *The Astrophysical Journal*, vol. 605, no. 1, pp. 321–339, 2004.

[296] M. E. Pessah and C. Chan, "Viscous, resistive magnetorotational modes," *The Astrophysical Journal*, vol. 684, no. 1, pp. 498–514, 2008.

[297] Y. Masada, T. Takiwaki, K. Kotake, and T. Sano, "Local simulations of the magneto-rotational instability in core-collapse supernovae," *The Astrophysical Journal*. In press, http://arxiv.org/abs/1209.2360.

[298] A. I. MacFadyen and S. E. Woosley, "Collapsars: gamma-ray bursts and explosions in 'failed supernovae'," *The Astrophysical Journal*, vol. 524, no. 1, article 262, 1999.

[299] A. I. MacFadyen, S. E. Woosley, and A. Heger, "Supernovae, jets, and collapsars," *The Astrophysical Journal*, vol. 550, no. 1, pp. 410–425, 2001.

[300] S. Harikae, T. Takiwaki, and K. Kotake, "Long-term evolution of slowly rotating collapsar in special relativistic magnetohydrodynamics," *The Astrophysical Journal*, vol. 704, no. 1, article 354, 2009.

[301] S. Harikae, K. Kotake, and T. Takiwaki, "Neutrino pair annihilation in collapsars: a ray-tracing method in special relativity," *The Astrophysical Journal*, vol. 713, no. 1, pp. 304–317, 2010.

[302] P. Meszaros, "Gamma-ray bursts," *Reports on Progress in Physics*, vol. 69, no. 8, article 2259, 2006.

[303] K. S. Thorne, "Multipole expansions of gravitational radiation," *Reviews of Modern Physics*, vol. 52, no. 2, pp. 299–339, 1980.

[304] A. Weinstein, "Advanced LIGO optical configuration and prototyping effort," *Classical and Quantum Gravity*, vol. 19, article 1575, 2002.

[305] H. Kudoh, A. Taruya, T. Hiramatsu, and Y. Himemoto, "Detecting a gravitational-wave background with next-generation space interferometers," *Physical Review D*, vol. 73, no. 6, Article ID 064006, 16 pages, 2006.

[306] S. Kawagoe, T. Takiwaki, and K. Kotake, "Neutrino oscillations in magnetically driven supernova explosions," *Journal of Cosmology and Astroparticle Physics*, vol. 2009, no. 9, article 33, 2009.

[307] J. Engel, G. C. McLaughlin, and C. Volpe, "What can be learned with a lead-based supernova-neutrino detector?" *Physical Review D*, vol. 67, no. 1, Article ID 013005, 2003.

[308] D. Autiero, J. Äystö, A. Badertscher et al., "Large underground, liquid based detectors for astro-particle physics in Europe: scientific case and prospects," *Journal of Cosmology and Astroparticle Physics*, vol. 2007, no. 11, article 011, 2007.

[309] K. Scholberg, "Future supernova neutrino detectors," *Journal of Physics Conference Series*, vol. 203, no. 1, Article ID 012079, 2010.

[310] M. Kachelrieß, R. Tomas, R. Buras, H. T. Janka, A. Marek, and M. Rampp, "Exploiting the neutronization burst of a galactic supernova," *Physical Review D*, vol. 71, no. 6, Article ID 063003, 14 pages, 2005.

[311] J. F. Beacom and M. R. Vagins, "Antineutrino spectroscopy with large water Čerenkov detectors," *Physical Review Letters*, vol. 93, no. 17, Article ID 171101, 4 pages, 2004.

[312] L. Hüdepohl, B. Müller, H. T. Janka, A. Marek, and G. G. Raffelt, "Neutrino signal of electron-capture supernovae from core collapse to cooling," *Physical Review Letters*, vol. 104, no. 25, Article ID 251101, 2010.

[313] T. Fischer, S. C. Whitehouse, A. Mezzacappa, F. K. Thielemann, and M. Liebendörfer, "Protoneutron star evolution and the neutrino-driven wind in general relativistic neutrino radiation hydrodynamics simulations," *Astronomy and Astrophysics*, vol. 517, article A80, 25 pages, 2010.

[314] E. O'Connor and C. D. Ott, "Black hole formation in failing core-collapse supernovae," *The Astrophysical Journal*, vol. 730, no. 2, article 70, 2011.

[315] K. Takahashi, K. Sato, A. Burrows, and T. A. Thompson, "Supernova neutrinos, neutrino oscillations, and the mass of the progenitor star," *Physical Review D*, vol. 68, no. 11, Article ID 113009, 8 pages, 2003.

[316] C. Lunardini and A.Y. Smirnov, "Supernova neutrinos: Earth matter effects and neutrino mass spectrum," *Nuclear Physics B*, vol. 616, no. 1-2, pp. 307–348, 2001.

[317] A. S. Dighe, M. Kachelrieß, G. G. Raffelt, and R. Tomàs, "Signatures of supernova neutrino oscillations in the Earth mantle and core," *Journal of Cosmology and Astroparticle Physics*, vol. 2004, article 4, 2004.

[318] R. C. Schirato and G. M. Fuller, "Connection between supernova shocks, flavor transformation, and the neutrino signal," in press, http://arxiv.org/abs/astro-ph/0205390.

[319] C. Lunardini and A. Y. Smirnov, "Probing the neutrino mass hierarchy and the 13-mixing with supernovae," *Journal of Cosmology and Astroparticle Physics*, vol. 2003, no. 6, article 9, 2003.

[320] R. Tomas, M. Kachelrieß, G. Raffelt, A. Dighe, H. T. Janka, and L. Scheck, "Neutrino signatures of supernova forward and reverse shock propagation," *Journal of Cosmology and Astroparticle Physics*, vol. 9, article 15, 2004.

[321] B. Dasgupta and A. Dighe, "Phase effects in neutrino conversions during a supernova shock wave," *Physical Review D*, vol. 75, no. 9, Article ID 093002, 10 pages, 2007.

[322] C. Lunardini and O. L. G. Peres, "Upper limits on the diffuse supernova neutrino flux from the SuperKamiokande data," *Journal of Cosmology and Astroparticle Physics*, vol. 8, article 33, 2008.

[323] A. S. Dighe and A. Y. Smirnov, "Identifying the neutrino mass spectrum from a supernova neutrino burst," *Physical Review D*, vol. 62, no. 3, Article ID 033007, 24 pages, 2000.

[324] K. Takahashi, K. Sato, H. E. Dalhed, and J. R. Wilson, "Shock propagation and neutrino oscillation in supernova," *Astroparticle Physics*, vol. 20, no. 2, pp. 189–193, 2003.

[325] G. L. Fogli, E. Lisi, A. Mirizzi, and D. Montanino, "Probing supernova shock waves and neutrino flavour transitions in next-generation water Cherenkov detectors," *Journal of Cosmology and Astroparticle Physics*, vol. 4, article 2, 2005.

[326] S. Chakraborty, T. Fischer, A. Mirizzi, N. Saviano, and R. Tomas, "No collective neutrino flavor conversions during the supernova accretion phase," *Physical Review Letters*, vol. 107, no. 15, Article ID 151101, 4 pages, 2011.

[327] B. Dasgupta, E. P. O'Connor, and C. D. Ott, "The role of collective neutrino flavor oscillations in core-collapse supernova shock revival," *Physical Review D*, vol. 85, no. 6, Article ID 065008, 16 pages, 2012.

[328] Y. Suwa, K. Kotake, T. Takiwaki, M. Liebendoerfer, and K. Sato, "Impacts of collective neutrino oscillations on core-collapse supernova explosions," *The Astrophysical Journal*, vol. 738, no. 2, article 165, 2011.

[329] O. Pejcha, B. Dasgupta, and T. A. Thompson, "Effect of collective neutrino oscillations on the neutrino mechanism of core-collapse supernovae," in press, http://arxiv.org/abs/1106.5718.

[330] G. G. Raffelt, *Stars As Laboratories for Fundamental Physics: The Astrophysics of Neutrinos, Axions, and Other Weakly Interacting Particles*, University of Chicago Press, 1996.

[331] J. Schechter and J. W. F. Valle, "Majorana neutrinos and magnetic fields," *Physical Review D*, vol. 24, no. 7, pp. 1883–1889, 1981.

[332] C. S. Lim and W. J. Marciano, "Resonant spin-flavor precession of solar and supernova neutrinos," *Physical Review D*, vol. 37, no. 6, pp. 1368–1373, 1988.

[333] E. K. Akhmedov and M. Y. Khlopov, "Resonant amplification of neutrino oscillations in longitudinal magnetic field," *Modern Physics Letters A*, vol. 3, no. 5, pp. 451–457, 1988.

[334] S. Ando and K. Sato, "Resonant spin-flavor conversion of supernova neutrinos: dependence on presupernova models and future prospects," *Physical Review D*, vol. 68, no. 2, Article ID 023003, 9 pages, 2003.

[335] T. Totani, K. Sato, H. E. Dalhed, and J. R. Wilson, "Future detection of supernova neutrino burst and explosion mechanism," *The Astrophysical Journal*, vol. 496, no. 1, article 216, 1998.

[336] C. Lunardini, B. Müller, and H. T Janka, "Neutrino oscillation signatures of oxygen-neon-magnesium supernovae," *Physical Review D*, vol. 78, no. 2, Article ID 023016, 13 pages, 2008.

[337] J. F. Cherry, M. R. Wu, J. Carlson, H. Duan, G. M. Fuller, and Y. Z. Qian, "Density fluctuation effects on collective neutrino oscillations in O–Ne–Mg core-collapse supernovae," *Physical Review D*, vol. 84, no. 10, Article ID 105034, 12 pages, 2011.

[338] S. Sarikas, G. G. Raffelt, L. Hüdepohl, and H. T. Janka, "Suppression of self-induced flavor conversion in the supernova accretion phase," *Physical Review Letters*, vol. 108, no. 6, Article ID 061101, 4 pages, 2012.

[339] A. Friedland and A. Gruzinov, "Neutrino signatures of supernova turbulence," *Astrophysics*. In press, http://arxiv.org/abs/astro-ph/0607244.

[340] S. Choubey, N. P. Harries, and G. G. Ross, "Turbulent supernova shock waves and the sterile neutrino signature in megaton water detectors," *Physical Review D*, vol. 76, no. 7, Article ID 073013, 20 pages, 2007.

[341] G. Reid, J. Adams, and S. Seunarine, "Collective neutrino oscillations in turbulent backgrounds," *Physical Review D*, vol. 84, no. 8, Article ID 085023, 32 pages, 2011.

[342] B. Dasgupta, A. Dighe, A. Mirizzi, and G. Raffelt, "Collective neutrino oscillations in nonspherical geometry," *Physical Review D*, vol. 78, no. 3, Article ID 033014, 14 pages, 2008.

[343] T. Lund, A. Marek, C. Lunardini, H. T. Janka, and G. Raffelt, "Fast time variations of supernova neutrino fluxes and their detectability," *Physical Review D*, vol. 82, no. 6, Article ID 063007, 13 pages, 2010.

[344] K. Kotake, S. Yamada, and K. Sato, "Anisotropic neutrino radiation in rotational core collapse," *The Astrophysical Journal*, vol. 595, no. 1, article 304, 2003.

[345] H. T. Janka and R. Mönchmeyer, "Anisotropic neutrino emission from rotating protoneutron stars," *Astronomy and Astrophysics*, vol. 209, no. 1-2, pp. L5–L8, 1989.

[346] G. Fogli, E. Lisi, A. Marrone, and A. Mirizzi, "Collective neutrino flavor transitions in supernovae and the role of trajectory averaging," *Journal of Cosmology and Astroparticle Physics*, vol. 2007, article 10, 2007.

[347] J.M. Lattimer and F.D. Swesty, "A generalized equation of state for hot, dense matter," *Nuclear Physics A*, vol. 535, no. 2, pp. 331–376, 1991.

[348] S. Shlomo, V. M. Kolomietz, and G. Colo, "Deducing the nuclear-matter incompressibilty coefficient from data on isoscalar compression modes," *European Physical Journal A*, vol. 30, no. 1, pp. 23–30, 2006.

[349] P. B. Demorest, T. Pennucci, S. M. Ransom, M. S. E. Roberts, and J. W. T. Hessels, "A two-solar-mass neutron star measured using Shapiro delay," *Nature*, vol. 467, no. 7319, pp. 1081–1083, 2010.

[350] K. Kiuchi and K. Kotake, "Equilibrium configurations of strongly magnetized neutron stars with realistic equations of state," *Monthly Notices of the Royal Astronomical Society*, vol. 385, no. 3, pp. 1327–1347, 2008.

[351] T. Fischer, I. Sagert, G. Pagliara et al., "Core-collapse supernova explosions triggered by a quark-hadron phase transition during the early post-bounce phase," *Astrophysical Journal Supplement Series*, vol. 194, article 39, 2011.

[352] T. K. Suzuki, K. Sumiyoshi, and S. Yamada, "Alfven wave-driven supernova explosion," *The Astrophysical Journal*, vol. 678, no. 2, pp. 1200–1206, 2008.

[353] B. Mueller, A. Marek, H. T. Janka, and H. Dimmelmeier, "General relativistic explosion models of core-collapse supernovae," ArXiv e-prints.

[354] J. Guilet, J. Sato, and T. Foglizzo, "The saturation of sasi by parasitic instabilities," *The Astrophysical Journal*, vol. 713, no. 2, article 1350, 2010.

[355] J. W. Murphy and C. Meakin, "A global turbulence model for neutrino-driven convection in core-collapse supernovae," *The Astrophysical Journal*, vol. 742, no. 2, article 74, 2011.

[356] M. Hempel and J. Schaffner-Bielich, "A statistical model for a complete supernova equation of state," *Nuclear Physics A*, vol. 837, no. 3-4, pp. 210–254, 2010.

[357] S. Furusawa, S. Yamada, K. Sumiyoshi, and H. Suzuki, "A new baryonic equation of state at sub-nuclear densities for core-collapse simulations," *The Astrophysical Journal*, vol. 738, no. 2, article 178, 2011.

[358] G. Shen, C. J. Horowitz, and E. O'Connor, "Second relativistic mean field and virial equation of state for astrophysical simulations," *Physical Review C*, vol. 83, no. 6, Article ID 065808, 11 pages, 2011.

[359] K. Sumiyoshi, S. Yamada, H. Suzuki, and S. Chiba, "Neutrino signals from the formation of a black hole: a probe of the equation of state of dense matter," *Physical Review Letters*, vol. 97, no. 9, Article ID 091101, 4 pages, 2006.

[360] K. Nakazato, K. Sumiyoshi, H. Suzuki, and S. Yamada, "Exploring hadron physics in black hole formations: a new promising target of neutrino astronomy," *Physical Review D*, vol. 81, no. 8, Article ID 083009, 6 pages, 2010.

[361] K. Nakazato, K. Sumiyoshi, and S. Yamada, "Impact of quarks and pions on dynamics and neutrino signal of black hole formation in non-rotating stellar core collapse," *The Astrophysical Journal*, vol. 721, no. 2, article 1284, 2010.

[362] K. Fukushima and T. Hatsuda, "The phase diagram of dense QCD," *Reports on Progress in Physics*, vol. 74, no. 1, Article ID 014001, 2011.

[363] I. Leonor, L. Cadonati, E. Coccia et al., "Searching for prompt signatures of nearby core-collapse supernovae by a joint analysis of neutrino and gravitational-wave data," ArXiv e-prints.

[364] S. Nishimura, K. Kotake, M. A. Hashimoto et al., "r-Process nucleosynthesis in magnetohydrodynamic jet explosions of core-collapse supernovae," *The Astrophysical Journal*, vol. 642, pp. 410–419, 2006.

[365] S. I. Fujimoto, M. A. Hashimoto, K. Kotake, and S. Yamada, "Heavy element nucleosynthesis in a collapsar," *The Astrophysical Journal*, vol. 656, no. 1, pp. 382–392, 2007.

[366] J. C. Wheeler, I. Yi, P. Hoflich, and L. Wang, "Asymmetric supernovae, pulsars, magnetars, and gamma-ray bursts," *The Astrophysical Journal*, vol. 537, no. 2, article 810, 2000.

[367] T. A. Thompson, P. Chang, and E. Quataert, "Magnetar spin-down, hyperenergetic supernovae, and gamma-ray bursts," *The Astrophysical Journal*, vol. 611, no. 1, pp. 380–393, 2004.

[368] D. A. Uzdensky and A. I. MacFadyen, "Magnetar-driven magnetic tower as a model for gamma-ray bursts and asymmetric supernovae," *The Astrophysical Journal*, vol. 669, pp. 546–560, 2007.

[369] Z. G. Dai and T. Lu, "Gamma-ray burst afterglows and evolution of postburst fireballs with energy injection from strongly magnetic millisecond pulsars," *Astronomy and Astrophysics*, vol. 333, pp. L87–L90, 1998.

[370] N. Bucciantini, E. Quataert, J. Arons, B. D. Metzger, and T. A. Thompson, "Relativistic jets and long-duration gamma-ray bursts from the birth of magnetars," *Monthly Notices of the Royal Astronomical Society*, vol. 383, no. 1, pp. L25–L29, 2007.

[371] B. D. Metzger, D. Giannios, T. A. Thompson, N. Bucciantini, and E. Quataert, "The protomagnetar model for gamma-ray bursts," *Monthly Notices of the Royal Astronomical Society*, vol. 413, no. 3, pp. 2031–3056, 2011.

[372] K. Kotake, K. Sumiyoshi, S. Yamada et al., "Core-collapse supernovae as supercomputing science: a status report toward 6D simulations with exact Boltzmann neutrino transport in full general relativity," *Progress of Theoretical and Experimental Physics*, vol. 2012, no. 1, Article ID 01A301, 2012.

Fab Four: When John and George Play Gravitation and Cosmology

J. P. Bruneton,[1] M. Rinaldi,[1] A. Kanfon,[2] A. Hees,[1, 3, 4] S. Schlögel,[1] and A. Füzfa[1, 5]

[1] *Namur Center for Complex Systems (naXys), University of Namur, 5000 Namur, Belgium*
[2] *Faculté des Sciences et Techniques, Université d'Abomey-Calavi, BP 526 Cotonou, Benin*
[3] *Royal Observatory of Belgium, Avenue Circulaire 3, 1180 Brussels, Belgium*
[4] *LNE-SYRTE, Observatoire de Paris, CNRS, UPMC, avenue de l'Observatoire 61, 75014 Paris, France*
[5] *Center for Cosmology, Particle Physics and Phenomenology (CP3), University of Louvain, 1348 Louvain-la-Neuve, Belgium*

Correspondence should be addressed to J.-P. Bruneton, jpbr@math.fundp.ac.be

Academic Editor: Elmetwally Elabbasy

Scalar-tensor theories of gravitation attract again a great interest since the discovery of the Chameleon mechanism and of the Galileon models. The former allows reconciling the presence of a scalar field with the constraints from Solar System experiments. The latter leads to inflationary models that do not need ad hoc potentials. Further generalizations lead to a tensor-scalar theory, dubbed the "Fab Four," with only first and second order derivatives of the fields in the equations of motion that self-tune to a vanishing cosmological constant. This model needs to be confronted with experimental data in order to constrain its large parameter space. We present some results regarding a subset of this theory named "John," which corresponds to a nonminimal derivative coupling between the scalar field and the Einstein tensor in the action. We show that this coupling gives rise to an inflationary model with very unnatural initial conditions. Thus, we include the term named "George," namely, a nonminimal, but nonderivative, coupling between the scalar field and Ricci scalar. We find a more natural inflationary model, and, by performing a post-Newtonian analysis, we derive the set of equations that constrain the parameter space with data from experiments in the Solar System.

1. Introduction

The Galileon theory has recently emerged as an effective theoretical realization of the Dvali-Gabadadze-Porrati model (DGP) [1]. The subsequent developments eventually led to the definition of the most general second order scalar-tensor theory that includes, besides the usual terms of scalar-tensor or $f(R)$ theories (as in, e.g., [2–10]), also nonminimal derivative couplings to the curvature [11–13]. This general theory, explored for the first time by the pioneering work of Horndeski many years ago [14], provides a wide framework that virtually encompasses all scalar models analyzed so far in the literature.

One interesting aspect of the nonlinearities of the scalar sector is that they trigger the Vainshtein mechanism that makes it possible to build viable cosmological models with sufficiently small effects at local scales to evade Solar System

constraints. Therefore, we have at hand an interesting alternative to chameleons [15], in which the parameter space for local gravity does not overlap with the one allowed by cosmic acceleration, as recently shown in [16]. In addition, inflationary phases are permitted without the introduction of ad hoc scalar potentials, making these models more "natural." Nonlinearities are also responsible for new phenomena in the dark sector including, for instance, sub/superluminality and/or effective violation of the Null Energy Condition, thus allowing for a stable and well-defined phantom-like phase [17–20]. Generally speaking, the Galileon model has opened the way to new models for cosmology, including inflationary or late-accelerated ones [21–24]. In fact, almost all sorts of cosmological scenarios are possible depending on which Galileon model is chosen; see, for example, [25].

In this paper, we first focus on a subclass of models dubbed "purely kinetic gravity" [25, 27–29] or also similar to "John Lagrangian" in the "Fab Four's" terminology. (Note that in the Fab Four Lagrangian, the scalar field has no standard kinetic term. It propagates however, via its derivative coupling to the curvature tensors.) [22, 30]. More generally speaking, it is a special case of the very general Galileon class considered in [23]. We mostly investigate the inflationary phase and its naturalness, taking into account no-ghost constraints and causality conditions. In a next step, a generalization of the theory including a coupling of the scalar field to the matter—via, in the Jordan frame, a Brans-Dicke-like coupling to the scalar curvature—is introduced. The analysis of the inflationary phase is then reconducted in such a more realistic theory, while a preliminary analysis of solar system constraints is discussed.

We begin in Section 2 with a review of John's equations of motion in a flat and empty Universe. We then complete the analysis made in [27] by computing the number of e-folds in Section 2.1. We show that a kinetically driven inflationary phase requires highly trans-Planckian values for the initial field velocity, which rule out the model. Moreover, in Section 2.2 we provide a detailed analysis of the no-ghost and causality conditions during cosmic evolution, and we show that the theory becomes acausal for such transplanckian values unless the coupling constant is vanishingly small or if the initial Hubble constant is trans-Planckian. In passing, we prove that some claims made in the literature regarding the sign of the coupling constant are wrong. We argue that this model is to be discarded also because there are no reasons why the scalar field should be generated at all in the first place. This is particularly clear in an astrophysical context where the theory is forced to reduce to pure GR ($\varphi = 0$), when the same equations are solved inside a compact body.

In Section 3, we thus extend our considerations to a more general model, in which we also include a coupling of the scalar field to the Ricci scalar. This corresponds to the "John plus George" combination in the Fab Four's terminology. We derive and study numerically the equations of motion for the matter-free cosmological background as well as the no-ghosts and causality conditions. We find that inflation is possible provided both coupling constants are positive. In Section 3.2 we derive the field equations in spherical and static symmetry, and we briefly sketch how to put some constraints on the parameter space using Solar System tests. We conclude in Section 4 with some remarks and perspectives.

2. John Lagrangian

We begin our analysis by considering the action

$$S = \int \sqrt{-g} d^4x \left[\frac{R}{2\kappa} - \frac{1}{2} (g^{\mu\nu} + \kappa\gamma G^{\mu\nu}) \partial_\mu \varphi \partial_\nu \varphi \right] + S_{\text{mat}} \left[\psi, g_{\mu\nu} \right], \tag{1}$$

where R is the curvature scalar, $G^{\mu\nu}$ the Einstein tensor, ψ collectively denotes the matter degrees of freedom coupled to the metric $g_{\mu\nu}$, and $\kappa = 8\pi G_N$. In this paper we use the mostly plus signature; γ is a dimensionless parameter whereas φ has the dimension of an inverse length. This action is a special

case of the generalized Galileon one presented in [23], as one can see by setting $K(X) = X$, $G_3 = 0$, $G_4 = 1/(2\kappa)$, $G_5 = \kappa\gamma\varphi/2$.

2.1. Inflation with John. As it was realized in [27], this model allows for a quasi de Sitter inflation with a graceful exit without the need for any specific scalar potential. Inflation is essentially driven by kinematics, and it crucially depends on the initial high velocity of the field, as we will shortly see. Although, in principle, the inflationary solutions begin at $t = -\infty$ (see [27]), we will consider the action as an effective model only valid from few Planck times after an unknown transplanckian phase. Our first concern is to establish whether the model accommodates an inflationary phase together with reasonable assumptions for the initial conditions at that time. This section thus completes the analysis found in [27] by providing the number of e-foldings as a function of the free parameters of the theory. The cosmological equations in vacuum derived from (1) read

$$3\dot{\alpha}^2 = \frac{\kappa\dot{\varphi}^2}{2} (1 - 9\kappa\gamma\dot{\alpha}^2), \tag{2a}$$

$$2\ddot{\alpha} + 3\dot{\alpha}^2 = -\frac{\kappa\dot{\varphi}^2}{2} (1 + \kappa\gamma[3\dot{\alpha}^2 + 2\ddot{\alpha} + 4\dot{\alpha}\ddot{\varphi}\dot{\varphi}^{-1}]), \tag{2b}$$

$$\frac{1}{a^3} \frac{d}{dt} (a^3 \dot{\varphi}(1 - 3\kappa\gamma\dot{\alpha}^2)) = 0, \tag{2c}$$

where a is the scale factor and $\alpha = \ln a$. The system can be partially decoupled to allow for a numerical integration whose results are shown in Section 2.3. Isolating the second order derivatives, we find

$$\ddot{\alpha} = \frac{(3\kappa\gamma\dot{\alpha}^2 - 1)}{2} \times \frac{3\dot{\alpha}^2 + (\kappa\dot{\varphi}^2/2)(1 - 9\kappa\gamma\dot{\alpha}^2)}{1 - 3\gamma\kappa\dot{\alpha}^2 + (\kappa^2\gamma\dot{\varphi}^2/2)(1 + 9\kappa\gamma\dot{\alpha}^2)}, \tag{3a}$$

$$\ddot{\varphi} = \frac{-3\dot{\alpha}\dot{\varphi}(1 + \kappa^2\gamma\dot{\varphi}^2)}{1 - 3\gamma\kappa\dot{\alpha}^2 + (\kappa^2\gamma\dot{\varphi}^2/2)(1 + 9\kappa\gamma\dot{\alpha}^2)}, \tag{3b}$$

which can be integrated numerically in a straightforward way. The effective equation of state (EoS) for the scalar field can be obtained from its stress-energy tensor or, more simply, by comparing our equations of motion directly to the standard Friedmann equations. After some algebra we find

$$\omega_\varphi = \frac{(2 + 3\kappa^2\gamma\dot{\varphi}^2)(1 - \kappa^2\gamma\dot{\varphi}^2)}{2 + 3\kappa^2\gamma\dot{\varphi}^2 + 3\kappa^4\gamma^2\dot{\varphi}^4}, \tag{4}$$

which is plotted in Figure 1 for both positive and negative γ. For both signs of γ, the EoS tends to -1 in the high energy limit ($\kappa|\dot{\varphi}| \gg 1$), so that a large initial velocity for the scalar field will result in a quasi de Sitter phase. However, only the case of positive γ can lead to inflation. Indeed, (2a) can be inverted to $3\dot{\alpha}^2 = \kappa\dot{\varphi}^2/(2 + 3\kappa^2\gamma\dot{\varphi}^2)$, which needs to be positive since the Hubble constant is a real number. Thus, $\gamma < 0$ implies that $|\dot{\varphi}| < \sqrt{-2/3\gamma}$, which, in turn, means that

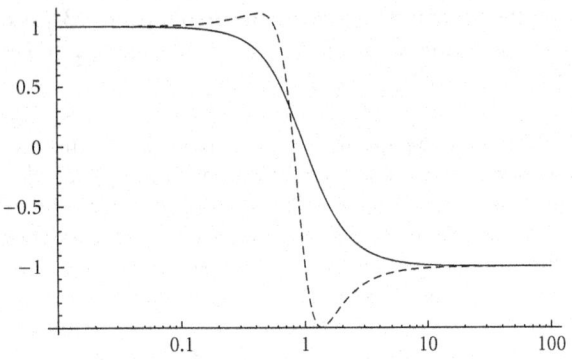

FIGURE 1: log linear plot of ω_φ as a function of the dimensionless variable $\kappa\dot\varphi$ for $\gamma = 1$ and $\gamma = -1$ (dashed).

we always have $w_\varphi > 0$. Therefore, the scalar field cannot even start in the $w_\varphi < 0$ region if $\gamma < 0$. More generally speaking, accelerated phases driven by a scalar field in this model require $\gamma > 0$.

In view of these considerations, in the following we shall assume that γ is positive. Equation (3b), together with the condition that $1 - 3\kappa\gamma\dot\alpha^2 > 0$ (see Section 2.2), shows that $\ddot\varphi < 0$ for an initially expanding Universe ($\dot\alpha > 0$). Hence, the velocity of the field decreases with time and ω_φ is driven towards $\omega = +1$. We characterize the end of inflation by the instant t_{end}, at which $\omega_\varphi = -1/3$ (throughout this paper we assume a vanishing cosmological constant). Under the assumption that $\kappa\gamma\dot\varphi^2$ is initially large, one may derive an analytical (approximate) solution for the scale factor $a(t) = e^{\alpha(t)}$ and the scalar field at early times. First, we write (2a) as $3\dot\alpha^2 = \kappa\dot\varphi^2/(2 + 3\kappa^2\gamma\dot\varphi^2)$ so that

$$H = \dot\alpha \simeq \frac{1}{3\sqrt{\kappa\gamma}}, \qquad (5)$$

where H is the Hubble constant. Integration yields the approximate scale factor

$$a \sim a_i \exp\left(\frac{t - t_i}{3\sqrt{\kappa\gamma}}\right). \qquad (6)$$

Using (5) in (3b), and expanding according to $\kappa^2\dot\varphi^2 \gg 1$, gives $\ddot\varphi/\dot\varphi \simeq -1/\sqrt{\kappa\gamma}$ and

$$\dot\varphi \sim \dot\varphi_i \exp\left(-\frac{t - t_i}{\sqrt{\kappa\gamma}}\right). \qquad (7)$$

Now recall that inflation ends at t_{end} such that $\omega_\varphi(t_{\text{end}}) = -1/3$. This corresponds to $\kappa^2\gamma\dot\varphi^2 = \zeta_{\text{end}}$, where

$$\zeta_{\text{end}} = \frac{1}{6}\left(3 + \sqrt{57}\right) \approx 1.76, \qquad (8)$$

as can be shown by solving (4). Using (7), the condition $\kappa^2\gamma\dot\varphi^2 = \zeta_{\text{end}}$ reduces to

$$\kappa^2\gamma\dot\varphi_i^2 \exp\left(-\frac{2\,(t_{\text{end}} - t_i)}{\sqrt{\kappa\gamma}}\right) \sim \zeta_{\text{end}}, \qquad (9)$$

from which one finds

$$t_{\text{end}} - t_i = \frac{\sqrt{\kappa\gamma}}{2}\ln\left(\frac{\kappa^2\gamma\dot\varphi_i^2}{\zeta_{\text{end}}}\right). \qquad (10)$$

Replacing this in the expression (6) for the scale factor leads to

$$\frac{a_{\text{end}}}{a_i} \sim \left(\frac{\kappa^2\gamma\dot\varphi_i^2}{\zeta_{\text{end}}}\right)^{1/6}. \qquad (11)$$

We finally impose that inflation lasts for a number of e-folds $N = \ln(a_{\text{end}}/a_i)$ greater than 60. This gives a relation between the initial velocity of the field $\dot\varphi_i$ and γ, namely:

$$\dot\varphi_i^2 \gtrsim \frac{\zeta_{\text{end}}}{\kappa^2\gamma}\exp(360), \qquad (12)$$

which is the crucial condition for a successful (purely kinetic-driven) inflationary phase. We see that it involves a rather unusual very large pure number. In order to discuss naturalness, (5) is also of interest, as it fixes the Hubble constant at the beginning of the inflationary phase $H_i \sim 1/3\sqrt{\kappa\gamma}$. Therefore, the last equation might also be written as

$$\kappa\frac{\dot\varphi_i^2}{H_i^2} \gtrsim 9\zeta_{\text{end}}\exp(360) \sim 10^{157}. \qquad (13)$$

It follows that the "natural" initial conditions $H_i = \mathcal{O}(1)$ and $\dot\varphi_i = \mathcal{O}(1)$ in Planckian units are not allowed. On the contrary, a natural value for the initial expansion $H_i = 1$ (and thus $\gamma \approx 0.11$) requires an extremely high transplanckian value for the initial velocity of the field $\dot\varphi_i \sim 10^{78}$ in natural units.

It is not even possible to obtain a Planckian value for the initial velocity in this model, since, in any event, the initial Hubble constant will be greater than the one today. This implies that in such an inflationary scenario, $\sqrt{\kappa\gamma}$ must be less than the Hubble radius today, still implying a very unnatural bound for the initial velocity, namely, $\dot\varphi_i \gtrsim 10^{51}$ in natural units.

2.2. Theoretical Constraints. The characteristic feature of Galileon models is the derivative coupling of the scalar field to the metric. This implies a direct coupling between scalar field and metric degrees of freedom or, in other words, that the scalar field propagation explicitly depends on the metric background and vice versa. Therefore, there might exist backgrounds for which the propagation becomes pathological (nonhyperbolic, i.e., noncausal, or carrying negative energy, i.e., ghosts). In the following, we restrict ourselves to the cosmological background in a flat universe with line element

$$ds^2 = -dt^2 + e^{2\alpha(t)}d\mathbf{x}^2, \qquad (14)$$

and we explore the conditions for the theory to be well defined, for both scalar field and metric perturbations. We start with the scalar field, whose action can be written as

$$S_\varphi = \int a^3 dt\, d^3x\, Q_\varphi\left(\dot\varphi^2 - \frac{c_\varphi^2}{a^2}\nabla\varphi^2\right), \qquad (15)$$

where

$$Q_\varphi = \frac{1}{2}(1 - 3\kappa\gamma\dot{\alpha}^2) > 0 \qquad (16)$$

as it needs to be positive for the scalar field to carry positive energy. The field propagates at a squared speed given by

$$c_\varphi^2 = \frac{1 - \kappa\gamma(2\ddot{\alpha} + 3\dot{\alpha}^2)}{1 - 3\kappa\gamma\dot{\alpha}^2} \geq 0. \qquad (17)$$

The condition $c_\varphi^2 \geq 0$ is necessary to ensure that the scalar field equation of motion remains hyperbolic, ultimately expressing its causal behavior regardless of whether the scalar field perturbations are sub- or super-luminal [17, 31]. The two conditions previously mentioned are indeed equivalent to the requirement that the effective metric

$$\tilde{g}^{\mu\nu} = g^{\mu\nu} + \kappa\gamma G^{\mu\nu},$$
$$\tilde{g}^{\mu\nu}\nabla_{\mu\nu}\varphi = 0, \qquad (18)$$

along which the scalar field propagates, is hyperbolic with the same (mostly +) signature than $g_{\mu\nu}$.

The conditions (16) and (17) are best analyzed in terms of the reduced dimensionless variables $x(t) = \kappa\dot{\varphi}$ and $y(t) = \sqrt{\kappa}\dot{\alpha}$. The first one requires $\gamma y^2 < 1/3$, and it is automatically satisfied if $\gamma < 0$, but also if $\gamma > 0$ provided that there is a maximal value for the Hubble constant $H = \dot{\alpha} < 1/\sqrt{3\kappa\gamma}$. The second one reduces to an algebraic condition with the help of (2a)–(3b), namely:

$$\frac{1 - 3\gamma y^2}{1 - 9\gamma y^2 + 54\gamma^2 y^4} \geq 0 \iff \frac{(1 + \gamma x^2)(2 + 3\gamma x^2)}{2 + 3\gamma x^2 + 3\gamma^2 x^4} \geq 0. \quad (19)$$

In summary, (16)-(17) require $\gamma > 0$ and $|y| < 1/\sqrt{3\gamma}$ or $\gamma < 0$ and the two possibilities $|x| > 1/\sqrt{-\gamma}$ or $|x| < \sqrt{-2/3\gamma}$. These conditions are therefore less restrictive than the one implied by the Friedmann equation (see previous section).

The curvature background implies nonstandard propagation for the scalar degree of freedom. In a quite similar fashion, the scalar field background modifies the standard spectrum of metric perturbations. Similar conditions for the avoidance of ghosts and euclidean metrics also exist. These have been derived in full generality in a very wide class of Galileon models in [32], whose conventions we follow. These conditions, namely, (23), (25)–(27) of [32], reduce to rather simple algebraic constraints in our case, after the necessary manipulation using the equations of motion (2a)–(3b):

$$Q_T > 0 \iff 1 + \frac{\gamma x^2}{2} > 0, \qquad (20a)$$

$$c_T^2 \geq 0 \iff 1 - \frac{\gamma x^2}{2} \geq 0, \qquad (20b)$$

where Q_T and c_T are defined as their scalar counterparts Q_φ and c_φ^2, but stand for the tensor perturbations of the metric

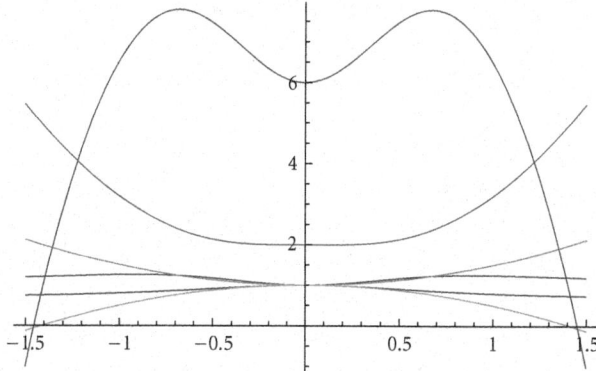

FIGURE 2: Plot of the six conditions Q_i and c_i derived previously, as a function x. Here $y = 1$. Allowed values for the field velocity are typically $|x| < \mathcal{O}(\gamma^{-1/2})$, as made clear by drawing similar plots while varying γ.

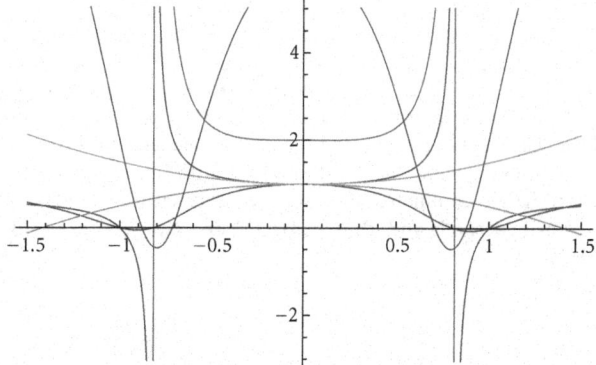

FIGURE 3: Plot of the six conditions Q_i and c_i derived previously, as a function x. Here $y = -1$. Allowed values for the field velocity are typically $|x| < \mathcal{O}(|\gamma|^{-1/2})$.

field. Similar conditions need to hold for the scalar part of the metric perturbations, namely:

$$Q_S > 0 \iff \frac{4 + 6\gamma x^2 + 6\gamma^2 x^4}{2 + 3\gamma x^2} > 0, \qquad (21a)$$

$$c_S^2 \geq 0 \iff \frac{12 + 36\gamma x^2 + 19\gamma^2 x^4 - 12\gamma^3 x^6 - 3\gamma^4 x^8}{2 + 3\gamma x^2 + 3\gamma^2 x^4} \geq 0. \qquad (21b)$$

This whole set of equations is difficult to reduce algebraically because of the last one. However, one might easily plot the six functions of x defined previously, and one typically finds that both positive and negative values for y are allowed on a given range $|x| < \xi_y$, where typically ξ_y behaves as $\mathcal{O}(1/\sqrt{|y|})$; see, for example, Figures 2 and 3. Hence, large (transplanckian) values for $|x|$ are only allowed for small $|y| \ll 1$. This means that the space for possible velocities of the field $x = \kappa\dot{\varphi}$ needs to be typically subplanckian, unless y is vanishingly small. This will be linked to the results found earlier, where transplanckian initial velocity was required for a successful

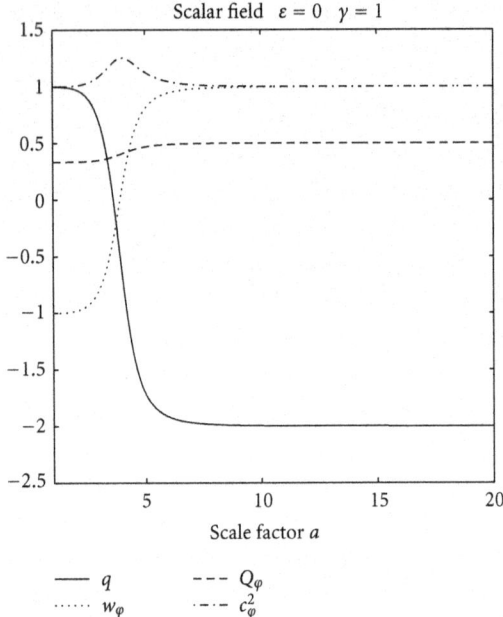

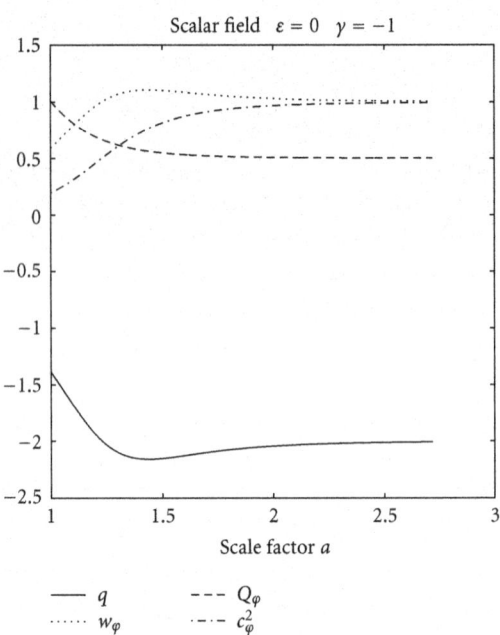

FIGURE 4: Evolution of the acceleration parameter $q = +\ddot{a}a/\dot{a}^2$ versus the scale factor, for $\gamma = 1$. Also shown in the graph: the evolution of the scalar field EoS w_φ and Q_φ and c_φ^2. The scalar field becomes superluminal during the transition from a de Sitter Universe to a to stiff matter-dominated one. However, as discussed in the text, it remains hyperbolic and, thus, causal. Moreover, it carries positive energy.

FIGURE 6: Evolution of the acceleration parameter q versus the scale factor, for $\gamma = -1$ with initial condition satisfying $|x_i| = 0.1 < \sqrt{-2/3\gamma}$. The field starts with an EoS $w_\varphi \sim 0.5$ and the Universe only decelerates.

inflation, leading to negative squared speeds c_S^2 and c_T^2 in that epoch. This is shown in Figure 5.

In passing, we note that the claim made in the literature (see, e.g., [24, 29, 33]) according to which only the subclass $\gamma < 0$ is a ghost-free theory is wrong (at least in the background considered here). Notice that the scalar field is well defined although being a phantom in certain regime (in the case $\gamma < 0$), a situation reminiscent of the one discussed in [18]. However, as shown previously, the Friedmann equation actually prevents the scalar field to enter this regime.

2.3. Numerical Results. In this section, we quickly show the cosmological behavior in the John model, for both positive and negative γ. As discussed before, the negative γ case leads only to a decelerating Universe: the phantom regime is not an acceptable initial condition (as it entails an imaginary Hubble constant), and neither can be reached. Only positive γ leads to acceleration and to an inflationary phase in the early Universe, a drawback being the presence of noncausal behavior for the scalar and tensor perturbations of the metric. These plots (Figures 4, 5, 6, and 7) have been obtained by numerical integration for an initial condition of $\dot{\varphi}_i = 10$ in natural units in the case $\gamma = 1$, and $\dot{\varphi}_i = 0.1$ for $\gamma = -1$.

2.4. Discussion. We have established that kinetically driven inflation in the Galileon theory involving the simplest coupling to Einstein tensor is not viable. It requires unnatural transplanckian values for the initial velocity of the field, which, in turn, implies various instabilities.

This model has anyway another serious drawback. In the absence of any direct coupling to the Ricci scalar, there is no

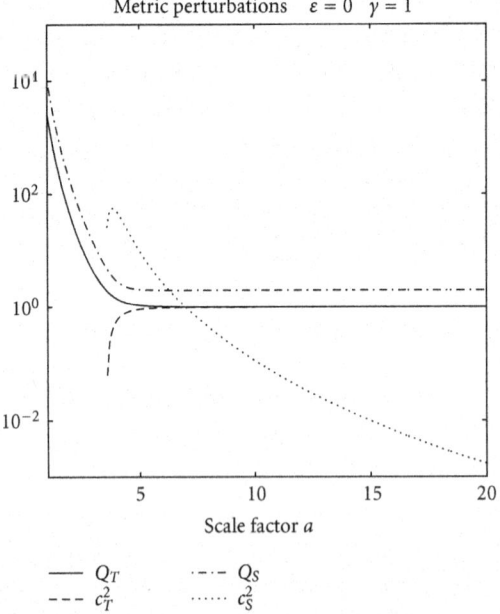

FIGURE 5: Evolution of the parameters for metric perturbations (scalar and tensorial parts), respectively, Q_S, c_S^2, Q_T, c_T^2, as defined in the text versus the scale factor, for $\gamma = 1$. As found theoretically, the initial high velocity of the field drives both the speed of (scalar and tensor) metric perturbations to imaginary values, thus signaling a breakdown of hyperbolicity for metric perturbations (here it happens when the corresponding curves terminate, since the y axis is in logarithmic scale).

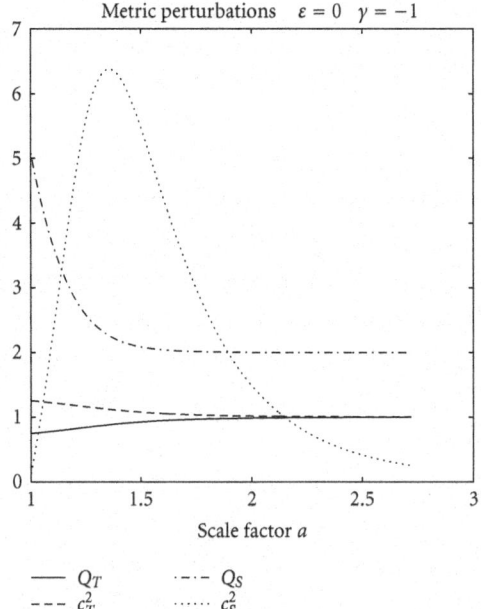

FIGURE 7: Evolution of the parameters for metric perturbations (scalar and tensorial parts), respectively, Q_S, c_S^2, Q_T, c_T^2, as defined in the text versus the scale factor, for $\gamma = -1$. This model with $\gamma < 0$ is well behaved but does not accommodate inflation.

reason why the scalar field should be generated at all (even in presence of cosmological matter fluid). In other words, $\varphi = 0$ is always a solution in this class of models, whatever the matter content is. At local scales, this problem appears in the following way. We checked numerically that a relativistic star (with flat asymptotic conditions) must be described by pure GR; that is, $\varphi = 0$ for all r is the only solution that is regular at the center of the star. To conclude, the model considered so far is trivial in the sense that it cannot be different than GR, except if one imposes nonvanishing initial conditions for the scalar field at early times.

This acts as a leitmotiv for a more realistic model, studied in the next section, where we add to the Lagrangian a direct coupling to the Ricci scalar: $F(\varphi)R$, namely, the "George" term in the Fab Four terminology, and where only a particular function F will be analyzed here.

3. George and John

We now consider the extended model given by

$$S = \int \sqrt{-g}d^4x$$

$$\times \left[\frac{R}{2\kappa}\left(1 + \epsilon\sqrt{\kappa}\varphi\right) - \frac{1}{2}\left(g^{\mu\nu} + \kappa\gamma G^{\mu\nu}\right)\partial_\mu\varphi\partial_\nu\varphi \right] \quad (22)$$

$$+ S_{\text{mat}}\left[\psi, g_{\mu\nu}\right],$$

where ϵ is a dimensionless, free parameter. Of course this is not the most general coupling one might consider but it is anyway reminiscent of Brans-Dicke coupling. Notice that one might worry that the effective gravitational constant

$G_{\text{eff}} = G/(1 + \epsilon\sqrt{\kappa}\varphi)$ might easily become negative in this model, meaning that the action chosen here shall trivially lead to dynamical pathologies for $\epsilon\varphi$ sufficiently large and negative. (In fact, what matters in the case $\gamma = 0$ is that the scalar field propagates positive energy in the Einstein frame. Performing a conformal transformation, this is equivalent to the usual Brans-Dicke condition $2w + 3 > 0$, where $w = \epsilon^2(1 + \epsilon\sqrt{\kappa}\varphi)$ here. Then, our model with $\gamma = 0$ would indeed be pathological if $\leq -(3/(2\epsilon^2) + 1)/(\epsilon\sqrt{\kappa})$. However the γ term introduces new terms in the equation of motion for the scalar field which invalidate such a conclusion in the general case $\gamma \neq 0$.).

Such an argument would call in favor of defining a better coupling function F in the Georges term $F(\varphi)R$. However, this would be a misleading conclusion here, since the John term couples the derivatives of the metric and of the scalar field, thus impacting their propagation. Therefore, only the entire set of no-ghost conditions together with causal propagation conditions (positivity of the squared velocities) for both the scalar and the metric perturbations can decide which regions of the configuration space are well behaved. This is done in the following sections (on a cosmological background), based on the conditions derived in Appendix B. In this light, the function F chosen above is just the simplest one could have chosen and might furthermore be understood as retaining only the first term in a series expansion of a more general function F.

The cosmological evolution in this theory is typically a function of four parameters, the initial value of the field, its velocity, and of the two dimensionless parameters γ and ϵ. It goes beyond the scope of the present paper to provide a comprehensive study of this parameter space. However we highlight some essential features of the model thanks to numerical results displayed in the next section. We provide both the cosmological evolution and the analysis of causality and positivity of energy within subclasses of the model, depending on the signs of γ and ϵ.

3.1. Cosmological Behavior. The equations of motion in a flat, empty Universe, derived from (22), are given in Appendix A; see (A.1). We extended the analysis of the no-ghost and causality conditions to this more general framework, and we also provide the scalar field EoS; see Appendices A and B.

The numerical results are the following. The case $\epsilon = 1$ and $\gamma = 1$ is pretty similar to the case John alone; see Figures 8 and 9. Inflation thus occurs in the case $\epsilon > 0$ and $\gamma > 0$, but the acausal behavior still shows up in the very early Universe. The number of e-folds is a function of the two initial conditions for the field and its velocity and the dimensionless parameters ϵ and γ. A further analysis that goes beyond the scope of this paper would determine whether the addition of the George term helps in solving the naturalness problem encountered with John alone in Section 2.

The case $\epsilon = -1$, $\gamma = 1$ is clearly pathological for the various no-ghosts and no-acausal conditions, as seen in Figure 10. Actually this theory leads to a double inflation scenario (see the acceleration parameter): the Universe transits from one de Sitter phase to another one and experiences in between a super acceleration phase. Finally, the case with

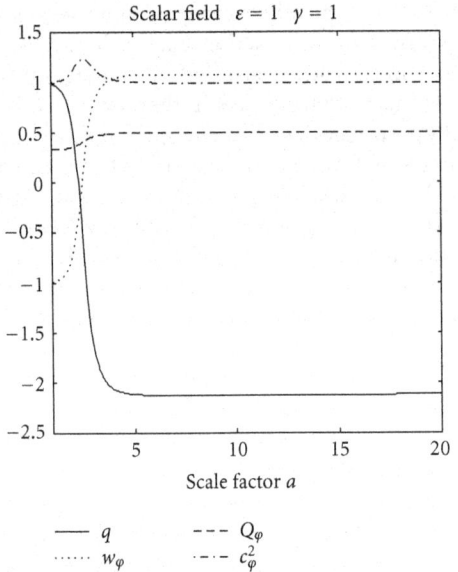

FIGURE 8: Evolution of q, w_φ, Q_φ, and c_φ^2 with the scale factor. We observe the same transition from inflation to stiff matter for the scalar field. Initial conditions are $\dot{\varphi}_i = 100$ and $\varphi_i = 1$.

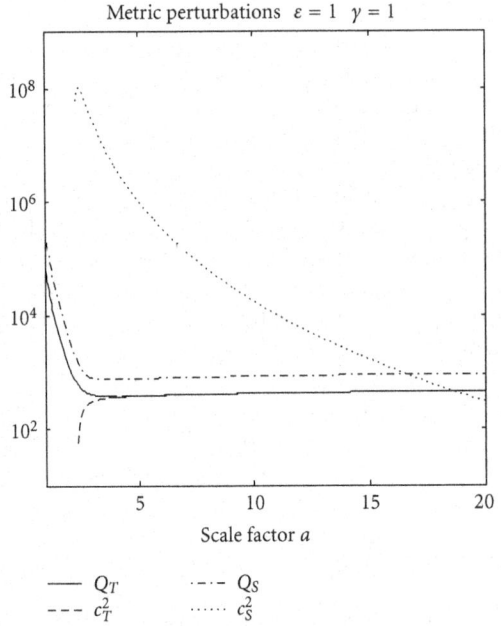

FIGURE 9: Evolution of Q_T, Q_φ, and c_T^2, c_φ^2 with the scale factor. These two speeds are negative in the early universe.

negative γ is similar to what we found for John alone: the theory is well defined, ghost-free, and causal, but fails to exhibit any acceleration at all; see Figure 10.

3.2. Solar System. In this section we show how to derive Solar System constraints on the free parameters of the model "John + George."

In a first step we solve asymptotically the field equations in spherical symmetric and isotropic coordinates. The line element reads

$$ds^2 = -A(r)^2 dt^2 + B(r)^2 (dr^2 + r^2 d^2\Omega), \qquad (23)$$

and the equations of motion for the components and for φ are given in Appendix C. The post-Newtonian analysis begins with the expansion of the metric components as $A^2 = 1 + \sum_i a_i/r^i$, $B^2 = 1 + \sum_i b_i/r^i$ and $\varphi = p_0 + \sum_i p_i/r^i$ and of the field equations in powers of $1/r$. By equating the coefficients of equal powers of r we find

$$A^2 = 1 - \frac{r_s}{r} + \frac{r_s^2}{2r^2} + \frac{\epsilon^2 p_1^2}{4M_p^2 z^2 r^2} + \frac{p_1 \epsilon r_s^2}{24 M_p zr^3}$$
$$- \frac{p_1^2 r_s}{24 M_p^2 zr^3} + \frac{3}{4} \frac{\overline{\gamma}\epsilon^2}{M_p^4 z^2 r^4} - \frac{r_s \overline{\gamma}}{8 M_p^4 zr^5}, \qquad (24a)$$

$$B^2 = 1 + \frac{r_s}{r} - 2\frac{\epsilon p_1}{M_p zr} - \frac{p_1^2}{4M_p^2 zr^2} - \frac{\overline{\gamma}}{4M_p^4 zr^4}, \qquad (24b)$$

where $r_s = 2GM/c^2$ is the Schwarzschild radius of the central body, $z = 1 + (\epsilon p_0/M_p)$, and $\overline{\gamma} = \gamma p_1^2$. In the expansion previously mentioned, we neglected higher order terms in r_s/r, in ϵ, in $p_1/(rM_p)$ and in $\overline{\gamma}/(rM_p)$ (which means we suppose these terms to be smaller than 1). We recall that, in our conventions, γ and ϵ are dimensionless parameters. The asymptotic scalar field value $p_0 = \varphi(r \to \infty)$ (in GeV) is a free parameter that can eventually be connected to the cosmological evolution of the scalar field.

The next step then amounts to relate the dimensionless parameter p_1 to the scalar charge of the central body and thus eventually to the couplings ϵ and γ. This could be determined numerically by solving the equations in the interior of the body [34] with suitable asymptotic and regularity conditions, together with, for example, a polytropic equation of state for the star's interior. Such a discussion about possible spontaneous scalarization goes however beyond the scope of the present paper and is left for future studies. In this section we only sketch how solar system constraints can be derived.

The final step consists then in computing observable effects from the metric Equations (23)–(24b), which show a deviation from standard General Relativity in the Solar System. A first constraint comes from the anomalous perihelion shift. First of all, we determine the geodesic equation (to first order in the metric deviation) to find the planetary equations of motion (when computing the geodesics, we suppose that matter is minimally coupled to the metric). The anomalous term is treated perturbatively to find the perihelion shift with the Gauss equations, which determines the evolution of the different orbital parameters due to perturbative forces. In particular, we derive the rate of change of the argument of the perihelion ω. Finally, the secular term of the change of

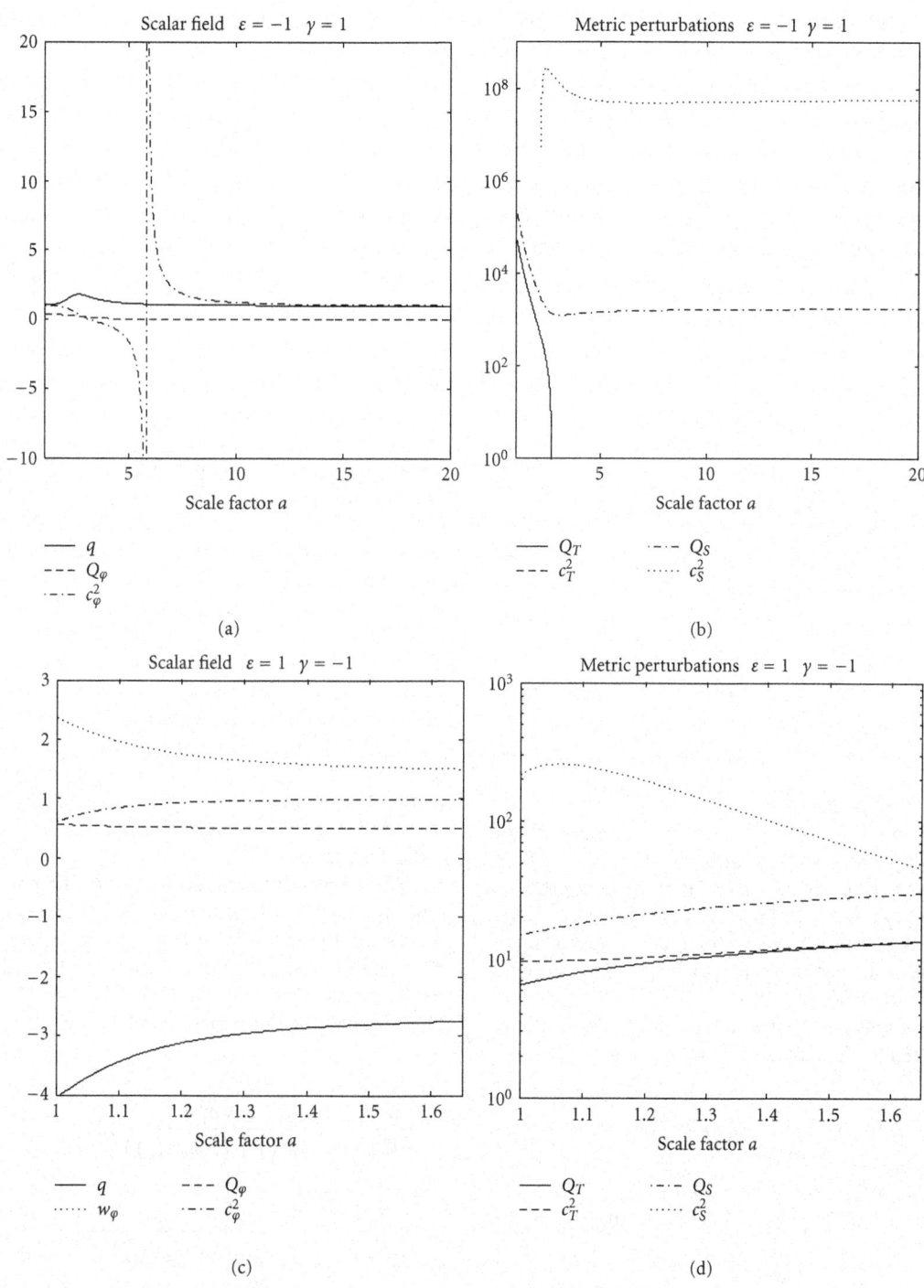

FIGURE 10: Cosmological evolution of q, Q_φ, c_φ^2, Q_S, c_S^2, Q_T, c_T^2 with the scale factor. Top: the case $\epsilon = -1$ and $\gamma = 1$ corresponds to a "double inflation." The scenario is pathological in many respects: singularity in the scalar field velocity (see (a)) while the scalar field EoS is imaginary. Also (see (b), the y axis is logarithmic): $c_T^2 < 0$ and there are periods for which $c_S^2 < 0$ and $Q_T < 0$. Bottom: the case $\epsilon = 1$ and $\gamma = -1$. The model is well behaved but does not accelerate the expansion, that is, as in the John alone model. The universe is actually in a super stiff regime and hence in a highly decelerating phase.

the argument of the perihelion is exhibited by a time average (over an orbital period T)

$$\left\langle \frac{d\omega}{dt} \right\rangle = \frac{1}{T} \int_0^T \frac{d\omega}{dt} dt. \qquad (25)$$

The results of this whole procedure are given by

$$\left\langle \frac{d\omega}{dt} \right\rangle = -\alpha_1 \frac{\bar{\gamma}}{M_p^4 z} - \alpha_2 \frac{\bar{\gamma}\epsilon^2}{M_p^4 z^2} - \alpha_3 \frac{p_1^2}{M_p^2 z}$$
$$- \alpha_4 \frac{p_1 \epsilon}{z M_p} - \alpha_5 \frac{p_1^2 \epsilon^2}{z^2 M_p^2}, \qquad (26)$$

TABLE 1: Values of the different α_i coeffcients (27a)–(27e) for planets and constraints given by INPOP10a for supplementary advances of perihelia (values given in [26]).

	α_1 (mas/cy/m^4)	α_2 (mas/cy/m^4)	α_3 (mas/cy/m^2)	α_4 (mas/cy/m)	α_5 (mas/cy/m^2)	$d\omega/dt$ (mas/cy) INPOP10a
Mercury	3.2×10^{-32}	4.85×10^{-24}	8.83×10^{-11}	19.4	8.2×10^{-4}	0.4 ± 0.6
Venus	7.69×10^{-34}	2.53×10^{-25}	9×10^{-12}	3.89	1.65×10^{-4}	0.2 ± 1.5
Earth	1.3×10^{-34}	5.9×10^{-26}	2.9×10^{-12}	1.73	7.33×10^{-5}	-0.2 ± 0.9
Mars	1.36×10^{-35}	9.13×10^{-27}	6.77×10^{-13}	6.1×10^{-1}	2.58×10^{-5}	-0.04 ± 0.15
Jupiter	1.51×10^{-38}	3.56×10^{-29}	9.06×10^{-15}	2.81×10^{-2}	1.19×10^{-6}	-41 ± 42
Saturn	5.33×10^{-40}	2.3×10^{-30}	1.08×10^{-15}	6.16×10^{-3}	2.61×10^{-7}	0.15 ± 0.65

with

$$\alpha_1 = \frac{n}{16a^4} \frac{8 + 24e^2 + 3e^4}{(1 - e^2)^4}, \tag{27a}$$

$$\alpha_2 = \frac{9n}{8a^3 r_s} \frac{4 - 3e^2 - e^4}{(1 - e^2)^4}, \tag{27b}$$

$$\alpha_3 = \frac{n}{8a^2} \frac{4 + e^2}{(1 - e^2)^2}, \tag{27c}$$

$$\alpha_4 = \frac{2n}{a(1 - e^2)}, \tag{27d}$$

$$\alpha_5 = \frac{n}{4a r_s (1 - e^2)}, \tag{27e}$$

where a is the semi major axis of the orbit, e its eccentricity, and n its mean motion $n = 2\pi/T$. The advance of perihelion of Solar System planets is very tightly constrained by planetary ephemerides. In particular, INPOP10a gives constraints on supplementary advances of perihelia (see Table 5 of [26]). Table 1 gives the value of the α_i coefficients appearing in the expression of the advance of perihelion (26) and the constraints coming from INPOP10a.

The most stringent constraints in the case considered here are obtained by data from Mercury and read

$$-3.12 \times 10^{31} \, \text{m}^4 < \frac{\bar{\gamma}}{M_p^4 z} < 6.25 \times 10^{30} \, \text{m}^4, \tag{28a}$$

$$-2.06 \times 10^{23} \, \text{m}^4 < \frac{\bar{\gamma} \epsilon^2}{M_p^4 z^2} < 4.12 \times 10^{22} \, \text{m}^4, \tag{28b}$$

$$-1.13 \times 10^{10} \, \text{m}^2 < \frac{p_1^2}{M_p^2 z} < 2.26 \times 10^9 \, \text{m}^2, \tag{28c}$$

$$-5.16 \times 10^{-2} \, \text{m} < \frac{p_1 \epsilon}{z M_p} < 1.03 \times 10^{-2} \, \text{m}. \tag{28d}$$

These constraints are obtained by considering only deviations on the dynamics (i.e., on the equations of motion). Other constraints can be derived using propagation of light rays in the solar system. For example, radioscience experiments include light propagation through a Shapiro-like term that can be derived from the expression of the metric (23)–(24b). One way to obtain such constraint is to follow the strategy and to use the software of [35–37]. The main idea presented in these papers is to simulate radioscience observables directly from the space-time metric so that the software includes both deviations on the dynamics and deviations on light propagation. In particular, we use this software to simulate a two-way Doppler link between Earth and Cassini spacecraft from May 2001 on the metric (23)–(24b) and to analyze them in GR by fitting the initial conditions of the spacecraft. The residuals that emerge are the incompressible signature produced by the alternative theory considered on Doppler signal for Cassini. Comparing this signal to the Doppler accuracy of the mission (10^{-14}) allows us to give order of magnitude of constraints on the theory. Figure 11 represents the incompressible signatures produced by parameters entering the metric (23)–(24b) on Cassini Doppler. The three sharp peaks occurred at solar conjunctions. The order of magnitude of the residuals observed in this figure is larger than Cassini accuracy, which means that the values of the parameters should be smaller than the indicated values.

We run a set of simulations with different values for the parameters appearing in the metric (23)–(24b). Figure 12 represents the evolution of the maximal Doppler residuals obtained in Cassini signal as function of metric parameters. Requiring the residuals to be lower than Cassini accuracy (10^{-14}) gives the boundary values

$$\left| \frac{\bar{\gamma}}{M_p^4 z} \right| = \left| \frac{\gamma p_1^2}{M_p^4 \left(1 + \left(\epsilon p_0/M_p\right)\right)} \right| < 3.65 \times 10^{26} \, \text{m}^4, \tag{29a}$$

$$\left| \frac{\bar{\gamma} \epsilon^2}{M_p^4 z^2} \right| = \left| \frac{\gamma p_1^2 \epsilon^2}{M_p^4 \left(1 + \left(\epsilon p_0/M_p\right)\right)^2} \right| < 1.15 \times 10^{26} \, \text{m}^4, \tag{29b}$$

$$\left| \frac{p_1^2}{M_p^2 z} \right| = \left| \frac{p_1^2}{M_p^2 \left(1 + \left(\epsilon p_0/M_p\right)\right)} \right| < 3.53 \times 10^8 \, \text{m}^2, \tag{29c}$$

$$\left| \frac{p_1 \epsilon}{z M_p} \right| = \left| \frac{p_1 \epsilon}{M_p \left(1 + \left(\epsilon p_0/M_p\right)\right)} \right| < 5.56 \times 10^{-2} \, \text{m}. \tag{29d}$$

It should be noted that radioscience constraints are significantly better for $\bar{\gamma}/M_p^2 z$ and $p_1^2/M_p^2 z$ as compared to perihelia advances. On the other hand, the constraint from

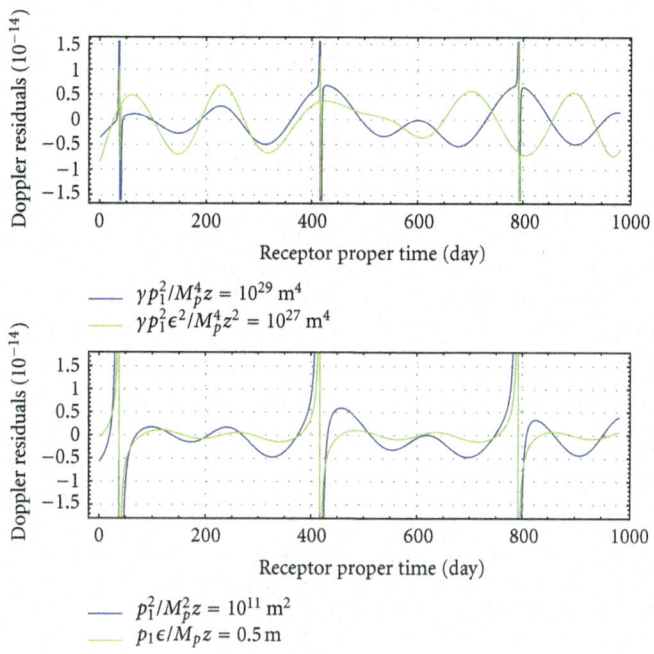

FIGURE 11: Incompressible anomalous signal due to the different terms in the metric (23)–(24b) on Cassini Doppler signal with indicated values. The produced residuals are larger than Cassini accuracy (which is 10^{-14}); therefore these values for the parameters are ruled out.

the ephemerides on $\bar{\gamma}\epsilon^2/M_p^4 z^2$ is significantly better while the constraint on $p_1\epsilon/zM_p$ is of the same order of magnitude.

We thus see that solar systems tests of GR will put severe constraints on the free parameters of the model. As already stressed; however, the full numerical study of Solar's interior is required, while a thorough analysis of possible spontaneous scalarization (depending also on the Georges coupling) is necessary.

4. Conclusions

In this paper, we have explored some phenomenology associated with a subset of the "Fab Four" scalar-tensor theory. The philosophy behind this preliminary work is that we cannot forget about solar system constraints on the parameter space, even when we deal with inflationary solutions. Traditional models of inflation rely upon the fact that the inflaton field decays, at some stage, into ordinary matter through some reheating mechanism. Therefore, the scalar-tensor nature of inflationary gravity is lost very soon in the evolution of the Universe. On the opposite, in the models studied in this paper the scalar field should live and show its effects until nowadays. Therefore, the parameter space determined by constraints from cosmological observations must overlap with the one determined by solar system tests. The "Fab Four" theory has many parameters with a very rich phenomenology, and with this paper we begin an ambitious plan for its systematic study.

In this work, we contained ourselves to the cases John and John plus George. The John case represents a theory with a nonminimal derivative coupling between scalar field and Einstein tensor. It was already known that this model

admits inflationary solutions with a graceful exit. Here, we constrained the coupling constant γ by showing that it must be positive in order to have successful inflation. If this is the case, however, very unnatural initial conditions are required. In particular the field velocity, which is related to the energy density, must be huge compared to the Planck scale, rendering the theory no longer trustworthy. Negative values for γ are permitted but do not allow for inflation while ghost states might appear. The most serious problem however comes from the fact that the model turns out to be trivial when one tries to solve the equations of motion inside a compact object. Indeed, we found that the only solution with a finite field at the centre is $\varphi = 0$ everywhere.

These facts have convinced us to extend the theory to include the term named "George," which is nothing but a coupling between the scalar field and the Ricci scalar. The parameter space is now two-dimensional, and we solved numerically the equations of motion for a cosmological background. The main result is that the sign of the two coupling constants must be positive in order to have an inflationary phase with graceful exit. We reserve the analysis of the naturalness problem, that is, the need for extreme initial conditions, for future work. We also performed a post-Newtonian analysis of the theory by solving the equations of motion by imposing static and spherical symmetry and expanding the fields for large radial distance. The aim was to put some constraint on the free parameters, which now include also the asymptotic value of the scalar field and the scalar charge, which now makes sense as there is nontrivial solution also for the interior of a compact object. The results still allow for a large parameter space; therefore future work is necessary in order to improve the constraints.

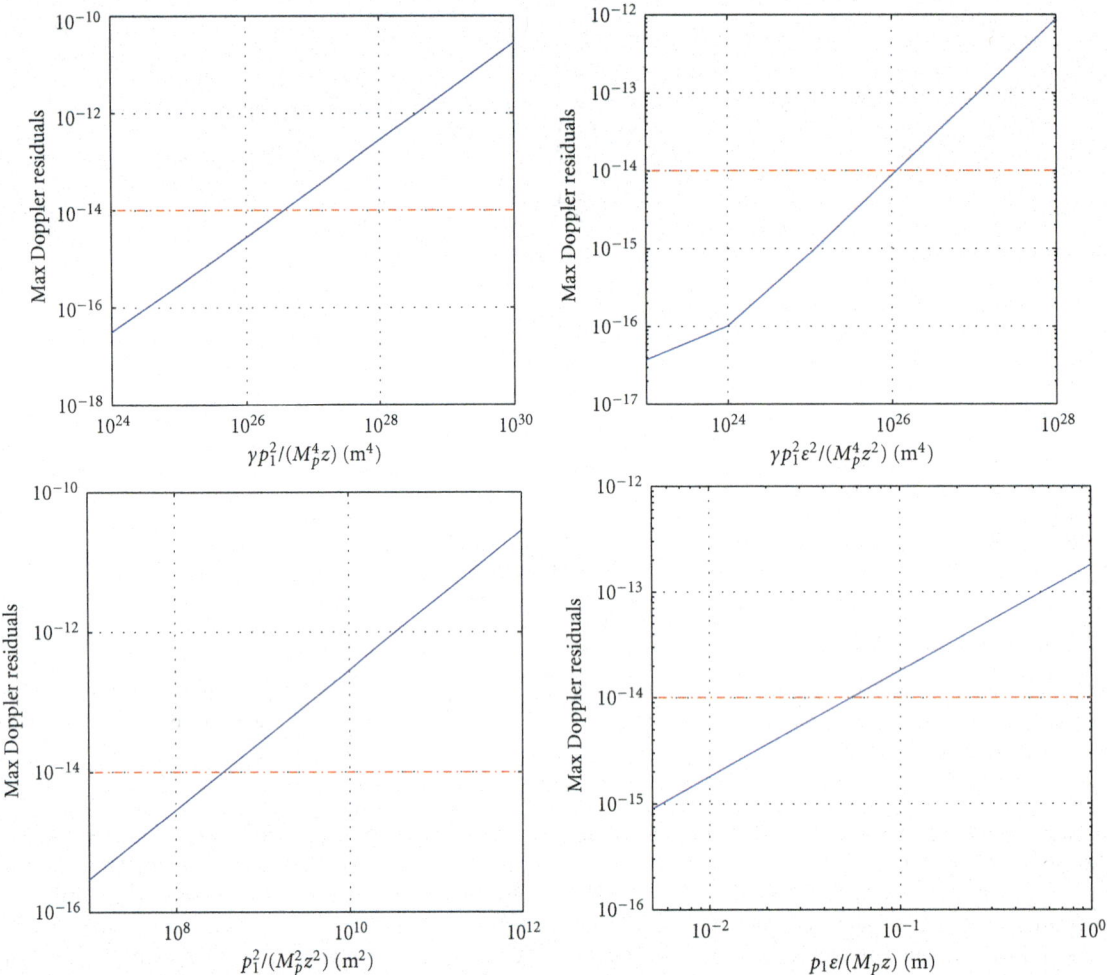

FIGURE 12: Representation of the maximal Doppler residuals signal due to John + George modification of gravity parametrized by the four parameters in the metric (23)–(24b) for the Cassini mission between Jupiter and Saturn. The red (dashed) lines represent the assumed Cassini accuracy.

As mentioned previously, this paper is the first of a series that aim at a systematic study of the "Fab Four" phenomenology, in order to isolate the experimentally viable sectors of the theory. Besides these aspects, there are several issues that deserve further investigations. For example, if these models truly lead to inflationary cosmology and graceful exit, we will need an alternative reheating mechanism. Maybe, the complexity of the theory acts as an effective potential for the scalar field, which resembles the usual power-law forms of usual inflation. But if this is not the case, then we need to find alternative explanations. Another aspect that needs to be studied is the relevance of these modifications of gravity in terms of late-time cosmology, as we expect modifications driven by the scalar field. This would include, at the background level, the study of tracking solutions and of the convergence mechanism towards GR, if any. This has not yet been addressed in the literature for the John Lagrangian (but see [38]). The study of cosmological perturbations, in particular CMB spectra and large-scale structures, might then further reduce the parameter space. Finally, gravitational effects might be relevant at galactic

scales and give rise to alternative explanations to the anomalous galactic rotation curves.

In conclusion, we believe that the recent developments in scalar-tensor theories of gravity have opened the door to new and intriguing research directions, and we are confident that many interesting results will be obtained in the near future.

Appendices

A. Cosmological Equations

In terms of the reduced variables $x(t) = \kappa\dot{\varphi}$, $y(t) = \sqrt{\kappa}\dot{\alpha}$ and $z(t) = 1 + \epsilon\sqrt{\kappa}\varphi(t)$, the equations of motion for a flat and empty universe derived from action Equation (22) are

$$6\epsilon xy + x^2(-1 + 9\gamma y^2) + 6y^2 z = 0,$$
$$4x(\epsilon + \gamma\sqrt{\kappa}\dot{x})y + x^2(1 + 3\gamma y^2 + 2\gamma\sqrt{\kappa}\dot{y})$$
$$+ 6y^2 z + 2\sqrt{\kappa}(\epsilon\dot{x} + 2\dot{y}z) = 0, \qquad (A.1)$$
$$3y(x - 2\epsilon y - 3\gamma xy^2)$$
$$+ \sqrt{\kappa}(\dot{x} - 3\gamma\dot{x}y^2 - 3(\epsilon + 2\gamma xy)\dot{y}) = 0,$$

which can be decoupled in the following way:

$$\dot{x} = \frac{-3x[\epsilon x + 4(\epsilon^2 + \gamma x^2)y + 7\epsilon\gamma xy^2] + 6y(-2x + \epsilon y)z}{2\kappa^{1/2}(3\epsilon^2 + 12\epsilon\gamma xy + \gamma x^2(1 + 9\gamma y^2) + (2 - 6\gamma y^2)z)}$$

$$\dot{y} = (2\epsilon xy(1 - 15\gamma y^2) + x^2[-1 + 3\gamma y^2(4 - 9\gamma y^2)]$$

$$- 6y^2(2\epsilon^2 + z - 3\gamma y^2 z))$$

$$\times \left(2\kappa^{1/2}[3\epsilon^2 + 12\epsilon\gamma xy + \gamma x^2(1 + 9\gamma y^2)\right.$$

$$\left. + (2 - 6\gamma y^2)z]\right)^{-1}.$$

(A.2)

The scalar field EoS is given by

$$\omega_\varphi = -x(3\gamma x^2 + 2z)\frac{N}{D}$$ (A.3)

with

$$N = 6x(\gamma x^2 - z)(3\gamma x^2 + 2z)^2 + 6\sqrt{3}\epsilon^3$$

$$\times \sqrt{x^2(3\epsilon^2 + 3\gamma x^2 + 2z)}(7\gamma x^2 + 2z)$$

$$- 18\epsilon^4(7\gamma x^3 + 2xz) - 2\sqrt{3}\epsilon\sqrt{x^2(3\epsilon^2 + 3\gamma x^2 + 2z)}$$

$$\times (33\gamma^2 x^4 + 16\gamma x^2 z - 4z^2)$$

$$+ 9\epsilon^2(15\gamma^2 x^5 + 4\gamma x^3 z - 4xz^2)$$

$$D = \left[-3\epsilon x + \sqrt{3}\sqrt{x^2(3\epsilon^2 + 3\gamma x^2 + 2z)}\right]^2$$

$$\times \left[18\gamma^3 x^6 + 30\gamma^2 x^4 z + 24\gamma x^2 z^2 + 8z^3\right.$$

$$+ 6\sqrt{3}\epsilon\gamma x(\gamma x^2 + 2z)\sqrt{x^2(3\epsilon^2 + 3\gamma x^2 + 2z)}$$

$$\left. + 3\epsilon^2(3\gamma^2 x^4 + 4z^2)\right].$$

(A.4)

B. Ghost Conditions

The coupling to the Ricci scalar, "George," does not change the analysis made for the scalar field sector of the theory. Thus the two following conditions still hold:

$$Q_\varphi = \frac{1}{2}(1 - 3\gamma y^2) > 0$$

$$c_\varphi^2 = \frac{1 - \gamma(3y^2 + 2\sqrt{\kappa}\dot{y})}{1 - 3\gamma y^2} \geq 0.$$

(B.1)

For the metric perturbations, we derive, based on equations (23), (25), (26), and (27) of [32]

$$Q_T > 0 \implies z + \frac{\gamma x^2}{2} > 0,$$

$$c_T^2 \geq 0 \implies z - \frac{\gamma x^2}{2} \geq 0$$

(B.2)

for the tensorial part, and also

$$Q_S > 0 \implies 3\epsilon^2 + 12\epsilon\gamma xy + 9\gamma^2 x^2 y^2 + 2z$$

$$+ \gamma(x^2 - 6y^2 z) > 0$$

(B.3)

for the scalar part of the metric perturbations, while their squared speed $c_s^2 \geq 0$ leads to

$$2y\left(\frac{\gamma x^2}{2} + z\right)^2 (\epsilon x + 3\gamma x^2 y + 2yz)$$

$$+ 2x(\epsilon + \gamma\sqrt{\kappa}\dot{x})(\gamma x^2 + 2z)(\epsilon x + 3\gamma x^2 y + 2yz)$$

$$+ \frac{1}{2}(\gamma x^2 - 2z)(\epsilon x + 3\gamma x^2 y + 2yz)^2$$

$$- 2\sqrt{\kappa}\left(\frac{\gamma x^2}{2} + z\right)^2$$

$$\times \left[\epsilon\left(\dot{x} + \frac{2xy}{\sqrt{\kappa}}\right) + 3\gamma x(2\dot{x}y + x\dot{y}) + 2\dot{y}z\right] \geq 0.$$

(B.4)

C. Spherically Symmetric Equations of Motion

We derive the equations of motion for the action (22) with a spherically symmetric and static field configuration. We consider the metric (23), and we replace its components in the Lagrangian. With the Noether theorem, we find the equations of motion for the fields A, B, R, and φ. Finally, we impose the gauge $R = r$, and we find three equations plus a Hamiltonian constraint that read

$$0 = \left(2\varphi'^2 Br^2 B'' + 4\varphi' B^2 r\varphi'' + 4\varphi' Br^2\varphi'' B'\right.$$

$$\left. - 5\varphi'^2 B'^2 r^2 + 2\varphi'^2 B^2\right)\gamma\kappa^2$$

$$+ \left(- 4B^3\varphi B'' r^2 - 2B^3 B' r^2\varphi' - 4B^4\varphi' r - 8B^3\varphi B' r\right.$$

$$\left. - 2B^4 r^2\varphi'' + 2B^2\varphi B' r^2\right)\epsilon\kappa^{1/2}$$

$$- 8B^3 rB' - 4B^3 r^2 B'' + 2B^2 r^2 B'^2 - \varphi'^2\kappa^2 B^4 r^2,$$

$$0 = \left(4\varphi'^2 B' ArB + 2\varphi'^2 B^2 A - 3\varphi'^2 B'^2 Ar^2\right.$$

$$+ 4\varphi' Br^2\varphi'' AB + 8\varphi'^2 B^2 rA' + 2\varphi'^2 B' A' Br^2$$

$$+ 2\varphi'^2 A'' r^2 B^2 + 4\varphi' Br^2\varphi'' A + 4\varphi' A' r^2\varphi'' B^2$$

$$\left. + 2\varphi'^2 ABr^2 B''\right)\gamma\kappa^2$$

$$+ \left(-4B^3 AB' r^2\varphi' + 2B^2 A\varphi B'^2 r^2 - 8B^4 A\varphi' r\right.$$

$$- 8\varphi rA' B^4 - 8B^3 A\varphi B' r - 4B^3 A\varphi B'' r^2$$

$$- 4\varphi B' A' r^2 B^3 - 6A' r^2\varphi' B^4 - 4\varphi A'' r^2 B^4$$

$$\left. - 4B^4 Ar^2\varphi''\right)\epsilon\kappa^{1/2}$$

$$- 4B^3 Ar^2 B'' - 4r^2 A'' B^4 - 8rA' B^4 - 8B^3 ArB'$$

$$- 4A' r^2 B' B^3 + 2B^2 Ar^2 B'^2 - \varphi'^2\kappa^2 B^4 Ar^2,$$

$$0 = \Big(4\varphi'A'B^3 + 4\varphi'AB'B^2 + 4\varphi'B'Ar^2B''B$$
$$+ 4\varphi'B''A'r^2B^2 + 4\varphi''rA'B^3 + 4\varphi'rA''B^3$$
$$+ 4\varphi''B'A'r^2B^2 + 8\varphi'B'A'rB^2 + 4\varphi'B''ArB^2$$
$$+ 4\varphi''B'ArB^2 - 4\varphi'B'^2ArB - 6\varphi'B'^2A'r^2B$$
$$- 6\varphi'B'^3Ar^2 + 4\varphi'B'A''r^2B^2 + 2\varphi''B'^2Ar^2B\Big)\gamma\kappa$$
$$+ \Big(- 8B'ArB^4 - 4B^5rA' - 4r^2AB''B^4 - 2B'A'r^2B^4$$
$$- 2B^5r^2A'' + 2B'^2Ar^2B^3\Big)\epsilon\kappa^{-1/2}$$
$$+ 2A'r^2\varphi'B^5 + 2B^5r^2\varphi''A + 4\varphi'B^5Ar$$
$$+ 2B'r^2\varphi'AB^4,$$

$$0 = \Big(2\varphi'^2A'B^2 + 2r\varphi'^2A''B^2 - 6\varphi'^2B'^2Ar$$
$$+ 4\varphi'A\varphi''B^2 - 2\varphi'^2B'AB - 4\varphi'^2A'B'rB$$
$$+ 4\varphi'rB'A\varphi''B + 4r\varphi'A'\varphi''B^2 + 2\varphi'^2rB''AB\Big)\gamma\kappa^2$$
$$+ \Big(-4\varphi B'AB^3 + 4r\varphi B'^2AB^2 - 4B^4rA'\varphi'$$
$$- 4B^4\varphi A''r - 4B^4\varphi''Ar - 4B^4\varphi A'$$
$$- 4B^4\varphi'A - 4\varphi B''ArB^3\Big)\epsilon\kappa^{1/2}$$
$$+ 4rB'^2AB^2 - 4rAB''B^3 - 4B'AB^3$$
$$- 4B^4A' - 4B^4A''r - 2B^4\varphi'^2Ar\kappa^2.$$

$$(C.1)$$

Acknowledgments

J.-P. Bruneton is FSR/COFUND postdoctoral researcher at naXys. M. Rinaldi is supported by a grant of ARC 11/15-040 convention. A. Kanfon thanks ARC 11/15-040 for travel support. A. Hees is FNRS Research Fellow. A. Hees thanks B. Lamine, P. Wolf, C. Le Poncin-Lafitte, V. Lainey, S. Reynaud, and M. T. Jaekel for useful discussions about radioscience simulations within Solar System.

References

[1] A. Nicolis, R. Rattazzi, and E. Trincherini, "Galileon as a local modification of gravity," *Physical Review D*, vol. 79, Article ID 064036, 21 pages, 2009.

[2] R. V. Wagoner, "Scalar-tensor theory and gravitational waves," *Physical Review D*, vol. 1, pp. 3209–3216, 1970.

[3] T. Damour and G. Esposito-Farese, "Tensor multiscalar theories of gravitation," *Classical and Quantum Gravity*, vol. 9, no. 9 , article 2093, 1992.

[4] T. Damour and K. Nordtvedt, "Tensor-scalar cosmological models and their relaxation toward general relativity," *Physical Review D*, vol. 48, no. 3436, p. 3450, 1993.

[5] F. Perrotta, C. Baccigalupi, and S. Matarrese, "Extended quintessence," *Physical Review D*, vol. 61, Article ID 023507, 12 pages, 1999.

[6] S. Nojiri and S. D. Odintsov, "Modified gravity with negative and positive powers of curvature: unification of inflation and cosmic acceleration," *Physical Review D*, vol. 68, Article ID 123512, 10 pages, 2003.

[7] T. Chiba, "1/R gravity and scalar-tensor gravity," *Physics Letters B*, vol. 575, no. 1-2, pp. 1–3, 2003.

[8] S. Capozziello and M. Francaviglia, "Extended theories of gravity and their cosmological and astrophysical applications," *General Relativity and Gravitation*, vol. 40, no. 2-3, pp. 357–420, 2008.

[9] T. P. Sotiriou and V. Faraoni, "f(R) theories of gravity," *Reviews of Modern Physics*, vol. 82, pp. 451–497, 2010.

[10] T. Clifton, P. G. Ferreira, A. Padilla, and C. Skordis, "Modified gravity and cosmology," *Physics Reports*, vol. 513, no. 1–3, pp. 1–189, 2012.

[11] C. Deffayet, G. Esposito-Farese, and A. Vikman, "Covariant Galileon," *Physical Review D*, vol. 79, Article ID 084003, 6 pages, 2009.

[12] C. Deffayet, X. Gao, D. Steer, and G. Zahariade, "From k-essence to generalized Galileons," *Physical Review D*, vol. 84, Article ID 064039, 15 pages, 2011.

[13] C. Deffayet, S. Deser, and G. Esposito-Farese, "Generalized Galileons: all scalar models whose curved background extensions maintain second-order field equations and stress tensors," *Physical Review D*, vol. 80, Article ID 064015, 5 pages, 2009.

[14] G. W. Horndeski, "Second-order scalar-tensor field equations in a four-dimensional space," *International Journal of Theoretical Physics*, vol. 10, no. 6, pp. 363–384, 1974.

[15] J. Khoury and A. Weltman, "Chameleon cosmology," *Physical Review D*, vol. 69, Article ID 044026, p. 15, 2004.

[16] A. Hees and A. Fuzfa, "Combined cosmological and solar system constraints on chameleon mechanism," *Physical Review D*, vol. 85, Article ID 103005, 21 pages, 2012.

[17] E. Babichev, V. Mukhanov, and A. Vikman, "k-Essence, superluminal propagation, causality and emergent geometry," *Journal of High Energy Physics*, vol. 2008, no. 02, article 101, 2008.

[18] P. Creminelli, A. Nicolis, and E. Trincherini, "Galilean genesis: an alternative to inflation," *Journal of Cosmology and Astroparticle Physics*, vol. 2010, no. 11, article 021, 2010.

[19] E. Babichev, "Galileon accretion," *Physical Review D*, vol. 83, Article ID 024008, 9 pages, 2011.

[20] C. Deffayet, O. Pujolas, I. Sawicki, and A. Vikman, "Imperfect dark energy from kinetic gravity braiding," *ournal of Cosmology and Astroparticle Physics*, vol. 2010, no. 10, article 026, 2010.

[21] N. Chow and J. Khoury, "Galileon cosmology," *Physical Review D*, vol. 80, Article ID 024037, p. 13, 2009.

[22] C. Charmousis, E. J. Copeland, A. Padilla, and P. M. Saffin, "General second order scalar-tensor theory, self tuning, and the Fab Four," *Physical Review Letters*, vol. 108, Article ID 051101, 2012.

[23] T. Kobayashi, M. Yamaguchi, and J. Yokoyama, "Generalized g-inflation," *Progress of Theoretical Physics*, vol. 126, pp. 511–529, 2011.

[24] S. Tsujikawa, "Observational tests of inflation with a field derivative coupling to gravity," *Physical Review D*, vol. 85, Article ID 083518, 2012.

[25] E. N. Saridakis and S. V. Sushkov, "Quintessence and phantom cosmology with nonminimal derivative coupling," *Physical Review D*, vol. 81, Article ID 083510, 8 pages, 2010.

[26] A. Fienga, J. Laskar, P. Kuchynka et al., "The INPOP10a planetary ephemeris and its applications in fundamental physics,"

Celestial Mechanics and Dynamical Astronomy, vol. 111, no. 3, pp. 363–385, 2011.

[27] S. V. Sushkov, "Exact cosmological solutions with nonminimal derivative coupling," *Physical Review D*, vol. 80, no. 10, Article ID 103505, 6 pages, 2009.

[28] G. Gubitosi and E. V. Linder, "Purely kinetic coupled gravity," *Physics Letters B*, vol. 703, no. 2, pp. 113–118, 2011.

[29] C. Germani and A. Kehagias, "Ultraviolet-protected inflation," *Physical Review Letters*, vol. 106, no. 16, Article ID 161302, 4 pages, 2011.

[30] C. Charmousis, E. J. Copeland, A. Padilla, and P. M. Saffin, "Self-tuning and the derivation of the Fab Four," *Physical Review D*, vol. 85, Article ID 104040, 2012.

[31] J. P. Bruneton, "Causality and superluminal behavior in classical field theories: applications to k-essence theories and modified-Newtonian-dynamics-like theories of gravity," *Physical Review D*, vol. 75, no. 8, Article ID 085013, 14 pages, 2007.

[32] A. De Felice and S. Tsujikawa, "Conditions for the cosmological viability of the most general scalar-tensor theories and their applications to extended Galileon dark energy models," *Journal of Cosmology and Astroparticle Physics*, vol. 2012, no. 02, article 007, 2012.

[33] C. Germani and A. Kehagias, "New model of inflation with nonminimal derivative coupling of standard model higgs boson to gravity," *Physical Review Letters*, vol. 105, Article ID 011302, 4 pages, 2010.

[34] T. Damour and G. Esposito-Farese, "Nonperturbative strong-field effects in tensor-scalar theories of gravitation," *Physical Review Letters*, vol. 70, pp. 2220–2223, 1993.

[35] A. Hees, B. Lamine, S. Reynaud et al., "Radioscience simulations in general relativity and in alternative theories of gravity," *Classical and Quantum Gravity*, vol. 29, no. 23, article 235027.

[36] A. Hees, P. Wolf, B. Lamine et al., "Radioscience simulations in General Relativity and in alternative theories of gravity," in *Proceedings of Les Rencontres de Moriond 2011*, 2011.

[37] A. Hees, P. Wolf, B. Lamine et al., "Testing Gravitation in the Solar System with Radio Science experiments," in *Proceedings of the Annual Meeting of the French Society of Astronomy and Astrophysics (SF2A '11)*, G. Alecian, K. Belkacem, R. Samadi, and D. Valls-Gabaud, Eds., pp. 653–658, 2011.

[38] E. J. Copeland, A. Padilla, and P. M. Saffin, "The cosmology of theFab-Four," http://arxiv.org/abs/1208.3373.

4

The Impact of Polarized Extragalactic Radio Sources on the Detection of CMB Anisotropies in Polarization

Marco Tucci[1] and Luigi Toffolatti[2,3]

[1] LAL, Université Paris-Sud and CNRS/IN2P3, 91400 Orsay, France
[2] Departamento de Física, Universidad de Oviedo, C. Calvo Sotelo s/n, 33007 Oviedo, Spain
[3] IFCA, Universidad de Cantabria, Avenida los Castros s/n, 39005 Santander, Spain

Correspondence should be addressed to Marco Tucci, tucci@lal.in2p3.fr

Academic Editor: Carlo Burigana

Recent polarimetric surveys of extragalactic radio sources (ERS) at frequencies $\nu \gtrsim 1$ GHz are reviewed. By exploiting all the most relevant data we study the frequency dependence of polarization properties of ERS between 1.4 and 86 GHz. For flat-spectrum sources the median (mean) fractional polarization increases from 1.5% (2–2.5%) at 1.4 GHz to 2.5–3% (3–3.5%) at $\nu > 10$ GHz. Steep-spectrum sources are typically more polarized, especially at high frequencies where Faraday depolarization is less relevant. Current data suggest that at high radio frequencies ($\nu \geq 20$) GHz the fractional polarization of ERS does not depend on total flux density and moderately increases with frequency. We estimate ERS number counts in polarization and the contribution of unresolved polarized ERS to angular power spectra. A first application is for the *Planck* satellite mission: we predict that only a dozen polarized ERS will be detected by the *Planck* LFI, and a few tens by the HFI. As for CMB power spectra, ERS should not be a strong contaminant to the CMB E-mode polarization at $\nu \gtrsim 70$ GHz, but they can become a relevant constraint for the detection of the cosmological B-mode polarization if the tensor-to-scalar ratio is $\lesssim 0.01$.

1. Introduction

The radiation emitted by extragalactic radio sources (ERS) at high radio frequencies, mainly synchrotron radiation from relativistic electrons in their jets and lobes, can be highly polarized, with an intrinsic degree of linear polarization as high as ~70–75% in homogeneous sources with an unidirectional magnetic field (see, e.g., [1, 2], for comprehensive discussions on the subject).

These have to be considered as maximum values, which could be detected only in the case of very homogeneous sources with highly aligned magnetic field lines, which are unlikely to be observed among ERS. Actually, a much lower degree of total linear polarization, P, is commonly observed in ERS at cm or mm wavelengths (see e.g., [3, 4]), with only very few ERS showing a total fractional polarization, $\Pi = P/S$, as high as ~10% of the total flux density, S.

Nevertheless, this low fractional polarization observed in ERS may constitute a problem for the detection of primordial polarization in the Cosmic Microwave Background (CMB), since ERS are the dominant polarized foreground at small angular scales and the CMB polarized signal is only a few percent of CMB temperature anisotropy. Therefore, CMB polarization studies need a careful determination of the polarized emission from foreground sources in general and of ERS in particular.

CMB polarized anisotropies can be decomposed into modes of even-parity (E-mode) and odd-parity (B-mode). As for the E-modes, generated by scalar perturbations in the primordial universe, they were first detected by [5] and all the following observations (e.g., [6, 7]) have confirmed measurements of CMB E-modes compatible with the concordance ΛCDM model. The more elusive B-modes are generated by tensor metric perturbations, that is, "primordial" gravitational waves generated during inflation, and, according to predictions of inflationary models, with an amplitude directly proportional to the energy scale at which the inflation occurred [8, 9]. A detection of these "primordial" B-modes would provide the first real measure of the energy scale of inflation; thus, it would produce a real breakthrough in modern cosmology. But the detection of the B-modes CMB polarization—parameterized by the tensor-to-scalar ratio, $T/S = r$—is still a great challenge. In fact, the *Planck* mission [10, 11] (detailed discussions of the *Planck* Low Frequency Instrument (LFI) and High Frequency Instrument (HFI) expected polarization capabilities and calibrations are given by [12, 13]) would be marginally able to detect tensor metric perturbations by the direct detection of the primordial CMB B-mode, but only in the case of very high values of the tensor-to-scalar ratio of primordial perturbations, $r \gtrsim 0.05$–0.1 [14]. Otherwise, we will have to rely on proposed future space experiments, like COrE [15], specifically designed to detect CMB polarization by virtue of a much higher sensitivity than current ones.

The above discussion clearly illustrates the importance of giving the best up-to-date estimate of the average/median fractional polarization as well as of the distributions of fractional polarization [16], Π_i, at least for each one of the different ERS populations [17–19] which are providing relevant contributions to temperature anisotropies of the CMB.

Statistical studies of polarized emission from ERS have shown that at ~1.4–5 GHz, the fractional polarization, Π, increases with decreasing flux density [3, 16, 20–22]. Different explanations have been proposed of this result, for example, a population change at fainter flux density [20, 22] or a changing fraction of radio-quiet AGN [3]. However, the cause of this increase in Π is still unknown. On the other hand, it has been shown by [16] that this increase of the degree of polarization with decreasing flux density is only observed, at cm wavelengths, for steep-spectrum ERS, that is, sources with $\alpha < -0.5$ if $S(\nu) \propto \nu^\alpha$, and not for flat-spectrum ERS ($\alpha \geq -0.5$).

Other very recent studies of the polarized emission in ERS have tried to analyze the dependence of the fractional polarization with luminosity, redshift, and the source environment. The current, still preliminary, results show no correlation between the fractional polarization and redshift, whereas a weak correlation is found between decreasing luminosity and increasing degree of polarization [23]. According to the analysis of highly polarized elliptical galaxies [24] no differences have been found in the source environments between low polarization and ultrahigh polarization sources. This result should indicate that a high polarization must be a result of intrinsic properties of ERS.

The outline of the paper is as follows: in Section 2 we briefly summarize the main processes giving rise to linear and circular polarization in ERS; in Section 3 we present current published data on the polarized emission from ERS; Section 4 is dedicated to discuss our current results on statistical properties of the polarized emission in ERS; in Section 5 we give a short review on the most recent cosmological evolution model for ERS; Section 6 presents our predictions on the contributions of ERS polarized radiation, given by unresolved sources, to the E- and B-modes; finally, in Section 7, we present our conclusions.

2. Polarized Emission from ERS

Observations as well as statistical studies of the fractional polarization of ERS are very interesting on their own, and not only because ERS constitute a major contaminant of the CMB polarized signal. In fact, they are an important tool in active galactic nuclei (AGN) research, as they give valuable information on the physical quantities which determine the characteristics of the synchrotron radiation emitted by relativistic electrons accelerated by homogeneous and/or random magnetic fields in AGN [1, 2].

On the one hand, precise measurements of the total synchrotron radiation emitted by a radio source give an estimate—under some assumption, that is, equipartition between the field and particle energy—of the total magnetic field strength. On the other hand, the degree of polarization of the radio wave provides information on the direction of the main magnetic field in the source environment and also on its degree of ordering. The magnetic field structure can, in turn, provide information on the relationship between the environments of the ERS and their properties.

As already noted, compact ERS typically show a degree of total linear polarization of a few percent of their radio total intensity. Therefore, magnetic fields in radio sources are believed to be highly inhomogeneous, or almost without ordering, although the observed nonvanishing linear polarization gives an indication of a certain degree of ordering of the field [25, 26].

The precise orientation of magnetic fields lines inside jets and lobes is still unknown, but theoretical arguments as well as observational evidence show that magnetic fields are indeed partially ordered. As for observations, these conclusions are based on measurements of the orientations of the linear polarization, revealing coherent structures across the images. From the theoretical point of view, an ordered magnetic field is expected when shocks compress an initially random field (with the field **B** perpendicular to the jet axis) or when such initial fields are sheared to lie in a plane, with **B** parallel to the jet axis [27–30].

Circular polarization (CP), which is a common feature of quasars and BL Lacertae objects (or simply BL Lacs), commonly known as blazars, (blazar sources are jet-dominated extragalactic objects—observed within a small angle of the jet axis—in which the beamed component dominates the observed emission [31]), is preferentially generated near synchrotron self-absorbed jet cores and is detected in about 30%–50% of these sources [25, 32]. In these inhomogeneous, optically thick, synchrotron sources, the emission of the electron population at lower energies is hidden by self-absorption. These invisible electrons produce *Faraday rotation* and conversion, which is the most likely mechanism capable of creating the observed CP in blazar sources [26, 32]. In any case, the measured degrees of CP are generally well below the levels of linear polarization [33], and thus negligible, at least at GHz frequencies.

The change in the position angle of the linearly polarized radiation which passes through a magnetoionic medium, that is, the *Faraday rotation*, can be expressed by the rotation angle $\Delta\phi[\text{rad}] = \text{RM}[\text{rad/m}^2] \, \lambda^2[\text{m}^2]$ [26, 34] experienced by the polarization vector, where RM indicates the *rotation measure*. The rotation measure, RM, is the line-of-sight integral $RM[rad/m^2] = 0.81 \int n_e[cm^{-3}]B_{\parallel}[\mu G]d\ell[pc]$, n_e being the electron density and $B_{\parallel}$ the component of the magnetic field along the line-of-sight [35].

If the rotation depth is the same for all the emission volumes of the radio source, the net result is a rotation equal to RM λ^2 of the direction of polarization, without any effect on the degree of polarization. To change the degree of polarization there must be a variation in depth, either along or transverse to the line-of-sight [26].

Faraday rotation is commonly observed towards extragalactic radio sources. It was first observed by [36], who found an RM of -60 rad/m^2 towards the center of Centaurus A. More recently, observations of the rest-frame RM in AGN cores and jets reported values from a few hundreds to several thousands rad/m^2 (e.g., [37–39]). On the other hand, ERS dominated by emission from radio lobes show lower values of RMs, that is, from a few tens to hundreds of rad/m^2 (e.g., [40]).

3. Polarization Data on ERS at cm and mm Wavelengths

3.1. Polarization Surveys at 1.4 GHz. A large-scale polarization catalogue of radio sources is provided by the NRAO VLA Sky Survey (NVSS) at 1.4 GHz [41]. This survey covers $\Omega \simeq 10.3$ sr of the sky with $\delta \geq -40°$. The catalogue contains the flux density S and the Stokes parameters Q and U of almost 2×10^6 discrete sources with $S \gtrsim 2.5$ mJy. Extensive analyses of these data were carried out by [20] and [16]. By correlating NVSS sources with sources from the Green Bank 4.85 GHz (GB6) catalogue [42], statistical polarization properties have been derived for a subsample of $\sim 30,000$ ERS with $S_{1.4GHz} \geq 100$ mJy, divided into steep- and flat-spectrum sources. Steep-spectrum sources are found to increase the degree of polarization (Π) with decreasing flux density: in fact, the median (mean) Π value increases from $\Pi \simeq 1.1\%$ ($\simeq 2\%$) for $S > 800$ mJy up to $\Pi \simeq 1.8\%$ ($\simeq 2.7\%$) for the faintest ERS of the sample, at $100 \leq S < 200$ mJy. On the other hand, flat-spectrum sources show a lower fractional polarization (median ~1.3% and mean $\simeq$ 2%) with no significant trend with flux density [16].

Two recent sub-mJy surveys at 1.4 GHz with polarization measurements, that is, the Dominion Radio Astrophysical Observatory (DRAO) *Planck* Deep Fields project [3, 23, 43] and the Australia Telescope Low-Brightness Survey (ATLBS) [22], seem to confirm the above result of a higher level of fractional polarization in steep-spectrum sources with fainter flux densities. However, there is no clear explanation for the origin of the anticorrelation between Π and flux density in this source population.

In Figure 1 we plot differential number counts at 1.4 GHz for the polarized intensity, computed from the surveys presented above. The continuous curve represents a fit to total number counts of AGNs from [44], whereas the dotted curve is obtained from the previous fit by assuming a constant fractional polarization of $\Pi = 3.3\%$ for *all* ERS.

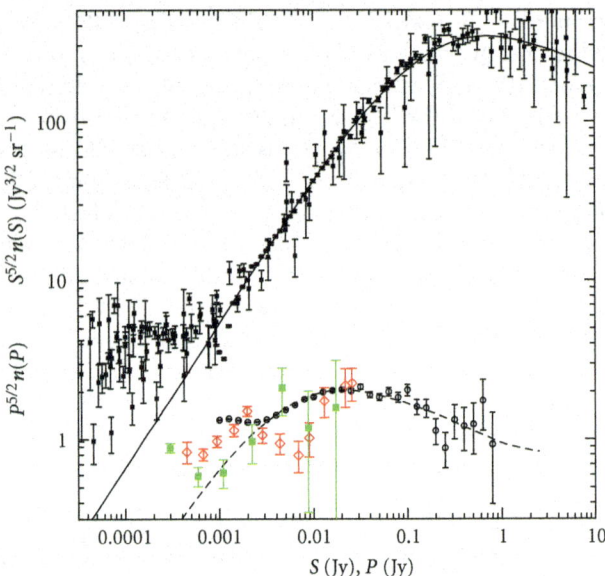

FIGURE 1: Differential number counts at 1.4 GHz as a function of total intensity (upper points) and of polarized intensity (lower points: black empty circles from NVSS; red empty diamonds from DRAO *Planck* Deep Fields; green squares from ATLBS). The continuous curve represents the number counts of AGNs from [44], whereas the dotted curve is obtained by the previous counts and assuming a constant fractional polarization of 3.3% for all ERS.

This value is chosen in order to fit NVSS data; the fit is extremely good down to polarized intensities of $P \simeq 3$ mJy. At fainter P fluxes the predicted curve bends down whereas observational data keep a flatter shape, in agreement with a higher fractional polarization for ERS at fainter flux density levels.

3.2. Multiwavelength Samples of Polarized ERS

3.2.1. Linear Polarization in a Sample of the B3-VLA Survey. Polarization measurements of sources in the B3-VLA survey [45] were carried out by [46] at 10.5, 4.85, and 2.7 GHz using the Effelsberg 100 m telescope. Taking into account only sources with $S_{10.5GHz} \geq 80$ mJy, a sample of 106 objects (out of 208) was defined with detected polarization. They found that flat-spectrum sources are significantly less polarized than steep-spectrum ones at 10.5 GHz. The latter ERS population shows a median fractional polarization that strongly depends on the frequency, from ~2% at 1.4 GHz to ~6% at 10.5 GHz, indicating a high Faraday depolarization at lower frequencies. Moreover, they notice that compact steep-spectrum sources exhibit much stronger depolarization than noncompact ones, and that sources showing larger linear size are more polarized. On the other hand, flat-spectrum sources are characterized by almost constant and low degrees of polarization (~2.5%) over the whole wavelength range considered.

3.2.2. The Australia Telescope 20 GHz (AT20G) Survey. The AT20G survey is a blind survey of the whole Southern sky at 20 GHz with follow-up observations at 4.8 and 8.6 GHz carried out with the Australia Telescope Compact Array (ATCA) [4]. The full source catalogue includes 5890 sources detected above a flux-density limit of 40 mJy. The AT20G catalogue is found to be 91% complete above 100 mJy and 79% above 50 mJy in regions south of declination $-15°$. Polarization was detected at 20 GHz for 768 sources, 467 of which also have simultaneous polarization detections at 5 and/or 8 GHz (upper limits on the polarized flux density are provided for nondetections, as described in [4]). Taking into account upper limits, a median (mean) fractional polarization of 2.6% (2.7%) at 20 GHz is found for sources with $S_{20\,GHz} > 250$ mJy, whereas an average 17% depolarization is observed at 5 GHz with respect to 20 GHz [47]. Compact sources with detected polarization are separated into flat- and steep-spectrum sources: the mean values of Π are 2.9% and 3.8%, respectively, thus confirming the higher fractional polarization in steep-spectrum sources.

3.2.3. VLA Polarization Measurements of WMAP Point Sources.

Using the VLA, [48] carried out polarization measurements at 8.4, 22 and 43 GHz of a complete sample of ERS brighter than 1 Jy in the 5-year WMAP catalogue and with declinations north of $-34°$. The sample consists of 203 objects: polarized emission was detected for 123, 169, and 167 sources at 8.4, 22, and 43 GHz, respectively, (at 8.4 GHz only a subset of 134 were observed) and 105 were detected at all the 3 frequencies. An accurate analysis of the statistical properties of the polarized intensity of the sample is done by [39], including an analysis of the correlations between the fractional polarization and the spectral indices. Here below, we summarize the main results obtained by them.

(i) The distribution of the fractional polarization varies slightly as a function of the frequency. Including sources undetected in polarization, the mean fractional polarization is $\langle \Pi \rangle = 2.9\ 3.0$ and 3.5 percent at 8.4, 22, and 43 GHz.

(ii) No correlation is found between the fractional polarization at 22 GHz and the intensity spectral indices, α_8^{22} and α_{22}^{43} (we remind readers that the sample, selected at 22 GHz, is dominated by flat-spectrum sources).

(iii) There is a significant change in the polarization angle between 8.4 and 22 GHz. For 45 sources, the position angle satisfies the λ^2 dependence, and intrinsic rotation measures in excess of $1000\ \mathrm{rad\,m^{-2}}$ were observed for a large number of them.

(iv) Polarization of 71 sources was also measured at 86 GHz by [49]. The fractional polarization is typically higher at 86 GHz than at 43 GHz.

3.2.4. VLA Polarization Measurements in an ACT Survey Field.

In [50], VLA observations in total intensity and polarization at 4.86, 8.46, 22.46, and 43.34 GHz are presented. The sample is selected from the AT20G survey and consists of 159 ERS, out of the almost 200 sources with flux density $S_{20\,GHz} > 40$ mJy in a field of the Atacama Comsmology Telescope

(ACT) survey. Polarized emission for about 60 sources was detected at all 4 frequencies, whereas the detections are 141, 146, 89, and 59 from low to high frequencies. Fractional polarization distributions are very similar at 5 and 8.5 GHz, whereas a trend of increasing polarization with increasing frequency is indicated at 22 and 43 GHz. In particular, a tail of strongly polarized ($\gtrsim 10\%$) sources is observed at 43 GHz. Data at 22 and 43 GHz suggest also that the polarization fraction in steep-spectrum sources is significant higher than in flat-spectrum sources.

3.3. Other Samples of Polarized ERS at High Frequencies

(i) Reference [51] presented polarization observations of 250 (out of 258) southern sources in the complete 5-GHz 1-Jy sample of [52] by using the ATCA facilities at 18.5 GHz. Polarized flux densities were measured for 170 sources (114 flat-spectrum and 56 steep-spectrum), upper limits were set for an additional subset of 27 sources (12 flat-spectrum and 15 steep-spectrum), and 53 sources were rejected (probably extended objects). The final flat-spectrum sample is almost complete (80%), while only 49% of steep-spectrum sources have reliable detections. In the flat-spectrum sample the median fractional polarization is $\simeq 2.7\%$ and the mean $\simeq 2.9\%$, and no sources have $\Pi > 10\%$. The median fractional polarization for the flat-spectrum sources included in the NVSS catalogue is about a factor 2 lower than that at 18.5 GHz. However, a relevant increase in the polarization fraction at 18.5 GHz is noticed only for sources with $\Pi \lesssim 1\%$ at 1.4 GHz.

(ii) Reference [49] presented a 3.5 mm polarimetric survey of ERS using the IRAM 30 m Telescope. Their sample consists of 145 flat-spectrum sources with $\delta > -30°$ and flux density $\gtrsim 1$ Jy at 86 GHz. Linear polarization is detected for 76% of the sample (110 objects). They found that BL Lacs ($\Pi_{median} \simeq 4.4\%$) are more strongly polarized than quasars ($\Pi_{median} \simeq 3.1\%$). This result seems to be in contradiction with the idea that quasars should be more polarized at high frequencies than BL Lacs because in the latter sources the synchrotron self-absorbed spectrum is maintained up to higher frequencies. A possible explanation provided by the authors comes from the recent evidence that the view angle of jets in quasars is smaller than that in BL Lacs [53, 54]. So, if the magnetic field is not homogeneous along the jet, a lower fractional polarization level is expected from sources better oriented to the line of sight (i.e., quasars).

Moreover, for those sources with detected polarization at 15 GHz, they found that $\Pi_{86\,GHz}$ is larger than $\Pi_{15\,GHz}$ by a median factor ≈ 2, and about 20% of sources have the $\Pi_{86\,GHz}/\Pi_{15\,GHz}$ ratio larger than 4. Reference [49] suggest that this increase may be explained by a combination of two phenomena: the 86-GHz emission comes from a region with

TABLE 1: Median, mean, and $\langle \Pi^2 \rangle^{1/2}$ of the fractional polarization for flat-spectrum sources in the almost complete subsample of the AT20G surveys at different observational frequencies and flux ranges. N_{tot} refers to the total number of objects in the corresponding flux range; N_{multi} to the number of objects with 5- and 8-GHz measurements; N_{fl} to the number of flat-spectrum sources; N_{det} to the number of flat-spectrum sources with polarization detection.

S(Jy)	N_{tot}	N_{multi}	N_{fl}	N_{det}	Π_{med}	$\langle \Pi \rangle$	$\langle \Pi^2 \rangle^{1/2}$
				20 GHz			
≥ 1.0	130	114	110	85	2.05	2.82	3.72
≥ 0.5	315	287	264	188	2.01	2.76	3.84
$[0.5, 1)$	185	173	154	103	1.92	2.72	3.92
				8.6 GHz			
≥ 1.0			110	87	2.00	2.52	3.00
≥ 0.5			264	180	1.76	2.34	2.85
$[0.5, 1)$			154	93	1.54	2.21	2.73
				4.8 GHz			
≥ 1.0			110	93	1.90	2.31	2.71
≥ 0.5			264	186	1.71	2.25	2.69
$[0.5, 1)$			154	93	1.59	2.20	2.68

greater degree of order in the magnetic field or/and, at 15 GHz, emission is still affected by Faraday depolarization.

(iii) Reference [55] looked for polarized sources in the *WMAP* five-year data, using a new technique named *filtered fusion*. They detected polarization in 13 ERS at a confidence level $\geq 99\%$ and polarized flux density higher than 300 mJy.

4. Statistical Properties of the Polarized Emission in ERS

In order to provide reliable estimates of ERS contamination to CMB anisotropy polarization measurements, we need to address the following main questions about polarization properties of ERS: (1) how the fractional polarization, Π, varies from cm to mm wavelengths; (2) *if* the fractional polarization depends on the flux density; (3) how polarization properties change among the different populations of ERS. In this section we address these questions on the basis of the observational data presented and discussed in the previous Section.

4.1. Flat-Spectrum Sources. We investigate more deeply the polarization of ERS observed in the AT20G survey. We consider the almost-complete sample at $\delta < -15°$ and, when available, we use 5- and 8-GHz measurements to separate ERS into steep- and flat-spectrum sources. In Table 1 we report statistical properties of the fractional polarization (i.e., the mean $\langle \Pi \rangle$, the median Π_{med}, and $\langle \Pi^2 \rangle^{1/2}$) for flat-spectrum sources as a function of the flux density range and the frequency. They are computed using the Survival Analysis techniques and the Kaplan-Meyer estimator as implemented in the ASURV code [56], which takes into account upper limits on the fractional polarization for estimating the above quoted quantities. In fact, when polarization is not detected,

an upper limit is provided. About 10% of sources in the AT20G sample we use here have measurements only at 20 GHz (see Table 1). Although most of them should be flat-spectrum sources, for a more consistent comparison with results at 5 and 8 GHz we prefer to exclude them from the analysis.

At 20 GHz the median and the mean fractional polarization do not present any significant variation between the subsamples defined by $S \geq 1$ Jy and $0.5 \leq S < 1$ Jy. At the lower frequencies, $\langle \Pi \rangle$ and Π_{med} show a moderate decrease at fainter flux densities. In particular, two-sample tests implemented in the ASURV code (e.g., the Gehan's generalized Wilcoxon test and the Peto & Prentice generalized Wilcoxon test) yield a probability of ∼10% and ∼1% that the distributions of the fractional polarization for $S \geq 1$ Jy and $0.5 \leq S < 1$ Jy at 5 and 8 GHz, respectively are drawn from the same parent distribution (compared to a probability of 30% at 20 GHz). At flux densities lower than 500 mJy, the high number of upper limits ($\gtrsim 50\%$) makes our estimates unreliable.

On the other hand, a larger fractional polarization is observed as the frequency increases, with, on average, ≈18% of depolarization at 5 GHz with respect to 20 GHz, in agreement with results from [47]. The ASURV two-sample tests yield only a 10% probability to have the same parent distribution for Π at 5 and 20 GHz. For a better comparison, we show in Figure 2 the distributions of the fractional polarization discussed above as obtained by using the Kaplan-Meyer estimator. These distributions are well fitted by a log-normal distribution with Π_{med} and $\langle \Pi^2 \rangle$ values taken from Table 1 (see, e.g., the case at 20 GHz in Figure 2). (Figure 2 clearly shows that the log-normal distribution, when averaged in each Π bin, is giving predictions always compatible with the observed Π values well inside the 1σ level, except for very few bins at high Π levels where the statistics is very poor. We have also verified that adopting other distribution functions, for example, a truncated gaussian, we are not able to reproduce equally well the observed distributions of Π.)

In Figure 3 we also study the correlation of $\Pi_{20\,GHz}$ with $\Pi_{4.8\,GHz}$ and $S_{20\,GHz}$: no clear correlation is found between the fractional polarization and the flux density at 20 GHz (the generalized Kendall's tau test yields a probability $P \sim 60\%$ of no correlation). On the other hand, as expected, we see a strong correlation between the fractional polarization at 20 and 4.8 GHz (with a $P < 0.01\%$ of no correlation). Using the Schmitt's method from the ASURV code, we find a linear regression $\Pi_{20\,GHz} = 1.58 + 0.46\Pi_{4.8\,GHz}$. This result seems to indicate that only sources with very low fractional polarization at 5 GHz (i.e., $\Pi_{4.8\,GHz} \lesssim 2\%$) have a significant increase of Π at 20 GHz (see also Figure 3). The large offset term in the linear regression we find, however, could be also partially due to Eddington bias in the AT20G catalogue and to the large number of sources with upper limits in fractional polarization at $\Pi \gtrsim 1\%$.

In Figure 5 we plot the mean and median values of the fractional polarization as obtained from the surveys presented in the previous section. These data allow us to cover frequencies from 1.4 to 86 GHz. Although there is large scatter in the data, we see a general increasing trend for the median fractional polarization: Π_{med} is $\sim1.5\%$ at 1.4 GHz, around 2–2.5% in the range 5–10 GHz, and finally 2-3% at $\nu > 10\,GHz$. The mean fractional polarization has a more linear increase with the frequency in all the samples, with $\langle\Pi\rangle$ varying from 2–2.5% at 1.4 GHz to 3–3.5% at $\nu \geq 20\,GHz$.

Most of the data presented in Figure 5 are coming from samples of bright sources, typically with $S \gtrsim 1\,Jy$. Samples with fainter sources are the B3-VLA ($S_c = 80\,mJy$) and the sample discussed in [50]. The latter one provides polarization for sources with $S \geq 40\,mJy$. However, since at the frequencies 20 and 43 GHz the number of polarized ERS detected in the sample is less than 50 percent, we have decided to consider only flat-spectrum sources with $S \geq 80\,mJy$. Spectral indices are estimated using flux densities at 4.86 and 8.46 GHz. Table 2 reports the number of detections and the corresponding values of the mean, median, and $\langle\Pi^2\rangle^{1/2}$ of the fractional polarization. For the two highest frequencies of the sample we estimate the median in two ways: firstly, we use the ASURV code, by taking into account the upper limits; secondly, we assume a fractional polarization $\leq1\%$ for those sources without measured polarization (values indicated in brackets in Table 2). The large spread between the two values at 43 GHz is an indication that the sample, also including upper limits, could provide biased values of statistical properties of Π at these frequencies.

If we compare results from the previous two samples (i.e., B3-VLA and [50]) with surveys of bright sources, we cannot find any clear evidence of higher fractional polarization in faint sources. However, larger and deeper samples of data are clearly required to settle the question.

In Figure 5 we also include the $\langle\Pi^2\rangle^{1/2}$ values. This quantity is important in order to estimate the angular power spectra of the polarized signal due to undetected ERS. At

$\nu > 20\,GHz$, that is, the most interesting frequencies for CMB data analyses, sources have $\langle\Pi^2\rangle^{1/2} \sim 4\%$. In [49] optical identifications are provided for ERS in the [50] sample. Over a total of 145 objects, 107 are identified as quasars and 26 as BL Lacs: we find $\Pi_{med} = 3.0\%$, 3.6% and $\langle\Pi^2\rangle^{1/2} = 3.8\%$, 4.5% for FSRQs and BL Lacs, respectively.

4.2. Steep-Spectrum Sources. The number of steep-spectrum sources with polarization measurements becomes very small at frequencies $\gtrsim 10\,GHz$, preventing any study of the correlation between fractional polarization and flux density (see, e.g., Figure 4). In the subsample of AT20G at $\delta < -15°$ there are 51 steep-spectrum sources with $S \geq 0.3\,Jy$ and only 25 of them have polarization detected at 20 GHz. We find that the median fractional polarization for these sources is less than 2% at all the frequencies (see Figure 5(b)). These very small values could be biased due to the small number of detected sources and to the incompleteness of the sample at faint flux densities. Nevertheless, from Figure 4 we can see a general increase of the fractional polarization between 4.8 and 20 GHz: the ratio of Π at 20 and 4.8 GHz is typically close to one (<1.5) for sources with $\Pi_{4.8\,GHz} > 2\%$, but become higher for sources weakly polarized at 4.8 GHz. In fact, the linear regression found by the Schmitt's method implemented in the ASURV code yields $\Pi_{20\,GHz} = 0.78 + 1.14\,\Pi_{4.8\,GHz}$.

From Figure 5 we can observe the strong increase of the fractional polarization from 1.4 GHz to 5 GHz, where Faraday depolarization is probably very relevant. The increase becomes more moderate up to 20 GHz. The large difference in the median values among different samples is perhaps related to the small samples considered and the large fraction of sources with upper limits in polarization. The largest sample of steep-spectrum sources is provided by B3-VLA (77 sources, all with detected polarization) and the Ricci et al. sample (71 sources, 15 of them with upper limits): in these samples the median fractional polarization is $\sim5\%$ between 5–20 GHz and the mean varies from 5 to 6.5%.

4.3. Number Counts in Polarization of ERS at 20 GHz. We provide a first estimate of number counts in polarization, $n(P)$, at 20 GHz by exploiting WMAP and AT20G polarization source catalogues. For very bright sources, we use the polarized source sample detected by [55] in the 5 yr WMAP CMB anisotropy maps (see their Table 2). We exclude from number counts Fornax A, Virgo A, and Centaurus A because they are all local objects. Then, we consider the nearly-complete subsample of AT20G at $\delta < -15°$ and $S \geq 50\,mJy$. This sample allows us to compute $n(P)$ down to polarized fluxes of $\sim10\,mJy$. Number counts, that are reported in Table 3 (see also Figure 6), have been corrected for the estimated incompleteness of the sample (i.e., the completeness is 0.91 for $S \geq 100\,mJy$ and 0.79 for $50 \leq S < 100\,mJy$ [4]). As displayed by Figure 6, number counts in total polarization, $n(P)$, are almost flat at 20 GHz, at least in the flux density range in which they can be estimated.

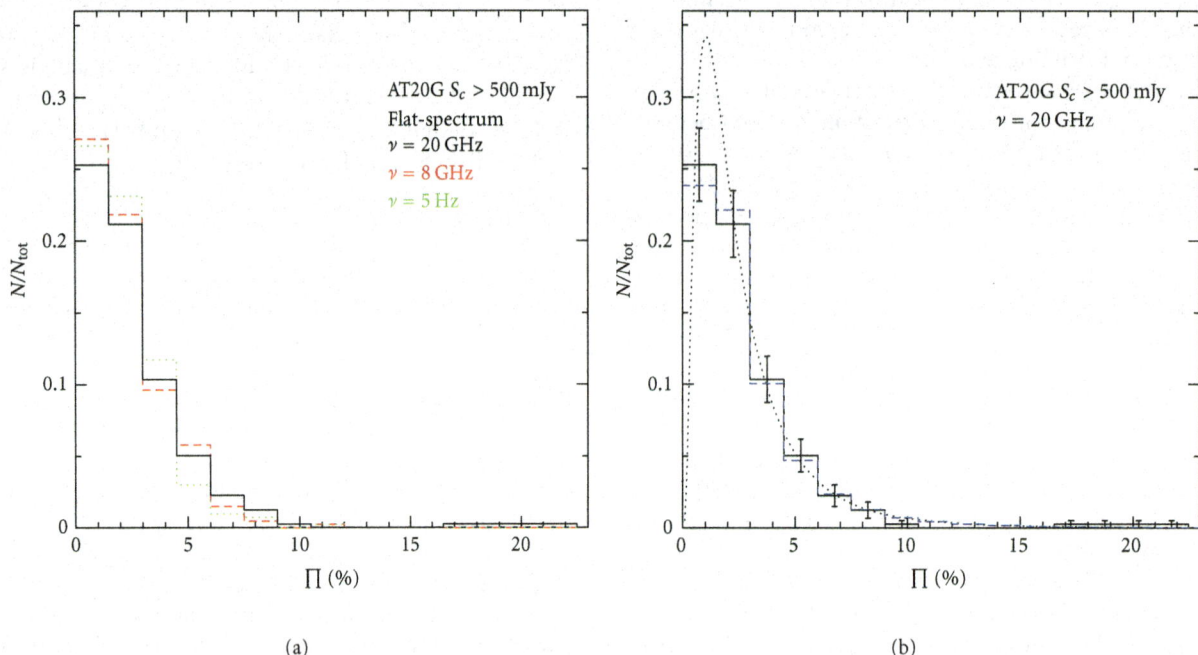

FIGURE 2: (a) Distribution of the fractional polarization at 4.8, 8.6, and 20 GHz (green dotted, red dashed, and black solid histograms, resp.) for flat-spectrum sources with $S \geq 500$ mJy in the almost-complete subsample of the AT20G survey. (b) Distribution of the fractional polarization at 20 GHz (black solid histogram) compared to the distribution (blue dashed histogram) produced by a log-normal distribution (see also Section 5.1) with $\Pi_{med} = 2.01$ and $\langle \Pi^2 \rangle = 3.84$ (black dotted line; see Table 2) at 20 GHz.

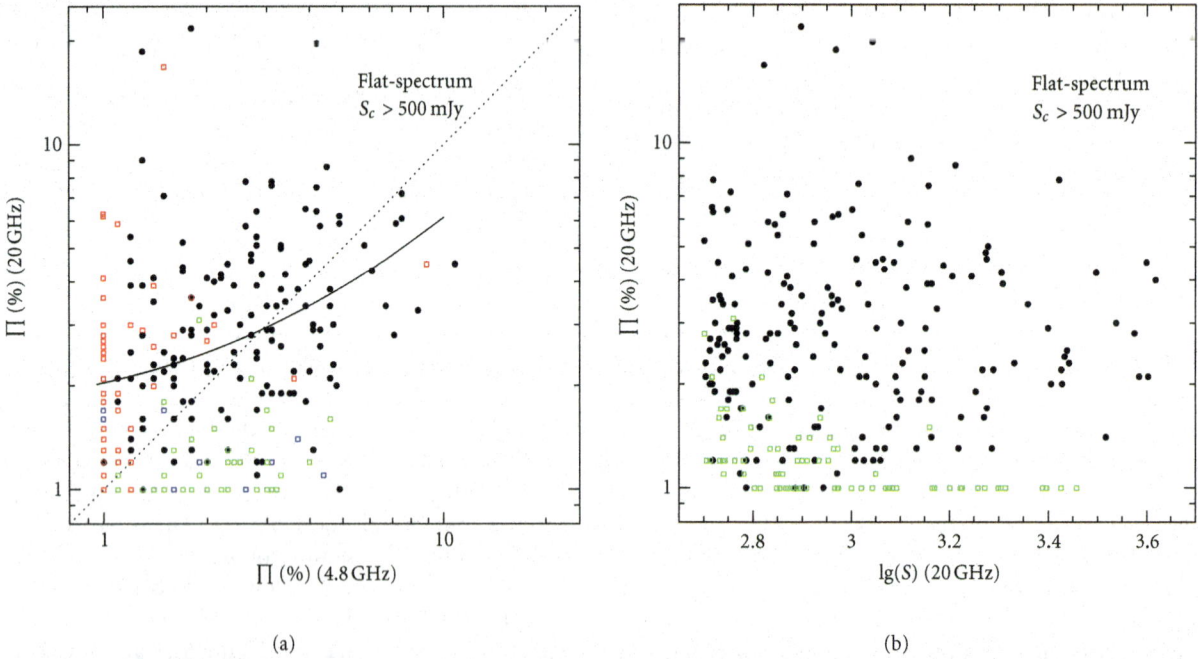

FIGURE 3: (a) Correlation between fractional polarizations at 4.8 and 20 GHz for flat-spectrum sources with $S \geq 500$ mJy. Green points indicate upper limits in Π at 20 GHz, red points at 4.8 GHz, and blue points at the both frequencies. The solid line is the linear regressions found by the ASURV code (see the text). (b) The fractional polarization at 20 GHz as a function of the flux density at the same frequency (green points indicate upper limits in Π).

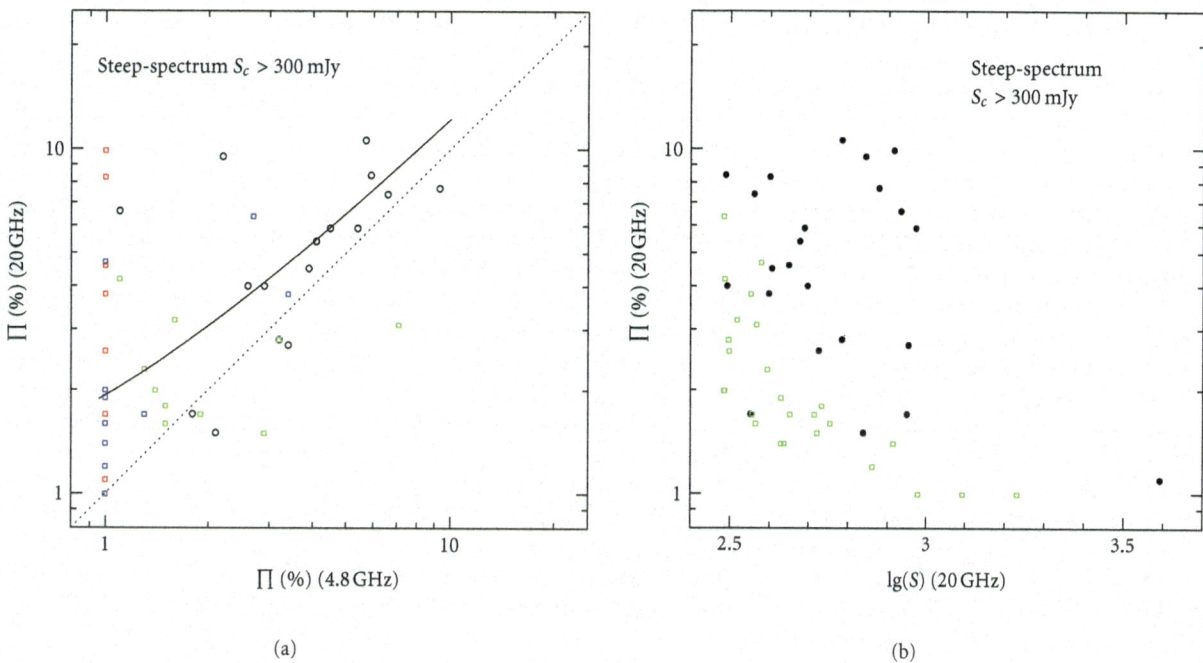

(a)   (b)

FIGURE 4: Fractional polarization for steep-spectrum sources at 20 GHz as a function of fractional polarization at 4.8 GHz (a) and of flux density at 20 GHz (b). Points of different colours have the same meaning as in Figure 3.

TABLE 2: Median and mean of the fractional polarization and $\langle \Pi^2 \rangle^{1/2}$ from the Sajina et al. [50] sample. For flat-spectrum sources we consider only objects with $S \geq 80$ mJy. Median values in brackets are computed by assigning a value of $\Pi < 1\%$ to sources without a detection in polarization.

ν[GHz]	N_{sou}	N_{det}	Π_{med}		$\langle \Pi \rangle$	$\langle \Pi^2 \rangle^{1/2}$
Flat-spectrum sources						
4.86	56	54	2.16		2.47	2.97
8.46	56	55	2.25		2.61	3.10
22.46	56	44	3.12	(2.88)	3.79	4.21
43.34	50	33	4.14	(2.64)	4.85	5.79
Steep-spectrum sources						
4.86	45	44	2.56		4.10	5.79
8.46	45	44	3.48		4.67	5.97
22.46	44	25	6.14	(3.40)	7.17	8.03
43.34	28	11	6.77		7.18	8.30

Given that an upper limit on the fractional polarization Π_{up} is always provided for the ERS in this sample (see [4], for more details), except for 5 very bright objects with no information on polarization, we also give a tentative estimate (Table 3, third column) of their contribution to the number counts $n_i^{up}(P)$ between P_i and $P_i + \Delta P$ by means of

$$n_i^{up}(P) = \sum_k \int_{\Pi_i}^{\Pi_f} \mathscr{P}(\Pi) d\Pi \quad \text{with } \Pi_i = \frac{P_i}{S_k}, \ \Pi_f = \frac{P_i + \Delta P}{S_k}.$$

(1)

The sum is done over all the sources without polarization detection and S_k is the flux density of the kth object. $\mathscr{P}(\Pi)$

is the probability function for the fractional polarization: we take a log-normal function with $\Pi_{med} = 2.0\%$ and $\langle \Pi^2 \rangle^{1/2} = 2.8\%$ (see Table 1 and Figure 3) if $\Pi \leq \Pi_{up}$ and zero otherwise. For the 5 objects without any polarization information, no upper limits are considered.

As shown by Table 3 and Figure 6, the contribution of sources undetected in polarization (displayed by empty squares) is generally negligible, except for bins at very faint polarization levels (i.e., less than 20 mJy), and for bins at $P \gtrsim 100$ mJy, due to the 5 sources without polarization information and flux density between 1 and 10 Jy. Only for the faintest bin n_i^{up} is larger than the uncertainty on the number counts. Therefore, we can conclude that number

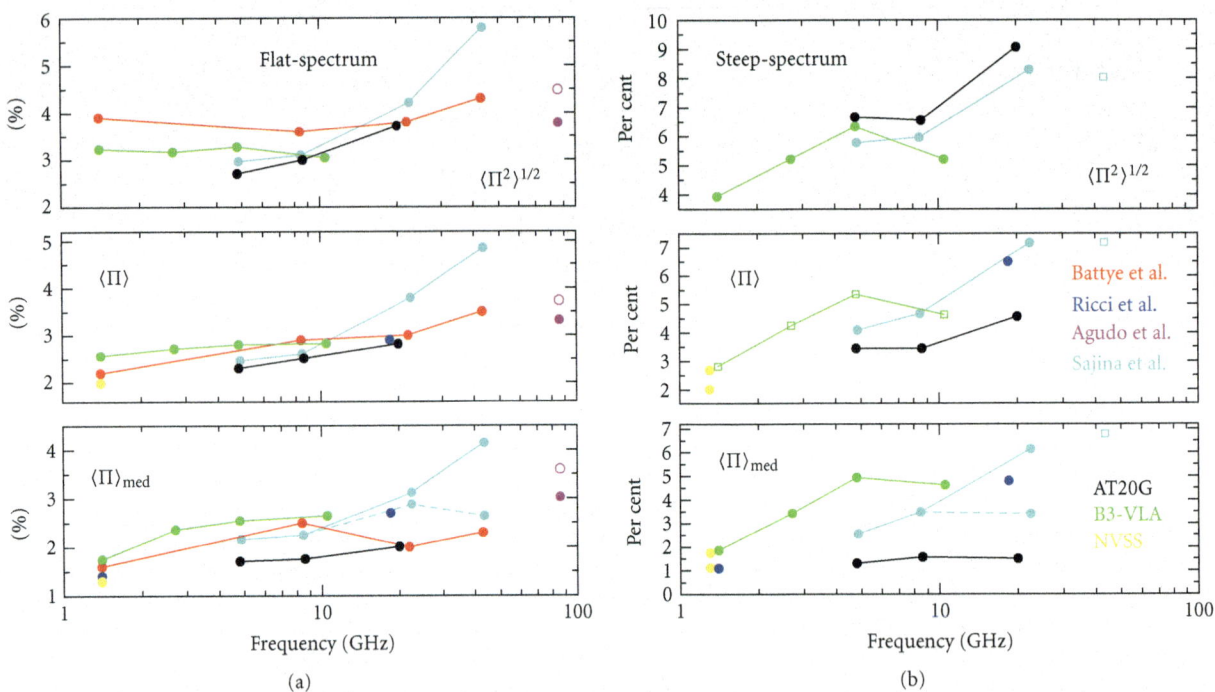

FIGURE 5: Median, mean, and $\langle \Pi^2 \rangle^{1/2}$ of fractional polarization for flat- (a) and steep-spectrum (b) sources. *Black points* refer to the almost complete subsample of the AT20G survey with $S_c = 500$ mJy for flat-spectrum sources and 300 mJy for steep-spectrum sources. *Red points* refer to data presented in [39]. *Green points* refer to the B3-VLA sample (with respect to [46], we distinguish flat- and steep-sources on the basis of their spectral index between 1.4 and 5 GHz; values shown in the plot are obtained using the ASURV code, taking into account upper limits in the fractional polarization). *Cyan points* refer to the sample from [50] (see text and Table 2; empty cyan points at 43 GHz for the steep-spectrum have to be considered as upper limits). *Blue points* refer to results from [51] and *magenta points* from [49] (solid and empty points are for FSRQs and BL Lacs, resp.). *Yellow points* refer to the NVSS survey (see [16]): for steep-spectrum sources we plot results for sources with $100 \le S < 200$ mJy and $S \ge 800$ mJy.

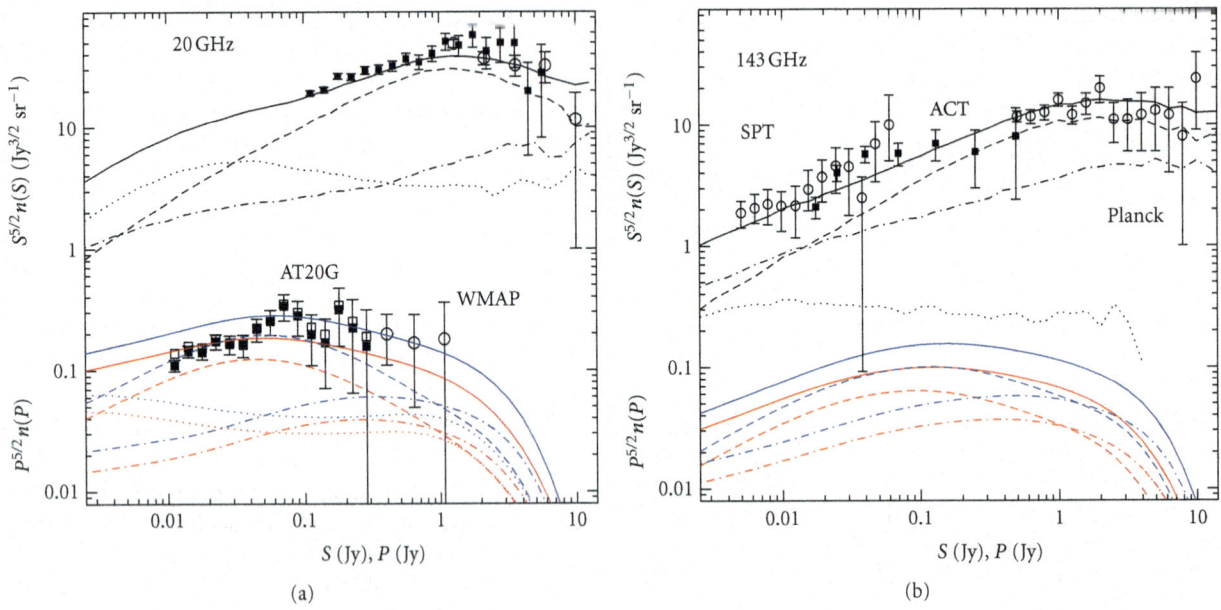

FIGURE 6: Normalized number counts at 20 (a) and at 143 GHz (b) in total intensity (upper curves and data points) and in polarized intensity (lower curves and data points; blue curves are for the more "conservative" case and red curves for the more "optimistic" case). Number counts are for the different source populations discussed in the text: solid lines represent total number counts; dotted lines are for steep-spectrum sources; long-dashed lines are for FSRQs; finally, dot-dashed lines are for BL Lacs. In (a), filled squares represent our current estimates from the AT20G subsample (empty squares include the contribution from sources without polarization detection; see Section 4.3) and empty circles are from WMAP 5-years data [55]. Data points in (b) are from published data [57–59].

counts in polarization estimated from the AT20G sample are not affected by upper limits in the polarized flux density.

5. Cosmological Evolution Models and High-Frequency Number Counts of ERS

Early evolutionary models of radio sources [17, 60–62] were able to give remarkably successful fits to the majority of data coming from surveys at $\nu \lesssim 10\,\mathrm{GHz}$, and down to flux densities of a few mJy. More recently, [18, 44] exploited the wealth of newly available data on luminosity functions, multifrequency source counts and redshift distributions to provide new cosmological evolution models of radio sources at frequencies $\gtrsim 5\,\mathrm{GHz}$ and $\lesssim 5\,\mathrm{GHz}$, respectively. These two models are based on the determination of the epoch-dependent luminosity functions for different source populations, starting from the local luminosity function and by adopting different luminosity evolution laws with free parameters fitted from observational data.

The predictions of high-frequency number counts of ERS provided by the above evolution models assume a simple power-law spectrum for ERS. Each source population is characterized by an "average," fixed, spectral index, or by two spectral indices (at most). This "classical" modeling has to be considered as a first-although successful-approximation, but it gives rise to an increasing mismatch with observed high-frequency (>30 GHz) number counts currently available.

Indeed, the ERS energy spectra can be quite different from a single power-law if analyzed in large frequency intervals. Different mechanisms can be responsible for this: (a) a spectral steepening due to the more rapid energy loss of high-energy electrons with source age, that is, "electron ageing"; (b) a transition from the optically thick to the optically thin regime at high radio frequencies; (c) at different wavelengths radio emission can be dominated by different components characterized by distinct spectral behaviors [63]. In particular, a clear steepening at mm wavelengths is theoretically expected for radio flat spectra of AGN core emission [64, 65]. This steepening has been already observed in blazars (see, e.g., [66]) and has also been statistically suggested by recent analyses of different ERS samples at $\nu > 30\,\mathrm{GHz}$ [19, 59, 67–69].

A first attempt of taking such steepening in blazar spectra into account has been done by [19]. In this work the spectral behaviour of blazars at mm wavelengths is statistically described by considering the main physical mechanisms responsible for the emission. In agreement with classical models of the synchrotron emission in the inner jets of blazars these authors interpret the high-frequency steepening observed in blazar spectra as caused, at least partially, by the transition from the optically-thick to the optically-thin regime in the observed spectra at mm wavelengths. Based on the published models of synchrotron emission from inhomogeneous, unresolved, relativistic jets [30, 65, 70], the value of the frequency ν_M at which the spectral break occurs is estimated as a function of the relevant physical parameters of AGNs: the redshift, the Doppler factor, and the linear dimension of the region (approximated as homogeneous and spherical) that is mainly responsible for the emission at the break frequency.

Recent high frequency data [57, 58], and in particular the first 1.6 *Planck* survey [59], have provided new important results on number counts and related statistics of ERS, and they can be used to constrain the different possible cases/models featuring a spectral break in the emission of AGN jets (see [19]). The most successful one (indicated in [19] as C2Ex, that is, the relevant emission is emitted from a more "extended" region in the inner jet of FSRQs) assumes different distributions of the break frequency for BL Lacs and FSRQs. According to this model, most of the FSRQs should show a bend in their otherwise flat spectra between 10 and 100 GHz, whereas in BL Lac spectral breaks should be typically observed at $\nu > 100\,\mathrm{GHz}$ (implying that the observed synchrotron radiation comes from more compact emitting regions). This dichotomy has been indeed found in the *Planck* ERCSC [66, 71]: almost all radio sources show very flat spectral indices at LFI frequencies, $\alpha_{\mathrm{LFI}} \gtrsim -0.2$; at HFI frequencies BL Lacs keep flat spectra ($\alpha_{\mathrm{HFI}} \gtrsim -0.5$) while most of FSRQs show steeper spectra, that is, $\alpha_{\mathrm{HFI}} < -0.5$. Moreover, the same model gives a remarkably good fit to all the observed data on number counts and spectral index distributions of ERS at frequencies above 5 and up to 220 GHz.

For the above reasons, hereafter we adopt the number counts provided by the "C2Ex" case in [19].

5.1. Number Counts of ERS in Polarization at mm Wavelengths. To assess the contamination due to undetected ERS in CMB polarization maps it is necessary to know how many sources can be found with polarized flux density $P = \sqrt{Q^2 + U^2}$ (typically, a debiased estimator is however used for the polarized flux density of point sources (as, for example, $P = \sqrt{Q^2 + U^2 - s^2}$, where s is the uncertainty on P; see [50] for more details)) above a given flux limit P_{lim}. Answering this question involves the determination of source counts in polarization. This estimate is quite difficult to perform directly, by the statistical analysis of observational data, since the polarized signal is weak and many samples are usually defined by their completeness in terms only of total intensity. However, this problem can be overcome by using the source number counts in total intensity complemented by information on the statistical properties of the fractional polarization Π, expressed in term of some probability function $\mathscr{P}(\Pi)$ [16].

Let us discuss this point in more detail. Polarization number counts $n(P)$ can be written as

$$n(P) = N \int_{S_0=P}^{\infty} \mathscr{P}(P, S)dS = N \int_{S_0=P}^{\infty} \mathscr{P}(\Pi, S)\frac{dS}{S}, \quad (2)$$

where N is the total number of sources with $S \geq S_0$ in the sample, $\mathscr{P}(P, S)$ and $\mathscr{P}(\Pi, S)$ are the probability functions of observing in a source of flux density S, a polarized intensity P and a fractional polarization Π, respectively. Assuming that Π is independent of S, $n(P)$ can be determined by

$$n(P) = \int_{S_0=P}^{\infty} \mathscr{P}\left(\Pi = \frac{P}{S}\right)n(S)\frac{dS}{S}. \quad (3)$$

TABLE 3: Number counts in polarization at 20 GHz from WMAP and AT20G source catalogues. N_{det} indicates the number of objects with detected polarization, while $n^{up}(P)$ is the estimated contribution of AT20G objects with no polarization information or with upper limits in the fractional polarization.

P (mJy)	N_{det}	AT20G subsample $n(P)$	$\sigma_{n(P)}$	$n^{up}(P)$
11.2	87	$0.828E + 04$	$0.889E + 03$	$0.205E + 04$
14.1	81	$0.607E + 04$	$0.675E + 03$	$0.636E + 03$
17.8	57	$0.338E + 04$	$0.447E + 03$	$0.263E + 03$
22.4	49	$0.231E + 04$	$0.329E + 03$	$0.102E + 03$
28.2	33	$0.123E + 04$	$0.215E + 03$	$0.242E + 02$
35.5	23	$0.683E + 03$	$0.142E + 03$	$0.112E + 02$
44.7	22	$0.519E + 03$	$0.111E + 03$	$0.104E + 02$
56.2	18	$0.337E + 03$	$0.795E + 02$	$0.938E + 01$
70.8	17	$0.253E + 03$	$0.613E + 02$	$0.842E + 01$
89.1	10	$0.118E + 03$	$0.374E + 02$	$0.752E + 01$
112.2	5	$0.469E + 02$	$0.210E + 02$	$0.664E + 01$
141.3	3	$0.224E + 02$	$0.129E + 02$	$0.385E + 01$
177.8	4	$0.237E + 02$	$0.118E + 02$	$0.184E + 01$
223.9	2	$0.941E + 01$	$0.665E + 01$	$0.121E + 01$
281.8	1	$0.374E + 01$	$0.370E + 01$	$0.755E + 00$
		5 yr WMAP sample		
398.1	5	$0.199E + 01$	$0.889E + 00$	
631.0	2	$0.530E + 00$	$0.375E + 00$	
1072.	1	$0.150E + 00$	$0.150E + 00$	

The probability function $\mathcal{P}(\Pi)$ can be constrained from the observed distributions of the fractional polarization. In agreement with [39] we model $\mathcal{P}(\Pi)$ by a log-normal distribution:

$$\mathcal{P}(\Pi) = \frac{1}{\sqrt{2\pi\sigma^2}\Pi} \exp\left\{-\frac{[\log(\Pi/\Pi_{med})]^2}{2\sigma^2}\right\}, \quad (4)$$

where Π_{med} is the median of the distribution and $\sigma^2 = 1/2\log(\langle\Pi^2\rangle/\Pi_{med}^2)$. These formulas are strictly valid only if $0 \leq \Pi < \infty$. However, because of the very low fractional polarization observed in ERS, the upper limit of $\Pi = 0.75$ can be effectively assumed as infinite. In Figure 2 we compared log-normal functions with the polarization distributions observed in the AT20G survey, confirming the very good fit of the model with data.

A critical point for our estimates is the variation with frequency of the fractional polarization observed in ERS. The data on ERS polarization discussed in this paper clearly show an higher fractional polarization at 10–20 GHz than at few GHz in both flat- and, more prominently, in steep-spectrum sources. At higher frequencies, data become scarce, but there are still indication of a possible further increase of the polarization fraction from 10–20 GHz to 40–90 GHz [39, 49]. For flat-spectrum sources, this increase could be due to the combination of two different effects: (1) the polarization degree actually increases with the frequency; and (2) BL Lacs, which are observed to be more polarized than quasars [49], become more relevant in number at higher frequencies.

In the following predictions we take into account both effects. We consider two possible values for the median and the dispersion $\langle\Pi^2\rangle^{1/2}$ of the log-normal distribution in order to provide a range of estimates for number counts and power spectra that could take into account uncertainties in observational data at $\nu > 10$ GHz. Table 4 reports our estimates for a more "optimistic" (lower) and a more "conservative" (upper) case. (Please note that the choice of the two adjectives, "optimistic" and "conservative", has been done under the point of view that polarized ERS do "contaminate" CMB maps in polarization and, thus, have to be removed from them.) The frequency dependence of the fractional polarization in ERS is simply modeled by means of two different sets of median and r.m.s. values at frequencies below and above about 40 GHz. Although somewhat arbitrary, this choice is motivated by the fact that this is the highest frequency at which multifrequency polarization samples of ERS are available.

(i) At $\nu \lesssim 40$ GHz we choose a lower and higher median and r.m.s. values of the fractional polarization based on the results displayed in Figure 5. Moreover, we require that number counts computed using values in Table 4 are (at least partially) compatible with AT20G/WMAP counts in polarization. As shown in Figure 6, the more "optimistic" case fits quite well with observational counts, especially at $P \lesssim 50$ mJy. On the other hand, the more "conservative" case tends to overestimate number counts at lower polarized fluxes, whereas it fits particularly well with

TABLE 4: Median and dispersion of the log-normal distribution for the fractional polarization, Π, as a function of frequency and for the different radio source populations relevant at CMB frequencies. Two cases are considered, a more "optimistic" one and a more "conservative" one (see Section 5.1).

| | $\nu \lesssim 40\,\text{GHz}$ | | | $\nu > 40\,\text{GHz}$ | | |
	Steep	FSRQ	BL Lac	Steep	FSRQ	BL Lac
	Lower case (more "optimistic")					
Π_{med} (%)	3.0	2.0	2.5	4.0	2.5	3.0
$\langle \Pi^2 \rangle^{1/2}$ (%)	5.0	3.0	3.5	6.0	3.5	4.2
	Upper case (more "conservative")					
Π_{med} (%)	4.0	3.0	3.6	5.0	3.6	4.3
$\langle \Pi^2 \rangle^{1/2}$ (%)	6.0	3.8	4.5	7.0	4.6	5.5

current data at high P fluxes. This is not unexpected because, in this latter case, we take as median and dispersion of the fractional polarization for blazars the values provided by [49] at 90 GHz.

(ii) At $\nu > 40\,\text{GHz}$, we make the assumption of an increase of the median fractional polarization of about 20% with respect to the corresponding cases at $\nu \lesssim 40\,\text{GHz}$. This choice is not firmly constrained by current data sets; however, we have been guided by the fractional polarization levels of ERS observed by [49] at 86 GHz which fall in the middle between our present lower and upper cases.

In Figure 6 we compare the predicted number counts, $n(P)$, of ERS in polarized intensity at 20 and 143 GHz. By comparing the two panels of the same Figure we can appreciate that $n(P)$ decreases by a factor $\simeq 1.5$ at 143 GHz at $P \sim 0.1\,\text{Jy}$, in comparison with the estimated $n(P)$ at 20 GHz. At still fainter polarized intensities, $P \sim 0.01\,\text{Jy}$, this decrement increases due to the fact that the contribution of steep-spectrum ERS, which are on average more polarized, becomes less and less relevant at high CMB frequencies. Finally, Figure 6(b) indicates a possible higher relative contribution to total polarized counts, $n(P)$, coming from BL Lacs with respect to FSRQs at 143 GHz. This is a direct consequence of the present analysis, which is based on the observations discussed by [49].

The good agreement with observational data at 20 GHz, as well as the reliability of our predictions on ERS number counts in flux density, S, gives us confidence in making extrapolations of integral number counts, $N(> P)$, in total linear polarization at *Planck* frequencies (see Table 5). It is expected that *Planck* LFI will be able to detect sources in polarization down to $\simeq 200$ mJy at 30 GHz and $\simeq 300$–400 mJy at 44 and 70 GHz, whereas *Planck* HFI, thanks to a better resolution and sensitivity, should reach polarized flux limits of $P \approx 100$ mJy. In both cases, however, a catalogue of only a few tens of compact ERS should be provided by *Planck* in polarization.

The above quoted detection limits in total polarization are estimated by the performances foreseen for the *Planck* LFI and HFI instruments and are based on prelaunch measurements [12, 13] of the detectors calibration and capabilities. These detection limits take also into account the future application to the *Planck* data—corresponding to the

end of the nominal mission, that is, in January 2012—of new detection techniques, specifically designed for detecting compact polarized sources in CMB maps (see, e.g., [72] for a recent discussion on the subject). These techniques have already been applied with success to WMAP 5 yr maps [55], improving on the results published by the WMAP team on the same data set.

6. Predictions on the Contribution of ERS to the CMB E- and B-Modes

As for temperature fluctuations, the analysis of CMB polarization measurements is usually made by the estimate of angular power spectra, that is, E- and B-modes spectra [8, 9]. B-mode polarization, that arises only from tensor perturbations, is expected to be extremely weak, and even for the most optimistic cases foreseen by inflationary models the rms signal is only a fraction of μK, less than 1 per cent of the level of temperature anisotropies at degree scales. On the other hand, polarization of foregrounds (and in particular of extragalactic sources) is equally shared between E- and B-modes [73]. ERS are therefore expected to dominate the sky B-mode polarization at subdegree angular scales at frequencies $\nu \lesssim 100$ GHz [74].

By using the statistical characterization of the fractional polarization described in the previous Sections, we are able to estimate the polarized angular power spectra given by undetected ERS in CMB anisotropy maps. First of all, we assume that ERS follow a Poisson distribution in the sky. The contribution of clustered ERS to the angular power spectrum of CMB temperature anisotropy is in fact small and can be neglected, if ERS are not subtracted down to faint flux limits (the signal due to clustered ERS becomes more relevant only at relatively low fluxes, that is, $S < 10$ mJy, [75, 76]).

It is well known that an ensemble of Poisson distributed point sources gives rise to a flat power spectrum of temperature fluctuations [77]. For a sample of sources with flux density below some cut-off S_c, the amplitude of this white noise spectrum is given by

$$C_{T\ell} = \left(\frac{dB}{dT}\right)^{-2} N\langle S^2 \rangle = \left(\frac{dB}{dT}\right)^{-2} \int_0^{S_c} n(S)S^2 dS, \quad (5)$$

where N and $n(S)$ are, respectively, the total and the differential number of sources per steradian, and dB/dT is

TABLE 5: Expected total numbers of ERS with polarized intensity $\geq P_{\mathrm{lim}}$ over the full sky at *Planck* frequencies. For each *Planck* channel, the minimum and maximum values here indicated refer to our predictions calculated by the more "optimistic" and by the more "conservative" cases previously discussed (see Table 4).

P_{lim}	ν [GHz]						
[mJY]	30	44	70	100	143	217	353
50	107–164	91–140	98–151	83–129	71–109	59–89	47–70
80	49–77	42–66	47–74	41–64	35–55	30–46	25–37
100	34–53	29–46	33–52	**28–45**	**25–39**	**21–33**	**18–27**
200	**10–16**	9–14	10–16	9–15	8–13	7–12	6–10
300	5–7	**4–7**	**5–8**	4–7	4–7	4–6	3–5
400	3-4	2–4	3–5	3-4	2–4	2–4	2-3
WMAP[1]	8	6	4				

[1]Number of polarized ERS detected in the WMAP 5 yr data at $|b| > 5°$ by [55] at 33, 41 and 61 GHz ($P_{\mathrm{lim}} \approx 300\,\mathrm{mJy}$).

the conversion factor from brightness to temperature, that is, $dB/dT \approx 10^{-2}\,\mu\mathrm{K}/(\mathrm{Jy\,sr}^{-1})(e^x - 1)^2/(x^4 e^x)$ and $x = \nu/56.8\,\mathrm{GHz}$.

As an analogy to the expression of the angular power spectrum in total intensity or CMB temperature, it is possible to define the angular power spectrum for the Stokes parameters Q and U. Because point sources contribute, on average, equally to Q, U and to the E-, B-mode power spectra, we assume $C_{E\ell} \simeq C_{B\ell} \simeq C_{Q\ell} \simeq C_{U\ell}$ (we generally refer to them as polarization spectra). Following the treatment given by [16], we have that

$$
\begin{aligned}
C_{Q\ell} &= \left(\frac{dB}{dT}\right)^{-2} N\langle Q^2\rangle \\
&= \left(\frac{dB}{dT}\right)^{-2} N\langle S^2\Pi^2\cos^2(2\phi)\rangle \\
&= \left(\frac{dB}{dT}\right)^{-2} N\langle S^2\rangle\langle\Pi^2\rangle\langle\cos^2(2\phi)\rangle \\
&= \frac{1}{2}\left(\frac{dB}{dT}\right)^{-2}\langle\Pi^2\rangle C_{T\ell},
\end{aligned}
\tag{6}
$$

where the Stokes parameter Q is written in terms of Π and of the polarization angle in the chosen reference system, ϕ. The factor 1/2 arises because of the uniform distribution of polarization angles. It is easy to demonstrate that the cross-correlation temperature-polarization spectra are null, for example, $C_{TQ\ell} = \langle S^2\Pi\cos(2\phi)\rangle = 0$.

In (3) and (6) we have assumed that the fractional polarization is independent of the total intensity of the source. Observations for flat-spectrum sources in the flux density range $S \gtrsim 100$ mJy seem to support this assumption (see Section 4.1). We expect that this is maintained also at fainter fluxes but only if FSRQs and BL Lacs are separated into two different populations. On the other hand, this hypothesis may not be true for steep-spectrum sources: a clear anticorrelation between Π and S is observed from data at 1.4 GHz, whereas, at higher frequencies, the lack of large samples of steep-spectrum sources does not allow to determine it. In any case, steep-spectrum ERS are giving a negligible contribution to number counts at $\nu \geq 100$ GHz

and, thus, this lack of information does not affect our current predictions.

In Figure 7 we present the results on ERS polarization power spectra for the six *Planck* frequencies where ERS are relevant. The value of $\langle\Pi^2\rangle^{1/2}$ is taken according to Table 4 for the different radio source populations. Moreover, we consider two cutoffs in flux density: $S_c = 1$, and 0.1 Jy. The former value is close to the completeness limit obtained by the *Planck* ERSCS for the frequency channels $\nu \leq 100$ GHz [59]; the latter one can be seen as a (somewhat optimistic) reference value for the *Planck* high frequency channels, or a reference value for future experiments.

In Figure 7 we also plot the CMB power spectrum for the E-mode and for the B-mode with a tensor-to-scalar ratio $r = 0.1, 0.01$ and 0.001. In this way we can have an indication of the level of ERS subtraction required to allow the detection of a gravitational wave induced primordial CMB B-mode signal. ERS should not be a strong constraint on detecting E-mode polarization or the B-modes with $r \gtrsim 0.01$. On the other hand, a primordial CMB B-mode signal corresponding to lower r values requires the subtraction of ERS down to flux detection limits of ~ 100 mJy, which will not be easy, or even possible, with the *Planck* sensitivity.

7. Conclusions

In this contribution we have reviewed recently available polarimetric surveys of extragalactic radio sources at frequencies $\nu \gtrsim 1$ GHz. These data point out that the typical intrinsic fractional polarization of ERS is around 2–5% of the total flux density, S, of the source even at frequencies as high as 20 GHz, and that in very few objects the fractional polarization is $\Pi \gtrsim 10\%$. This may be due to the low degree of uniformity of magnetic fields in the internal part of AGN jets and in lobes. Faraday depolarization is probably the cause of the large number of sources with a very low level of polarization, that is, $\Pi \lesssim 1\%$, at GHz frequencies, which also explains the strong increase of the fractional polarization observed at $\nu \gtrsim 10$ GHz in those objects. This conclusion is supported by high or extreme values of rotation measure RM $\gg 1000\,\mathrm{rad\,m}^{-2}$ observed in some blazars (e.g., [37, 39]).

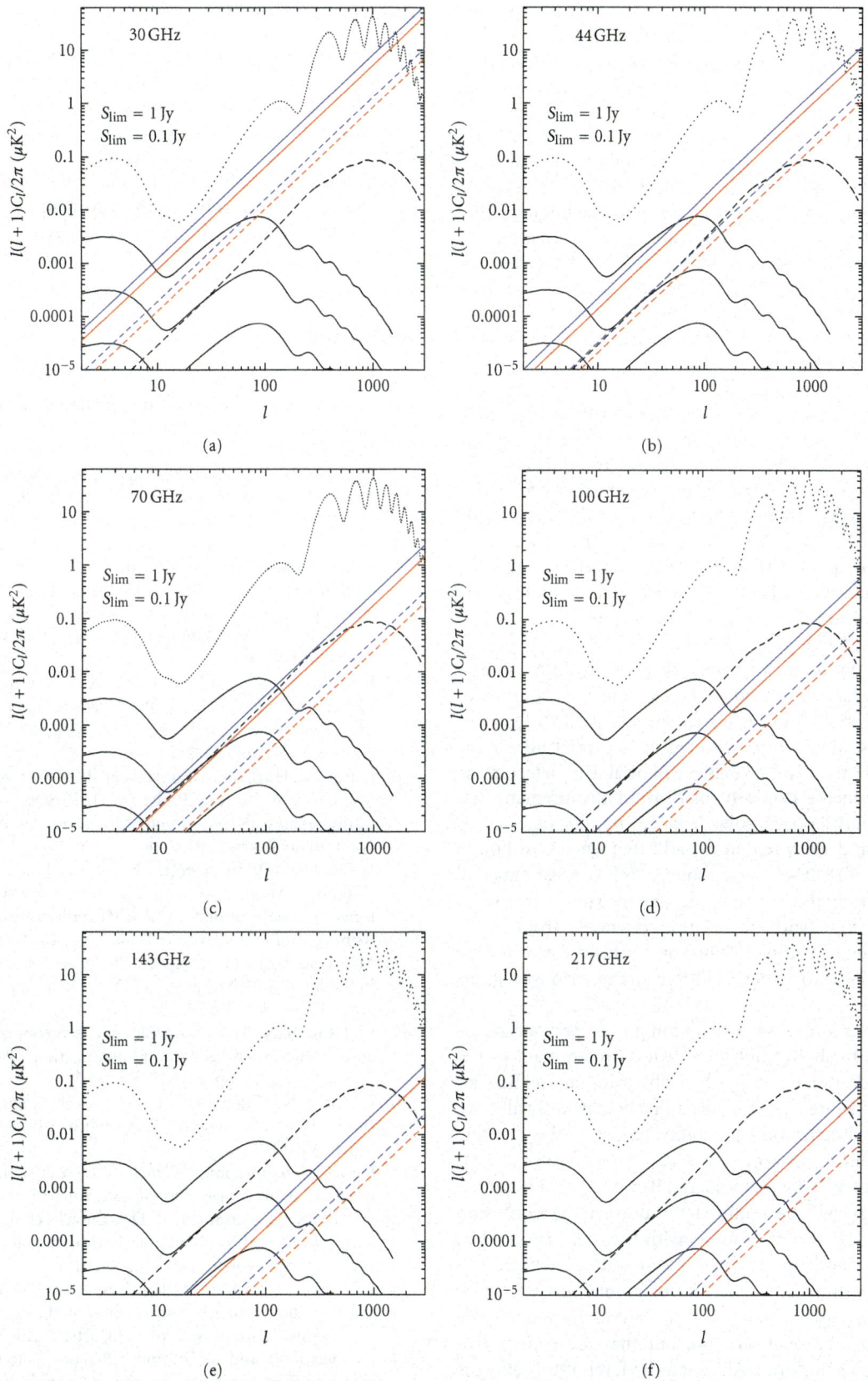

FIGURE 7: Polarization power spectra at six *Planck* frequencies for the CMB radiation (black lines: dotted lines for the E-mode; solid lines for the gravitational wave B-mode with $r = 0.1$, 0.01 and 0.001; dashed lines for the lensing-induced B-mode) and for ERS (solid lines are for $S_c = 1$ Jy and dashed lines for $S_c = 0.1$ Jy; blue lines correspond to the more "conservative" case and red lines to the more "optimistic" case).

Moreover, we have studied how polarization properties of ERS change from cm to mm wavelengths. For flat-spectrum sources a weak but constant increase of fractional polarization is observed, with median (mean) values varying from 1.5% (2–2.5%) at 1.4 GHz, to 2–2.5% (2.5–3%) at 5–10 GHz and 2-3% (3–3.5%) at 10–40 GHz. Indications that fractional polarization in blazars could further increase above 40 GHz come from the recent works by [39, 49]. On the other hand, a significantly higher fractional polarization is typically found in steep-spectrum sources, especially at high frequencies (median is 4-5% at 10–20 GHz and mean between 5 and 6.5%). However, because of incompleteness of the samples and of the small number of steep-spectrum sources in surveys at $\nu \gtrsim 10$ GHz, current observations could be biased by high-polarized objects.

In general, we do not find any dependence of the fractional polarization of ERS with the flux density at high radio frequencies. However, more conclusive evidences require larger and deeper surveys. Nevertheless, an anticorrelation between Π and S in blazars is expected when very faint flux densities are considered (see Section 4). For the flux-density ranges covered by available surveys, flat-spectrum sources are dominated by quasars (FSRQs). These objects are typically less polarized than BL Lacs [49], which become increasingly relevant at fainter flux densities and become the dominant population at $S \lesssim 10$ mJy [18, 78].

We also discuss a formalism to estimate ERS number counts in polarization and to predict the contribution of unresolved ERS to angular power spectra at CMB frequencies. As a first application, we attempt to predict how many polarized ERS the *Planck* Satellite will be able to detect in the different channels sensible to polarization measurements: we expect that only a dozen polarized ERS could be detected by *Planck* LFI, and a few tens at the HFI frequencies. Although the number of *Planck* detected sources is low, these data will allow us to study the frequency dependence of the fractional polarization in a wide range of frequencies, from 30 to 353 GHz, thus providing original and valuable information on polarization properties of ERS in the innermost regions of AGN jets.

Finally, our results on polarization power spectra demonstrate that ERS should not be a strong contaminant to the CMB E-mode polarization when observing at frequencies $\nu \gtrsim 70$ GHz. Moreover, it seems unlikely that ERS will have a significant impact on our ability to detect the B-mode polarization from primordial graviational waves if $r > 0.01$. On the contrary, if the cosmological B-mode signal is fainter, some strategy will be required to subtract the confusion noise produced by radio sources with flux-density $S < 1$ Jy. At sub-mm wavelengths, where radio sources become less and less relevant, confusion noise of point sources may be dominated by dusty galaxies (e.g., [79]). Polarization of these objects should be relatively low, but they are expected to be significantly clustered. Although the level of this effect is still quite uncertain, polarization spectra from dusty galaxies could begin to dominate the one produced by ERS already at $\nu \gtrsim 200$–300 GHz, at least at angular scales relevant for the detection of the B-modes.

Acknowledgments

The authors thank the referee, R. B. Partridge, for his insightful comments and criticisms that helped a lot in clarifying the main assumptions of the paper and also in improving its final presentation. M. Tucci acknowledges financial support from the French "Centre National d' Études Spatiales" (CNES). L. Toffolatti acknowledges partial financial support from the Spanish Ministry of Science and Innovation (MICINN), under project AYA2010-21766-C03-01.

References

[1] V. L. Ginzburg and S. I. Syrovatski, "Cosmic Magneto-bremsstrahlung (synchrotron Radiation)," *Annual Review of Astronomy & Astrophysics*, vol. 3, pp. 297–350, 1965.

[2] V. L. Ginzburg and S. I. Syrovatski, "Developments in the theory of synchrotron radiation and its reabsorption," *Annual Review of Astronomy & Astrophysics*, vol. 7, pp. 375–420, 1969.

[3] A. R. Taylor, J. M. Stil, J. K. Grant et al., "Radio polarimetry of the ELAIS N1 field: polarized compact sources," *The Astrophysical Journal*, vol. 666, p. 201, 2007.

[4] T. Murphy, E. M. Sadler, R. D. Ekers et al., "The Australia Telescope 20 GHz survey: the source catalogue," *Monthly Notices of the Royal Astronomical Society*, vol. 402, no. 4, pp. 2403–2423, 2010.

[5] J. M. Kovac, E. M. Leitch, C. Pryke, J. E. Carlstrom, N. W. Halverson, and W. L. Holzapfel, "Detection of polarization in the cosmic microwave background using DASI," *Nature*, vol. 420, no. 6917, pp. 772–787, 2002.

[6] L. Page, G. Hinshaw, E. Komatsu et al., "Three-year Wilkinson Microwave Anisotropy Probe (WMAP) observations: polarization analysis," *The Astrophysical Journal Supplement Series*, vol. 170, no. 2, article 335, 2007.

[7] J. Dunkley, D. N. Spergel, E. Komatsu et al., "Five-year Wilkinson Microwave Anisotropy Probe (WMAP) observations: bayesian estimation of CMB polarization maps," *The Astrophysical Journal*, vol. 701, no. 2, pp. 1804–1813, 2009.

[8] M. Zaldarriaga and U. Seljak, "All-sky analysis of polarization in the microwave background," *Physical Review D*, vol. 55, no. 4, pp. 1830–1840, 1997.

[9] M. Kamionkowski, A. Kosowsky, and A. Stebbins, "Statistics of cosmic microwave background polarization," *Physical Review D*, vol. 55, no. 12, pp. 7368–7388, 1997.

[10] J. Tauber, N. Mandolesi, J.-L. Puget et al., "*Planck* pre-launch status: the *Planck* mission," *Astronomy & Astrophysics*, vol. 520, article A1, 2010.

[11] Planck Collaboration, "*Planck* early results. I. The *Planck* mission," *Astronomy & Astrophysics*, vol. 536, article A1, 2011.

[12] J. P. Leahy, M. Bersanelli, O. D'Arcangelo et al., "*Planck* pre-launch status: expected LFI polarisation capability," *Astronomy & Astrophysics*, vol. 520, article A8, 2010.

[13] C. Rosset, M. Tristram, N. Ponthieu et al., "*Planck* pre-launch status: high frequency instrument polarization calibration," *Astronomy & Astrophysics*, vol. 520, article A13, 2010.

[14] G. Efstathiou and S. Gratton, "B-mode detection with an extended Planck mission," *Journal of Cosmology and Astroparticle Physics*, vol. 6, p. 11, 2009.

[15] The COrE Collaboration, C. Armitage-Caplan et al., "Cosmic origin explorer," a proposal submitted to ESA for the 2015–2025 Horizon Programme, 2011.

[16] M. Tucci, E. Martinez–Gonzalez, L. Toffolatti, J. Gonzalez-Nuevo, and G. De Zotti, "Predictions on the high-frequency polarization properties of extragalactic radio sources and implications for polarization measurements of the cosmic microwave background," *Monthly Notices of the Royal Astronomical Society*, vol. 349, no. 4, pp. 1267–1277, 2004.

[17] L. Toffolatti, F. Argueso, G. De Zotti et al., "Extragalactic source counts and contributions to the anisotropies of the cosmic microwave background: predictions for the *Planck* surveyor mission," *Monthly Notices of the Royal Astronomical Society*, vol. 297, no. 1, pp. 117–127, 1998.

[18] G. De Zotti, R. Ricci, D. Mesa et al., "Predictions for high-frequency radio surveys of extragalactic sources," *Astronomy & Astrophysics*, vol. 431, no. 3, pp. 893–903, 2005.

[19] M. Tucci, L. Toffolatti, G. De Zotti, and E. Martinez-Gonzalez, "High-frequency predictions for number counts and spectral properties of extragalactic radio sources. New evidence of a break at mm wavelengths in spectra of bright blazar sources," *Astronomy & Astrophysics*, vol. 533, article A57, 2011.

[20] D. Mesa, C. Baccigalupi, G. De Zotti et al., "Polarization properties of extragalactic radio sources and their contribution to microwave polarization fluctuations," *Astronomy & Astrophysics*, vol. 396, no. 2, pp. 463–471, 2002.

[21] E. M. Sadler, R. Ricci, R. D. Ekers et al., "The properties of extragalactic radio sources selected at 20 GHz," *Monthly Notices of the Royal Astronomical Society*, vol. 371, no. 2, pp. 898–914, 2006.

[22] R. Subrahmanyan, R. D. Ekers, L. Saripalli, and E. M. Sadler, "ATLBS: the Australia telescope low-brightness survey," *Monthly Notices of the Royal Astronomical Society*, vol. 402, no. 4, pp. 2792–2806, 2010.

[23] J. K. Banfield, S. J. George, A. R. Taylor et al., "Polarized radio sources: a study of luminosity, redshift and infrared colors," *The Astrophysical Journal*, vol. 733, no. 1, article 69, 2011.

[24] H. Shi, H. Liang, J. L. Han, and R. W. Hunsted, "Radio source with ulttrahigh polarization," *Monthly Notices of the Royal Astronomical Society*, vol. 409, no. 2, pp. 821–838, 2010.

[25] M. Ruszkowski and M. C. Begelman, "Circular polarization from stochastic synchrotron sources," *Astrophysical Journal Letters*, vol. 573, no. 2 I, pp. 485–495, 2002.

[26] F. F. Gardner and J. B. Whiteoak, "The polarization of cosmic radio waves," *Annual Review of Astronomy & Astrophysics*, vol. 4, pp. 245–292, 1966.

[27] J. A. Hogbom, "A study of the radio galaxies 3C111, 192, 219, 223, 315 and 452," *Astronomy & Astrophysics Supplement Series*, vol. 36, pp. 173–192, 1979.

[28] R. A. Laing, "A model for the magnetic-field structure in extended radio sources," *Monthly Notices of the Royal Astronomical Society*, vol. 193, pp. 439–449, 1980.

[29] R. A. Laing, "Magnetic fields in extragalactic radio sources," *The Astrophysical Journal*, vol. 248, pp. 87–104, 1981.

[30] A. P. Marscher and W. K. Gear, "Models for high-frequency radio outbursts in extragalactic sources, with application to the early 1983 millimeter-to-infrared flare of 3C 273," *The Astrophysical Journal*, vol. 298, pp. 114–127, 1985.

[31] J. R. P. Angel and H. S. Stockman, "Optical and infrared polarization of active extragalactic objects," *Annual Review of Astronomy & Astrophysics*, vol. 18, pp. 321–361, 1980.

[32] T. Beckert and H. Falcke, "Circular polarization of radio emission from relativistic jets," *Astronomy & Astrophysics*, vol. 388, no. 3, pp. 1106–1119, 2002.

[33] D. P. Rayner, R. P. Norris, and R. J. Sault, "Radio circular polarization of active galaxies," *Monthly Notices of the Royal Astronomical Society*, vol. 319, no. 2, pp. 484–496, 2000.

[34] R. G. Strom, "Faraday depolarization of radio galaxies and quasars with simple spectra," *Astronomy & Astrophysics*, vol. 25, article 303, 1973.

[35] C. Burigana, L. La Porta, W. Reich et al., "Polarized synchrotron emission," in *Proceedings of the CMB and Physics of Early Universe*, pp. 20–22, Ischia, Italy, April 2006.

[36] B. F. C. Cooper and R. M. Price, "Faraday rotation effects associated with the radio source Centaurus A," *Nature*, vol. 195, no. 4846, pp. 1084–1085, 1962.

[37] R. T. Zavala and G. B. Taylor, "A View through Faraday's fog: parsec scale rotation measures in active galactic nuclei," *The Astrophysical Journal*, vol. 589, no. 1, article 126, 2003.

[38] R. T. Zavala and G. B. Taylor, "A view through Faraday's Fog 2: parsec scale rotation measures in 40 AGN," *The Astrophysical Journal*, vol. 612, no. 2, pp. 749–779, 2004.

[39] R. A. Battye, I. W. A. Browne, M. W. Peel, N. J. Jackson, and C. Dickinson, "Statistical properties of polarized radio sources at high frequency and their impact on cosmic microwave background polarization measurements," *Monthly Notices of the Royal Astronomical Society*, vol. 413, no. 1, pp. 132–148, 2011.

[40] A. R. Taylor, J. M. Stil, and C. Sunstrum, "A rotation measure image of the sky," *Astrophysical Journal Letters*, vol. 702, no. 2, pp. 1230–1236, 2009.

[41] J. J. Condon, W. D. Cotton, E. W. Greisen et al., "The NRAO VLA sky survey," *Astronomical Journal*, vol. 115, no. 5, pp. 1693–1716, 1998.

[42] P. C. Gregory, W. K. Scott, K. Douglas, and J. J. Condon, "The GB6 catalog of radio sources," *Astrophysical Journal Supplement*, vol. 103, pp. 427–432, 1996.

[43] J. K. Grant, A. R. Taylor, J. M. Stil et al., "The drao *Planck* deep fields: the polarization properties of radio galaxies at 1.4 GHz," *Astrophysical Journal Letters*, vol. 714, no. 2, pp. 1689–1701, 2010.

[44] M. Massardi, A. Bonaldi, M. Negrello, S. Ricciardi, A. Raccanelli, and G. de Zotti, "A model for the cosmological evolution of low-frequency radio sources," *Monthly Notices of the Royal Astronomical Society*, vol. 404, no. 1, pp. 532–544, 2010.

[45] M. Vigotti, G. Grueff, R. Perley, B. G. Clark, and A. J. Bridle, "Structures, spectral indexes, and optical identifications of radio sources selected from the B3 catalogue," *Astronomical Journal*, vol. 98, pp. 419–499, 1989.

[46] U. Klein, K. H. Mack, L. Gregorini, and M. Vigotti, "Multifrequency study of the B3-VLA sample. III. Polarisation properties," *Astronomy & Astrophysics*, vol. 406, no. 2, pp. 579–592, 2003.

[47] M. Massardi, R. D. Ekers, T. Murphy et al., "The Australia Telescope 20 GHz (AT20G) survey: analysis of the extragalactic source sample," *Monthly Notices of the Royal Astronomical Society*, vol. 412, no. 1, pp. 318–330, 2011.

[48] N. Jackson, W. A. Browne, R. A. Battye, D. Gabudza, and A. C. Taylor, "High-frequency radio polarization measurements of *WMAP* point sources," *Monthly Notices of the Royal Astronomical Society*, vol. 401, no. 2, pp. 1388–1389, 2010.

[49] I. Agudo, C. Thum, H. Wiesemeyer, and T. P. Krichbaum, "A 3.5 mm polarimetric survey of radio-loud active galactic nuclei," *Astrophysical Journal, Supplement Series*, vol. 189, no. 1, pp. 1–14, 2010.

[50] A. Sajina, B. Partridge, T. Evans et al., "High-frequency radio spectral energy distributions and polarization fractions of sources in Atacama Cosmology Telescope survey field," *The Astrophysical Journal*, vol. 732, no. 1, article 45, 2011.

[51] R. Ricci, I. Prandoni, C. Gruppioni, R. J. Sault, and G. De Zotti, "High-frequency polarization properties of southern Kühr sources," *Astronomy & Astrophysics*, vol. 415, no. 2, pp. 549–558, 2004.

[52] H. Kuhr, A. Witzel, I. I. K. Pauliny-Toth, and U. Nauber, "A catalogue of extragalactic radio sources having flux densities greater than 1 Jy at 5 GHz," *Astronomy & Astrophysics*, vol. 45, pp. 367–430, 1981.

[53] T. Hovatta, E. Valtaoja, M. Tornikoski, and A. Lähteenmäki, "Doppler factors, Lorentz factors and viewing angles for quasars, BL Lacertae objects and radio galaxies," *Astronomy & Astrophysics*, vol. 494, no. 2, pp. 527–537, 2009.

[54] A. B. Pushkarev, Y. Y. Kovalev, M. L. Lister, and T. Savolainen, "Jet opening angles and gamma-ray brightness of AGN," *Astronomy & Astrophysics*, vol. 507, no. 2, pp. L33–L36, 2009.

[55] M. Lopez-Caniego, M. Massardi, J. Gonzalez–Nuevo et al., "Polarization of the *WMAP* point sources," *The Astrophysical Journal*, vol. 705, no. 1, artkicle 868, 2009.

[56] M. Lavalley, T. Isobe, and E. Feigelson, "ASURV: astronomy survival analysis package," in *Proceedings of the Astronomical Data Analysis Software and Systems I, A.S.P. Conference Series*, D. M. Worrall, C. Biemesderfer, and J. Barnes, Eds., vol. 25, p. 245, 1992.

[57] J. D. Vieira, T. M. Crawford, E. R. Switzer et al., "Extragalactic millimeter-wave sources in South Pole Telescope survey data: source counts, catalog, and statistics for an 87 square-degree field," *The Astrophysical Journal*, vol. 719, no. 1, article 763, 2010.

[58] T. A. Marriage, J. B. Juin, Y.-T. Lin et al., "The Atacama Cosmology telescope: extragalactic sources at 148 GHz in the 2008 survey," *The Astrophysical Journal*, vol. 731, no. 2, article 100, 2011.

[59] Planck Collaboration, "*Planck* Early results. XIII. Statistical properties of extragalactic radio sources in the *Planck* early release compact source catalogue," *Astronomy & Astrophysics*, vol. 536, article A13, 2011.

[60] L. Danese, A. Franceschini, L. Toffolatti, and G. De Zotti, "Interpretation of deep counts of radio sources," *The Astrophysical Journal*, vol. 318, pp. L15–L20, 1987.

[61] J. S. Dunlop and J. A. Peacock, "The redshift cut-off in the luminosity function of radio galaxies and quasars," *Monthly Notices of the Royal Astronomical Society*, vol. 247, no. 1, article 19, 1990.

[62] C. A. Jackson and J. V. Wall, "Extragalactic radio-source evolution under the dual-population unification scheme," *Monthly Notices of the Royal Astronomical Society*, vol. 304, no. 1, pp. 160–174, 1999.

[63] M. Tucci, J. A. Rubino-Martin, R. Rebolo et al., "Multifrequency spectral analysis of extragalactic radio sources in the 33-GHz VSA catalogue: sources with flattening and upturn spectrum," *Monthly Notices of the Royal Astronomical Society*, vol. 386, no. 3, pp. 1729–1738, 2008.

[64] K. I. Kellermann, "On the interpretation of radio-source spectra and the evolution of radio galaxies and quasi-stellar sources," *The Astrophysical Journal*, vol. 146, article 621, 1966.

[65] R. D. Blandford and A. Konigl, "Relativistic jets as compact radio sources," *The Astrophysical Journal*, vol. 232, pp. 34–48, 1979.

[66] Planck Collaboration, "*Planck* early results. XV. Spectral energy distributions and radio continuum spectra of northern extragalactic radio sources," *Astronomy & Astrophysics*, vol. 536, article A15, 2011.

[67] E. M. Waldram, R. C. Bolton, G. G. Pooley, and J. M. Riley, "Some estimates of the source counts at *Planck* Surveyor frequencies, using the 9C survey at 15 GHz," *Monthly Notices of the Royal Astronomical Society*, vol. 379, no. 4, pp. 1442–1452, 2007.

[68] J. Gonzalez-Nuevo, M. Massardi, F. Argueso et al., "Statistical properties of extragalactic sources in the New Extragalactic *WMAP* Point Source (NEWPS) catalogue," *Monthly Notices of the Royal Astronomical Society*, vol. 384, no. 2, pp. 711–718, 2008.

[69] M. Massardi, A. Bonaldi, L. Bonavera et al., "The *Planck*-ATCA Co-eval observations project: the bright sample," *Monthly Notices of the Royal Astronomical Society*, vol. 415, no. 2, pp. 1597–1610, 2011.

[70] A. Konigl, "Relativistic jets as X-ray and gamma-ray sources," *The Astrophysical Journal*, vol. 243, pp. 700–709, 1981.

[71] P. Giommi, G. Polenta, A. Lahteenmaki et al., "Simultaneous *Planck*, Swift, and Fermi observations of X-ray and gamma-ray selected blazars," *Astronomy & Astrophysics*. In press. http://arxiv.org/abs/1108.1114.

[72] D. Herranz, F. Argueso, and P. Carvalho, "Compact source detection in multi-channel microwave surveys: from SZ clusters to polarized sources," *Advances in Astronomy*. In press. Astrophysical Foregrounds in Microwave Surveys.

[73] U. Seljak, "Measuring polarization in the cosmic microwave background," *Astrophysical Journal Letters*, vol. 482, no. 1, pp. 6–16, 1997.

[74] M. Tucci, E. Martinez-Gonzalez, P. Vielva, and J. Delabrouille, "Limits on the detectability of the CMB B-mode polarization imposed by foregrounds," *Monthly Notices of the Royal Astronomical Society*, vol. 360, no. 3, pp. 935–949, 2005.

[75] J. Gonzalez-Nuevo, L. Toffolatti, and F. Argueso, "Predictions of the angular power spectrum of clustered extragalactic point sources at cosmic microwave background frequencies from flat and all-sky two-dimensional simulations," *The Astrophysical Journal*, vol. 621, no. 1, article 1, 2005.

[76] L. Toffolatti, M. Negrello, J. Gonzalez-Nuevo et al., "Extragalactic source contributions to arcminute-scale cosmic microwave background anisotropies," *Astronomy & Astrophysics*, vol. 438, no. 2, pp. 475–480, 2005.

[77] M. Tegmark and G. Efstathiou, "A method for subtracting foregrounds from multifrequency CMB sky maps," *Monthly Notices of the Royal Astronomical Society*, vol. 281, no. 4, pp. 1297–1314, 1996.

[78] P. Padovani, P. Giommi, H. Landt, and E. S. Perlman, "The deep x-ray radio blazar survey. III. radio number counts, evolutionary properties, and luminosity function of blazars," *The Astrophysical Journal*, vol. 662, no. 1 I, pp. 182–198, 2007.

[79] M. Negrello, F. Perrotta, J. Gonzalez-Nuevo et al., "Astrophysical and cosmological information from large-scale submillimetre surveys of extragalactic sources," *Monthly Notices of the Royal Astronomical Society*, vol. 377, no. 4, pp. 1557–1568, 2007.

5

CMB Map Restoration

J. Bobin,[1] J. L. Starck,[1] F. Sureau,[1] and J. Fadili[2]

[1] Laboratoire AIM, IRFU, SEDI-SAP, Service d'Astrophysique, Orme des Merisiers, Bat 709, Piece 282, 91191 Gif-Sur-Yvette, France
[2] GREYC CNRS, ENSICAEN, Université de Caen 6, Boulevard du Maréchal Juin, 14050 Caen Cedex, France

Correspondence should be addressed to J. Bobin, jbobin@cea.fr

Academic Editor: Carlo Burigana

Estimating the cosmological microwave background is of utmost importance for cosmology. However, its estimation from full-sky surveys such as WMAP or more recently Planck is challenging: CMB maps are generally estimated via the application of some source separation techniques which never prevent the final map from being contaminated with noise and foreground residuals. These spurious contaminations whether noise or foreground residuals are well known to be a plague for most cosmologically relevant tests or evaluations; this includes CMB lensing reconstruction or non-Gaussian signatures search. Noise reduction is generally performed by applying a simple Wiener filter in spherical harmonics; however, this does not account for the nonstationarity of the noise. Foreground contamination is usually tackled by masking the most intense residuals detected in the map, which makes CMB evaluation harder to perform. In this paper, we introduce a novel noise reduction framework coined LIW-Filtering for Linear Iterative Wavelet Filtering which is able to account for the noise spatial variability thanks to a wavelet-based modeling while keeping the highly desired linearity of the Wiener filter. We further show that the same filtering technique can effectively perform foreground contamination reduction thus providing a globally cleaner CMB map. Numerical results on simulated Planck data are provided.

1. Introduction

In mid-2009, European Spatial Agency (ESA) put in orbit the latest space observatory Planck to investigate the Cosmological Microwave Background (CMB). These data are of particular scientific importance as it will provide more insight into the understanding of the birth of our universe. Most cosmological parameters can be derived from the study of these CMB data. After a series of successful CMB experiments (Archeops, Boomerang, Maxima, COBE, and WMAP [1]), the Planck ESA mission is providing more accurate data which will be the reference full-sky high-resolution CMB data for the next decades. More precisely, recovering useful scientific information requires disentangling the CMB from the contribution of several astrophysical components, namely, Galactic emissions from dust and synchrotron, Sunyaev-Zel'dovich (SZ) clusters, Free-Free emissions, CO emission to only name a few, see [2]. Classically, the CMB map is estimated via the application of some source separation methods. However, the application of very sophisticated source separation methods (see [3–6]) does not prevent the

final estimated map from being contaminated with various spurious components: (i) instrumental noise and (ii) foreground residuals. Dealing with these various sources of contaminations is of paramount importance for most cosmologically relevant tests or evaluations: this includes CMB lensing reconstruction [7] or non-Gaussian signatures search. Tackling the problem of noise and foreground residual reduction is therefore important and is at the heart of this paper.

(a) Instrumental Noise. The very specific scanning pattern in full-sky CMB experiments leads to an instrumental noise closely Gaussian that exhibits significant spatial variation: Planck noise is far from being homogeneous. More precisely, the statistics of the noise vary spatially; in the field of statistics, it is said that the noise is nonstationary. As an illustration, Figure 1 displays a Planck simulated root mean square (RMS) map of instrumental noise at 217 GHz. These data originate from the publicly available simulations described in [3]. Furthermore, while noise is insignificant at large scale, it is largely predominant at small scale (e.g., typically for $\ell > 1500$ for Planck). Most source separation methods

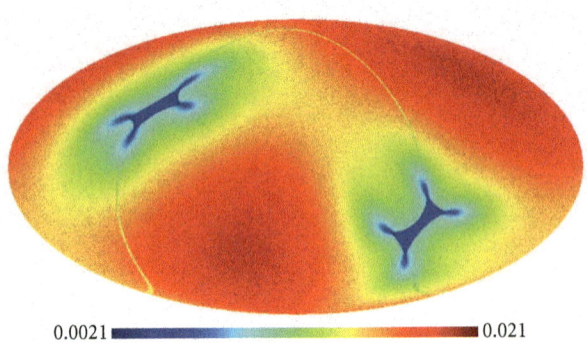

FIGURE 1: *Instrumental noise in Planck*—simulated noise root mean square map at 217 GHz in mK (Antenna temperature). These simulated Planck data are described in [3].

applied so far to CMB data are linear methods: the estimated CMB map can be expressed as a linear combination of all the pixels of the input observations. As such, the propagated instrumental noise in the CMB maps is also typically assumed nonstationary Gaussian.

Noise can be a significant limitation for most cosmological tests or reconstructions. High noise level can hamper non-Gaussianity tests as the noisy map is likely to be "more Gaussian" than the noise-free map. The classical approach in the field of cosmology generally consists in reducing the noise of the CMB via a Wiener filter in spherical harmonics. If C_ℓ stands for the theoretical power spectrum of the CMB and C_ℓ^n the power spectrum of the noise, the Wiener filter is defined as follows:

$$\forall \ell, m > 0; \quad a_{\ell m}^{\hat{x}} = \frac{C_\ell}{C_\ell + C_\ell^n} \, a_{\ell m}^{y}, \qquad (1)$$

where $\{a_{\ell m}^{\hat{x}}\}_{\ell,m}$ (resp., $\{a_{\ell m}^{y}\}_{\ell,m}$) stands for the spherical harmonics coefficients of the filtered map $\hat{x}$ (resp, the input map y). While this approach makes profit of some knowledge of the energy distribution of the noise in frequency, it does not account for its spatial variations or nonstationarity. It is very likely that such a nonstationarity may create global non-Gaussian signatures at high frequency when such variations are sharp. Strong nonstationarities are known to have a strong impact on CMB lensing reconstruction (see [8]), creating an undesired reconstruction bias.

Another less traditional but commonly encountered noise reducing technique is the local Wiener filtering. It boils down to apply the Wiener filter in the Fourier space on patches in the pixel domain. The main drawback of this method is that while large patches should be favored to capture the nonstationarity of noise, they are enable to capture correlations at scales larger than the patch size.

(b) Foreground Residual. The first evaluation of already sophisticated CMB-dedicated source separation methods has been performed in [3]. Since, novel very effective methods have been proposed Needlet ILC [4], Generalized ILC [9], SMICA [10] or GMCA [6]. However, due to the very high complexity of the data, none of these very effective methods is able to provide a CMB map that is guaranteed to be free of foreground contaminations. If the residual contamination is generally low at large scales (e.g., for $\ell < 500$ for Planck), it is generally significant at higher frequencies. It is likely that the remaining contamination mainly originates from point sources or strong features in the galactic plane. Obviously the presence of these residuals highly biases any cosmological tests to be performed on contaminated CMB map. The usual approach consists in masking the known spurious pixels of the CMB prior to any post-processing (e.g., 20 to 40% of the sky are usually masked). However masking raises several issues: (i) it limits the number of samples to which any test can be applied, limiting its statistical relevance and (ii) masking generally makes any post-processing much more complicated to perform properly (e.g., inpainting of the mask is generally needed prior to any CMB lensing reconstruction [7]). Avoiding or at least limiting the extent of the mask to be applied should greatly help improving any cosmological test to be made on the estimated CMB map. Furthermore, masking does not prevent from the presence of residuals of undetected point sources which are likely be be numerous all over the sky. Further reducing the amount of foreground residuals should clearly be helpful.

(c) Contribution of This Paper. The contribution of this paper is twofold:

(i) It introduces a novel noise reduction framework that, in opposition to the classical spherical-harmonics based Wiener filter, accounts for the nonstationarity of the noise. This new method preserves the linearity of the filtering. In fact, linearity is crucial in this field as it makes the study of error propagation via Monte-Carlo simulations much more convenient to carry out. In this framework, we make use of a simple, but efficient, wavelet-based modeling of the noise that allows for the modeling of nonstationarities at different scales. This will be discussed in Section 3.

(ii) We discuss an extension of this method to further estimate an approximate contribution of the foreground residuals per wavelet scales by means of RMS maps. Thereby contamination reduction could be tackled within the same filtering framework.

The noise/contamination reduction problem, recast as a quadratically regularized least square problem, is then effectively solved using recently introduced algorithms based on proximal calculus. Extensive numerical experiments are provided that show the effectiveness of the proposed method in reducing noise in Section 2.3 and foreground residuals in Section 3.3.

2. CMB Map Filtering with Nonstationary Noise

2.1. CMB Map Filtering. In a very general context and more precisely in the context of Planck, it is customary to assume that the input noise data, denoted by y in the following, is modeled as the linear combination of the sought after noise-free signal x and the noise term n:

$$y = x + n. \tag{2}$$

Decorrelation between x and n is also classically assumed. In what follows, we will assume that each of these signals is sampled on the sphere according to the Healpix [11] sampling grid. The sampled CMB signal x is assumed to be composed of N samples $\{x[k]\}_{k=1,...,N}$. Recovering an estimate $\hat{x}$ of x can then be formulated as an inverse problem. According to the Bayesian way of solving inverse problems, some *a priori* knowledge about the signal of interest x has to be expressed. This inference framework is particularly well suited to the CMB denoising issue as strong *prior* assumptions can be formulated. Indeed, since [12], it is well established that the cosmological microwave background can be approximated as an isotropic Gaussian random field with a high accuracy. This particularly means that it is fully characterized by its power spectrum C_ℓ.

More precisely, the CMB x can be defined by its expansion in spherical harmonics:

$$x = \sum_{\ell=0}^{\infty} \sum_{m=-\ell}^{\ell} a_{\ell m} Y_{\ell m}. \tag{3}$$

Assuming that the CMB is fully Gaussian with power spectrum C_ℓ, its spherical harmonics $\{a_{\ell m}\}_{\ell m}$ are independently distributed following a Gaussian distribution with mean 0 and variance C_ℓ:

$$a_{\ell m} \sim \mathcal{N}(0, C_\ell). \tag{4}$$

The power spectrum of the CMB is generally estimated with the pseudo-power spectrum defined as follows:

$$\hat{C}_\ell = \frac{1}{2\ell + 1} \sum_{m=-\ell}^{\ell} a_{\ell m}^\star a_{\ell m}, \tag{5}$$

where $a_{\ell m}^\star$ stands for the conjugate of $a_{\ell m}$. The *a priori* probability distribution $\mathcal{P}$ of the CMB map x is a normal distribution with covariance matrix $\mathcal{F}^H \mathbf{C} \mathcal{F}$ where $\mathcal{F}$ stands for the spherical harmonics basis and $\mathcal{F}^H$ its adjoint. The matrix $\mathbf{C}$ is diagonal with $\mathbf{C} = \mathrm{diag}(C_\ell)$. This yields the following CMB *prior* probability distribution:

$$\mathcal{P}(x) \propto \exp\left\{-x^T \mathcal{F}^H \mathbf{C}^{-1} \mathcal{F} x\right\}. \tag{6}$$

In this section, we assume that the instrumental noise can be modeled as a nonstationary but uncorrelated Gaussian field in the pixel domain such that

$$\forall k = 1,\ldots,N, \quad n[k] \sim \mathcal{N}\left(0, \sigma_k^2\right). \tag{7}$$

Written differently, n is a realization of a multivalued Gaussian variable with mean 0 and covariance matrix $\Sigma = \mathrm{diag}(\sigma_1^2 \cdots \sigma_N^2)$. The case of correlated noise will be discussed later on in this section. From the Gaussian

modeling of n, the likelihood function is trivially obtained as follows:

$$\mathcal{L}(y \mid x, \Sigma) \propto \exp\left\{-(y - x)^T \Sigma^{-1} (y - x)\right\}. \tag{8}$$

Following Bayes rule, the posterior distribution for x is the following:

$$P(x \mid y) \propto \mathcal{L}(y \mid x, \Sigma)\mathcal{P}(x) \tag{9}$$

$$\propto \exp -\left[x^T \mathcal{F}^H \mathbf{C}^{-1} \mathcal{F} x + (y - x)^T \Sigma^{-1}(y - x)\right]. \tag{10}$$

The CMB map can be estimated using the maximum *a posteriori* (MAP) estimator with the following analytical solution:

$$\hat{x} = \underset{x}{\mathrm{Argmax}} P(x \mid y) \tag{11}$$

$$= \left(\Sigma^{-1} + \mathcal{F}^H \mathbf{C}^{-1} \mathcal{F}\right)^{-1} \Sigma^{-1} y. \tag{12}$$

The reader would here recognize a classical Wiener filter. Contrary to the classically used Wiener filter in spherical harmonics, the solution of (11) explicitly account for the nonstationarity of the noise. However, the computational evaluation of this Wiener-like solution raises some numerical issues: the noise covariance matrix is diagonal in the pixel domain while the CMB covariance matrix is diagonal in the spherical harmonics domain. The solution filter that appears in (11) is formulated in the pixel domain:

$$\left(\Sigma^{-1} + \mathcal{F}^H \mathbf{C}^{-1} \mathcal{F}\right)^{-1} \Sigma^{-1}. \tag{13}$$

In this expression, the matrix $\mathcal{F}^H \mathbf{C}^{-1} \mathcal{F}$ is the inverse covariance matrix of the CMB in the pixel domain. This matrix is clearly not diagonal. Recalling that N^2 is of the order of $1e13$ for WMAP and $2e15$ for Planck, storing and handling this matrix is way too unrealistic. As an alternative, the most effective approach in this situation would amount to solve the linear system of equations $(\Sigma^{-1} + \mathcal{F}^H \mathbf{C}^{-1} \mathcal{F})\hat{x} = \Sigma^{-1} y$ using a conjugate gradient (CG, [13]) for instance. However, in the next section, we will rather make use of recently introduced proximal algorithms. This optimization framework leads to algorithms which exhibits a lower speed of convergence than CG but with the gain of more flexibility to handle different filtering techniques or better constrain the solution.

2.1.1. Iterative Filtering. As said clearly in the previous paragraph, computing the MAP solution requires inverting a system of equations which turns to be intractable due to the large scale of Planck data. The most straightforward numerical solver is the well-known conjugate gradient. However, while this solver is known to be a fast and accurate numerical method for solving linear systems of equations, it lacks the flexibility of more modern optimization techniques. For this reason we rather opt for the more flexible optimization framework of proximal calculus (see [14] and references therein).

---

(1) **Initialization:** set the number of iterations I_{max}, the step size $\gamma = \min_k \sigma_k^2$, and the starting point x^0.
(2) **Main loop:** apply iteratively the following steps at each iteration k:
    (a) Forward step of the FPS:
      $u = x^k + 2\gamma \Sigma^{-1}(y - x^k)$
    (b) Backward step of the FPS:
      $x^{k+1} = \mathcal{F}^H(I + 2\gamma\mathbf{C}^{-1})^{-1}\mathcal{F}u$
(3) **Stop** when $k = I_{max}$.

ALGORITHM 1

(a) A Flexible Optimization Framework: Forward-Backward Splitting. The problem in (12) can be written as minimization of the sum of two operators f_1 and f_2:

$$\hat{x} = \underset{x}{\text{Argmin}}\, f_1(x) + f_2(x) \qquad (14)$$

$$= \underset{x}{\text{Argmin}}\, x^T \mathcal{F}^H \mathbf{C}^{-1} \mathcal{F} x + (y - x)^T \Sigma^{-1}(y - x). \qquad (15)$$

The function $f_1(x) = x^T \mathcal{F}^H \mathbf{C}^{-1} \mathcal{F} x$ is convex and lower semicontinuous and differentiable; the function $f_2(x) = (y - x)^T \Sigma^{-1}(y - x)$ is strictly convex, continuous and differentiable. Such problem as therefore a unique solution (see [14]). If $\nabla f_2(x)$ denotes the gradient of f_2, we will also assumed that this gradient is L-Lipschitz. In the specific setting of interest in this article, $\nabla f_2(x) = -2\Sigma^{-1}(y - x)$ and $L = 2\|\Sigma^{-1}\|_2$. In this context, an effective algorithmic framework is the forward-backward splitting (FPS) technique defined in [14]. In the paradigm of proximal calculus, any convex lower semicontinuous function (see [14]) admits a well-defined proximal operator:

$$\text{prox}_{\gamma f}(z) = \underset{u}{\text{Argmin}}\, \gamma f(u) + \frac{1}{2}\|z - u\|_{\ell_2}^2. \qquad (16)$$

In the FBS framework, the solution to the problem in (14) is computed via an iterative fixed point algorithm such that at iteration k

$$x^{k+1} = \text{prox}_{\gamma f_1}\left(x^k - \gamma\nabla f_2\left(x^k\right)\right) \qquad (17)$$

$$= \text{prox}_{\gamma f_1}\left(x^k + 2\gamma\Sigma^{-1}\left(y - x^k\right)\right). \qquad (18)$$

As proved in [14], the convergence of the above fixed point algorithm is guaranteed under the condition $\gamma < 2/L$; note that in the setting of CMB map recovery with uncorrelated nonstationary noise, $L = 1/\min_k \sigma_k^2$.

(b) The Proximal Operator of f_1. In the problem of interest in this paper, $f_1(x) = x^T \mathcal{F}^H \mathbf{C}^{-1} \mathcal{F} x$. As this function is continuous and differentiable, it is *a fortiori* lower semicontinuous. Following the definition in (16), the proximal operator of f_1 is defined as follows:

$$\text{prox}_{\gamma f_1}(z) = \underset{u}{\text{Argmin}}\, \gamma u^T \mathcal{F}^H \mathbf{C}^{-1} \mathcal{F} u + \frac{1}{2}\|z - u\|_{\ell_2}^2. \qquad (19)$$

The solution to this problem has a simple closed-form expression:

$$\text{prox}_{\gamma x^T \mathcal{F}^H \mathbf{C}^{-1}\mathcal{F}x} = \left(I + 2\gamma\mathcal{F}^H\mathbf{C}^{-1}\mathcal{F}\right)^{-1} z \qquad (20)$$

$$= \mathcal{F}^H\left(I + 2\gamma\mathbf{C}^{-1}\right)^{-1}\mathcal{F}z. \qquad (21)$$

The proximal operator of f_1 is a mere Wiener filter in the spherical harmonics space. To sum everything up, the resulting iterative Wiener-like filtering is as we can see in Algorithm 1.

(c) Computational Cost. The computational cost of this iterative Wiener algorithm is particularly low. The first step (2-a) only requires entrywise multiplications and additions of vectors. It is therefore of the order of $\mathcal{O}(N)$. The second step (2-b) is by far the most computationally demanding; it requires entrywise multiplications and additions of vectors of the order of $\mathcal{O}(\ell_{max})$ (where generally $\ell_{max} \simeq 3000$) and single forward/backward spherical harmonics transforms. In the numerical experiments of Sections 2.3 and 3.3, the starting point $x^{(0)}$ will be chosen as having entries equal to zero. The algorithm stops when the relative variance—in euclidian norm—between consecutive estimates of the CMB map is lower than a certain threshold: $\|x^{(k+1)} - x^{(k)}\|_{\ell_2}/\|x^{(k)}\|_{\ell_2} \leq 10^{-6}$. The total number of iterations is generally no greater than $I_{max} = 250$.

(d) Linearity and Bias of the Iterative Wiener Filter. The proposed iterative filtering technique preserves the linearity of the overall CMB map processing. In fact, in every step of the proposed algorithm, each output depends linearly on their inputs. This property is essential as it allows for a simple way to study the propagation of errors via Monte-Carlo simulations; each simulation then has to undergo exactly the same processing step described previously. It is worthwhile to outline that the solution to the proposed iterative Wiener filter is exactly the solution of the Wiener filter described in (11). Therefore, the proposed method exhibits exactly the same bias and variance as the Wiener estimator. Similarly to the global Wiener filter, the bias of this estimator does not vanish and can be shown to depend on the noise statistics and the signal itself. The same holds for the variance as well.

2.2. The Case of Correlated Noise. Assuming a perfect calibration of the data, the nonstationarities of instrumental noise mainly originate from the nonuniform scanning of the sky. The lower the number of scanning time in one area is, the higher the noise level is and vice versa. Figure 1 shows the RMS map of noise at channel 217 GHz; mathematically this is no more than the square root of the diagonal of Σ. However this is also well known that, contrary to what has been assumed in the previous section, the noise is far from being uncorrelated. There are mainly two origins for the correlation of noise:

(i) The way the sky is scanned by Planck makes the noise correlated along the scanning direction. This means that the noise is correlated along elongated patterns in the pixel domain; for numerical reasons, this can hardly be accounted for in the pixelwise covariance matrix Σ.

(ii) Most modern CMB-dedicated source separation methods, Needlet ILC [4], Generalized ILC [9], SMICA [10] or GMCA [6] rely on a local analysis of the data in the wavelet or needlet domain. Furthermore, the final CMB map is generally obtained as the mixture of observations that do not share the same resolution. This means that the noise that finally contaminates the CMB is, in the pixel domain, correlated.

It then seems natural to model non-stationary-correlated noise: (i) in the wavelet domain to capture its correlation and (ii) locally to capture its nonstationarity.

(1) Wavelets: the Basics. Wavelets are well known to be the tool of choice to analyze nonstationary signals [15]. When considering spherical data, there exist several implementations of wavelets on the sphere. In this paper, we will make use of the isotropic undecimated wavelet transform (IUWT) introduced in [16]. The IUWT can be seen as an extension of the celebrated à trous algorithm to spherical data (see the seminal book [15] and references therein).As any wavelet transform, the IUWT is first defined by a scaling function $\phi_{\ell_c}(\theta, \varphi)$ defined by a fixed frequency cut-off ℓ_c. As emphasized in [16], this scaling function is built so that it is azimuth-invariant on the sphere (i.e., it does not de-pend on φ). Classically, each rescaled version of the scaling function $\phi_{\ell_c/2^j}$ is associated to a low-pass filter h_j or equivalently $H_j(\ell)$ in spherical harmonics. Traditionally, the wavelet-based multiscale analysis is performed by computing successive smooth approximations of x. These smooth approximations are obtained by convolving x with dyadically rescaled versions of the scaling function:

$$\forall j = 0, \ldots, J; \quad c_j = h_j \star x, \tag{22}$$

where J is the total number of scales. The so-called wavelet function at scale j, $\psi_{\ell_c/2^j}(\theta, \varphi)$, is defined as the difference of two consecutive rescaled scaling functions:

$$\psi_{\ell_c/2^j}(\theta, \varphi) = \phi_{\ell_c/2^{j-1}}(\theta, \varphi) - \phi_{\ell_c/2^j}(\theta, \varphi). \tag{23}$$

The wavelet function at scale j can be equivalently associated to a high-pass filter g_j or equivalently $G_j(\ell) = 1 - H_j(\ell)$ in

spherical harmonics. The wavelet coefficients at scale j are then computed by convolving the smooth approximation of x at scale j with the jth wavelet function $\psi_{\ell_c/2^j}(\theta, \varphi)$.

$$\forall j = 0, \ldots, J; \quad w_{j+1} = g_j \star c_j. \tag{24}$$

In spherical harmonics, this amounts to a simple multiplication of the signal's spherical harmonic coefficients with the wavelet filter G_j. The decomposition being redundant, several strategies can be designed to reconstruct x from the wavelet coefficients and the coarse resolution c_J. The simplest is the one advocated by the à trous algorithm:

$$x = c_J + \sum_{j=1}^{J} w_j. \tag{25}$$

In the following, noise modeling will be performed in each wavelet scale; that is, on each set of coefficients $\{w_j\}_{j=1,\ldots,J}$.

(2) Wavelet-Based Statistical Modeling of Correlated Noise. As emphasized previously, one elegant way to model the instrumental noise is to consider the distribution of its expansion coefficients in the spherical wavelet domain. The basic idea consists in assuming that the wavelet coefficients of noise are approximately decorrelated. This idea takes its roots in the field of multiscale statistical modeling: it has been shown that wavelets exhibit (almost)-decorrelating properties for some classes of nonstationary stochastic processes. To give only one example, this is the case for fractional Brownian motion [17] which has been extensively used to model power-law following stochastic processes. In the following, $w_j^n[k]$ will denote the wavelet coefficient of the noise term n in scale j at pixel k with the convention: $j = 1$ corresponds to the finest scale and $j = J - 1$ to the coarse resolution. From this definition, we will assume that these wavelet coefficients verify:

$$\mathbb{E}\left\{w_j^n[k]w_j^n[i]\right\} = \sigma_j^n[k]^2 \delta_{k,i}, \tag{26}$$

where $\sigma_j^n[k]^2$ stands for the local variance of the noise at pixel k and $\delta_{k,i}$ for the Kronecker delta function. This entails that the covariance matrix of the noise in wavelet scale j is diagonal and equal to:

$$\Sigma_{n,j} = \text{diag}\left(\sigma_j^n[k]^2\right). \tag{27}$$

The local standard deviation of the noise, $\Sigma_{n,j}$, has to be estimated from the data beforehand. In the numerical experiments in Section 2.3, these parameters are estimated from a single noise realization. It is commonly assumed that one has access to noise realizations which generally originate from simulations or from the data themselves. Assuming that the local variances $\sigma_j^n[k]^2$ of the noise vary slowly from one pixel to another, it can be approximated at pixel k by the empirical estimate of the variance in a fixed surrounding neighborhood; in the following this neighborhood will be a patch of size $\sqrt{P} \times \sqrt{P}$ centered about the pixel k:

$$\hat{\sigma}_j^n[k]^2 = \frac{1}{P} \sum_{i \in \mathcal{N}[k]} w_j^n[i]^2, \tag{28}$$

where $\mathcal{N}[k]$ stands for the indices of the pixels that compose the patch that surrounds the pixel k. Note that the patch size $\sqrt{P}$ will highly depend on how fast the noise variance varies spatially. This particularly means that $\sqrt{P}$ should be smaller in the finest scale and larger when j increases. Following the natural dyadic scaling of the wavelet transform, the patch size in scale j will scale as follows: $P_j = 4^j P_1$ where P_1 stands for the patch size at the finest scale.

2.3. Numerical Experiments

2.3.1. Experimental Dataset. In the following, numerical experiments will be performed on simulated Planck data described in [3]; to our knowledge, these are the only full-sky simulated data that are publicly available. The input CMB map has been computed by applying a recently introduced source separation technique coined Local-GMCA (L-GMCA, see [6, 18]) to Planck simulated data. These data were already deployed in [3] to evaluate the performances of source separation methods. For more technical details on these data, we refer the interested reader to [3]. To our knowledge this dataset is the only full-sky realistic Planck simulated dataset that is publicly available. It contains contributions as various as synchrotron and free-free emission, anomalous dust, dust, SZ effect and CMB. The instrumental noise on each of these Planck simulated observations was assumed to be uncorrelated but nonstationarity.

The noise that contaminates the raw CMB map at the output of most component separation techniques is nonstationary and correlated. The CMB map estimates being estimated from several observations at different resolutions, the output noise is obviously correlated. Source separation techniques like L-GMCA [6, 18] or Needlet ILC [4] also analyze the data on small patches in different wavelet or needlet scales. Such local/multiscale process is also at the origin of the correlation of the noise that contaminates the CMB map.

2.4. Experimental Results.

In this section, we particularly focus on the noise reduction aspect of the proposed method. The reference denoising method in the field of astrophysics and more specifically CMB data analysis is the *global* Wiener filter applied in spherical harmonics. In opposition to this classical filtering technique, the proposed iterative filtering approach accounts for the nonstationarity of the noise; thereby, it should particularly provide better solution than global Wiener.

As detailed in the previous section, the modeling of the noise contribution is performed in the wavelet domain: the noise is assumed to be nonstationary but decorrelated at each wavelet scale j. In each wavelet scale, the noise covariance matrix is estimated locally from a single noise realization. The noise covariance per wavelet scale is evaluated locally on patches of size $4^{(j+1)}$ where, by convention, $j = 1$ is defined as the finest wavelet scale. As an illustration, Figure 8 displays the estimated noise RMS map at scale $j = 1$.

The global Wiener solution is computed by filtering the original CMB map y with the linear filter defined in spherical harmonics by (11). The CMB power spectrum C_ℓ is a WMAP7 best-fit estimate; the noise power spectrum C_ℓ^n is estimated via the pseudo-power spectrum of the noise realization.

Iterative Wiener has been applied to the original noisy CMB map y. The single parameter to be tuned is the number of iterations I_{max}; in the following experiments, it has been fixed to 250. Figure 2 shows the noisy map, the LIW-Filtering solution and the noise realization used to estimate the noise local variance in the finest wavelet scale. Figure 3 displays the residual maps of the *global* Wiener and LIW-Filtering solution. As expected the nonstationarity footprint of the noise appears clearly on the residual LIW-Filtering map where the bias of the estimated map is lower.

Figure 4 features the power spectra of the original and filtered maps. We would like to point out that these power spectra have been computed from the raw maps; no masking has been applied. This particularly means that these estimates are likely to be biased by strong foreground from the galactic plane. The first observation is that the global Wiener filter does not fully remove the noise at high ℓ contrary to the iterative filtering technique. This can be explained by the presence of significant nonstationarities at high ℓ which can produce a substantial bias at high frequency. Interestingly, the iterative technique succeed in largely reducing noise at high ℓ thus supporting the significant contribution of the noise nonstationarities in producing a bias on the power spectrum. More technically, it is worth noting that the power spectrum of a stochastic process is properly defined as long as this process is wide sense stationary [19]. Obviously, the nonstationarities of the noise make it depart from this assumption; thereby the power spectrum has no rigorous statistical meaning in this context. Said differently, the power spectrum of the noise C_ℓ^n does not contain all the information that a nonstationary noise does in the pixel domain.

It is well known that the Wiener filter provides a CMB map that has a biased power spectrum. Thereby the Wiener filter—in red in Figure 4—yield a power spectrum that is negatively biased at high ℓ when the noise highly predominates upon the CMB—typically for $\ell > 2500$.

Figure 5 shows the power spectrum of the residual: $x - \hat{x}$. This quantity is particularly well suited to visualize the power discrepancy in frequency between the original CMB map x and its estimate $\hat{x}$. As expected, the *global* Wiener filtering reduces the contribution of noise at high ℓ (see the blue line in Figure 5). Still, the filtered map largely departs from the original map. Interestingly, the proposed iterative filtering technique (in red in Figure 5) largely outperforms the classical Wiener filter from $\ell = 900$ to $\ell = 3000$. More precisely, the residual power is reduced by up to an order of magnitude for $\ell > 2500$.

A more complete measure of discrepancy between the original map x and one of its estimate $\hat{x}$ consists in evaluating the mean squared error (MSE) in the wavelet domain. To that end, Figure 6 displays the MSE of different estimators of the CMB map in 5 wavelet scales in logarithmic scale. The normalized MSE between some CMB map estimate $\hat{x}$ and the "true" CMB map is computed as follows: $\sum_k (\hat{x}[k] - x[k])^2 / \sum_k x[k]^2$. Again, this figure

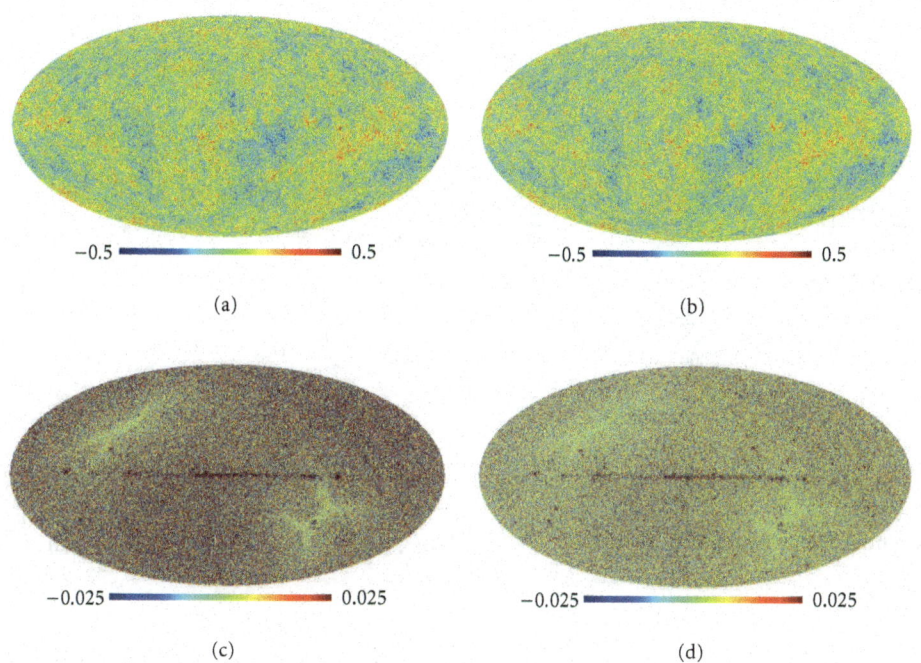

FIGURE 2: *CMB maps*—(a) Noisy image. (b) LIW-Filtering solution. (c) Residual map (true map-noisy map) in the finest wavelet scale. (d) Residual map (true map-LIW-Filtering solution) in the finest wavelet scale. Units in mK.

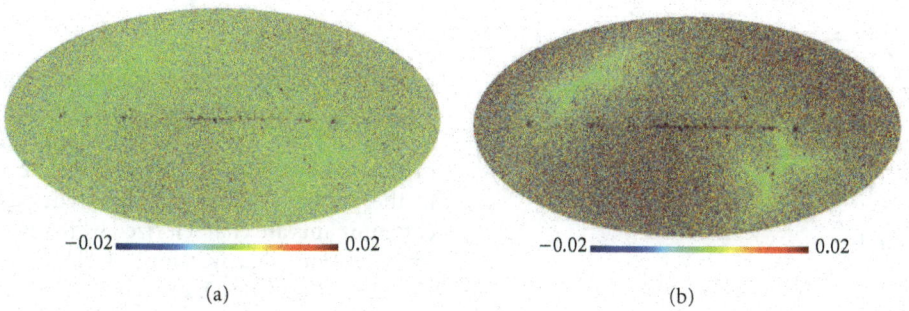

FIGURE 3: *Removed noise map* (i.e., *noisy data-filtered data*) in the finest wavelet scale—(a) from the *global* Wiener filter. (b) from the LIW-Filtered solution. *These maps correspond to what has been filtered from the noisy map*. The amount of noise removed by LIW-Filtering is more realistic. Units in mK.

shows that the iterative technique which directly account for the nonstationarity of the noise, performs better than the *global* Wiener filter. The impact of the noise tends to vanish for larger scales (*typically* for $\ell < 1500$).

To better illustrate the differences between the *global* Wiener filter and its iterative counterparts, Figure 7 displays the MSE of the different estimators of the CMB map which has been computed in the wavelet domain per bands of latitude of width 10°. By convention, the galactic plane is centered around the latitude 0°. Figure 7(a) displays the MSE which has been evaluated from the finest wavelet scale $j = 1$; this corresponds to an analysis window that ranges from $\ell = 1500$ to $\ell = 3000$. Again, taken globally, one can point out the clear improvement between the *global* Wiener filter and its iterative version. As expected the Wiener filter is optimal in the sense of the MSE. More interestingly, the MSE of the

two filtering techniques seems to be rather constant for high latitudes and sharply increases in the range $[-15°, 15°]$. As already emphasized, the impact of any of these linear filtering techniques is the more significant when the signal to noise ratio (SNR) decrease. This is obviously the case at smaller scales. At larger scales, the SNR increases and the impact of the filtering is much less significant. This is exactly what can be observed on the plot on the right of Figure 7(b): the MSE of any of the filtered CMB maps at scale $j = 2$, which grossly correspond to an analysis window ranging from $\ell = 750$ to $\ell = 1500$, are approximately the same.

As a brief conclusion for these experimental results, it appears clearly that the nonstationarity of the noise has an impact on the CMB estimate. The most classical noise reducing technique in the field, that is, the *global* Wiener filter, only relies on the power spectra of the CMB C_ℓ

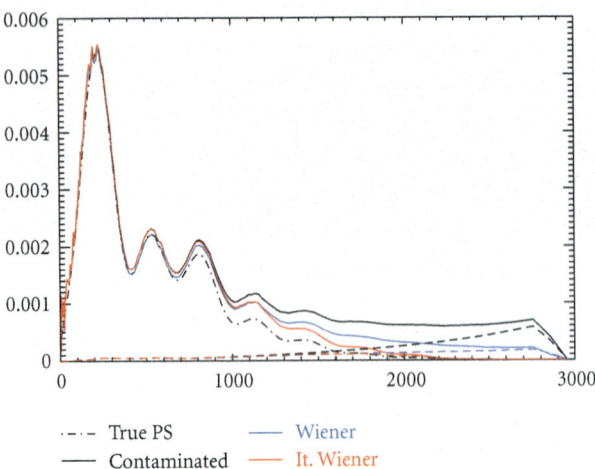

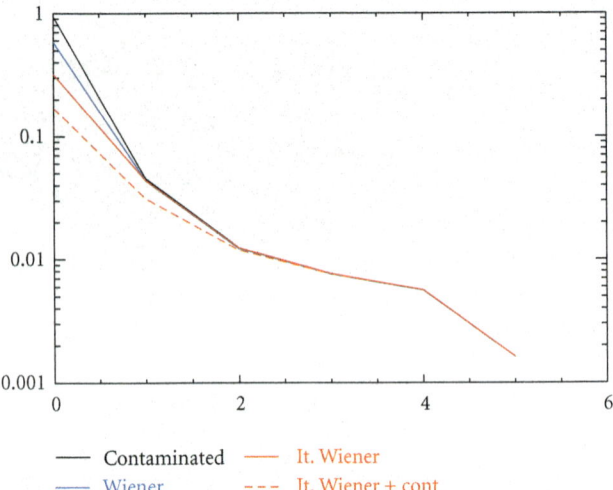

FIGURE 4: *Power spectra.* In abscissa: spherical harmonics coefficient. In ordinate: CMB power in mK^2. Dashed line show the power spectra of the noise for the different maps.

FIGURE 6: *MSE per wavelet scale.* In abscissa: wavelet scale—0 corresponds to $j = 1$. In ordinate: normalized mean squared error.

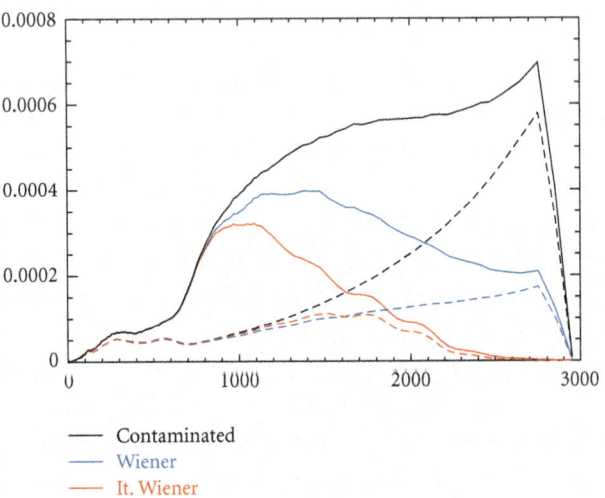

FIGURE 5: *Residual power spectra.* In abscissa: spherical harmonics coefficient. In ordinate: CMB power in mK^2. Dashed line show the power spectra of the noise for the different maps.

and the noise C_ℓ^n. Obviously, the behavior of the noise power spectrum contains almost no information about its spatial behavior. These experiments have shown that the noise nonstationary has a clear impact on the CMB. Further accounting for the spatial variations of the noise variance helps improving the noise reduction process by a dramatic amount.

3. From Noise Reduction to Contamination Reduction

As stated in the introduction in Section 1, one of the purposes of this paper is to tackle the problem of post-separation contamination reduction. In this setting, contamination generally means instrumental noise and residuals of galactic foregrounds. In the case of Planck, CMB maps are generally

obtained by applying sources separation methods. Most of these methods provide a solution that can be formulated as a linear combination of the original Planck multichannel observations. Therefore, it is relevant to assume that these contaminants essentially contribute additively to the CMB map:

$$y = x + f + n, \qquad (29)$$

where f stands for the foreground contamination and n for the instrumental noise. Furthermore, it is also natural to assume that these three components x, f and n are mutually uncorrelated. The filtering technique introduced in the previous sections mainly relies on the fact that the contaminants are characterized by their covariance matrix Σ. The questions that we aim at tackling in this paragraph is: *How far do we know Σ and how can it be estimated from the data ?*

3.1. Foreground Residuals Modeling. Unlike noise, the foreground residual are generally non-Gaussian, with the exception of the CIB (see [20]). We further restrict the modeling of residuals to their second order statistics thus opting for a Gaussian approximation of their contribution. The contribution of foreground residuals are obviously not known in advance; this means that their covariance matrix has to be estimated from the data directly. From the uncorrelation of the different components, the additive contamination model in (29) then reads at the level of second statistics:

$$\Sigma = \Sigma_x + \Sigma_f + \Sigma_n. \qquad (30)$$

In this equation, the correlation of foreground pixels implies that Σ_f is non-diagonal. Similarly to the noise, second order dependencies of foreground pixels are way too complex to be modeled in the pixel domain.

So as to capture the correlation of foreground pixels, a natural and simple strategy boils down to adopting the wavelet-based statistical modeling used to model correlated

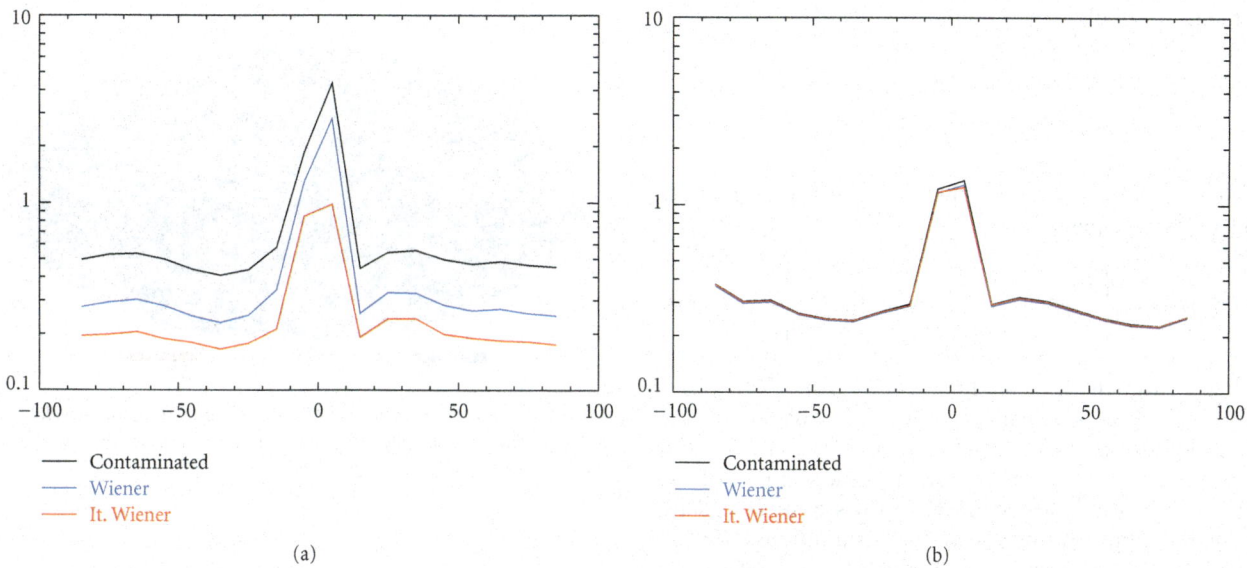

FIGURE 7: *MSE per latitude at wavelet scale* (a) $j = 1$ *and* (b) $j = 2$. In abscissa: latitude. The galactic plane corresponds to $0°$. In ordinate: normalized mean squared error.

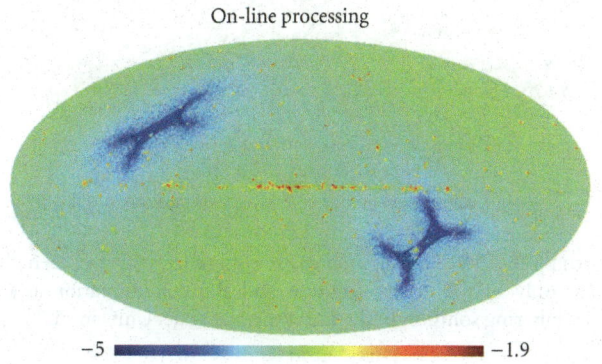

FIGURE 8: *Estimated noise covariance map at wavelet scale $j = 1$*: this covariance map has been estimated from a single realization of the noise that contaminates the CMB estimated by GMCA on simulated Planck data—see [3] for a description of these data. Units in mK^2.

noise in the previous Section 2.2.(2). To this end, $w_j^f[k]$ will denote the wavelet coefficient of f in scale j at pixel k with the convention: $j = 1$ corresponds to the finest scale and $j = J - 1$ to the coarse resolution. From this definition, we will assume that these wavelet coefficients verify:

$$\mathbb{E}\left\{w_j^f[k]w_j^f[i]\right\} = \sigma_j^f[k]^2 \delta_{k,i}, \qquad (31)$$

where $\sigma_j^f[k]^2$ stands for the local variance of the foreground at pixel k. The covariance matrix of the foreground residual in wavelet scale j is diagonal and equal to:

$$\Sigma_{f,j} = \mathrm{diag}\left(\sigma_j^f[k]^2\right). \qquad (32)$$

In practical situations, $\Sigma_{f,j}$ has to be estimated from the data y themselves. Similarly to the modeling of noise, the local noise variance of the foreground will be approximated

by the empirical estimate of the variance in a surrounding neighborhood of size $\sqrt{P} \times \sqrt{P}$ centered about the pixel k:

$$\hat{\sigma}_j^f[k]^2 = \frac{1}{P}\sum_{i \in \mathcal{N}[k]} w_j^f[i]^2. \qquad (33)$$

3.2. Contamination Filtering. In the rest of this paper, we will generally use the term contamination for the contribution of both the foreground residuals and the instrumental noise. In the following, as suggested previously, we also opt for exactly the same model to capture the pixel dependencies of the instrumental noise. Therefore, we will now define locally at pixel k and scale j the variance of contamination (noise and foreground) as $\hat{\sigma}_j^c[k]^2 = \hat{\sigma}_j^f[k]^2 + \hat{\sigma}_j^n[k]^2$. The joint contribution of noise and foreground residuals will be evaluated jointly from the data.

From the estimation of the local variance $\hat{\sigma}_j^y[k]^2$; one has to recall that the contaminants' variance is not directly accessible but is equal to:

$$\hat{\sigma}_j^c[k]^2 = \left[\hat{\sigma}_j^y[k]^2 - \hat{\sigma}_j^2\right]_+, \qquad (34)$$

where $\hat{\sigma}_j^2$ stands for the variance of the CMB and $[x]_+ = \max(0, x)$. Let use note that, from the stationarity of the CMB map, $\hat{\sigma}_j^2$ does not explicitly depend on k; this holds as long as the patch size at scale j is large enough compared to the CMB fluctuations (i.e., correlation length) in the jth wavelet scale. The value of the CMB variance at scale jth is directly related to the CMB power within this scale; its value can be derived from the CMB power spectrum C_ℓ and the wavelet analysis filters. To that end, let us denote by $\psi_{j,\ell}$ the spherical harmonics filter from which the wavelet scale j is defined. As emphasized in Section 1 the spherical harmonics coefficients of the CMB map should be mutually uncorrelated. This particularly entails that the power of the CMB in the jth wavelet scale is exactly:

$$\sigma_j^2 = \frac{1}{4\pi}\sum_\ell (2\ell+1)\psi_{j,\ell}^2 C_\ell. \qquad (35)$$

It follows that the variance of the contamination at pixel k is approximately equal to:

$$\hat{\sigma}_j^c[k]^2 = \left[\hat{\sigma}_j^y[k]^2 - \frac{1}{4\pi}\sum_\ell(2\ell+1)\psi_{j,\ell}^2 C_\ell\right]_+. \qquad (36)$$

Thereby, from the *a priori* knowledge of the CMB power spectrum C_ℓ, an approximate contribution of the residual can be evaluated at each pixel and each wavelet scale. The reader has to keep in mind that the measurement of this contribution is only approximate and has to be considered under the light of the following assumptions: (1) the contribution of the residual is modeled as a nonstationary Gaussian field where the covariance matrix is diagonal in each wavelet scales and (2) its variance being evaluated on small patches of size $\sqrt{P} \times \sqrt{P}$, it is assumed to be approximately stationary in a given analysis patches. However, even with these approximations, our model offers a much more realistic modeling of the complexity of the data than the traditional Wiener method. Improvements of this modeling will be discussed in Section 4.1.

Let $\Sigma_{f,j} = \mathrm{diag}(\hat{\sigma}_j^c[k]^2)$. Then, in the algorithm introduced in Section 2, the noise matrix Σ is substituted with the following covariance matrix:

$$\Sigma = \mathbf{W}^t \begin{bmatrix} \Sigma_{f,1} & 0 & 0 \\ 0 & \ddots & 0 \\ 0 & 0 & \Sigma_{f,J} \end{bmatrix} \mathbf{W}, \qquad (37)$$

where $\mathbf{W}$ stands for the isotropic wavelet transform and $\mathbf{W}^T$ its adjoint. Along with the iterative technique introduced in Section 2, the resulting wavelet-based filtering will be coined LIW-Filtering for Linear Iterative Wiener Filtering in the following.

3.3. Contamination Reduction on Planck Simulated Data. The dataset used in this section is exactly the same that we used to study noise reduction. It is described in Section 2.3.

As emphasized in Section 3, the wavelet-based noise/contamination variance modeling used so far makes it possible to account for the contribution of the foreground contamination in the filtering process. Estimating the contribution of the foreground contamination, as detailed in Section 3, makes the reduction of contamination possible. In the following experiments, the contamination, which includes the contribution of the noise and an approximate contribution of the foreground residuals, has been modeled in the wavelet domain [16] using 6 scales. Within each wavelet scale, with the exception of the coarsest scale to which no filtering is applied, the contribution of the contamination is assumed to be Gaussian and locally stationary on patches

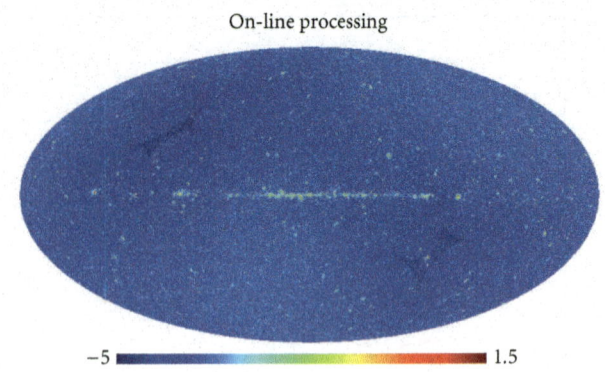

On-line processing

FIGURE 9: *Estimated contamination covariance map at wavelet scale* $j = 1$: this covariance map has been estimated from the CMB estimated by GMCA. Units in mK^2.

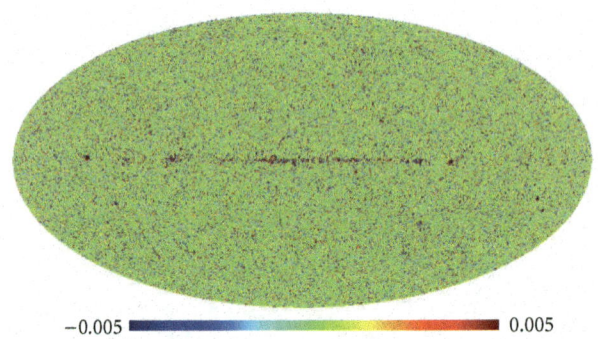

FIGURE 10: *CMB and residual maps* Difference map between the noise-only LIW-Filtering solution and the contamination-aware LIW-Filtering solution in the finest wavelet scale. Units in mK.

of size $\{4^{j+1}\}_{j=1,\dots,5}$. Figure 9 features the estimated variance map in logarithmic scale. As a remark, if the morphology of this contaminant variance map resembles at first glance the noise-only variance map, the dynamic range of the former is clearly larger by 3 orders of magnitudes. These differences are mostly significant on the galactic plane and on local features that are likely point sources or residuals of SZ clusters.

The proposed Wiener-based iterative method has been applied on the aforementioned Planck simulated data. Figure 10 shows the difference map between the noise-only LIW-Filtering solution and the proposed contamination-aware LIW-Filtering solution in the finest wavelet scale. It appears clearly that the difference between the map are mainly concentrated on the galactic plane and on likely point sources at high latitude.

Figure 11 displays the power spectra of the original and filtered CMB maps along with the theoretical power spectrum of the simulated map in black dash-dotted line. At first glance, accounting for the contribution of the foreground residuals, even approximately, yields improvements from $\ell = 750$. Perhaps more interesting, Figure 12 shows the power spectrum of the residual maps $x - \hat{x}$: the contaminant-aware filtering technique clearly outperforms the previous methods

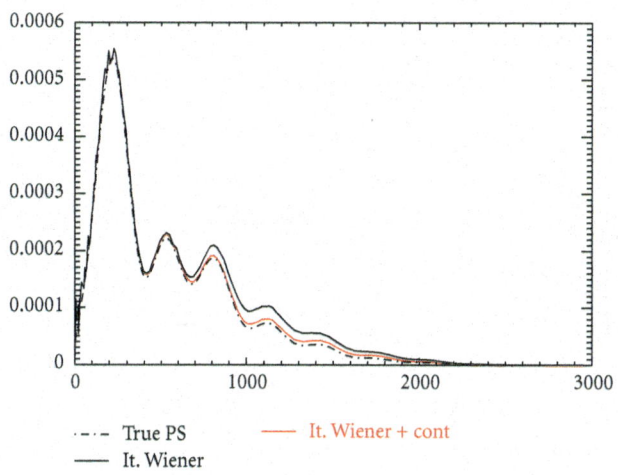

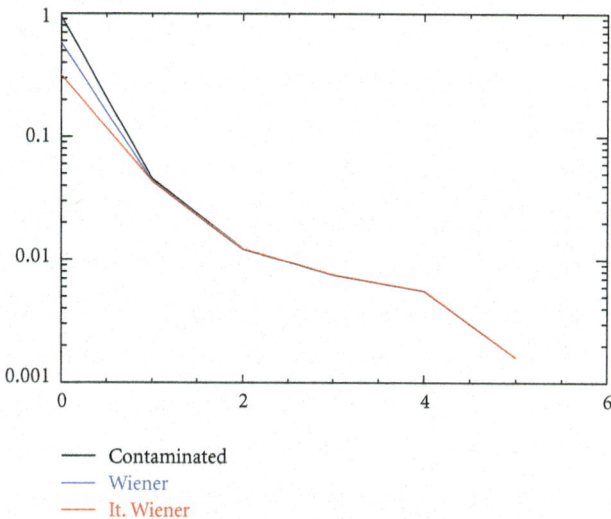

FIGURE 11: *Power spectra.* In abscissa: spherical harmonics coefficient. In ordinate: CMB power in mK^2. Dashed line show the power spectra of the noise for the different maps.

FIGURE 13: *Normalized MSE per wavelet scale.* In abscissa: wavelet scale—0 corresponds to $j = 1$. In ordinate: normalized mean squared error.

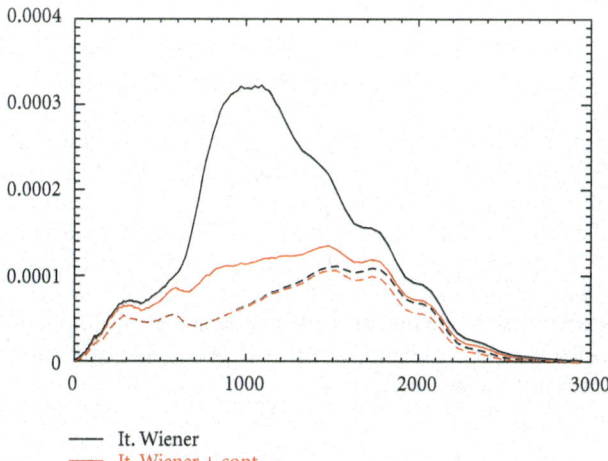

FIGURE 12: *Residual power spectra.* In abscissa: spherical harmonics coefficient. In ordinate: CMB power in mK^2. Dashed line show the power spectra of the noise for the different maps.

from $\ell = 250$ to $\ell = 3000$ with a dramatic improvement in the range $[750, 1500]$.

Figure 13 displays the wavelet-based MSE we described in the previous paragraph. The results of the contaminant-aware filtering technique have been further added to the plot in Figure 6. Let us recall that the noise reducing techniques described in the previous section mainly help reducing the noise in the noise-dominant regime. This particularly explained why the MSE remains almost unaltered in wavelet scales $j > 1$. Interestingly, accounting for an approximate contribution of the foregrounds helps reducing their impact in larger scales, typically for $j < 3$.

Figure 14 presents the MSE computed at different latitudes in the finest wavelet scale $j = 1$— plot on Figure 14(a)—and $j = 2$—plot on Figure 14(b). Accounting for the contaminant contribution improves the MSE of the filtered map at all latitudes. Interestingly, the MSE is

decreased by a large amount—a factor of approximately 5—on the galactic (latitude 0°). The same holds in a slighter amount in wavelet scale $j = 2$.

Assuming that the Planck instrumental noise is Gaussian is widely considered as a reasonable assumption. This is obviously not the case for foreground residuals the presence of which is likely to largely distort the search for non-Gaussian features in the CMB. Thereby, reducing the amount of contaminant should help preventing non-Gaussianity test from the non-Gaussian impact of foreground residuals. In the field of CMB non-Gaussianity evaluation, a classical technique boils down to measuring higher-order statistics in the wavelet domain [21]. One of the statistics of choice is the statistics of order 4, denoted κ_4, a.k.a. the kurtosis. The kurtosis of the wavelet coefficients of the estimated CMB map has been evaluated in 5 scales; non-Gaussianity test results are shown in Figure 15. The first observation is that the proposed method helps reducing non-Gaussian contamination even when only the noise is modeled as a spurious component. This suggests that the nonstationary behavior of the noise can generate undesired non-Gaussian signatures. Thereby, accounting for the nonstationary behavior of the noise in the noise reduction process decreases the kurtosis of the filtered map by an order of magnitude in the finest wavelet scale.

The red dashed line in Figure 15 shows a clear reduction of the kurtosis of the filtered map when the contribution of the foreground residuals is modeled. More precisely, the value of κ_4 is grossly reduced by 4 orders of magnitude for $j = 1$, 2 orders of magnitude for $j = 2$ and 1 order of magnitude for $j = 3$. While noise-only filtering yields map that departs from the Gaussian assumption for $J < 2$, the contamination-aware filtered map is now compatible with the Gaussian assumption in all wavelet scales with $\kappa_4 < 1$.

Figure 16 presents more detailed non-Gaussianity tests; the three plots that this figure displays show the evolution of κ_4, in the three finest wavelet scales $j = 1, \ldots, 3$, which

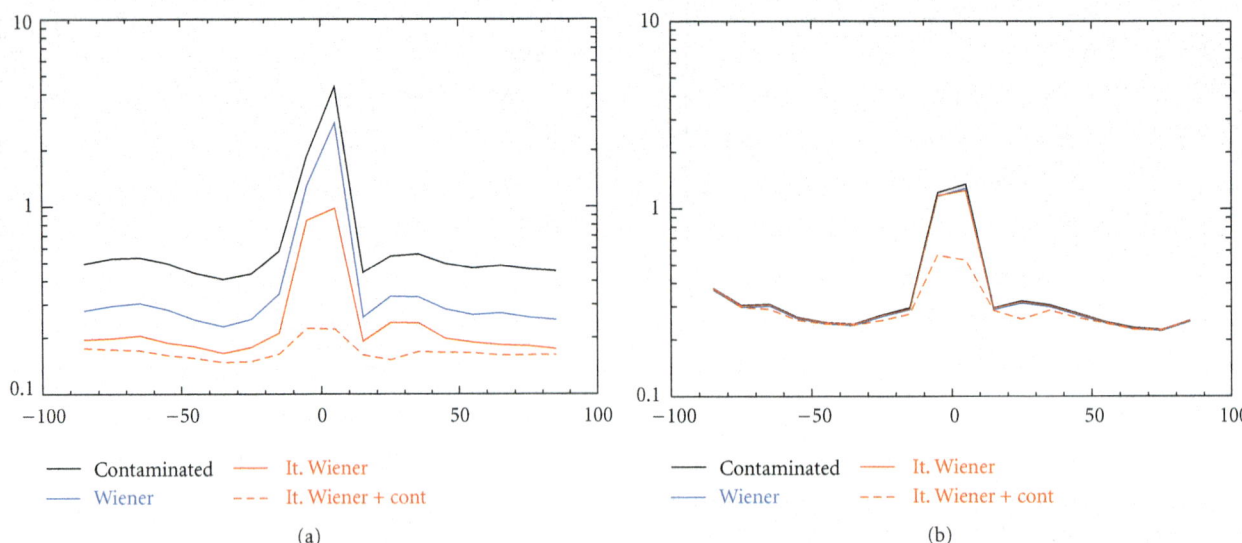

(a)   (b)

FIGURE 14: (a) *Normalized MSE per latitude at wavelet scale* $j = 1$ *and* (b) $j = 2$. In abscissa: latitude. The galactic plane corresponds to $0°$. In ordinate: normalized mean squared error.

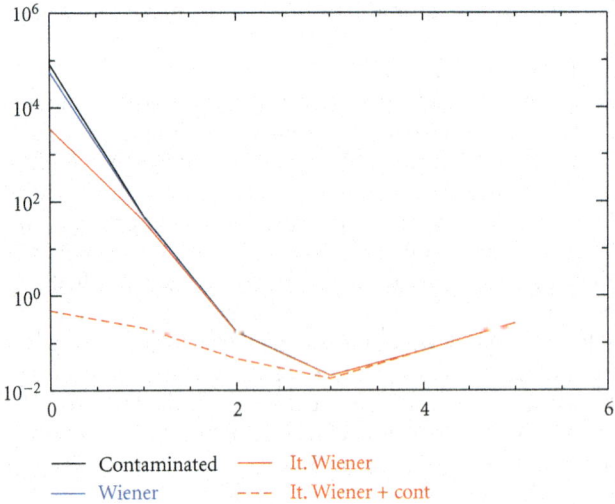

FIGURE 15: *Kurtosis* κ_4 *per wavelet scale*. In abscissa: wavelet scale—0 corresponds to $j = 1$. In ordinate: value of the kurtosis κ_4.

has been evaluated in bands of latitude of width $10°$. The galactic plane is centered about the point $0°$. The non-Gaussian contamination mainly originates from the presence of galactic foreground residuals which are mainly concentrated on the galactic plane. This explains the very large value of κ_4 in the range of latitude $[-15°, 15°]$. The second well-known source of non-Gaussian signatures at spatial high frequency are the radio and infrared point sources. These foregrounds are roughly uniformly scattered all over the sky. Point sources are likely to be the predominant source of non-Gaussian signatures at high latitude thus explaining the relatively large value of κ_4 in this range of latitudes. As shown in Figure 16(a), further filtering the contamination (see the red dashed line) yields a dramatic reduction of κ_4 up to 5 orders of magnitude on the galactic plane. The filtered CMB map is likely to be compatible with the Gaussian assumption

at every latitude, even on the galactic plane. In the second wavelet scale $j = 2$, the same reduction of non-Gaussian signatures is also visible by a lesser extent, mainly on the galactic plane in the range $[-15°, 15°]$. In the third wavelet scale, the proposed filtering is likely to reduce the non-Gaussian signatures in the galactic plane; in this range of scales (i.e., the third wavelet scale roughly corresponds to an analysis window that spans the range $[375, 750]$), the CMB is largely predominant which entails that: (i) the input map is already quite compatible with the Gaussian assumption and (ii) the impact of the filtering may be less visible when non-Gaussianity is measured.

4. Discussion, Prospects, and Conclusion

4.1. Discussion and Prospects

4.1.1. Filtering the Maps, for What Use? Important cosmological tests and evaluations are performed on estimated CMB maps; this is particularly the case for CMB lensing reconstruction [7] and non-Gaussian signature detection to only name two. These tests are particularly sensitive to spurious components that contaminate the estimated CMB map, whether it is noise or foreground residuals. This is therefore crucial to estimate a CMB map that is the less contaminated possible. The second point we would like to emphasize is that the aforementioned cosmological tests or evaluations are generally performed on the full-sky data; prior to any post-processing the estimated map is masked to get rid of the impact of galactic foregrounds and point sources. This has two major consequences: (i) it limits the size of the samples to which these tests/evaluations are performed thus increasing the statistical variance of these tests and (ii) any of these tests has to deal with this mask (e.g., via an inpainting procedure in [7]) thus making the analysis of the CMB map generally more complex.

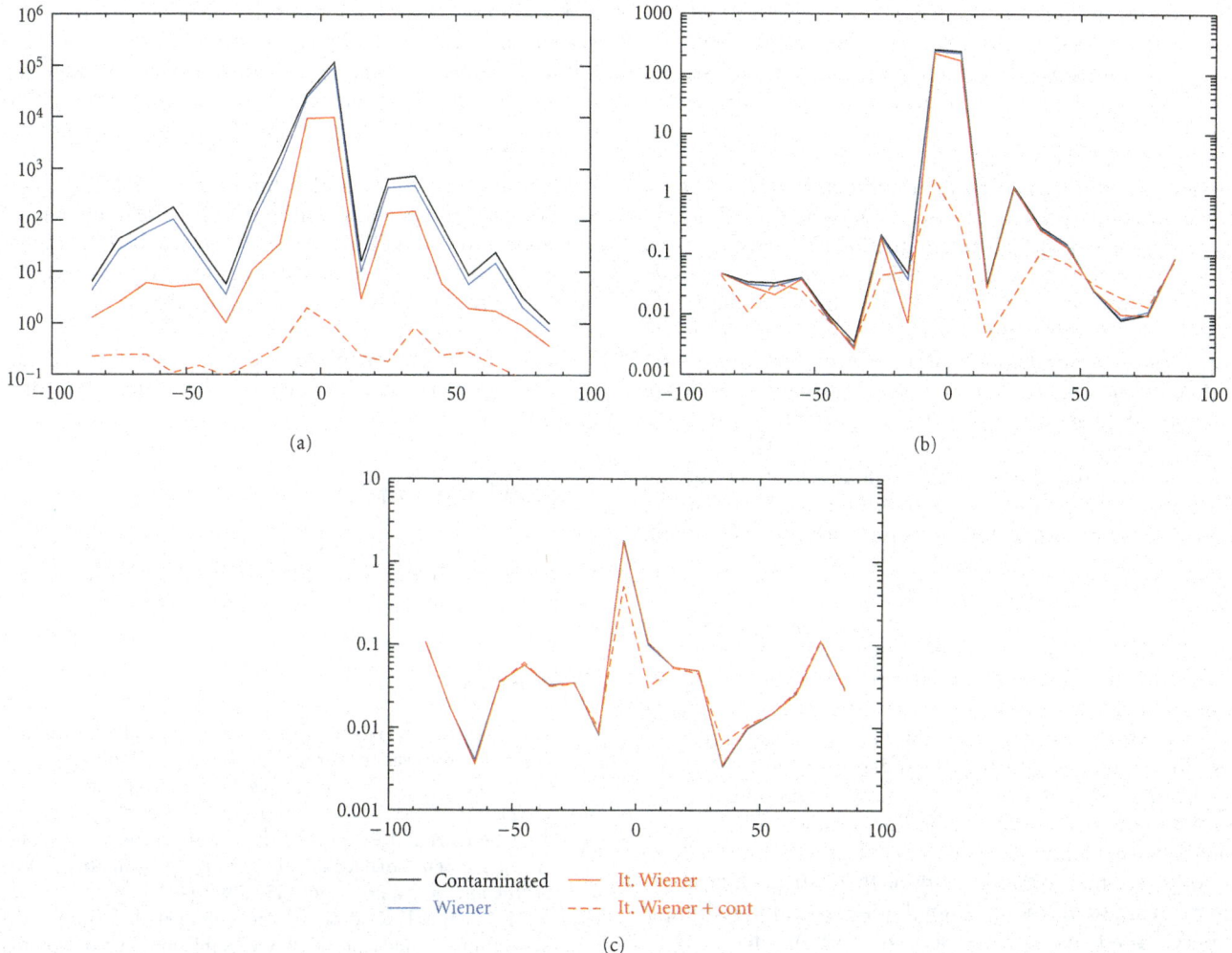

FIGURE 16: *Kurtosis κ_4 per latitude at wavelet scale* (a) $j = 1$, (b) $j = 2$ and (c) $j = 3$. In abscissa: latitude. The galactic plane corresponds to $0°$. In ordinate: value of the kurtosis κ_4.

As shown in the previous experiments, the proposed filtering clearly limits the impact of noise by taking into account its nonstationary behavior. Furthermore, results are presented that clearly show that modeling the contribution of foreground residuals make contamination reduction effective even on the galactic plane. This has two major consequences: (i) it makes it possible the use of a much smaller mask prior to any analysis of the estimated CMB map and (ii) it helps reducing the non-Gaussian features that originate from the presence of foreground residuals.

4.1.2. Beyond the Proposed Methods. Whether noise or foreground residual is at stake, the central assumption that is at the heart of the modeling of contamination is the decorrelation hypothesis we made in Section 3. It is assumed that the contaminants have potentially nonstationary but decorrelated samples in each wavelet scales. The basic idea that supports this assumption is the well-known decorrelating property of wavelets bases. If this assumption almost holds for rather theoretical stochastic processes (e.g., fractional Brownian motion [17] to only name one), this is only roughly approximate for more general signals such as galactic foreground or complex correlated noise. The basic idea behind this assumption is that the multiscale analysis basis is able to capture the *correlation morphology* of the component to be modeled. In surveys involving raster scanning such as WMAP or Planck, it is very likely that instrumental noise will be highly correlated along the scanning direction. This means that the correlation of the Planck instrumental noise is better modeled as a decorrelated stochastic process in a multiscale signal representation that best represent elongated structures such as the Curvelet tight frame [16]. We can extrapolate the same argument to the modeling of foreground residuals.

The contamination modeling used so far in this paper makes profit of the decorrelation assumption; it particularly helps simplifying the filtering process by only requiring the handling of diagonal matrices (i.e., root mean squared maps). Departing from the decorrelation assumption will largely increase the complexity of the proposed filtering techniques. However, a straightforward way of extending the proposed methods is to choose multiscale signal representations which are better adapted to the morphology or structure of the signal to be modeled.

4.1.3. Potential Impact of Contamination Reduction on CMB Non-Gaussianities. It is important to wonder whether the contamination reduction filtering introduced in Section 3 might affect or reduce potential CMB-related non-Gaussianities. One important point is that the way the local variance of the contamination is estimated in Section 3 leads, in practice, to an estimation of foreground contaminations that largely exceeds the average CMB power in each wavelet scale. Therefore, the filtering should have a much lesser impact on non-Gaussianities whose magnitude is of the order or lower than the magnitude of the CMB. It is worth pointing out that the rule in (36) to estimate the local con-tamination variance can also be chosen differently to only account for non-Gaussianities that are guaranteed to largely exceed the magnitude of the CMB thus avoiding for sure any effect on potential CMB-related non-Gaussianities. Future work will focus on evaluating the potential impact of the proposed filtering technique on CMB lensing reconstruction and non-Gaussianity detection.

4.2. Conclusion. Cosmological microwave background maps estimated from full-sky surveys such as WMAP or more recently Planck generally suffers from various sources of contaminations: (i) instrumental noise is generally nonstationary which may generate non-Gaussian signatures and (ii) foreground residuals generally remain even after the application of state-of-the-art source separation methods. In this context, the most classical denoising technique, aka. *global* Wiener filter; despite its simplicity, it is not able to account for the nonstationarity of the instrumental noise. To that end, we introduce a novel noise reduction technique coined LIW-Filtering for Linear Iterative Wavelet Filtering which combines the linearity of the *global* Wiener filter while accounting for the potential nonstationarity of the noise. In this framework, the noise is modeled as a nonstationary but decorrelated process in the wavelet domain. The denoising problem then takes the very simple form of a quadratically regularized least square where the noise covariance matrix is diagonal in the wavelet domain and the signal covariance matrix is also diagonal in the spherical harmonic domain (i.e., CMB power spectrum). The solution is computed using recently introduced proximal algorithms. When noise reduction is at stake, we showed that the proposed iterative technique succeeds in reducing the mean squared error (MSE) of the filtered solution with respect to the *global* Wiener filter classically applied in the field. Moreover, the modeling/estimation framework introduced in this paper makes the reduction of foreground contamination possible. Similar to the nonstationary instrumental noise, foreground contamination is modeled as a nonstationary but decorrelated process in the wavelet domain. Numerical experiments show that: (i) the MSE of the filtered map is improved; specifically on the galactic plane, and (ii) very interestingly, non-Gaussian signatures are also dramatically reduced. This particularly provides arguments supporting the application of the proposed filtering technique as post-processing step to be applied to the CMB with the crucial aim of reducing noise and perhaps more importantly foreground contamination.

It is also important to point out that, similarly to the *global* Wiener filter, LIW-Filtering is—at this acronym indicates—a linear filtering technique. This means that LIW-Filtering is also relevant when studying errors and their propagation on the estimated CMB map via Monte-Carlo simulations are unavoidable.

Future work will focus on refining the modeling of noise and foreground residuals. Without losing the simplicity of this approach, we will more specifically focus on studying more adapted signal representation to better model the noise/foreground contamination spatial behavior.

The developed IDL code will be released with the next version of ISAP (Interactive Sparse astronomical data Analysis Packages) via the following web site: http://jstarck.free .fr/isap.html.

Acknowledgments

This work was supported by the French National Agency for Research (ANR-08-EMER-009-01) and the European Research Council Grant SparseAstro (ERC-228261).

References

[1] C. L. Bennett, M. Halpern, G. Hinshaw et al., "First-year Wilkinson Microwave Anisotropy Probe (WMAP) observations: preliminary maps and basic results," *Astrophysical Journal, Supplement Series*, vol. 148, no. 1, pp. 1–27, 2003.

[2] F. R. Bouchet and R. Gispert, "Foregrounds and CMB experiments: I. Semi-analytical estimates of contamination," *New Astronomy*, vol. 4, no. 6, pp. 443–479, 1999.

[3] S. M. Leach, J. F. Cardoso, C. Baccigalupi et al., "Component separation methods for the PLANCK mission," *Astronomy and Astrophysics*, vol. 491, no. 2, pp. 597–615, 2008.

[4] J. Delabrouille, J. F. Cardoso, M. Le Jeune, M. Betoule, G. Fay, and F. Guilloux, "A full sky, low foreground, high resolution CMB map from WMAP," *Astronomy and Astrophysics*, vol. 493, no. 3, pp. 835–837, 2009.

[5] J. Bobin, J. L. Starck, J. Fadili, and Y. Moudden, "Sparsity and morphological diversity in blind source separation," *IEEE Transactions on Image Processing*, vol. 16, no. 11, pp. 2662–2674, 2007.

[6] J. Bobin, F. Sureau, and J.-L. Starck, "Source separation in cosmology, from global to local models," in *Proceedings of the IEEE International Conference on Image Processing (ICIP '11)*, pp. 1297–1300, 2011.

[7] L. Perotto, J. Bobin, S. Plaszczynski, J.-L. Starck, and A. Lavabre, "Reconstruction of the cosmic microwave background lensing for *Planck*," *Astronomy & Astrophysics*, vol. 519, article A4, 14 pages, 2010.

[8] D. Hanson, G. Rocha, and K. Górski, "Lensing reconstruction from Planck sky maps: inhomogeneous noise," *Monthly Notices of the Royal Astronomical Society*, vol. 400, no. 4, pp. 2169–2173, 2009.

[9] M. Remazeilles, J. Delabrouille, and J.-F. Cardoso, "Foreground component separation with generalized Internal Linear Combination," *Monthly Notices of the Royal Astronomical Society*, vol. 418, no. 1, pp. 467–476, 2011.

[10] J. Delabrouille, J. F. Cardoso, and G. Patanchon, "Multidetector multicomponent spectral matching and applications for cosmic microwave background data analysis," *Monthly Notices*

of the Royal Astronomical Society, vol. 346, no. 4, pp. 1089–1102, 2003.

[11] M. Górski, E. Hivon, A. J. Banday et al., "HEALPix: a framework for high-resolution discretization and fast analysis of data distributed on the sphere," *Astrophysical Journal*, vol. 622, no. 2 I, pp. 759–771, 2005.

[12] E. Komatsu, J. Dunkley, M. R. Nolta et al., "Five-year wilkinson microwave anisotropy probe observations: cosmological interpretation," *Astrophysical Journal, Supplement Series*, vol. 180, no. 2, pp. 330–376, 2009.

[13] G. H. Golub and C. F. Van Loan, *Matrix Computations*, The Johns Hopkins University Press, Baltimore, Md, USA, 3rd edition, 1996.

[14] P. L. Combettes and V. R. Wajs, "Signal recovery by proximal forward-backward splitting," *Multiscale Modeling and Simulation*, vol. 4, no. 4, pp. 1168–1200, 2005.

[15] S. Mallat, *A Wavelet Tour of Signal Processing*, Academic Press, San Diego, Calif, USA, 1998.

[16] J. L. Starck, Y. Moudden, P. Abrial, and M. Nguyen, "Wavelets, ridgelets and curvelets on the sphere," *Astronomy and Astrophysics*, vol. 446, no. 3, pp. 1191–1204, 2006.

[17] P. Flandrin, "Wavelet analysis and synthesis of fractional Brownian motion," *IEEE Transactions on Information Theory*, vol. 38, no. 2, pp. 910–917, 1992.

[18] J. Bobin, Y. Moudden, J. L. Starck, J. Fadili, and N. Aghanim, "SZ and CMB reconstruction using generalized morphological component analysis," *Statistical Methodology*, vol. 5, no. 4, pp. 307–317, 2008.

[19] T. Chonavel, *Statistical Signal Processing: Modelling and Estimation*, Springer, New York, NY, USA, 2002.

[20] F. R. Bouchet, R. Gispert, and J. Puget, "Unveiling the cosmic infrared background," in *Proceedings of the IP Conference*, E. Dwek, Ed., pp. 255–258, 1995.

[21] J. L. Starck, N. Aghanim, and O. Forni, "Detection and discrimination of cosmological non-gaussian signatures by multi-scale methods," *Astronomy and Astrophysics*, vol. 416, no. 1, pp. 9–17, 2004.

6

Clustering of X-Ray-Selected AGN

N. Cappelluti,[1,2] V. Allevato,[3] and A. Finoguenov[2,4]

[1] Osservatorio Astronomico di Bologna, INAF, Via Ranzani 1, 40127 Bologna, Italy
[2] University of Maryland, Baltimore County, 1000 Hilltop Circle, Baltimore, MD 21250, USA
[3] Max-Planck-Institut für Plasmaphysik and Excellence Cluster Universe, Boltzmannstrasse 2, 85748 Garching, Germany
[4] Max-Planck-Institute für Extraterrestrische Physik, Giessenbachstrasse 1, 85748 Garching, Germany

Correspondence should be addressed to N. Cappelluti, nico.cappelluti@oabo.inaf.it

Academic Editor: Angela Bongiorno

The study of the angular and spatial structure of the X-ray sky has been under investigation since the times of the *Einstein* X-ray Observatory. This topic has fascinated more than two generations of scientists and slowly unveiled an unexpected scenario regarding the consequences of the angular and spatial distribution of X-ray sources. It was first established from the clustering of sources making the CXB that the source spatial distribution resembles that of optical QSO. It then became evident that the distribution of X-ray AGN in the Universe was strongly reflecting that of Dark Matter. In particular, one of the key results is that X-ray AGNs are hosted by dark matter halos of mass similar to that of galaxy groups. This result, together with model predictions, has lead to the hypothesis that galaxy mergers may constitute the main AGN-triggering mechanism. However, detailed analysis of observational data, acquired with modern telescopes, and the use of the new halo occupation formalism has revealed that the triggering of an AGN could also be attributed to phenomena-like tidal disruption or disk instability and to galaxy evolution. This paper reviews results from 1988 to 2011 in the field of X-ray-selected AGN clustering.

1. Introduction

After about 50 years from the opening of the X-ray window on the Universe with the discovery of Sco-X1 and the Cosmic X-ray background (CXB, [1]), our knowledge of high-energy processes in the Universe has dramatically improved. One of the leading mechanisms for the production of X-ray in the Universe is accretion onto compact objects. For this reason, the study of astrophysical X-ray sources is a powerful tool for studying matter under the effects of extreme gravity. As the efficiency of converting matter into energy in accretion processes is proportional to the "compactness" of the object (i.e., $\propto M/R$), it is clear that the strongest sources powered by accretion are super-massive black holes (SMBH). It also became a cornerstone of astrophysics that every galaxy with a bulge-like component hosts a SMBH at its centre and that the BH mass and the bulge velocity dispersion are strictly related [2]. It is also believed that black holes reach those high masses via one or more phases of intense accretion activity and therefore shining as active galactic nuclei (AGN). It is

believed that an AGN basically shines mostly from the power emitted by a thin, viscous, accretion disk orbiting the central SMBH Shakura and Sunyaev [3]. Such a disk produces a high amount of X-rays both from its hot inner regions (as far as the soft X-ray emission is concerned) and from a nonthermal source which is supposed to be the primary source of X-rays (both soft and hard).

Since its discovery, the nature of the CXB has been strongly debated, but soon the community converged into interpreting most of the CXB as the integrated emission of AGN across the cosmic time. While the discrete nature of the CXB has been proposed [4] and rapidly unveiled by experiments like *Einstein* [1] and ROSAT (see, e.g., [5]), little cosmological information has been obtained from samples of AGN because of the scarce number of detected sources in the X-ray band. Structure formation models and numerical simulations have shown that structures in the Universe have undergone a hierarchical growth starting from the denser peaks in the primordial Gaussian matter distribution. The large-scale structures (LSS) of the Universe are gravitationally dominated by dark matter (DM), and we can consider it as the responsible and one of the main drivers of the Cosmological structures evolution. Dark matter is believed to clump in large-scale halos (DMH Navarro, [6]) which are populated by galaxies. Thus, galaxies can be considered as tracers of the DM distribution in the Universe, and the study of their spatial clustering led us to a most comprehensive view of the LSS. On the other hand AGN/Quasar, as phase of the galactic evolution, is a quite rare phenomenon in the Universe as their space density of these objects is about 1/100–1/1000 lower than that of galaxies. This means that AGN/Quasar survey requires large field of view and/or deep exposure to provide statistically significant samples.

The study of their clustering and its evolution is a powerful tool to understand, from a statistical point of view, what kind of environment is more likely to host AGN. This is not just an academic question, but this is strictly related to the mechanism of AGN activation. We know that one of the candidate mechanisms for triggering an AGN is galaxy merger (see, e.g., [7–10]). The probability of such an event is definitely dependent on the environment inhabited by the host galaxy. Even if the mean distance between galaxies is relatively small, in high-density (mass) environments, they have a high velocity dispersion, and, therefore, the likelihood of a major merger is very low. On the contrary, in the field, the likelihood of galaxy mergers is low because of the large average distance between galaxies. The most favorable place to detect a merger is therefore a moderately low-density (mass) environment like a group (see, e.g., [11]).

In fact, merger-driven models (see, e.g., [7]) accurately predict the observed large-scale clustering of quasars as a function of redshift up to z~4. The clustering is precisely that predicted for small group halos in which major mergers of gas-rich galaxies should proceed most efficiently. Thus, it is well established empirically and with theoretical predictions that quasar clustering traces a characteristic host halo mass $\sim 4 \times 10^{12} h^{-1} M_{\odot}$, supporting the scenario in which major mergers dominate the bright quasar populations.

In addition, other phenomena like secular processes may become dominant at lower luminosities as suggested by Milosavljević et al. [12]; Hopkins et al. [13]; Hopkins and Henquist [9]. Low-luminosity AGN could be triggered in more common nonmerger events, like stochastic encounters of the black holes and molecular clouds, tidal disruption, or disk instability. This leads to the expectation of a characteristic transition to merger-induced fueling around the traditional quasar-Seyfert luminosity division (growth of BH masses above/below $\sim 10^7 M_{\odot}$). However, the triggering mechanism of the SMBH growth must be compliant with M_{BH}-σ relation that links the growth of the SMBH with growth of the bulge of the host galaxy [2].

As shown in Hopkins et al. [8], the predicted large-scale bias of quasars triggered by secular processes is, at all redshifts, lower than the bias estimated for quasars fueled by major mergers. This implies that low-luminosity Seyfert galaxies live in DMHs that never reach the characteristic mass associated with small group scales.

On the other hand, the majority of the results on the clustering of X-ray-selected AGN suggest a picture where moderate-luminosity AGN live in massive DMHs ($12.5 < \log M_{DMH} [h^{-1} M_{\odot}] < 13.5$) up to $z \sim 2$, that is, X-ray-selected AGN samples appear to cluster more strongly than bright quasars. The reason for this is not completely clear, but several studies argued that these large bias and DMH masses could suggest a different AGN-triggering mechanism respect to bright quasars characterized by galaxy merger-induced fueling.

This paper reviews results of clustering of X-ray-selected AGN from the first *Einstein* to the most recent *Chandra* and XMM-*Newton* surveys. We give a detailed description of the methods used in this kind of analysis from simple power-law to halo models. In addition, we discuss the results of X-ray AGN clustering in the framework of AGN evolution and triggering. We adopt a ΛCDM cosmology with $\Omega_{\Lambda} = 0.7$, $\Omega_m = 0.3$, $H_0 = 100 \, h^{-1}$ km/s/Mpc with h = 0.7 and $\sigma_8 = 0.8$ ([14], WMAP-7).

2. Previous Measures of X-Ray Clustering Amplitude

As far as the X-ray source clustering results are concerned, the development of the field has always been driven by the performance of the telescopes. In particular, while first results studied the angular distribution of the unresolved CXB under the assumption that Quasars were its main contributors, recent *Chandra* and XMM-*Newton* surveys sample clustering of AGN with a precision comparable to that achievable with redshift galaxy surveys.

In the following section, we will use the following convention for reporting results of clustering analysis in the case of power-law representation of the auto(cross)-correlation function: if the clustering is measured in the angular space, we will use

$$w(\theta) = \left(\frac{\theta}{\theta_0}\right)^{1-\gamma}, \qquad (1)$$

where θ_0 is the angular correlation length. If the measurements have been performed in the real (redshift) space this becomes

$$\xi(r) = \left(\frac{r}{r_0}\right)^{-\gamma}, \quad \left(\xi(s) = \left(\frac{s}{s_0}\right)^{-\gamma}, \quad \text{in } z - \text{space}\right), \qquad (2)$$

where γ is the 3D correlation slope and r_0 or s_0 are the correlation lengths. Barcons and Fabian [15] measured with *Einstein* a clustering signal of the CXB on scales ≤5′ corresponding to an angular correlation length $\theta_0 \sim 4'$. They have shown the importance of studying the angular structure of the CXB by pointing out that a large fraction of the CXB could have been attributed to sources with a redshift distribution similar to optical QSOs. In addition, the first prediction was not consistent with the hypothesis that the CXB was also partly produced by a diffuse hot intergalactic medium (IGM) component. It was also proposed that these sources were actually clustered on comoving scales of the order of $\sim 10\,h^{-1}$ Mpc.

Carrera and Barcons [16], Georgantopoulos et al. [17], and Soltan and Hasinger [18] observed that the CXB was highly isotropic on scales of the order of $2°-25°$. The first attempt of measuring the clustering of X-ray-selected AGN was performed by Boyle and Mo [19] that measured a barely significant signal by using a sample of 183 EMSS sources, mostly local AGN ($z < 0.2$). These evidences have brought the attention to the study of the clustering of the CXB down to the arcminute scale. The first significant upward turn for the measurement of AGN clustering in the X-ray band has been brought to light by ROSAT. By using a set of ROSAT-PSPC pointing on an area of $\sim 40\,\deg^2$, Vikhlinin and Forman [20] measured, for the first time, an angular correlation signal of faint (ROSAT) X-ray sources on scales $<10'$. By using the Limber equation (see Appendix B and [21]) they have deprojected their angular correlation function into a real-space correlation function and found that, under the assumption that the redshift distribution of the sources was the same as that of optical QSOs, the spatial correlation length was in the range $6-10\,h^{-1}$ Mpc. With such a result, they confirmed the hypothesis that the CXB was mostly produced by sources with a redshift distribution comparable to that of optically selected QSO, though with almost double source density. By using the results of Vikhlinin and Forman [20] and Akylas et al. [22] (who obtained similar results), Barcons et al. [23] have shown for the first time that X-ray-selected AGNs are highly biased tracers of the underlying LSS at $z < 1$ by showing a redshift evolving bias factor as large as $b \sim 2$.

However, it is worth to consider that the deprojection of the angular correlation function into a 3D correlation function relies on several assumptions, like the model-dependent expected redshift distribution, which may lead to a biased estimate of the real-space clustering. It is, however, worth noticing that angular correlation can be very useful to provide a first overview in the early phase of surveys, when optical identifications are not available, especially sampling new part of the parameter space of sources, like that is, new unexplored luminosity/flux limits and therefore source classes. Detailed physical models are, however, much better investigated by more sophisticated techniques as shown in the following parts.

The first firm detection of 3D spatial clustering of X-ray-selected AGN has been claimed by Mullis et al. [24] by using data of the ROSAT-NEP survey. They detected on an area of $\sim 81\,\deg^2$ a 3σ significant signal in the redshift space autocorrelation function of soft X-ray-selected sources at $\langle z \rangle \sim 0.22$. They have shown that, at that redshift AGN cluster with a typical correlation length, $r_0 = 7.4 \pm 1.9\,h^{-1}$ Mpc. Their results suggest that the population of AGN in such a sample is consistent with an unbiased population with respect to the underlying matter. Their result suggested that, at that redshift, AGNs were hosted in DMHs of mass of the order of $10^{13}\,h^{-1}\,M_\odot$.

With the development of *Chandra* and XMM-*Newton* surveys and thanks to the high source surface densities (i.e., $>400-1000\,\deg^{-2}$), our capabilities in tracing the LSS have dramatically increased. One of the first evidences that AGNs are highly correlated with the underlying LSS has been pointed out by Cappi et al. [25] and Cappelluti et al. [26] and references therein, who showed that, around massive high-z galaxy clusters, the source surface density of *Chandra* point sources is significantly, up to two times, higher than that of the background. More recently, Koulouridis and Plionis [27] showed that, although the X-ray source surface density of AGN around galaxy clusters is larger than in the background, the amplitude of their overdensities is about 4 times lower than that of galaxies in the same fields. This has been interpreted as a clear indication of an environmental influence on the AGN activity. Silverman et al. [10] in the COSMOS field and Koss et al. [28] in the *Swift*-BAT all-sky survey have shown that the AGN fraction in galaxy pairs is higher relative to isolated galaxies of similar stellar mass providing an additional evidence of the influence of the environment on AGN activity.

Chandra and XMM-*Newton* performed several blanck sky extragalactic surveys, and most of them dedicated part of their efforts in the study of the LSS traced by AGN to unveil their coevolution. Basilakos et al. [29, 30] by using data of the XMM-*Newton* 2dF-survey have measured an unexpected high correlation length both in the angular ($\theta_0 \sim 10''$) and, by projection, in the real space ($r_0 \sim 16\,h^{-1}$ Mpc). Such a high correlation length has been detected in this field only, thus one can explain such a measurement as a statistical fluctuation. With the same technique, Gandhi et al. [31] obtained a marginal 2-3σ detection of angular clustering in the XMM-LSS survey and obtained $\theta_0 = 6.3(42) \pm 3(^{+7}_{-13})$ in the 0.5-2 (2-10) keV bands and a slope $\gamma \sim 2.2$. Puccetti et al. [32] measured the clustering of X-ray sources in the XMM-*Newton* ELAIS-S1 survey in the soft and hard energy bands with a sample of 448 sources. They obtained $\theta_0 = 5.2 \pm 3.8\,4''$ and $\theta_0 = 12.8 \pm 7.8\,4''$ in the two bands, respectively. These measurements have been deprojected with the Limber's inversion in the real space and obtained $r_0 = 9.8-12.8\,h^{-1}$ Mpc and $r_0 = 13.4-17.9\,h^{-1}$ Mpc in the two bands, respectively.

In the *Chandra* era, Gilli et al. [33] measured the real space autocorrelation function of point sources in the CDFS-CDFN. They have measured, in the CDFS, $r_0 = 8.6 \pm 1.2\,h^{-1}$ Mpc at $z = 0.73$, while, in the CDFN, they obtained $r_0 = 4.2 \pm 0.4\,h^{-1}$ Mpc. The discrepancy of these measurements has been explained with variance introduced by the relatively small field of view and the consequent random sampling of LSSs in the field. In the CLASXS survey, Yang et al. [34] obtained a measurement of the clustering at $z = 0.94$ with $r_0 = 8.1^{+1.2}_{-2.2}\,h^{-1}$ Mpc which proposes that AGNs are hosted by DMH of mass of $10^{12.1}\,h^{-1}\,M_\odot$ (see Section 3). In addition, they proposed that AGN clustering evolves with luminosity and they found that the bias factor evolves with the redshift. Such a behavior is similar to that found in optically selected quasars. The XMM-*Newton* [35-37] and *Chandra* [38, 39] survey of the COSMOS field have provided a leap forward to the field of X-ray AGN clustering by surveying a 2 deg^2 field of view. The key of the success of this project is a redshift survey zCOSMOS [40] performed simultaneously with the X-ray survey, together with observations in more than 30 energy bands from radio to X-ray that allowed to measure either the spectroscopic or the photometric redshift of every source. In the X-ray band, the survey covers $2\,\mathrm{deg}^2$ with XMM-*Newton* with a depth of ~60 ks with the addition of a central $0.9\,\mathrm{deg}^2$ observed by *Chandra* with ~150 ks exposure. The first sample of ~1500 X-ray sources [36] has been used by Miyaji et al. [41] to determine their angular correlation function, without knowing their distance, and just assuming a theoretical redshift distribution for the purpose of Limber's deprojection. Significant positive signals have been detected in the 0.5–2 band, in the angular range of 0.5′–24′, while the positive signals were at the ~2σ and 3σ levels in the 2–4.5 and 4.5–10 keV bands, respectively. With power-law fits to the ACFs without the integral constraint term, they have found correlation lengths of $\theta_0 = 1.9 \pm 0.3''$, $0.8^{+0.5}_{-0.4}$, and $6 \pm 2''$ for the three bands, respectively, for a fixed slope $\gamma = 1.8$. The inferred comoving correlation lengths were $r_0 = 9.8 \pm 0.7$, $5.8^{+1.4}_{-1.7}$, and $12 \pm 2\,\mathrm{h}^{-1}$ Mpc at the effective redshifts of $z = 1.1$, 0.9, and 0.6, respectively. Comparing the inferred rms fluctuations of the spatial distribution of AGNs $\sigma_{8,\mathrm{AGN}}$ (see Appendix D) with those of the underlying dark matter, the bias parameters of the X-ray source clustering at these effective redshifts were found in the range $b = 1.5$–4. Such a result leads to the conclusion that the typical mass of the DMH hosting an AGN is of the order $M_{\mathrm{DMH}} \sim 10^{13}\,M_\odot\,\mathrm{h}^{-1}$. Similar results have been found by Ebrero et al. [42] using the angular correlation function of 30000 X-ray sources in the AXIS survey. In the XMM-LSS survey, Elyiv et al. [43] measured the clustering of ~5000 AGN and computed via Limber's deprojection the obtained $r_0 = 7.2 \pm 0.8\,\mathrm{Mpc/h}$ and $r_0 = 10.1 \pm 0.8\,\mathrm{Mpc/h}$ and $\gamma \sim 2$ in the 0.5–2 keV and 2–10 keV energy bands, respectively. In the XMM-COSMOS field, Gilli et al. [44] measured the clustering of 562 X-ray selected and spectroscopically confirmed AGN. They have obtained that the correlation length of these source, $r_0 = 8.6 \pm 0.5\,\mathrm{h}^{-1}$ Mpc, and slope of $\gamma = 1.88 \pm 0.07$. They also found that, if source in redshift spikes removed, the correlation length decreases to about 5-6 h^{-1} Mpc. Even if not conclusively, they also showed that narrow-line AGN and broad-line AGN cluster in the same way, indicating that both classes of sources share the same environment, an argument in favor of the unified AGN model which predicts that obscuration, and therefore the Type-I/Type II dichotomy is simply a geometrical problem. However, it is worth noticing that such a procedure may artificially reduce the clustering signal and the effects of such a cut in the sample may lead to an unreliable estimate of the clustering signal.

Even if the results of Gilli et al. [44] provide a quite complete overview of the environments of the AGN in the COSMOS field, Allevato et al. [45] analyzed the same field by using the halo model formalism (see Section 3). Their results show that AGNs selected in the X-ray band are more biased than the more luminous optically selected QSO. This observation significantly deviates from the prediction of models of merger-driven AGN activity [13, 46], indicating that other mechanisms like disk/bar instability of tidal disruptions may trigger an AGN. They also found that Type

1 AGN are more biased than Type 2 AGNs up to redshift of ~1.5.

In the Böotes field, Hickox et al. [47] explored the connection between different classes of AGN and the evolution of their host galaxies, by deriving host galaxy properties, clustering, and Eddington ratios of AGN selected in the radio, X-ray, and infrared (IR) wavebands from the wide-field ($9\,\mathrm{deg}^2$) Böotes survey. They noticed that radio and X-ray AGNs reside in relatively large DMHs ($M_{\mathrm{DMH}} \sim 3 \times 10^{13}$ and $10^{13}\,M_\odot\,\mathrm{h}^{-1}$, resp.) and are found in galaxies with red and green colors. In contrast, IR AGNs are in less luminous galaxies, have higher Eddington ratios, and reside in halos with $M_{\mathrm{DMH}} < 10^{12}\,M_\odot\,\mathrm{h}^{-1}$.

On the same line, Coil et al. [48] measured the clustering of nonquasar X-ray active galactic nuclei at $z = 0.7$–1.4 in the AEGIS field. Using the cross-correlation of Chandra-selected AGN with 5000 DEEP2 galaxies, they measured a correlation length of $r_0 = 5.95 \pm 0.90\,\mathrm{h}^{-1}$ Mpc and slope $\gamma = 1.66 \pm 0.22$. They also concluded that X-ray AGNs have a similar clustering amplitude as red, quiescent, and "green" transition galaxies at $z \sim 1$ and are significantly more clustered than blue, star-forming galaxies. In addition, they proposed a "sequence" of X-ray AGN clustering, where its strength is primarily determined by the host galaxy color; AGNs in red host galaxies are significantly more clustered than AGNs in blue host galaxies, with a relative bias that is similar to that of red to blue DEEP2 galaxies. They did not observe any dependence of clustering on optical brightness, X-ray luminosity, or hardness ratio. In addition, they obtained evidence that galaxies hosting X-ray AGN are more likely to reside in groups and more massive DMHs than galaxies of the same color and luminosity without an X-ray AGN. Allevato et al. [45], Coil et al. [48] and Mountrichas and Georgakakis [49] concluded that DEEP2 X-ray AGN at $z \sim 1$ are more clustered than optically selected quasars (with a 2.6σ significance) and therefore may reside in more massive DMHs. In an evolutionary picture, their results are consistent with galaxies undergoing a quasar phase while in the blue cloud before settling on the red sequence with a lower-luminosity X-ray AGN, if they are similar objects at different evolutionary stages [47]. At lower redshift, Krumpe et al. [50] confirmed the results of Coil et al. [48]. Various recent works have presented indications and/or evidences, of varying significance, regarding a correlation between the X-ray Luminosity and the AGN clustering amplitude, based either on the spatial [34, 44, 48, 50–52] or the angular [53] correlation function.

Note that luminosity-dependent clustering is one of the key features of merger-triggered AGN activity and is one of the prime motivations for AGN clustering analyses. Low L_X AGNs have been found to cluster in a similar way as blue star forming galaxies while high L_X AGN cluster like red passive galaxies. Such a result has been confirmed by Cappelluti et al. [51] using the Swift-BAT all-sky survey at $z \sim 0$. They detected both a L_X dependence of AGN clustering amplitude and a larger clustering of Type I AGN than that of Type II AGN. Krumpe et al. [50, 52] confirm the weak dependence of the clustering strength on AGN X-ray luminosity at a 2σ level for $z < 0.5$.

TABLE 1: Cappelluti, Allevato, and Finoguenov.

Survey	Band keV	N_{obj}	z	θ_0 arcsec	r_0 h^{-1} Mpc	γ	$b(z)^a$	Log(M$_{DMH}$)^b $M/(M_\odot h)$
EMSS	0.5–2	183	<0.2	X	<10	X	X	X
RASS	0.1–2.4	2158	1–1.5	~10	<10	1.7 ± 0.3	X	X
RASS	0.1–2.4	2096	0.1	~3.7	6.0 ± 1.6	1.9 ± 0.31	X	X
ROSAT-NEP	0.1–2.4	220	0.22	X	$7.5^{+2.7}_{-4.2}$	$1.85^{+1.90}_{-0.80}$	$1.83^{+1.88}_{-0.61}$	$13.51^{+0.91}_{-0.79}$
AXIS[1]	0.5–2	31288	0.96	22.9 ± 2.0	6.54 ± 0.12	1.12 ± 0.04	2.48 ± 0.07	$13.20^{+0.11}_{-0.12}$
AXIS[1]	2–10	9188	0.94	$29.2^{+5.1}_{-5.7}$	9.9 ± 2.4	$2.33^{+0.10}_{-0.11}$	2.38 ± 0.51	$13.14^{+0.28}_{-0.41}$
AXIS[1]	5–10	1259	0.77	$40.9^{+19.6}_{-29.3}$	5.1 ± 4.1	$1.47^{+0.43}_{-0.57}$	2.14 ± 1.88	$13.17^{+0.84}_{-2.44}$
ELAIS-S1	0.5–2	392	0.4	5.2 ± 3.8	$9.8^{+2.7}_{-4.3}$	1.8	X	X
ELAIS-S1	2–10	205	0.4	12.8 ± 7.8	$13.4^{+2.7}_{-4.3}$	1.8	X	X
CDFS	0.5–2	97	0.84	X	8.6 ± 1.2	1.33 ± 0.11	$2.64^{+0.29}_{-0.30}$	$13.41^{+0.55}_{-0.18}$
CDFN[2]	0.5–2	164	0.96	X	4.2 ± 0.4	1.42 ± 0.07	$1.87^{+0.14}_{-0.16}$	$12.73^{+0.12}_{-0.17}$
XMM-2dF[3]	0.5–2	432	1.2	10.8 ± 1.9	~16	1.8	1.9–2.7	12.5–13.1
XMM-LSS	0.5–2	1130	0.7	6.3 ± 3	6 ± 3	2.2 ± 0.2	X	X
XMM-LSS	2–10	413	0.7	4.2^{+7}_{-13}	6 ± 3	$3.1^{+1.1}_{-0.5}$	X	X
CLASXS	0.5–8	233	1.2	X	$8.1^{+1.2}_{-2.2}$	2.1 ± 0.5	$3.58^{+2.49}_{-1.38}$	$12.86^{+0.61}_{-0.16}$
CDFN[4]	0.5–8	252	0.8	X	$5.8^{+1.0}_{-1.5}$	$1.38^{+0.12}_{-0.14}$	$1.77^{+0.80}_{-0.15}$	$13.53^{+0.63}_{-0.71}$
XMM-COSMOS[5]	0.5–2	1037	1.1	2.9 ± 0.6	$11.8 \pm 1.1,$	1.8	3.7 ± 0.3	13.6 ± 0.1
XMM-COSMOS[5]	2–4.5	545	0.9	$1.2^{+1.1}_{-0.9}$	$6.9^{+2.2}_{-3.1},$	1.8	$2.5^{+0.7}_{-1.0}$	$13.3^{+0.3}_{-0.7}$
XMM-COSMOS[5]	4.5–10	151	0.6	$6.5^{+3.0}_{-2.7}$	$12.7^{+2.3}_{-2.7}$	1.8	$3.8^{+0.6}_{-0.8}$	13.9 ± 0.2
XMM-COSMOS[6]	0.5–2	538	0.98	X	$8.65^{+0.41}_{-0.48}$	$1.88^{+0.06}_{-0.07}$	3.08 ± 0.14	$13.51^{+0.05}_{-0.07}$
XMM-COSMOS[7]	0.5–2	593	1.21	X	$7.12^{+0.28}_{-0.18}$	$1.81^{+0.04}_{-0.03}$	2.71 ± 0.14	$13.10^{+0.06}_{-0.07}$
SWIFT-BAT	15–55	199	0.045	X	$5.56^{+0.49}_{-0.43}$	$1.64^{+0.07}_{-0.08}$	$1.21^{+0.06}_{-0.07}$	$13.15^{+0.09}_{-0.13}$
AEGIS	0.5–2	113	0.9	X	5.95 ± 0.90	1.66 ± 0.22	$1.97^{+0.26}_{-0.25}$	$13.0^{+0.1}_{-0.4}$
AGES	0.5–2	362	0.51	X	4.5 ± 0.6	1.6 ± 0.1	$1.35^{+0.06}_{-0.07}$	$12.60^{+0.1}_{-0.1}$
ROSAT + SDSS	0.1 2.4	1552	0.27	X	$4.28^{+0.41}_{-0.54}$	$1.67^{+0.13}_{-0.12}$	$1.11^{+0.10}_{-0.12}$	$12.58^{+0.20}_{-0.33}$
XMM-LSS	0.5–2	4360	1.1	3.2 ± 0.5	7.2 ± 0.8	1.93 ± 0.03	2.7 ± 0.3	13.2 ± 0.3
XMM-LSS	2–10	1712	1.0	9.9 ± 0.4	10.1 ± 0.9	1.98 ± 0.04	3.3 ± 0.3	13.7 ± 0.3

X: Unconstrained or undetermined, ^{a}: Bias factors converted to a common cosmology ($\Omega_\Lambda = 0.7$, $\Omega_m = 0.3$, $\sigma_8 = 0.8$), ^{b}: DMH masses estimated using van den Bosch [54] and Sheth et al. [55], [1]: Ebrero et al. [42], fit ID = 2, assuming no redshift evolution of the correlation length, [2]: Gilli et al. [33], [3]: Basilakos et al. [30], using the LDDE model, [4]: Yang et al. [34], [5]: Miyaji et al. [41], fit ID = 6 with integral constrain, assuming redshift evolution of the correlation length, [6]: Gilli et al. [44], [7]: Allevato et al. [45].

Table 1 summarizes all the discussed results on the clustering of AGN in X-ray surveys with bias factors converted to a common cosmology ($\Omega_\Lambda = 0.7$, $\Omega_m = 0.3$, $\sigma_8 = 0.8$) in the EMSS, Boyle and Mo [19]; RASS, Vikhlinin and Forman [20], Akylas et al. [22]; ROSAT-NEP, Mullis et al. [24]; AXIS, Ebrero et al. [42]; ELAIS-S1, Puccetti et al. [32]; CDFS, Gilli et al. [33]; CDFN, Gilli et al. [33], Yang et al. [34]; XMM-2dF, Basilakos et al. [30]; XMM-LSS, Gandhi et al. [31]; CLASXS, Yang et al. [34]; COSMOS, Gilli et al. [44] Allevato et al. [45]; Swift-BAT, Cappelluti et al. [51]; AEGIS, Coil et al. [48]; AGES, Hickox et al. [47]; ROSAT-SDSS, Krumpe et al. [50], while Figure 3 shows the redshift evolution of the correlation length r_0 as estimated in previous works, according to the legend.

2.1. Techniques of Investigation. The continuously increasing volume and quality of data allowed a parallel improvement of the techniques of investigation. The first surveys of *Einstein*

(see, e.g., [15]) used the autocorrelation function of the unresolved CXB and linked it to the clustering properties of the clustering of X-ray source that produced it.

Modern surveys have mostly estimated correlation function with estimators that use random samples and real data pairs and then estimating physical clustering properties by fitting the correlation function functions with simple power-law models in the form of (2). A detailed description of the method to estimate correlation functions is given in the appendix A. Considering its power, here we give a detailed description of halo modeling which is by far the most reliable formalism to describe clustering of AGN/Galaxies and to determine the environment of a specific DMH tracer.

3. Halo Model

In the hierarchical model of cosmological structure formation, galaxies, group of galaxies, clusters, and so on are built

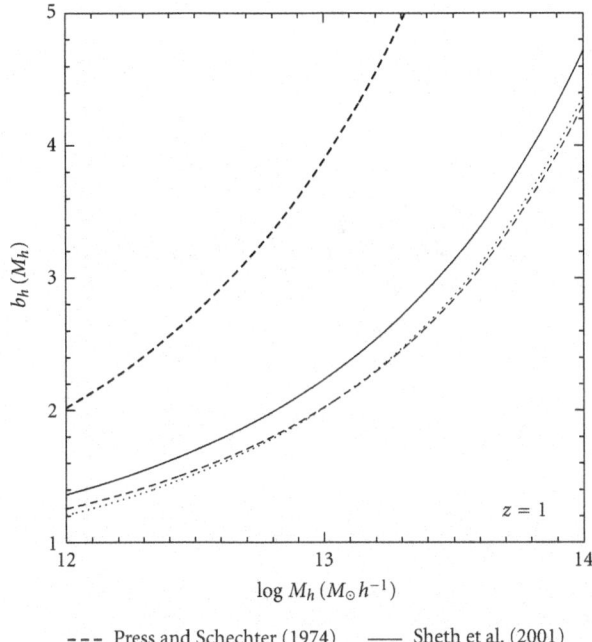

FIGURE 1: Halo bias as function of halo masses for a fixed redshift $z = 1$ and the corresponding predictions of Press and Shechter [56] (long-dashed line), Sheth and Tormen [57] (dashed line), Sheth et al. [55] (solid line), and Tinker et al. [58] (dotted line).

from the same initial perturbation in the underlying dark matter density field. Regions of dark matter denser-than-average collapse to form halos in which structures form. Galaxies and AGN, as well as, groups and clusters are believed to populate the collapsed DMHs.

The theoretical understanding of galaxy clustering has been greatly enhanced through the framework of the *halo model* [58, 64–67]. One can fill DMHs with objects based on a statistical *halo occupation distribution* (HOD), allowing one to model the clustering of galaxies within halos (and thus at nonlinear scales) while providing a self-consistent determination of the bias at linear scales. Similarly the problem of discussing the abundance and spatial distribution of AGN can be reduced to studying how they populate their host halos.

The HOD analysis recasts AGN-clustering measurements into a form that is more physically informative and conducive for testing galaxy/AGN formation theories.

Thus, one can use measurements of AGN two-point correlation functions to constrain the HOD of different sets of AGN and gain information on the nature of DMH in which they live. In fact, the power of the HOD modeling is the capability to transform data on AGN pair counts at small scales into a physical relation between AGN and DMH at the level of individual halos.

The key ingredient needed to describe the clustering properties of AGN is their *halo occupation distribution function* $P_N(M_h)$, which gives the probability of finding N AGN within a single halo as a function of the halo mass, M_h. In the most general case, $P_N(M_h)$ is entirely

specified by all its moments which, in principle, could be observationally determined by studying AGN clustering at any order. Regrettably, AGNs are so rare that their two-point function is already poorly determined, so that it is not possible to accurately measure higher-order statistics. One overcomes this problem by assuming a predefined functional form for the lowest-order moments of $P_N(M_h)$, defining the *halo occupation number* $N(M_h)$ which is the mean value of the halo occupation distribution $N(M_h) = \langle N \rangle(M_h) = \sum_N N P_N(M_h)$. It is convenient to describe $N(M_h)$ in terms of a few parameters whose values will then be constrained by the data.

An accurate description of matter clustering on the basis of the halo approach requires three major ingredients: these halo mass function $n(M_h)$ (the number of DMHs per unit mass and volume), the mass-dependent biasing factor $b(M_h)$, and the density profile of halos. These terms, along with a parametrization of $N(M_h)$, allow us to calculate some useful quantities; the number density of AGN:

$$n_{\text{AGN}} = \int n(M_h)N(M_h)dM_h, \tag{3}$$

the large-scale bias:

$$b = \frac{\int b_h(M_h)N(M_h)n(M_h)dM_h}{\int N(M_h)n(M_h)dM_h}, \tag{4}$$

and the average mass of the host dark halo:

$$M = \frac{\int M_h N(M_h)n(M_h)dM_h}{\int N(M_h)n(M_h)dM_h}. \tag{5}$$

The number density and clustering properties of the DMHs can be easily computed, at any redshift, by means of a set of analytical tools which have been tested and calibrated against numerical simulations [55, 57, 58, 68–72]. Popular choices for both $n(M_h)$ and $b(M_h)$ are the analytical spherical collapse [57] or an ellipsoidal collapse model ([55], see Section 4 for more details). A detailed description of HOD mathematical formalism is given in Appendix B.

3.1. Occupation Number. In the past ten years, a very successful framework for modeling the nonlinear clustering properties of galaxies has been developed and a number of halo models have been presented in the literature. These have been successfully used to describe the abundance and clustering properties of galaxies at both low [58, 65, 73–82] and high [83–87] redshifts, as well as whether these galaxies occupy the centers of the DMH or are satellite galaxies [67, 88].

Partially due to the low number density of AGN, there have been few results in the literature interpreting AGN correlation function using HOD modeling, where the small-scale clustering measurements are essential. Porciani et al. [89] studied the clustering of 2QZ QSO with the halo model to infer the mean number of optically selected quasars which are harboured by a virialized halo of given mass and the characteristic quasar lifetime. Padmanabhan et al. [90] discussed qualitative HOD constraints on their LRG-optical

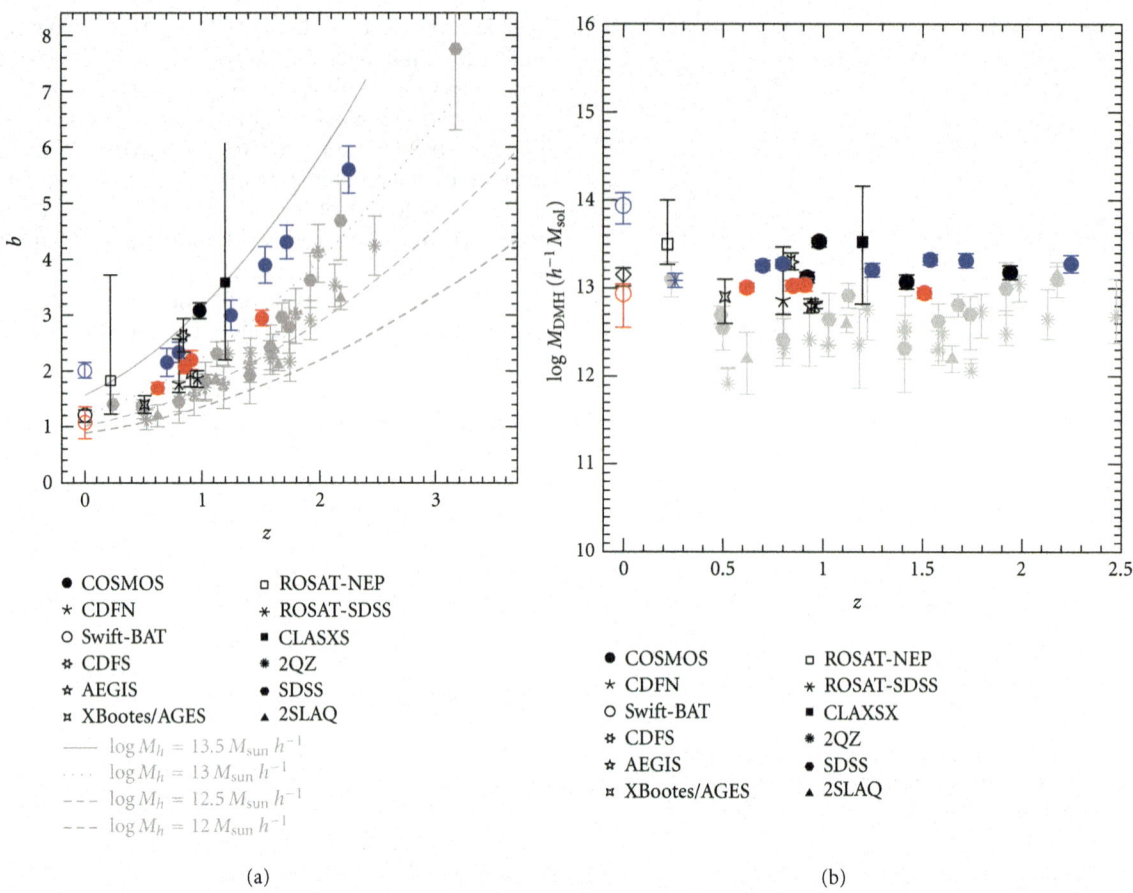

(a)   (b)

FIGURE 2: Bias factor (a) and mass of AGN hosting halos (b) as a function of redshift for X-ray-selected AGN (black data points), X-ray-selected Type 1 AGN (blue data points), and X-ray selected Type 2 AGN (red data points) as estimated in different surveys (COSMOS, Gilli et al. [44], Allevato et al. [45]; CDFN, Gilli et al. [33], Yang et al. [34]; Swift-BAT, Cappelluti et al. [51]; CDFS, Gilli et al. [33]; AEGIS, Coil et al. [48]; AGES, Hickox et al. [47]; ROSAT NEP, Mullis et al. [24]; ROSAT-SDSS, Krumpe et al. [50], CLASXS, Yang et al. [34]). The dashed lines show the expected $b(z)$ of DMHs with different masses according to the legend, based on Sheth et al. [55]. The grey points show results from quasar-quasar correlation measurements using spectroscopic samples from SDSS [59, 60], 2QZ [61, 62], and 2SLAQ [63]. All the previous studies infer the picture that X-ray-selected AGN which are moderate luminosity AGN compared to bright quasars inhabit more massive DMHs than optically selected quasars in the range $z = 0.5$–2.25.

QSO cross-correlation function (CCF), and Shen et al. [91] modelled with the HOD the observed two-point correlation function of 15 binary quasars at $z > 2.9$.

The standard halo approach used for quasars and galaxies is based on the idea that the elements of HOD can be effectively decomposed into two components, separately describing the properties of central and satellite galaxies within the DMH. A simple parametric form used to describe the galaxy HOD is to model the mean occupation number for central galaxies as a step function, that is, $\langle N_{cen} \rangle = 1$ for halos with mass $M \geq M_{min}$ and $\langle N_{cen} \rangle = 0$ for $M < M_{min}$, while the distribution of satellite objects can be well approximated by a Poisson distribution with the mean following a power law, $\langle N_{sat} \rangle = (M/M_1)^{\alpha}$. Previously derived HOD of galaxies show α values ~ 1–1.2 which imply a number of satellite galaxies approximately proportional to M_h.

The clustering properties of X-ray-selected AGN have been modelled with the HOD in two previous works for sources in the *Bootes* field Starikova et al. [92] and in the

ROSAT All-Sky Survey Miyaji et al. [93]. Starikova et al. [92] used the the projections of the two-point correlation function both on the sky plane and in the line of sight to show that *Chandra/Bootes* AGNs are located at the center of DM halos with $M > M_{min} = 4 \times 10^{12} \, h^{-1} M_{\odot}$, assuming a halo occupation described by a step function (zero AGN per halo/subhalo below M_{min} and one above it). They also showed that Chandra/Boötes AGNs are located at the centers of DMHs, limiting the fraction of AGN in noncentral galaxies to be <0.09 at the 95% CL. The central locations of the AGN host galaxies are expected in the merger trigger model because mergers of equally sized galaxies preferentially occur at the centers of DMH [8].

Miyaji et al. [93] modelled the AGN HOD testing the effects of having or not AGN in central galaxies by using the RASS AGN-LRG cross-correlation. In the first scenario, they assumed that all the AGNs are satellites and they visualized the HOD of the LRG as a step function with a step at $\log M_h[h^{-1} M_{\odot} = 13.5]$. While formally they assumed that

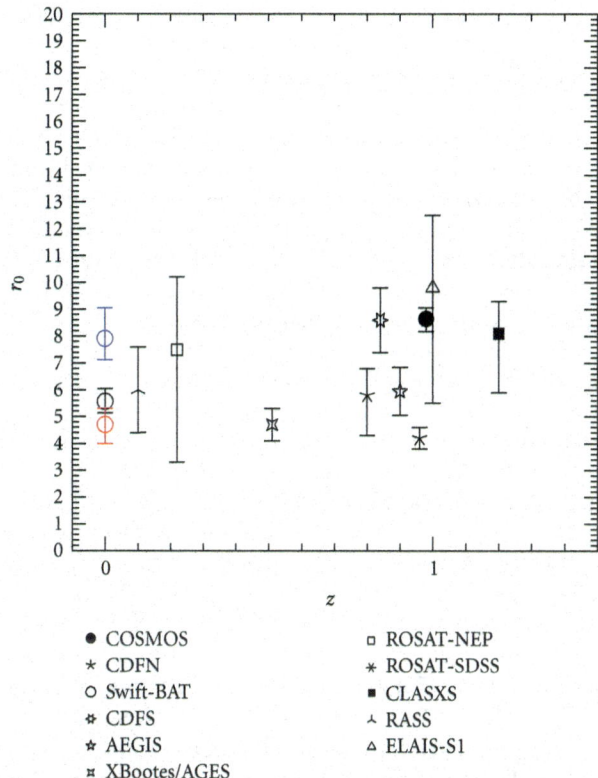

FIGURE 3: Redshift evolution of the correlation length r_0 as estimated in different X-ray surveys (COSMOS, Gilli et al. [44], Allevato et al.[45]; CDFN, Gilli et al. [33], Yang et al. [34]; Swift-BAT, Cappelluti et al. [51]; CDFS, Gilli et al. [33]; AEGIS, Coil et al. [48]; AGES, Hickox et al. [47]; ROSAT-NEP, Mullis et al. [24]; ROSAT-SDSS, Krumpe et al. [50]; CLASXS, Yang et al. [34]; RASS, Akylas et al. [22]; ELAIS-S1, Puccetti et al. [32]).

all AGNs are not in central galaxies, the HOD constraints obtained from this assumption can be applied to satellite and central AGN if the AGN activity in central galaxies of high-mass halos ($\log M_h[h^{-1} M_\odot > 13.5]$) is suppressed. In particular, they used a truncated power-law satellite HOD, with two parameters: the critical DMH mass below which the AGN HOD is zero and the slope α of the HOD for $M_h > M_{cr}$. They also investigated a model where the central HOD is constant and the satellite HOD has a power-law form, both at masses above M_{min}. In all the cases, they rejected $\alpha \sim 1$, finding a marginal preference for an AGN fraction among satellite galaxies which decreases with increasing M_h. They argued that this result might be explained by a decrease of the cross-section for galaxy merging in the environment of richer groups or clusters. In fact, previous observations infer that the AGN fraction is smaller in clusters than in groups [27, 94–96].

It is important to stress that the small number statistics has so far limited the accuracy of correlation function of X-ray AGN at small-scales, especially through the autocorrelation function of the AGN themselves. The situation can be improved by measuring the cross-correlation function of AGN with a galaxy sample that has a much higher space

density, with common sky and redshift coverage as the AGN redshift surveys. The AGN clustering through cross-correlation function with galaxies is emerging in the last years [47, 48, 90, 97–99] and can be used to improve our understanding of how AGNs populate DMH [52, 93].

4. Bias and DMH Mass

In the literature, the bias parameter is often calculated with the power-law fits [24, 34, 41, 48, 50, 100] over scales of 0.1–0.3 $< r_p <$ 10–20 h^{-1} Mpc. The power-law models of the ACF are usually converted to the rms fluctuation over 8 h^{-1} Mpc spheres or are averaged up to the distance of 20 h^{-1} Mpc. While some authors use only large scales ($r_p > 1$-2 h^{-1} Mpc) to ensure that the linear regime is used, others include smaller scales to have better statistics. As an example, Hickox et al. [47] fitted their data with a biased DMH-projected correlation function.

In the HOD analysis, the bias factor only comes from the 2-halo term ($r_p > 1$-2 h^{-1} Mpc). Miyaji et al. [93] compared the bias of RASS-AGN from the full HOD model (D.4) with the one estimated using the power-law best fits parameters, finding that the bias estimates are consistent within 1σ. Moreover, using (D.1), one introduces large statistical errors. Allevato et al. [45] found a similar results in comparing the bias of X-ray AGN in COSMOS field from the 2-halo term with (D.3) and the one estimated from the power-law best fits parameters. In Appendix C, we describe the mathematical procedures for the bias parameter calculation commonly used in the literature.

Most of the authors [47, 50, 51] used an analytical expression (as the one described in [55, 57, 58, 69]) to assign a characteristic DMH mass to the hosting halos. The large-scale bias is directly related to the mass function of halos, so that the mass of a halo dictates the halo clustering and the number of such halos. The halo mass can be quantified in terms of the peak height $\nu = \delta_c/\sigma(M_h, z)$, which characterizes the amplitude of density fluctuations from which a halo of mass M_h forms at a given redshift. In generals one assumes $\delta_c = 1.686$ and $\sigma(M_h, z)$ is the linear overdensity variance in spheres enclosing a mean mass M_h. The traditional choice of the mass function and then of the bias has been that of Press and Schechter [56]:

$$b^{PS} = 1 + \frac{\nu^2 - 1}{\delta_c}.$$ (6)

A commonly used prescription was derived by Sheth and Tormen [57]:

$$b^{ST} = 1 + \frac{a\nu^2 - 1}{\delta_c} + \frac{2p/\delta_c}{1 + (a\nu^2)^p},$$ (7)

where $a = 0.707$ and $p = 0.3$ or the ellipsoidal collapse formula of Sheth et al. [55]:

$$b^{SMT} = 1 + \frac{1}{\sqrt{a}\delta_c}\left[\sqrt{a}(a\nu^2) + \sqrt{a}b(a\nu^2)^{1-c}\right.$$
$$\left. - \frac{(a\nu^2)^c}{(a\nu^2)^c + b(1-c)\left(\frac{1-c}{2}\right)}\right],$$ (8)

where $a = 0.707$, $b = 0.5$, $c = 0.6$ or the recalibrated parameters $a = 0.707$, $b = 0.35$, $c = 0.8$ of Tinker et al. [58]. The ν parameter can be estimated following the appendix of Van den Bosch [54]. Figure 1 shows the bias as function of the halo mass M_h, at $z = 1$, following the predictions of Press and Schechter [56], Sheth and Tormen [57], Sheth et al. [55], and Tinker et al. [58].

Allevato et al. [45] argued that this approach reveals an incongruity due to the fact that the AGN bias used in the formulas above is the average bias of a given AGN sample at a given redshift. In fact, following this approach, one cannot take into account that the average bias is sensitive to the entirety of the mass distribution; different mass distributions with different average masses can give rise to the same average bias.

On the contrary, by using the halo model, the average bias and the average mass of the sample, (D.4), and (5) properly account for the shape of the mass distribution: the average bias depends on the halo number density and on the AGN HOD, integrated over the mass range of the particular AGN sample. They introduced a new method that uses the 2-halo term in estimating the AGN bias factor assuming an AGN HOD described by a δ-function. Following this approach, they properly took into account for the sample variance and the growth of the structures over time associated with the use of large redshift interval of the AGN sample.

On the other hand, Miyaji et al. [93] and Krumpe et al. [52] applied the HOD modeling technique to the RASS AGN-LRG CCF in order to move beyond determining the typical DMH mass based on the clustering signal strength and instead constrain the full distribution of AGN as a function of DMH mass. Along with a parametrization of $N(M_h)$, they estimated the large-scale bias and the typical mass of hosting DM halos using (D.4) and (5). This method improves the clustering analysis because it properly uses the nonlinear growth of matter in the *1-halo* term through the formation and growth of DMHs. These results are significant improvements with respect to the standard method of fitting the signal with a phenomenological power law or using the 2-halo term (see Appendix C).

4.1. X-Ray-Selected AGN Bias, Bias Evolution, and Mass of the Hosting Halos.

The majority of the X-ray surveys agree with a picture where X-ray AGNs are typically hosted in DM halos with mass of the order of $12.5 < \log M_{DMH}[h^{-1} M_\odot] < 13.5$, at low ($z < 0.4$) and high ($z \sim 1$) redshift [33, 34, 44, 47, 48, 50–52, 92, 93].

At high redshift, Gilli et al. [33] measured the clustering of X-ray AGN with $z = 0$–4 in both the $\sim 0.1\,\mathrm{deg}^2$ CDFs, finding $b = 1.87^{+0.14}_{-0.16}$ for 240 sources in the northern field and $b = 2.64^{+0.29}_{-0.30}$ for 124 sources in the southern field. At $z \sim 1$, Yang et al. [34] measured the clustering of 233 spectroscopic sources in the $0.4\,\mathrm{deg}^2$ *Chandra* CLASXS area and of 252 spectroscopic sources from the CDFN, both at $z = 0.1$–3. They found $b = 3.58^{+2.49}_{-1.38}$ for the CLASXS AGN and $b = 1.77^{+0.80}_{-0.15}$ for the CDFN field. Gilli et al. [44] studied 538 XMM-COSMOS AGN with $0.1 < z < 3$, and they found a bias factor $b = 3.08^{+0.14}_{-0.14}$ at $\bar{z} \sim 1$. Using

the Millennium simulations, they suggested that XMM-COSMOS AGNs reside in DMH with mass $M_{DMH} > 2.5 \times 10^{12}\,h^{-1}\,M_\odot$. Coil et al. [48] measured the clustering of X-ray AGN at $z = 0.7$–1.4 in the AEGIS field, and they estimated $b = 1.85^{+0.28}_{-0.28}$. Following Zheng et al. [87], they infer from the bias factor that, at $z = 0.94$, the minimum DM halo mass of the X-ray AGN is $>10^{12}\,M_\odot\,h^{-1}$. These results combined with Mountrichas and Georgakakis [49] show that moderate luminosity X-ray-selected AGN live in DMHs with masses $M_h \sim 10^{13}\,h^{-1}\,M_\odot$ at all redshifts since $z \sim 1$. At lower redshift, Hickox et al. [47] analysed 362 AGES X-ray AGN at $\langle z \rangle = 0.51$. The bias factor equal to $b = 1.40 \pm 0.16$ indicates that X-ray AGNs inhabit DM halos of typical mass $\sim 10^{13}\,M_\odot\,h^{-1}$.

In the local Universe, Cappelluti et al. [51] estimated for ~ 200 Swift-BAT AGN a bias equal to $b = 1.21^{+0.07}_{-0.06}$ which corresponds to $\log M_{DM} = 13.15^{+0.09}_{-0.13}\,h^{-1}\,M_\odot$.

Allevato et al. [45] estimated an average mass of the XMM-COSMOS AGN hosting halos equal to $\log M_0[h^{-1}Mpc] = 13.10 \pm 0.06$ at $z \sim 1.2$. They also measured the bias of Type 1 and Type 2 AGN, finding that the latter resides in less massive halos than Type 1 AGN. Only two other works [50, 51] analysed the clustering properties of X-ray-selected Type 1 AGN and Type 2 AGN. Cappelluti et al. [51] estimated the typical DM halo mass hosting type 1 and type 2 Swift-BAT AGN at $z \sim 0$. They measured that these two different samples are characterized by halos with mass equal to $\log M_{DM}[h^{-1}\,M_\odot] \sim 13.94^{+0.15}_{-0.21}$ and $\sim 12.92^{+0.11}_{-0.38}$, respectively. However, the lack of small separation pair of Type I AGN in the local Universe may have produced systematic deviations which were not accounted in their fits. In Krumpe et al. [50], the bias factor of BL RASS AGN at $z = 0.27$ is consistent with BL AGN residing in halos with mass $\log M_{DM}[h^{-1}\,M_\odot] = 12.58^{+0.20}_{-0.33}$.

Using the HOD model, Starikova et al. [92] suggested that X-ray Chandra/Bootes AGN is located at the center of DM halos with $M > M_{min} = 4 \times 10^{12}\,h^{-1}\,M_\odot$, while Miyaji et al. [93] estimated for RASS AGN at $z = 0.25$, $b = 1.32 \pm 0.08$, and a typical mass of the host halos of 13.09 ± 0.08.

The redshift evolution of the clustering of X-ray-selected AGN has been first studied by Yang et al. [34] in the CLAXS+CDFN fields. They measured an increase of the bias factor with redshift, from $b = 0.95 \pm 0.15$ at $z = 0.45$ to $b = 3.03 \pm 0.83$ at $z = 2.07$, corresponding to an average halo mass of $\sim 12.11\,h^{-1}\,M_\odot$.

Allevato et al. [45] studied the redshift evolution of the bias for a sample of XMM-COSMOS AGN at $z < 2$. They found a bias evolution with time from $b(z = 0.92) = 1.80 \pm 0.19$ to $b(z = 1.94) = 2.63 \pm 0.21$ with a DM halo mass consistent with being constant at $\log M[h^{-1}\,M_\odot] \sim 13.1$ at all redshifts $z < 2$. They also found evidence of a redshift evolution of the bias factor of XMM-COSMOS Type 1 AGN and Type 2. The bias evolves with redshift at constant average halo mass $\log M_0[h^{-1}\,M_\odot] \sim 13.3$ for Type 1 AGN and $\log M_0[h^{-1}\,M_\odot] \sim 13$ for Type 2 AGN at $z < 2.25$ and $z < 1.5$, respectively. In particular, Allevato et al. [45] argued that X-ray selected Type 1 AGNs reside in more massive DMHs compared to X-ray-selected Type 2 AGN at all redshifts at

~2.5σ level, suggesting that the AGN activity is a mass-triggered phenomenon and that different AGN classes are associated with the DM halo mass, irrespective of redshift z.

Krumpe et al. [52] measured the clustering amplitudes of both X-ray RASS and optically selected SDSS broad-line AGNs, as well as for X-ray-selected narrow-line RASS/SDSS AGNs through cross-correlation functions with SDSS galaxies and derive the bias by applying the HOD model directly to the CCFs. They estimated typical DMH masses of broad-line AGNs in the range $\log(M_h/[h^{-1} M_\odot]) = 12.4$–13.4, consistent with the halo mass range of typical non-AGN galaxies at low redshifts, and they found no significant difference between the clustering of X-ray-selected narrow-line AGNs and broad-line AGNs up to $z \sim 0.5$.

Figure 2(a) shows the bias parameter and Figure 2(b) the mass of the AGN hosting halos as a function of redshift for X-ray-selected AGN (black data points), X-ray-selected Type 1 AGN (blue data points), and X-ray-selected Type 2 AGN (red data points) as estimated for different surveys (see the legend). The dashed lines show the expected b(z) of typical DM halo masses M_{DMH} based on Sheth et al. [55]. The masses are given in $\log M_{DMH}$ in units of $h^{-1} M_\odot$.

There have been several studies of the bias evolution of optical quasar with the redshift as shown in Figure 2 (grey data points), based on large survey samples such as 2QZ, 2SLAQ, and SDSS [59–63]. These previous studies infer the picture that X-ray-selected AGNs which are moderate luminosity AGN compared to bright quasars inhabit more massive DMHs than optically selected quasars in the range $z = 0.5$–2.25.

Recently, Krumpe et al. [52] verified that the clustering properties between X-ray and optically selected AGN samples are not significantly different in three redshift bins below $z = 0.5$ (the differences are 1.5σ, 0.1σ, and 2.0σ). The reason for the fact that X-ray-selected AGN samples appear to cluster more strongly than optically selected AGNs is still unclear. Allevato et al. [45] and Mountrichas and Georgakakis [49] suggested that the difference in the bias and then in the host DMH masses is due to the different fueling mode of those sources from that of the X-ray-selected moderate luminosity AGN. On the contrary, Krumpe et al. [52] suggested that some of the X-ray clustering studies significantly underestimate their systematic uncertainties and then it may turn out that these measurements are consistent with optical AGN clustering measurements. More high-z AGN clustering measurements based on larger samples are needed to gain a clearer picture.

4.2. AGN Life Time. One of the most important tests for studying the evolution models of AGN is understanding their lifetime. It is widely accepted that AGN is phase of the galaxy life necessary to explain the coevolution of the bulge and the black hole. After a triggering event of which we do not know the nature, yet, the central black hole begins its accretion phase and it is believed that it undergoes several regimes of Eddington rates and bolometric luminosity. Martini and Weinberg [101] proposed a method to derive the AGN life time by knowing their space density and their DMH host mass.

By knowing the AGN and DMH halo space density at a given luminosity and mass (n_{AGN}, n_{DMH}), one can estimate the duty cycle of the AGN, $\tau_{AGN}(z) = (n_{AGN}(L, z)/n_{DMH}(M, z))(\tau_H(z))$, where $\tau(H(z))$ is the Hubble time at a given redshift. Actually, this method provides only an upper limit since it assumes that the life of halo of a given mass is similar to the Hubble time. A more exhaustive formulation would be $\tau_{AGN}(z) = (n_{AGN}(L, z)/n_{DMH}(M, z))(\tau_{DMH}(z))$, where $\tau_{DMH}(z)$ is the age of a DMH at given redshift. Unfortunately, this quantity cannot be estimated analytically but could be estimated in a statistical way by using hydrodynamic simulations. Several results can be mentioned for these quantities, but their dispersion is very large, therefore we report only some example. At $z = 1$, Gilli et al. [44] obtains that the typical duty cycle of AGN is <1 Gyr. At $z = 0$, Cappelluti et al. [51] have measured a duty cycle in the range 0.2 Gyr–5 Gyr with an expectation value of 0.7 Gyr. Both the measurements are fairly larger than the 40 million years determined by Martini and Weinberg [101] at $z = 2$-3. These differences, however, are not surprising if we assume that the different populations of AGN grow with a different Eddington rate as function of their typical luminosities and/or redshifts [102].

5. Discussion

In this paper, we reviewed the results in the field of X-ray AGN clustering, for energies between 0.1 keV to 55 keV over a period of more than 20 years. The literature has produced an increasingly convincing and consistent picture of the physical quantities derivable from this kind of study. Most of the advancements in the field have been achieved with the improvement of survey capabilities and instruments sensitivity. The availability of simultaneously wide and deep fields, coupled with multiwavelength information, has produced larger and larger samples of spectroscopically confirmed sources. This allowed several teams to refine the techniques needed to estimate the two-point ACF and the quantities derived form it. In particular, we are entering a phase where, at least at $z < 2$, AGN clustering studies will not probably provide any new result unless evaluated with the HOD formalism. Open questions as what is the AGN occupation number and the evolution of HOD define a new barrier which is necessary to break in order to understand the history of X-ray emission from accretion onto AGN. In this respect, samples of X-ray-selected AGN always need a spectroscopical followup to provide a solid base to compute clustering in the real space rather than in the angular space.

Summarizing, the current picture is that X-ray-selected AGNs are highly biased objects with respect to the underlined matter distribution. Such an evidence is clearer when measuring the redshift dependence of AGN bias. At every redshift from $z = 0$ to $z = 2$, AGNs cluster in way similar to DMH of mass of the order of $\log(M_\odot h^{-1}) = 13$. The spread of such a value is of the order 0.3–0.5 dex at 1σ. This means that the determination of what kind of environment is inhabited by AGN is relatively well constrained and identical at every redshift sampled by X-ray surveys. This allows us to formulate the hypothesis that every phase of AGN activity is mass-triggered phenomenon (i.e., each AGN evolutionary phase is characterized by a critical halo mass).

It is believed that major mergers of galaxies is one of the dominant mechanisms for fueling quasars at high redshift and bright luminosities, while minor interactions, bar instabilities, or tidal disruptions are important at low redshift ($z \lesssim 1$) and low luminosities ($L \lesssim 10^{44}\, \mathrm{erg\, s^{-1}}$) [9, 13, 103, 104]. In the local Universe, for example, the study of the environment of Swift BAT Seyfert galaxies [28] finds a larger fraction of BAT AGNs with disturbed morphologies or in close physical pairs (<30 kpc) compared to matched control galaxies or optically selected AGNs. The high rate of apparent mergers (25%) suggests that AGN activity and merging are critically linked for the moderate luminosity AGN in the BAT sample. Moreover, models of major mergers appear to naturally produce many observed properties of quasars, as the quasar luminosity density, the shape, and the evolution of the quasar luminosity function and the large-scale quasar clustering as a function of L and z (e.g., [8, 46, 105–109]). Quasar clustering at all redshift is consistent with halo masses similar to group scales, where the combination of low velocity dispersion and moderate galaxy space density yields to the highest probability of a close encounter [8, 11]. Moreover, recent detections of an L_X-dependent clustering play in favor of major mergers being the dominant AGN triggering mechanism.

On the other hand, it has became clear that many AGNs are not fueled by major mergers and only a small fraction of AGNs are associated with morphologically disturbed galaxies. Georgakakis et al. [110] and Silverman et al. [96] found that AGNs span a broad range of environments, from the field to massive groups and thus major mergers of galaxies, possibly relevant for the more luminous quasar phenomenon, may not be the primary mechanism for fueling these moderate luminosity AGN.

Georgakakis et al. [111] suggest that bar instabilities and minor interactions are more efficient in producing luminous AGN at $z \lesssim 1$ and not only Seyfert galaxies and low-luminosity AGN as the Hopkins and Henquist [9] model predicts. Cisternas et al. [112] analysed a sample of X-ray-selected AGN host galaxies and a matched control sample of inactive galaxies in the COSMOS field. They found that mergers and interactions involving AGN hosts are not dominant and occur no more frequently than for inactive galaxies. Over 55% of the studied AGN sample that is characterized by $L_{BOL} \sim 10^{45}\, \mathrm{erg\, s^{-1}}$ and by mass of the host galaxies $M_* \gtrsim 10^{10}\, M_\odot$ are hosted by disk-dominated galaxies, suggesting that secular fuelling mechanisms can be highly efficient.

Moreover, several works on the AGN host galaxies [113–118] show that the morphologies of the AGN host galaxies do not present a preference for merging systems.

At high redshift ($z \sim 2$), recent findings of Schlegel et al. [119] and Rosario et al. [120], who examined a smaller sample of AGN in the ERS-II region of the GOODS-South field, inferred that late-type morphologies are prevalent among the AGN hosts. The role that major galaxy mergers play in triggering AGN activity at $1.5 < z < 2.5$ was also studied in the CDF-S. At $z = 1.5$–3, Schawinski et al. [121] showed that, for X-ray-selected AGN in the Chandra Deep Field South and with typical luminosities of $10^{42}\, \mathrm{erg\, s^{-1}} < L_X < 10^{44}\, \mathrm{erg\, s^{-1}}$, the majority (80%) of the host galaxies of these AGNs have low Srsic indices indicative of disk-dominated light profiles, suggesting that secular processes govern a significant fraction of the cosmic growth of black holes. That is, many black holes in the present-day Universe grew much of their mass in disk-dominated galaxies and not in early-type galaxies or major mergers.

Later, Kocevski et al. [122] found that X-ray-selected AGNs at $z \sim 2$ do not exhibit a significant excess of distorted morphologies while a large fraction reside in late-type galaxies. They also suggested that these late-type galaxies are fueled by the stochastic accretion of cold gas, possibly triggered by a disk instability or minor interaction.

Allevato et al. [45] argued that for moderate luminosity X-ray AGN secular processes such as tidal disruptions or disk instabilities might play a much larger role than major mergers up to $z \sim 2.2$.

It becomes important to study the clustering properties of AGN at high redshift when we assume the peak of the merger-driven accretion. Moreover, given the complexity of AGN triggering, a proper selection of AGN samples, according to the luminosity or the mass of the host galaxies, can help to test a particular model boosting the fraction of AGN host galaxies associated with morphologically disturbed galaxies.

From the evolutionary point of view, the evidence of a bias segregation of optically and X-ray-selected AGN might be a sufficient proof to claim that the two phenomena are sensitive to different environments and therefore likely driven by different triggering mechanisms. A more comprehensive picture will be available when the clustering of different phases of AGN activity will be studied and compared.

Hickox et al. [47] interpreted their clustering results in terms of a general picture for AGN and galaxy evolution which is reproduced in Figure 4. The picture consists of an evolutionary sequence that occurs at different redshifts for halos with different masses. In this scenario, luminous AGN accretion occurs preferentially (through a merger or some secular process) when a host DMH reaches a critical M_{DMH} between 10^{12} and $10^{13}\, M_\odot\, h^{-1}$ (this phase is indicated by the solid ovals). Once a large halo reaches this critical mass, it becomes visible as a ULIRG or SMG (owing to a burst of dusty star formation) or (perhaps subsequently) as a luminous, unobscured quasar. The ULIRG/quasar phase is associated with rapid growth of the SMBH and formation of a stellar spheroid and is followed by the rapid quenching of star formation in the galaxy. Subsequently, the young stellar population in the galaxy ages (producing "green" host galaxy), and the galaxy experiences declining nuclear accretion that may be associated with an X-ray AGN. Eventually, the aging of the young stars leaves a "red" and "dead" early-type galaxy, which experiences intermittent

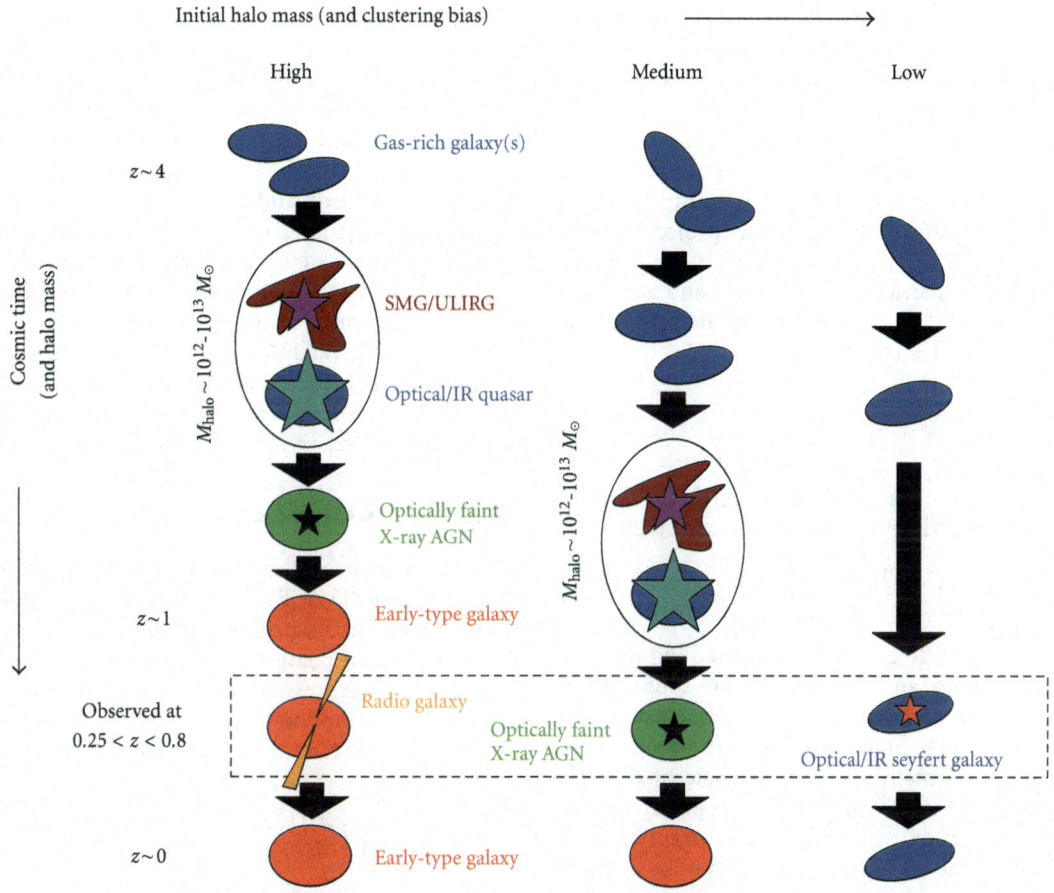

FIGURE 4: Schematic for a simple picture of AGN and host galaxy evolution, taken from Hickox et al. [47] and motivated by the AGN host galaxy and clustering results presented in that study.

"radio-mode" AGN outbursts that heat the surrounding medium. For "medium" initial DMHs, the quasar phase and formation of the spheroid occur later than for the systems with high halo mass, so that, at $z \sim 0.5$, we may observe the green X-ray AGN phase. Even smaller halos never reach the threshold mass for quasar triggering; these still contain star-forming disk galaxies at $z \lesssim 0.8$, and we observe some of them as optical or IR-selected Seyfert galaxies. The dashed box indicates the AGN types (in their characteristic DMH) that would be observable in the redshift range $0.25 < z < 0.8$.

Further steps in the field will require the study of clustering of AGN from $z = 3$ to $z = 6$-7. This will likely lead to the determination of the mass of early DM spheroids who hosted primordial black holes seeds. However, this is a very challenging task since it requires a very deep and wide survey with an almost complete optical followup.

BOSS [123] and BigBOSS [119] will detect high redshift AGNs at $z \sim 2.2$, which will improve AGN clustering measurements at higher redshifts. The only approved mission that at the moment will allow to study the $z = 3$–5 X-ray Universe is eROSITA ([124], launch Dec. 2013) for which an estimate of the completeness of the typical followup is still unavailable. Additionally, the Large Synoptic Survey Telescope ([125], LSST) is expected to identify ~ 2 million

AGNs in optical bands. eROSITA and LSST have the potential to significantly improve AGN clustering measurements at low and high redshifts, though only if there are dedicated large spectroscopic follow-up programs. Another strong contribution will come from either Nustar that will likely provide a better view of AGN clustering without the selection biases introduced by photoelectric absorption. Athena, the proposed ESA new generation telescope that will mount a wide field imager on a very large collecting area telescope, will provide a further view on the deep X-ray sky and likely push our knowledge of the high-z X-ray Universe.

In addition to better model the evolution of SMBH environments, a fundamental point to start is to establish the nature of BH seeds at $z = 10$. Such a determination will likely come with the new generation of telescope like JWST and ESO-ELT.

Appendices

A. Deriving the Two-Point Autocorrelation Function

The two-point autocorrelation function ($\xi(r)$, ACF) describes the excess probability over random of finding a pair with an object in the volume dV_1 and another in the volume dV_2, separated by a distance r so that $dP = n^2[1 + \xi(r)]dV_1dV_2$, where n is the mean space density. A known effect when measuring pairs separations is that the peculiar velocities combined with the Hubble flow may cause a biased estimate of the distance when using the spectroscopic redshift. To avoid this effect, it is usually computed the projected ACF [126]: $w(r_p) = 2\int_0^{\pi_{max}} \xi(r_p, \pi)d\pi$, where r_p is the distance component perpendicular to the line of sight and π parallel to the line of sight [127]. It can be demonstrated that, if the ACF is expressed as $\xi(r) = (r/r_0)^{-\gamma}$, then

$$w(r_p) = A(\gamma)r_0^\gamma r_p^{1-\gamma}, \tag{A.1}$$

where $A(\gamma) = \Gamma(1/2)\Gamma[(\gamma - 1)/2]/\Gamma(\gamma/2)$ [21].

The ACF is mostly estimated by using the minimum variance estimator described by Landy and Szalay [128]:

$$\xi(r_p, \pi) = \frac{DD - 2DR + RR}{RR}, \tag{A.2}$$

where DD, DR, and RR are the normalized number of data-data, data-random, and random-random source pairs, respectively. Equation (A.2) indicates that an accurate estimate of the distribution function of the random samples is crucial in order to obtain a reliable estimate of $\xi(r_p, \pi)$. Note that other estimators have been proposed in the literature, but the Landy and Szalay [128] one has been shown to provide the smallest statistical variance. Such a formalism can be easily adopted when computing the angular or the redshift space correlation function, with the only difference that the evaluation is made on a single dimension. Several observational biases must be taken into account when generating a random sample of objects in a X-ray-flux limited survey. In particular, in order to reproduce the selection function of the survey, one has to carefully reproduce the space and flux distributions of the sources, since the sensitivity in X-ray surveys is not homogeneous on the detector and therefore on the sky. This points out the necessity of creating a random sample which includes as many selection effects as possible since the estimate of $\xi(r)$ (or $w(\theta)$) is strongly dependent on RR (see (A.2)). Moreover, in several cases, optical followup of the X-ray source is not 100% complete, therefore one must carefully reproduce the mask effect. What is usually done is that to create random samples in 3D, sources are placed at the same angular position of the real sources and redshift are randomly drawn from a smoothed redshift distribution of the real sources. If instead the spectral completeness is close to 100%, then the right procedures are to occupy the survey volume with random sources drawn from a L-z dependent luminosity function and accept check if they would be observable using a sensitivity map. An important choice for obtaining a reliable estimate of $w(r_p)$ is to set π_{max} in the calculation of the integral above. One should avoid values of π_{max} too large since they would add noise to the estimate of $w(r_p)$. If, instead, π_{max} is too small one could not recover all the signal. Uncertainties in the ACF are usually evaluated with a bootstrap resampling technique, but it is worth noting

that, in the literature, several methods are adopted for errors estimates in two-point statistics, (See, [129] for a detailed description). It is known that Poisson estimators generally underestimate the variance because they do consider that points in ACF are not statistically independent. Jackknife resampling method, where one divides the survey area in many sub fields and iteratively recomputes correlation functions by excluding one subfield at a time, generally gives a good estimate of errors. But it requires that sufficient number of almost statistically independent subfields, this is not the case for most of X-ray surveys where the source statistics is moderately low. Coil et al. [48] estimated the error bars on the two-point correlation function including both Poisson and cosmic variance errors estimated, using DEEP2 mock catalogs derived from the Millenium Run simulations.

B. Limber's Deprojection

The 2D angular correlation function (ACF) is a projection of the real-space 3D ACF of the sources along the line of sight. In the following discussions and thereafter, r is in comoving coordinates. The relation between the 2D (angular) ACF and the 3D ACF is expressed by the Limber equation (e.g., [21]). Under the assumption that the scale length of the clustering is much smaller than the distance to the object, this reduces to

$$w(\theta)N^2 = \int \left(\frac{dN}{dZ}\right)^2$$
$$\int \xi\left(\sqrt{[d_A(z)\theta^2] + l^2(1+z)}\right)\left(\frac{dl}{dz}\right)^{-1} dl\, dz, \tag{B.1}$$

where $d_A(z)$ is the angular distance, N is the total number of sources, and dN/dz is the redshift distribution (per z) of the sources. The redshift evolution of the 3D correlation function is customarily expressed by

$$\xi(r, z) = \left(\frac{r}{r_0}\right)^{-\gamma}(1+z)^{-3-\epsilon+\gamma}, \tag{B.2}$$

where $\epsilon = -3$ and $\epsilon = \gamma - 3$ correspond to the case where the correlation length is constant in physical and comoving coordinates, respectively. In these notations, the zero-redshift 3D correlation length r_0 can be related to the angular correlation length θ_0 by

$$r_0^\gamma = \left(\frac{N^2}{S}\right)\theta_0^{\gamma-1},$$

$$S = H_\gamma \int \left(\frac{dN}{dZ}\right)^2 \left[\frac{cd\tau(z)}{dz}\right]^{-1}$$
$$d_A^{1-\gamma}(1+z)^{-3-\epsilon} dz, \tag{B.3}$$

$$H_\gamma = \frac{\Gamma[(\gamma - 1)/2]\Gamma(1/2)}{\Gamma(1/2)},$$

where $\tau(z)$ is the look-back time. We also define the comoving correlation length

$$r_0(z_{eff}^-) = r_0(1 + z_{eff}^-)^{-3-\epsilon+\gamma}, \tag{B.4}$$

at the effective redshift z_{eff}^-, which is the median redshift of the contribution to the angular correlation (the integrand of the second term). An essential ingredient of the deprojection process is the redshift distribution of the sources, and, when individual redshifts are not available, this is derived from integration of the luminosity function.

C. 1-Halo and 2-Halo Terms in the HOD Formalism

In the halo model approach, the two-point correlation function of AGN is the sum of two contributions: the first term (1-halo term) is due to the correlation between objects in the same halo and the second term (2-halo term) arises because of the correlation between two distinct halos:

$$\xi(r) = \xi_{1h}(r) + \xi_{2h}(r). \tag{C.1}$$

Recent articles prefer to express $w = (1 + \xi_{1h}) + \xi_{2h}$ [58, 67, 130], instead of $\xi = \xi_{1h} + \xi_{2h}$, as used in older articles. This is because $1 + \xi$ represents a quantity that is proportional to the number of pairs $\propto [1 + \xi_{1h}] + [1 + \xi_{2h}]$. In this new convention, the projected correlation function ξ_{1h} represents the projection of $1 + \xi_{1h}$ rather than ξ_{1h}.

Similarly, one expresses the power spectrum of the distribution of the AGN in terms of the 1- and 2-halo term contributions:

$$P(k) = P_{1h}(k) + P_{2h}(k), \tag{C.2}$$

and then the projected correlation function as

$$w_{p,1h}\left(r_p\right) = \int \frac{k}{2\pi} P_{1h}(k) J_0\left(kr_p\right) dk,$$
$$w_{p,2h}\left(r_p\right) = \int \frac{k}{2\pi} P_{2h}(k) J_0\left(kr_p\right) dk, \tag{C.3}$$

where $J_0(x)$ is the zeroth-order Bessel function of the first kind.

Several parametrizations exist in literature for representing the DMH profile [66, 131, 132], and the Navarro et al. [6] (NFW) profile is a popular choice. If $y(k, M_h)$ expresses the Fourier transform of the NFW profile of the DMH with mass M_h, normalized such that volume integral up to the virial radius is unity, then the one-halo term of the power spectrum can be written as

$$P_{1h}(k) = \frac{1}{n_{AGN}^2} \int n(M_h) N(M_h) |y(k, M_h)|^2 dM_h. \tag{C.4}$$

Assuming the linear halo bias model [68], the two-halo term of the power spectrum reduces to

$$P_{2h}(k) = P_m(k) \left[\frac{1}{n_{AGN}} \int n(M_h) b(M_h) y(k, M_h) dM_h \right]^2. \tag{C.5}$$

Since the clustering on large scales is dominated by the two-halo term, it is fairly insensitive to the assumption of AGN distribution inside the hosting halo [75]. It should be noted that since $y \sim 1$ on large scales (e.g., scales much larger than the virial radius of halos), on such scales the two-halo term can be rewritten as

$$P_{2h}(k) \approx b^2 P_m(k, z), \tag{C.6}$$

or, in terms of projected correlation function,

$$w_{p,2h}\left(r_p\right) = b^2 w_{m,2h}\left(r_p\right), \tag{C.7}$$

where b is the bias parameter of the sample and $w_{m,2h}$ is the DM-projected correlation function. For the matter power spectrum, $P_m(k)$, one can use the primordial power spectrum with a fixed n_s and a transfer function calculated using the fitting formula of Eisenstein and Hu [133] or the nonlinear form given by Smith et al. [134] and Tinker et al. [58].

D. Bias Parameter Calculation

In the majority of works on clustering of X-ray AGN [24, 33, 34, 48, 50, 51], the standard approaches used to estimate the bias are based on the power-law fit parameters of the AGN correlation function. This method assumes that the projected correlation function is well fitted by a power-law and the bias factors are derived from the best fit parameters r_0 and γ of the clustering signal at large scale. Using the power-law fit, one can estimate the AGN bias factor using the power-law best fit parameters:

$$b_{PL} = \frac{\sigma_{8,AGN}(z)}{\sigma_{DM}(z)}, \tag{D.1}$$

where $\sigma_{8,AGN}(z)$ is the rms fluctuations of the density distribution over the sphere with a comoving radius of $8\,\mathrm{Mpc}\,h^{-1}$, $\sigma_{DM}(z)$ is the dark matter correlation function evaluated at $8\,\mathrm{Mpc}\,h^{-1}$, normalized to a value of σ_{DM} ($z = 0$) = 0.8. For a power-law correlation function, this value can be calculated by [21]:

$$(\sigma_{8,AGN})^2 = J_2(\gamma) \left(\frac{r_0}{8\,\mathrm{Mpc}\,h^{-1}} \right)^\gamma, \tag{D.2}$$

where $J_2(\gamma) = 72 / [(3 - \gamma)(4 - \gamma)(6 - \gamma) 2^\gamma]$.

Differently in the halo model approach, the 2-halo term of the projected correlation function, which dominates at large scales, can be considered in the regime of linear density fluctuations. In the linear regime, AGNs are biased tracers of the dark matter distribution and the bias factor is described by:

$$b = \left(\frac{w_{p,1h}\left(r_p\right)}{w_{m,2h}\left(r_p\right)} \right)^{1/2}. \tag{D.3}$$

HOD modeling is currently the optimal method to establish the large-scale bias parameter, provided the parametrization of $N(M_h)$, by using

$$b = \frac{\int b_h(M_h) N(M_h) n(M_h) dM_h}{\int N(M_h) n(M_h) dM_h} \tag{D.4}$$

assuming the halo mass function $n(M_h)$ and the halo bias factor $b(M_h)$.

In fact, power-law fit bias measurements commonly use smaller scales (<1-$2\,h^{-1}$ Mpc) that are in the 1-halo term in order to increase the statistical significance. If power-law fits are restricted only to larger scales, the method suffers from the problem that the lowest scale, where the linear biasing scheme can still be applied, varies from sample to sample and remains ambiguous.

HOD modeling allows, in principle, the use of the full range of scales since the method first determines the 1- and 2-halo terms and then constrains the linear using data down to the smallest r_p values that are dominated by the 2-halo term for each individual sample.

Krumpe et al. [52] estimated the RASS-AGN bias following the power-law (D.1) and the HOD (D.4) approach, pointing out that, using the first method, the errors on the bias are much larger, but the values are statistically consistent which those derived from the HOD model fits. Allevato et al. [45] found similar results in estimating the COSMOS-AGN bias following (D.1) and (D.3).

In order to derive a reliable picture of AGN clustering, bias parameters should be inferred from HOD modeling, or at least from the comparison of the correlation function with that of the DM only in the linear regime, because systematic errors based on power-law bias parameters will be larger than the statistical uncertainties of the clustering measurement.

Acknowledgments

N. Cappelluti thanks the INAF-Fellowship program for support. N. Cappelluti thanks the Della Riccia and Blanceflor-Lodovisi-Boncompagni foundation for partial support. V. Allevato is supported by the DFG cluster of excellence Origin and Structure of the Universe (http://www.universe-cluster.de/). N. Cappelluti, V. Allevato, and A. Finoguenov thank the referees, Ryan Hickox and Manolis Plionis, for valuable suggestions for improving the paper.

References

[1] R. Giacconi, J. Bechtold, G. Branduardi et al., "A high-sensitivity X-ray survey using the Einstein Observatory and the discrete source contribution to the extragalactic X-ray background," *Astrophysical Journal*, vol. 234, pp. L1–L17, 1979.

[2] J. Magorrian, S. Tremaine, D. Richstone et al., "The demography of massive dark objects in galaxy centers," *Astronomical Journal*, vol. 115, no. 6, pp. 2285–2305, 1998.

[3] N. I. Shakura and R. A. Sunyaev, "A theory of the instability of disk accretion on to black holes and the variability of binary X-ray sources, galactic nuclei and quasars," *Monthly Notices of the Royal Astronomical Society*, vol. 175, pp. 613–32, 1976.

[4] R. Bergamini, P. Londrillo, and G. Setti, "The cosmic black-body radiation and the inverse compton effect in the radio galaxies: the X-ray background," *Il Nuovo Cimento B Series*, vol. 52, no. 2, pp. 495–506, 1967.

[5] G. Hasinger, R. Burg, R. Giacconi et al., "A deep X-ray survey in the lockman-hole and the soft X-ray N-log," *Astronomy*

and *Astrophysics*, vol. 275, no. 1, p. 1, 1993.

[6] J. F. Navarro, C. S. Frenk, and S. D. M. White, "A universal density profile from hierarchical clustering," *Astrophysical Journal*, vol. 490, no. 2, pp. 493–508, 1997.

[7] P. F. Hopkins, A. Lidz, L. Hernquist et al., "The co-formation of spheroids and quasars traced in their clustering," *Astrophysical Journal*, vol. 662, no. 1, pp. 110–130, 2007.

[8] P. F. Hopkins, L. Hernquist, T. J. Cox, and D. Kerbs, "A cosmological framework for the co-evolution of quasars, supermassive black holes, and elliptical galaxies. i. galaxy mergers and quasar activity," *Astrophysical Journal, Supplement Series*, vol. 175, no. 2, pp. 356–389, 2008.

[9] P. F. Hopkins and L. Hernquist, "A characteristic division between the fueling of quasars and seyferts: five simple tests," *Astrophysical Journal*, vol. 694, no. 1, pp. 599–609, 2009.

[10] J. D. Silverman, P. Kampczyk, K. Jahnke et al., "The impact of galaxy interactions on active galactic nucleus activity in zCOSMOS," *The Astrophysical Journal*, vol. 743, no. 1, article 2, 2011.

[11] D. H. McIntosh, Y. Guo, H. J. Mo, F. van den Bosch, and X. Yan, "The SDSS view of galaxy mergers and their environments," *Bulletin of the American Astronomical Society*, vol. 41, p. 244, 2009.

[12] M. Milosavljević, D. Merritt, and L. C. Ho, "Contribution of stellar tidal disruptions to the X-ray luminosity function of active galaxies," *Astrophysical Journal*, vol. 652, no. 1, pp. 120–125, 2006.

[13] P. F. Hopkins, L. Hernquist, T. J. Cox, T. Di Matteo, B. Robertson, and V. Springel, "A unified, merger-driven model of the origin of starbursts, quasars, the cosmic X-ray background, supermassive black holes, and galaxy spheroids," *Astrophysical Journal, Supplement Series*, vol. 163, no. 1, pp. 1–49, 2006.

[14] D. Larson, J. Dunkley, G. Hinshaw et al., "Seven-year wilkinson microwave anisotropy probe (WMAP) observations: power spectra and WMAP-derived parameters," *Astrophysical Journal, Supplement Series*, vol. 192, article 16, 2011.

[15] X. Barcons and A. C. Fabian, "Fluctuations in the X-ray background and the large-scale structure of the universe," *Monthly Notices of the Royal Astronomical Society*, vol. 230, pp. 189–206, 1988.

[16] F. J. Carrera and X. Barcons, "The spatial distribution of cosmic X-ray sources from the isotropy of the soft X-ray background," *Monthly Notices of the Royal Astronomical Society*, vol. 257, no. 3, pp. 507–512, 1992.

[17] I. Georgantopoulos, G. C. Stewart, T. Shanks, R. E. Griffiths, and B. J. Boyle, "A deep ROSAT survey. II—observations of the isotropy of the 1-2 keV X-ray background," *Monthly Notices of the Royal Astronomical Society*, vol. 262, no. 3, pp. 619–626, 1993.

[18] A. Soltan and G. Hasinger, "The angular correlation function of the soft X-ray background," *Astronomy and Astrophysics*, vol. 288, no. 1, pp. 77–88, 1994.

[19] B. J. Boyle and H. J. Mo, "The clustering of QSOs at low redshift," *Monthly Notices of the Royal Astronomical Society*, vol. 260, no. 4, pp. 925–928, 1993.

[20] A. Vikhlinin and W. Forman, "Detection of the angular correlation of faint X-ray sources," *Astrophysical Journal*, vol. 455, no. 2, pp. L109–L113, 1995.

[21] P. J. E. Peebles, *The Large Scale Structure of the Universe*, Princeton University Press, Princeton, NJ, USA, 1980.

[22] A. Akylas, I. Georgantopoulos, and M. Plionis, "The angular correlation function of the ROSAT all-sky survey bright source catalogue," *Monthly Notices of the Royal Astronomical*

Society, vol. 318, no. 4, pp. 1036–1040, 2000.

[23] X. Barcons, F. J. Carrera, M. T. Ceballos, and S. Mateos, "X-ray sources as tracers of the large-scale structure in the universe ," in *Proceedings of the X-Ray Astronomy: Stellar Endpoints,AGN, and the Diffuse X-ray Background*, vol. 599, pp. 3–12, 2001.

[24] C. R. Mullis, J. P. Henry, I. M. Gioia et al., "Spatial correlation function of X-ray-selected active galactic nuclei," *Astrophysical Journal*, vol. 617, no. 1, pp. 192–208, 2004.

[25] M. Cappi, P. Mazzotta, M. Elvis et al., "Chandra study of an overdensity of x-ray sources around two distant (z ∼ 0.5) clusters," *Astrophysical Journal*, vol. 548, no. 2, pp. 624–638, 2001.

[26] N. Cappelluti, M. Cappi, M. Dadina et al., "X-ray source overdensities in *Chandra* distant cluster fields: a new probe to map the cosmic tapestry?" *Astronomy and Astrophysics*, vol. 430, no. 1, pp. 39–45, 2005.

[27] E. Koulouridis and M. Plionis, "Luminous X-ray active galactic nuclei in clusters of galaxies," *Astrophysical Journal*, vol. 714, no. 2, pp. L181–L184, 2010.

[28] M. Koss, R. Mushotzky, S. Veilleux, and L. Winter, "Merging and clustering of the swift bat AGN sample," *Astrophysical Journal*, vol. 716, no. 2, pp. L125–L130, 2010.

[29] S. Basilakos, M. Plionis, A. Georgakakis et al., "The XMM-Newton/2dF survey—III. Comparison between optical and X-ray cluster detection methods," *Monthly Notices of the Royal Astronomical Society*, vol. 351, no. 3, pp. 989–996, 2004.

[30] S. Basilakos, M. Plionis, A. Georgakakis, and I. Georgantopoulos, "The XMM-Newton/2dF survey—VI. Clustering and bias of the soft X-ray point sources," *Monthly Notices of the Royal Astronomical Society*, vol. 356, no. 1, pp. 183–191, 2005.

[31] P. Gandhi, O. Garcet, L. Disseau et al., "The XMM large scale structure survey: properties and two-point angular correlations of point-like sources," *Astronomy and Astrophysics*, vol. 457, no. 2, pp. 393–404, 2006.

[32] S. Puccetti, F. Flore, V. D'Elia et al., "The XMM-Newton survey of the ELAIS-S1 field I. Number counts, angular correlation function and X-ray spectral properties," *Astronomy and Astrophysics*, vol. 457, no. 2, pp. 501–515, 2006.

[33] R. Gilli, E. Daddi, G. Zamorani et al., "The spatial clustering of X-ray selected AGN and galaxies in the Chandra Deep Field South and North," *Astronomy and Astrophysics*, vol. 430, no. 3, pp. 811–825, 2005.

[34] Y. Yang, R. F. Mushotzky, A. J. Barger, and L. L. Cowie, "Spatial correlation function of the Chandra-selected active galactic nuclei," *Astrophysical Journal*, vol. 645, no. 1, pp. 68–82, 2006.

[35] G. Hasinger, N. Cappelluti, H. Brunner et al., "The XMM-Newton wide-field survey in the COSMOS field. I. Survey description," *Astrophysical Journal, Supplement Series*, vol. 172, no. 1, pp. 29–37, 2007.

[36] N. Cappelluti, G. Hasinger, M. Brusa et al., "The XMM-Newton wide-field survey in the COSMOS field. II. X-ray data and the log N-log S relations," *Astrophysical Journal, Supplement Series*, vol. 172, no. 1, pp. 341–352, 2007.

[37] N. Cappelluti, M. Brusa, G. Hasinger et al., "The XMM-Newton wide-field survey in the COSMOS field," *Astronomy and Astrophysics*, vol. 497, no. 2, pp. 635–648, 2009.

[38] M. Elvis and Chandra-COSMOS Team, "The chandra-COSMOS survey: first results," *Bullettin of the American Astronomical Society*, vol. 39, no. 4, p. 899, 2007.

[39] S. Puccetti, C. Vignali, and N. Cappelluti, "The chandra survey of the cosmos field. II. source detection and photometry," *The Astrophysical Journal Supplement Series*, vol. 185, no. 2, article 586, 2009.

[40] S. J. Lilly, O. Le Fèvre, A. Renzini et al., "zCOSMOS: a large VLT/VIMOS redshift survey covering 0 < z < 3 in the COSMOS field 1," *Astrophysical Journal, Supplement Series*, vol. 172, no. 1, pp. 70–85, 2007.

[41] T. Miyaji, G. Zamorani, N. Cappelluti et al., "The XMM-Newton wide-field survey in the COSMOS field. V angular clustering of the X-ray point sources," *Astrophysical Journal, Supplement Series*, vol. 172, no. 1, pp. 396–405, 2007.

[42] J. Ebrero, S. Mateos, G. C. Stewart, F. J. Carrera, and M. G. Watson, "High-precision multi-band measurements of the angular clustering of X-ray sources," *Astronomy and Astrophysics*, vol. 500, no. 2, pp. 749–762, 2009.

[43] A. Elyiv, N. Clerc, M. Plionis et al., "Angular correlation functions of X-ray point-like sources in the full exposure XMM-LSS field," *Astronomy and Astrophysics*, vol. 537, article A131, 2012.

[44] R. Gilli, G. Zamorani, T. Miyaji et al., "The spatial clustering of X-ray selected AGN in the XMM-COSMOS field," *Astronomy and Astrophysics*, vol. 494, no. 1, pp. 33–48, 2009.

[45] V. Allevato, A. Finoguenov, N. Cappelluti et al., "The XMM-newton wide field survey in the cosmos field: redshift evolution of agn bias and subdominant role of mergers in triggering moderate-luminosity AGNs at redshifts up to 2.2," *Astrophysical Journal*, vol. 736, no. 2, article 99, 2011.

[46] S. Bonoli, F. Marulli, V. Springel, S. D. M. White, E. Branchini, and L. Moscardini, "Modelling the cosmological co-evolution of supermassive black holes and galaxies— II. the clustering of quasars and their dark environment," *Monthly Notices of the Royal Astronomical Society*, vol. 396, no. 1, pp. 423–438, 2009.

[47] R. C. Hickox, C. Jones, W. R. Forman et al., "Host galaxies, clustering, Eddington ratios, and evolution of radio, X-ray, and infrared-selected AGNs," *Astrophysical Journal Letters*, vol. 696, no. 1, pp. 891–919, 2009.

[48] A. L. Coil, A. Georgakakis, J. A. Newman et al., "Aegis: the clustering of x-ray active galactic nucleus relative to galaxies at z ∼ 1," *Astrophysical Journal Letters*, vol. 701, no. 2, pp. 1484–1499, 2009.

[49] G. Mountrichas and A. Georgakakis, "The clustering of X-ray-selected active galactic nuclei at z = 0.1," *Monthly Notices of the Royal Astronomical Society*, vol. 420, no. 1, pp. 514–525, 2012.

[50] M. Krumpe, T. Miyaji, and A. L. Coil, "The spatial clustering of rosat all-sky survey AGNs. I. the cross-correlation function with sdss luminous red galaxies," *Astrophysical Journal*, vol. 713, no. 1, pp. 558–572, 2010.

[51] N. Cappelluti, M. Burlon D. Aiello et al., "Active galactic nuclei clustering in the local universe: an unbiased picture from swift-BAT," *The Astrophysical Journal Letters*, vol. 716, no. 2, pp. L209–L213, 2010.

[52] M. Krumpe, T. Miyaji, A. L. Coil, and H. Aceves, "The spatial clustering of ROSAT All-Sky Survey active galactic nuclei. III. Expanded sample and comparison with optical active galactic nuclei," *Astrophysical Journal*, vol. 746, article 1, 2012.

[53] M. Plionis, M. Rovilos, S. Basilakos, I. Georgantopoulos, and F. Bauer, "Luminosity-dependent X-ray active galactic nucleus clustering?" *The Astrophysical Journal*, vol. 674, no. 1, pp. L5–L8, 2008.

[54] F. C. van den Bosch, "The universal mass accretion history

of cold dark matter haloes," *Monthly Notices of the Royal Astronomical Society*, vol. 331, no. 1, pp. 98–110, 2002.

[55] R. K. Sheth, H. J. Mo, and G. Tormen, "Ellipsoidal collapse and an improved model for the number and spatial distribution of dark matter haloes," *Monthly Notices of the Royal Astronomical Society*, vol. 323, no. 1, pp. 1–12, 2001.

[56] W. H. Press and P. Schechter, "Formation of galaxies and clusters of galaxies by self-similar gravitational condensation," *Astrophysical Journal*, vol. 187, pp. 425–438, 1974.

[57] R. K. Sheth and G. Tormen, "Large-scale bias and the peak background split," *Monthly Notices of the Royal Astronomical Society*, vol. 308, no. 1, pp. 119–126, 1999.

[58] J. L. Tinker, D. H. Weinberg, Z. Zheng, and I. Zehavi, "On the mass-to-light ratio of large-scale structure," *Astrophysical Journal*, vol. 631, no. 1, pp. 41–58, 2005.

[59] N. P. Ross, Y. Shen, M. A. Strauss et al., "Clustering of low-redshift ($z \leq 2.2$) quasars from the sloan digital sky survey," *Astrophysical Journal*, vol. 697, no. 2, pp. 1634–1655, 2009.

[60] Y. Shen, M. A. Strauss, N. P. Ross et al., "Quasar clustering from SDSS DR5: dependences on physical properties," *Astrophysical Journal Letters*, vol. 697, no. 2, pp. 1656–1673, 2009.

[61] S. M. Croom, B. J. Boyle, T. Shanks et al., "The 2dF QSO redshift survey - XIV. Structure and evolution from the two-point correlation function," *Monthly Notices of the Royal Astronomical Society*, vol. 356, no. 2, pp. 415–438, 2005.

[62] C. Porciani and P. Norberg, "Luminosity- and redshift-dependent quasar clustering," *Monthly Notices of the Royal Astronomical Society*, vol. 371, no. 4, pp. 1824–1834, 2006.

[63] J. Da Ângela, T. Shanks, S. M. Croom et al., "The 2dF-SDSS LRG and QSO survey: QSO clustering and the L-z degeneracy," *Monthly Notices of the Royal Astronomical Society*, vol. 383, no. 2, pp. 565–580, 2008.

[64] G. Kauffmann, A. Nusser, and M. Steinmetz, "Galaxy formation and large-scale bias," *Monthly Notices of the Royal Astronomical Society*, vol. 286, no. 4, pp. 795–811, 1997.

[65] J. A. Peacock and R. E. Smith, "Halo occupation numbers and galaxy bias," *Monthly Notices of the Royal Astronomical Society*, vol. 318, no. 4, pp. 1144–1156, 2000.

[66] A. Cooray and R. Sheth, "Halo models of large scale structure," *Physics Report*, vol. 372, no. 1, pp. 1–129, 2002.

[67] Z. Zheng, A. A. Berlind, D. H. Weinberg et al., "Theoretical models of the halo occupation distribution: separating central and satellite galaxies," *Astrophysical Journal*, vol. 633, no. 2, pp. 791–809, 2005.

[68] H. J. Mo and S. D. M. White, "An analytic model for the spatial clustering of dark matter haloes," *Monthly Notices of the Royal Astronomical Society*, vol. 282, no. 2, pp. 347–361, 1996.

[69] S. Basilakos, M. Plionis, and C. Ragone-Figueroa, "The halo mass-bias redshift evolution in the ΛCDM cosmology," *Astrophysical Journal*, vol. 678, no. 2, pp. 627–634, 2008.

[70] J. L. Tinker, B. E. Robertson, A. V. Kravtsov et al., "The large-scale bias of dark matter halos: numerical calibration and model tests," *Astrophysical Journal*, vol. 724, no. 2, pp. 878–886, 2010.

[71] A. Pillepich, C. Porciani, and O. Hahn, "Halo mass function and scale-dependent bias from N-body simulations with non-Gaussian initial conditions," *Monthly Notices of the Royal Astronomical Society*, vol. 402, no. 1, pp. 191–206, 2010.

[72] C.-P. Ma, M. Maggiore, A. Riotto, and J. Zhang, "The bias and mass function of dark matter haloes in non-Markovian

extension of the excursion set theory," *Monthly Notices of the Royal Astronomical Society*, vol. 411, no. 4, pp. 2644–2652, 2011.

[73] Seljak, "Analytic model for galaxy and dark matter clustering," *Monthly Notices of the RoyalAstronomical Society*, vol. 318, p. 2035, 2000.

[74] R. Scoccimarro, R. K. Sheth, L. Hui, and B. Jain, "How many galaxies fit in a halo? Constraints on galaxy formation efficiency from spatial clustering," *Astrophysical Journal*, vol. 546, no. 1, pp. 20–34, 2001.

[75] A. A. Berlind and D. H. Weinberg, "The halo occupation distribution: toward an empirical determination of the relation between galaxies and mass," *Astrophysical Journal*, vol. 575, no. 2, pp. 587–616, 2002.

[76] C. Marinoni and M. J. Hudson, "The mass-to-light function of virialized systems and the relationship between their optical and X-ray properties," *Astrophysical Journal*, vol. 569, no. 1, pp. 101–111, 2002.

[77] M. Magliocchetti and C. Porciani, "The halo distribution of 2dF galaxies," *Monthly Notices of the Royal Astronomical Society*, vol. 346, no. 1, pp. 186–198, 2003.

[78] F. C. van den Bosch, X. Yang, and H. J. Mo, "Linking early- and late-type galaxies to their dark matter haloes," *Monthly Notices of the Royal Astronomical Society*, vol. 340, no. 3, pp. 771–792, 2003.

[79] X. Yang, H. J. Mo, and F. C. Van den Bosch, "Constraining galaxy formation and cosmology with the conditional luminosity function of galaxies," *Monthly Notices of the Royal Astronomical Society*, vol. 339, no. 4, pp. 1057–1080, 2003.

[80] I. Zehavi, D. H. Weinberg, Z. Zheng et al., "On departures from a power law in the galaxy correlation function," *Astrophysical Journal Letters*, vol. 608, no. 1, pp. 16–24, 2004.

[81] S. Phleps, J. A. Peacock, K. Meisenheimer, and C. Wolf, "Galaxy clustering from COMBO-17: the halo occupation distribution at z = 0.6," *Astronomy and Astrophysics*, vol. 457, no. 1, pp. 145–155, 2006.

[82] Z. Zheng, I. Zehavi, D. J. Eisenstein, D. H. Weinberg, and Y. P. Jing, "Halo occupation distribution modeling of clustering of luminous red galaxies," *Astrophysical Journal*, vol. 707, no. 1, pp. 554–572, 2009.

[83] J. S. Bullock, R. H. Wechsler, and R. S. Somerville, "Galaxy halo occupation at high redshift," *Monthly Notices of the Royal Astronomical Society*, vol. 329, no. 1, pp. 246–256, 2002.

[84] L. A. Moustakas and R. S. Somerville, "The masses, ancestors, and descendants of extremely red objects: constraints from spatial clustering," *Astrophysical Journal*, vol. 577, no. 1, pp. 1–10, 2002.

[85] T. Hamana, M. Ouchi, K. Shimasaku, I. Kayo, and Y. Suto, "Properties of host haloes of Lyman-break galaxies and Lyman α emitters from their number densities and angular clustering," *Monthly Notices of the Royal Astronomical Society*, vol. 347, no. 3, pp. 813–823, 2004.

[86] Z. Zheng, "Interpreting the observed clustering of red galaxies at z ~ 3," *Astrophysical Journal Letters*, vol. 610, no. 1, pp. 61–68, 2004.

[87] Z. Zheng, A. L. Coil, and I. Zehavi, "Galaxy evolution from halo occupation distribution modeling of DEEP2 and SDSS galaxy clustering," *Astrophysical Journal*, vol. 667, no. 2, pp. 760–779, 2007.

[88] A. V. Kravtsov, A. A. Berlind, R. H. Wechsler et al., "The dark side of the halo occupation distribution," *Astrophysical Journal*, vol. 609, no. 1, pp. 35–49, 2004.

[89] C. Porciani, M. Magliocchetti, and P. Norberg, "Cosmic evolution of quasar clustering: implications for the host haloes," *Monthly Notices of the Royal Astronomical Society*, vol. 355, no. 3, pp. 1010–1030, 2004.

[90] N. Padmanabhan, M. White, P. Norberg, and C. Porciani, "The real-space clustering of luminous red galaxies around z < 0.6 quasars in the Sloan Digital Sky Survey," *Monthly Notices of the Royal Astronomical Society*, vol. 397, no. 4, pp. 1862–1875, 2009.

[91] Y. Shen, J. F. Hennawi, F. Shankar et al., "Binary quasars at high redshift. II. Sub-Mpc clustering at z ~ 3-4," *Astrophysical Journal Letters*, vol. 719, no. 2, pp. 1693–1698, 2010.

[92] S. Starikova, R. Cool, D. Eisenstein et al., "Constraining halo occupation properties of x-ray active galactic nuclei using clustering of Chandra sources in the Botes survey region," *Astrophysical Journal*, vol. 741, no. 1, article 15, 2011.

[93] T. Miyaji, M. Krumpe, A. L. Coil, and H. Aceves, "The spatial clustering of rosat all-sky survey AGNs. II. halo occupation distribution modeling of the cross-correlation function," *Astrophysical Journal Letters*, vol. 726, no. 2, article 83, 2011.

[94] T. J. Arnold, P. Martini, J. S. Mulchaey, A. Berti, and T. E. Jeltema, "Active galactic nuclei in groups and clusters of galaxies: detection and host morphology," *Astrophysical Journal*, vol. 707, no. 2, pp. 1691–1706, 2009.

[95] P. Martini, G. R. Sivakoff, and J. S. Mulchaey, "The evolution of active galactic nuclei in clusters of galaxies to redshift 1.3," *Astrophysical Journal*, vol. 701, no. 1, pp. 66–85, 2009.

[96] J. D. Silverman, K. Kova, and C. Knobel, "The environments of active galactic nuclei within the zCOSMOS density field," *The Astrophysical Journal*, vol. 695, no. 1, p. 171, 2009.

[97] C. Li, G. Kauffmann, L. Wang, S. D. M. White, T. M. Heckman, and Y. P. Jing, "The clustering of narrow-line AGN in the local Universe," *Monthly Notices of the Royal Astronomical Society*, vol. 373, no. 2, pp. 457–468, 2006.

[98] A. L. Coil, J. F. Hennawi, J. A. Newman, M. C. Cooper, and M. Davis, "The DEEP2 galaxy redshift survey: clustering of quasars and galaxies at z = 1," *Astrophysical Journal Letters*, vol. 654, no. 1, pp. 115–124, 2007.

[99] G. Mountrichas, U. Sawangwit, T. Shanks et al., "QSO-LRG two-point cross-correlation function and redshift-space distortions," *Monthly Notices of the Royal Astronomical Society*, vol. 394, no. 4, pp. 2050–2064, 2009.

[100] R. C. Hickox, A. D. Myers, M. Brodwin et al., "Clustering of obscured and unobscured quasars in the Boötes field: placing rapidly growing black holes in the cosmic web," *Astrophysical Journal Letters*, vol. 731, no. 2, article 117, 2011.

[101] P. Martini and D. H. Weinberg, "Quasar clustering and the lifetime of quasars," *Astrophysical Journal*, vol. 547, no. 1, pp. 12–26, 2001.

[102] A. C. Fabian, R. V. Vasudevan, R. F. Mushotzky, L. M. Winter, and C. S. Reynolds, "Radiation pressure and absorption in AGN: results from a complete unbiased sample from *Swift*," *Monthly Notices of the Royal Astronomical Society*, vol. 394, no. 1, pp. L89–L92, 2009.

[103] P. F. Hopkins, T. J. Cox, D. Kerbš, and L. Hernquist, "A cosmological framework for the co-evolution of Quasars, supermassive black holes, and elliptical galaxies. II. Formation of red ellipticals," *Astrophysical Journal, Supplement Series*, vol. 175, no. 2, pp. 390–422, 2008.

[104] G. Hasinger, "Absorption properties and evolution of active galactic nuclei," *Astronomy and Astrophysics*, vol. 490, no. 3, pp. 905–922, 2008.

[105] Y. Shen, "Supermassive black holes in the hierarchical universe: a general framework and observational tests," *Astrophysical Journal Letters*, vol. 704, no. 1, pp. 89–108, 2009.

[106] F. Shankar, D. H. Weinberg, and J. Miralda-Escudé, "Self-consistent models of the AGN and black hole populations: duty cycles, accretion rates, and the mean radiative efficiency," *Astrophysical Journal*, vol. 690, no. 1, pp. 20–41, 2009.

[107] F. Shankar, C. Martin, M.-E. Jordi, F. Pablo, H. W. David et al., "On the radiative efficiencies, eddington ratios, and duty cycles of luminous high-redshift quasars," *The Astrophysical Journal*, vol. 718, no. 1, article 231, 2010.

[108] F. Shankar, "Merger-induced quasars, their light curves, and their host halos," in *Proceedings of the International Astronomical Union (IAUS '10)*, vol. 267, pp. 248–253, 2010.

[109] E. Treister, C. M. Urry, K. Schawinski, C. N. Cardamone, and D. B. Sanders, "Heavily obscured active galactic nuclei in high-redshift luminous infrared galaxies," *Astrophysical Journal*, vol. 722, no. 2, pp. L238–L243, 2010.

[110] A. Georgakakis, K. Nandra, E. S. Laird et al., "AEGIS: the environment of X-ray sources at z ~ 1," *Astrophysical Journal Letters*, vol. 660, no. 1, pp. L15–L18, 2007.

[111] A. Georgakakis, A. L. Coil, E. S. Laird et al., "Host galaxy morphologies of X-ray selected AGN: assessing the significance of different black hole fuelling mechanisms to the accretion density of the Universe at z ~ 1," *Monthly Notices of the Royal Astronomical Society*, vol. 397, no. 2, pp. 623–633, 2009.

[112] M. Cisternas, K. Jahnke, K. J. Inskip et al., "The bulk of the black hole growth since z ~ 1 occurs in a secular universe: no major merger-AGN connection," *Astrophysical Journal Letters*, vol. 726, no. 2, article 57, 2011.

[113] J. S. Dunlop, R. J. McLure, M. J. Kukula, S. A. Baum, C. P. O'Dea, and D. H. Hughes, "Quasars, their host galaxies and their central black holes," *Monthly Notices of the Royal Astronomical Society*, vol. 340, no. 4, pp. 1095–1135, 2003.

[114] N. A. Grogin, C. J. Conselice, E. Chatzichristou et al., "AGN host galaxies at z ~ 0.4-1.3: bulge-dominated and lacking merger-AGN connection," *Astrophysical Journal*, vol. 627, no. 2, pp. L97–L100, 2005.

[115] C. M. Pierce, J. M. Lotz, E. S. Laird et al., "AEGIS: host galaxy morphologies of X-ray-selected and infrared-selected active galactic nuclei at 0.2 ≤ z < 1.2," *Astrophysical Journal Letters*, vol. 660, no. 1, pp. L19–L22, 2007.

[116] J. M. Gabor, C. D. Impey, K. Jahnke et al., "Active galactic nucleus host galaxy morphologies in cosmos," *The Astrophysical Journal*, vol. 691, no. 1, article 705, 2009.

[117] T. A. Reichard, T. M. Heckmas, and G. Rudnick, "The lopsidedness of present-day galaxies: connections to the formation of stars, the chemical evolution of galaxies, and the growth of black holes," *The Astrophysical Journal*, vol. 691, no. 2, article 1005, 2009.

[118] T. Tal, P. G. Van Dokkum, J. Nelan, and R. Bezanson, "The frequency of tidal features associated with nearby luminous elliptical galaxies from a statistically complete sample," *Astronomical Journal*, vol. 138, no. 5, pp. 1417–1427, 2009.

[119] D. J. Schlegel, F. Abdalla, and T. Abraham, "The BigBOSS experiment," *Astrophysics*, vol. 1, no. 11, p. 2865, 2011.

[120] D. J. Rosario, R. C. McGurk, and C. E. Max, "Adaptive optics imaging of QSOs with double-peaked narrow lines: are they dual AGNs?" *The Astrophysical Journal*, vol. 739, no. 1, article 44, 2011.

[121] K. Schawinski, M. Urry, E. Treister, B. Simmons, P. Natarajan, and E. Glikman, "Evidence for three accreting black holes in a galaxy at z ~ 1.35: a snapshot of recently formed black hole seeds?" *Astrophysical Journal Letters*, vol. 743, no. 2, article L37, 2011.

[122] D. D. Kocevski, S. M. Faber, M. Mozena et al., "Candels: constraining the AGN-merger connection with host morphologies at z ~ 2," *Astrophysical Journal*, vol. 744, no. 2, article 148, 2012.

[123] D. J. Eisenstein, D. H. Weinberg, and E. Agol, "SDSS-III: massive spectroscopic surveys of the distant universe, the milky way galaxy, and extra-solar planetary systems," *The Astronomical Journal*, vol. 142, no. 3, article 72, 2011.

[124] P. Predehl, R. Andritschke, W. Bornemann et al., "eROSITA," in *UV, X-Ray, and Gamma-Ray Space Instrumentation for Astronomy XV*, vol. 6686 of *Proceedings of SPIE*, August 2007.

[125] Z. Ivezic, J. A. Tyson, E. Acosta et al., "LSST: from science drivers to reference design and anticipated data products," *Bulletin of the American Astronomical Society*, vol. 41, p. 366, 2008.

[126] M. Davis and P. J. E. Peebles, "A survey of galaxy redshifts. V—the two-point position and velocity correlations," *Astrophysical Journal*, vol. 267, pp. 465–482, 1983.

[127] K. B. Fisher, M. Davis, M. A. Strauss, A. Yahil, and J. P. Huchra, "Clustering in the 1.2 Jy IRAS galaxy redshift survey II: redshift distortions and $\xi(r_p, \pi)$," *Monthly Notices of the Royal Astronomical Society*, vol. 267, pp. 927–948, 1994.

[128] S. D. Landy and A. S. Szalay, "Bias and variance of angular correlation functions," *Astrophysical Journal*, vol. 412, no. 1, pp. 64–71, 1993.

[129] P. Norberg, C. M. Baugh, E. Gaztañaga, and D. J. Croton, "Statistical analysis of galaxy surveys—I. Robust error estimation for two-point clustering statistics," *Monthly Notices of the Royal Astronomical Society*, vol. 396, no. 1, pp. 19–38, 2009.

[130] C. Blake, A. Collister, and O. Lahav, "Halo-model signatures from 380 000 Sloan Digital Sky Survey luminous red galaxies with photometric redshifts," *Monthly Notices of the Royal Astronomical Society*, vol. 385, no. 3, pp. 1257–1269, 2008.

[131] S. R. Knollmann, C. Power, and A. Knebe, "Dark matter halo profiles in scale-free cosmologies," *Monthly Notices of the Royal Astronomical Society*, vol. 385, no. 2, pp. 545–552, 2008.

[132] J. Stadel, D. Potter, B. Moore et al., "Quantifying the heart of darkness with GHALO—a multibillion particle simulation of a galactic halo," *Monthly Notices of the Royal Astronomical Society*, vol. 398, no. 1, pp. L21–L25, 2009.

[133] D. J. Eisenstein and W. Hu, "Power spectra for cold dark matter and its variants," *Astrophysical Journal Letters*, vol. 511, no. 1, pp. 5–15, 1999.

[134] R. E. Smith, J. A. Peacock, A. Jenkins et al., "Stable clustering, the halo model and non-linear cosmological power spectra," *Monthly Notices of the Royal Astronomical Society*, vol. 341, no. 4, pp. 1311–1332, 2003.

7

The Role of Gravitational Instabilities in the Feeding of Supermassive Black Holes

Giuseppe Lodato

Dipartimento di Fisica, Università degli Studi di Milano, Via Celoria 16, Milano, Italy

Correspondence should be addressed to Giuseppe Lodato, giuseppe.lodato@unimi.it

Academic Editor: Francesco Shankar

I review the recent progresses that have been obtained, especially through the use of high-resolution numerical simulations, on the dynamics of self-gravitating accretion discs. A coherent picture is emerging, where the disc dynamics is controlled by a small number of parameters that determine whether the disc is stable or unstable, whether the instability saturates in a self-regulated state or runs away into fragmentation, and whether the dynamics is local or global. I then apply these concepts to the case of AGN discs, discussing the implications of such evolution on the feeding of supermassive black holes. Nonfragmenting, self-gravitating discs appear to play a fundamental role in the process of formation of massive black hole seeds at high redshift ($z \sim 10$–15) through direct gas collapse. On the other hand, the different cooling properties of the interstellar gas at low redshifts determine a radically different behaviour for the outskirts of the accretion discs feeding typical AGNs. Here the situation is much less clear from a theoretical point of view, and while several observational clues point to the important role of massive discs at a distance of roughly a parsec from their central black hole, their dynamics is still under debate.

1. Introduction

The accretion discs surrounding the growing supermassive black holes (SMBH) in active galactic nuclei (AGN) are expected to become gravitationally unstable at a distance of ~ 0.01 pc from the black hole [1, 2]. Traditionally, this occurrence has been interpreted in relation to star formation: a self-gravitating disc, in this picture, would rapidly fragment and form stars [3, 4]. At the same time, it has been noted very early that the development of gravitational instability may also act as an efficient mechanism to produce torques through the effect of the resulting spiral structure and thus might be very effective in redistributing angular momentum within the disc and promote accretion [5–7]. As we shall see, the modern debate about these issues still concentrates on these two extreme cases. While we now have a much clearer understanding of the mechanism of growth and saturation of the instability in gaseous discs, and—especially though the use of high-resolution numerical simulations— we have clarified what are the main parameters regulating the disc structure and evolution, some questions are still unanswered. Are massive discs effectively truncated by star

formation at the radius where they become self-gravitating, thus preventing accretion beyond these scales? Or does accretion proceed effectively through gravitational torques even in fragmenting discs, allowing the central black holes to be fed by gas on parsec scales?

From the observational point of view, on the one hand, it has now become quite clear that fragmentation in massive discs can be very important for the formation of compact, young stellar clusters in AGNs, and in particular in our own Galaxy [8, 9]. On the other hand, it is also clear that rotating gaseous discs exist on parsec scales in AGN [10–12], often displaying a clear Keplerian rotation [10].

All the issues discussed above bear important consequences not just for the dynamics of the disc itself, but, in a broader context, relate to the overall process of coevolution between the supermassive black hole and the host galaxy. The efficiency of star formation in the disc, the efficiency of the accretion process, and the related timescales and duty cycle of AGN activity are all often assumed as subgrid physics in simulations of galactic evolution on larger scales [17–19], which turn out to be quite sensitive to the chosen subgrid prescriptions.

In this contribution, I will not try to give an exhaustive answer to the questions above. I will rather give an account of the progresses we have made in recent years in our understanding of these phenomena and highlight their importance in several contexts related to the feeding of supermassive black holes. I will first summarize, in Section 2, the state of the art about the evolution of gravitational instabilities in gaseous discs, from a purely theoretical point of view. In Section 3, I will describe the possible importance of gravitational torques in the formation of the seeds of supermassive black holes by direct collapse in the early evolution of pregalactic discs. In Section 4, I will address the issue of feeding the SMBH in AGN through gravitational torques, and the related issue of fragmentation of AGN discs. Finally, in Section 5 I will draw some conclusions.

2. Gravitational Instabilities in Gaseous Discs

The issue of the nonlinear evolution of gravitational instabilities in gaseous discs has been studied in great detail over the last 10–15 years [13, 20–25]. As a result, despite the differences in the numerical methods adopted and in the setup used, a coherent picture of the overall dynamics is emerging. This issue has also been covered in several reviews, see for example Lodato [26] and Durisen et al. [27], and the reader is referred to these papers for further details and for an application of these concepts to different astrophysical systems (such as protostellar and protoplanetary discs), which share similar characteristics.

Consider an accretion disc with surface density $\Sigma(R)$, where R is the cylindrical distance to the central object of mass M, around which the disc is rotating in approximately centrifugal balance with angular velocity $\Omega(R)$. Let us also define the epicyclic frequency κ, which is equal to Ω in the case in which the rotation curve of the disc is Keplerian, $\Omega^2 = GM/R^3$. If the disc mass is high enough ($M_{\text{disc}} \approx M$), deviations from Keplerian rotation might arise [28] and κ is not going to be exactly equal to Ω. The disc midplane temperature is $T(R)$, and the sound speed is $c_s \propto T^{1/2}$. The disc thickness is $H = c_s/\Omega$ for a non-self-gravitating disc and $H = c_s^2/\pi G\Sigma$ for a self-gravitating disc: we shall see that for gravitationally unstable discs the two definitions are equivalent. For most cases, we will consider thin discs, for which $H/R \ll 1$.

Fundamentally, the dynamics of self-gravitating accretion discs depends on three dimensionless parameters. Firstly, there is the well-known axisymmetric stability parameter $Q = c_s\kappa/\pi G\Sigma$ [29]. The second important parameter is the ratio between the cooling time t_{cool} and the dynamical time $t_{\text{dyn}} = \Omega^{-1}$, a parameter often called $\beta = \Omega t_{\text{cool}}$. Thirdly, we have the ratio between the disc mass and the central object mass M_{disc}/M. As we shall see, each of these parameters controls some important features about the evolution of the gravitational instability.

2.1. The Role of Q: Linear Stability. As mentioned above, the basic, and most widely used, criterion to determine the

stability of a massive disc against gravitational perturbations is related to the linear dispersion relation in the WKB approximation for an infinitesimally thin disc [30]:

$$(\omega - m\Omega)^2 = c_s^2 k^2 - 2\pi G\Sigma|k| + \kappa^2, \quad (1)$$

where ω is the frequency of the perturbation, k is the radial wave number, and m is the azimuthal wave number. The above dispersion relation is quadratic in k from which one easily sees that, for $m = 0$ (axisymmetric perturbations), ω^2 is positive (and the perturbation is stable) at all wavelengths if

$$Q = \frac{c_s\kappa}{\pi G\Sigma} > 1. \quad (2)$$

Marginal stability occurs at $Q = 1$.

Here we should note that the above (local) dispersion relation is strictly speaking only appropriate for infinitesimally thin discs and for tightly wound perturbations ($m/kR \ll 1$) for which the WKB approximation holds. Finite thickness effects generally act so as to dilute the effect of self-gravity, thus making the disc more stable and decreasing the marginal stability value of Q below unity (i.e., allowing a colder disc to remain stable). On the contrary, global perturbations are more unstable [31] thus effectively increasing the marginal stability value of Q.

For most cases considered here, the disc is close to being in Keplerian rotation, for which $\kappa = \Omega$. In this case, it is easy to show that the requirement of marginal stability ($Q \approx 1$) is equivalent to

$$\frac{M_{\text{disc}}}{M} \approx \frac{H}{R}, \quad (3)$$

where $M_{\text{disc}} = \pi\Sigma R^2$ is a measure of the enclosed disc mass within radius R. Thus, for marginally stable discs, "thin" and "light" on the one hand and "thick" and "massive" on the other hand are equivalent. Also note that, as mentioned above, when $Q \approx 1$, the two expressions for the disc thickness in the non-self-gravitating and in the self-gravitating regime are indeed equivalent.

AGN discs are generally quite thin, with $H/R \approx 10^{-3}$, and thus even a relatively light disc, much less massive than the central black hole, can be marginally stable. It is then easy, based on standard models of accretion discs around supermassive black holes [32], to calculate the distance from the black hole at which the disc first becomes gravitationally unstable [2, 33]. This turns out to be of the order of $10^3 R_\bullet$ (where $R_\bullet$ is the Schwarzschild radius of the black hole), or 0.01 pc, for a $10^8 M_\odot$ black hole. Thus, discs that extend beyond this radius are going to be gravitationally unstable: in order to determine their evolution, we need to understand the behaviour of the instability at the nonlinear stage: this is addressed in the next subsection.

2.2. The Role of β: Fragmentation versus Self-Regulation. The details of the nonlinear evolution of the gravitational instability are best understood through the use of hydrodynamical simulations, which include the disc self-gravity. However, before discussing such simulations, let us make some preliminary remarks to guide us in the interpretation of the results of the simulation.

The very same fact that the linear stability of the disc depends on Q, which is directly proportional to the sound speed $c_s \propto T^{1/2}$ (where T is the disc temperature), offers a possible way to predict the nonlinear evolution of the system. In fact, the development of the instability will act as to feed back energy into the disc and to heat it up, thus making it more stable. In practice, the linear stability condition works as a "thermostat" for the disc, so that heating turns on only when Q drops below the marginal stability value, which we have seen is of order unity. If the thermostat works, we would expect the disc to be always close to marginal stability, at least under some conditions, in a so-called "self-regulated state" [34, 35].

From a numerical point of view, it is clear then that if we want to catch the dynamics associated with self-regulation, we need to make sure that the instability is able to feed back energy into the disc, and we should not then use isothermal simulations (such as the pioneering ones of [36]), which by constraint do not allow the disc to heat up. At the same time, we need to make sure that the disc is able to cool; otherwise, once the instability sets in, it will stabilize the disc forever (cf. the "perennial heating" problem for the spiral structure in galaxies), and we should thus also avoid pure N-body simulations, unless special arrangements are made to artificially cool the disc down [37].

One such approach has been taken by Gammie [20], who ran local, shearing-sheet simulations of razor-thin discs, which were allowed to heat up through shocks and pdV work and to cool down, according to a simple cooling prescription, such that

$$\frac{dT}{dt} = -\frac{T}{t_{\text{cool}}}, \tag{4}$$

where the cooling time t_{cool} is a free input parameter for the simulation. While more complex approaches, which consider the details of the radiative transfer within the disc [38], can certainly be adopted, such an approach should be considered as a useful "numerical experiment," in order to evaluate the disc response as a function of the main parameters, rather than as a "realistic" simulation of some particular system. Having clarified the main dependencies from the physical parameters, we may then establish the disc response in any particular system. Following this approach, a number of papers have considered the details of the process [13, 21, 24, 25], extending the simulations to full 3D and considering thus global and potentially thick configurations, as a function of the main parameters of the system, such as the disc mass and thickness. Here, I will present a summary of the main results concerning the issue of fragmentation and self-regulation of the instability. In the next subsection, I will address the important issue of the locality of the induced transport.

It turns out that the behaviour of the disc is actually determined by the ratio of t_{cool} to the dynamical time in the disc,

$$\beta = \Omega t_{\text{cool}}. \tag{5}$$

It should be noted that, in most of the simulations described here [13, 20, 21, 24, 25], the parameter β is taken to be a single-free parameter for each simulation, with no dependance on either time or position in the disc. This is certainly not realistic, as in fact the cooling time should and will depend on the local microphysics associated with the disc opacity and radiative properties. These simulations should thus be regarded as simple "numerical experiments," where we test the disc response in a controlled configuration, as a function of the main parameters. For an actual, astrophysically relevant disc, we would thus calculate at any given radius the cooling properties and thus infer the disc behaviour from our controlled experiments. In doing this, care should be taken that the results are not affected by global effects (see below) or by nonlinear effects induced by a temperature dependence of the cooling rate, which has been studied by Johnson and Gammie [39] and Cossins et al. [40].

If the cooling timescale is larger than a few dynamical timescales, an initially stable (large Q) disc cools down until Q becomes of the order of unity. At this stage, the disc becomes gravitationally unstable and develops a spiral structure which provides a heating source, through compressional heating and shock dissipation, able to balance the externally imposed cooling. Once in thermal equilibrium, the disc is characterized by an approximately constant value of Q very close to marginal stability. In such a state, a spiral structure persists in the disc, to provide the required heating. Therefore, the self-regulation mechanism described above determines the disc structure and evolution. Figure 1(a) shows the result of one such simulations, where in this case $\beta = 10$ and the total disc mass $M_{\text{disc}} = 0.1 M$ [13]. The colour plot shows the disc surface density, in which a spiral structure is clearly seen. Figure 1(b) shows the azimuthally and vertically averaged value of Q as a function of radius, for several simulations with the same mass ratio but with different values of β, as indicated. The disc in this case extends from $R = 0.25$ to $R = 25$ in code units. It is then seen that far from the boundaries (where the density drops and Q correspondingly grows) the disc is self-regulated, with $Q \approx 1$ over a wide radial range. Cossins et al. [13] have also computed the amplitude of the perturbed surface density as a function of β. Analysis of the disc structure showed that while the cooling rate β does not influence the spectrum of wavenumbers that are excited, it does affect that amplitude of the density perturbations, such that

$$\frac{\delta \Sigma}{\Sigma} \approx \frac{1}{\sqrt{\beta}}, \tag{6}$$

which is shown in Figure 2. Thus as the cooling becomes more rapid (and thus as β decreases), the amplitude of the density perturbation increases. Similarly, it was found [13] that the spectrum of the radial wavenumber k peaks strongly where $kH = 1$, a result that can be predicted from the dispersion relation (1) but has now also been demonstrated numerically. This result is independent of both the cooling rate and the disc to star mass ratio.

The behaviour described above changes when the cooling time is decreased to smaller values [20]. In this case,

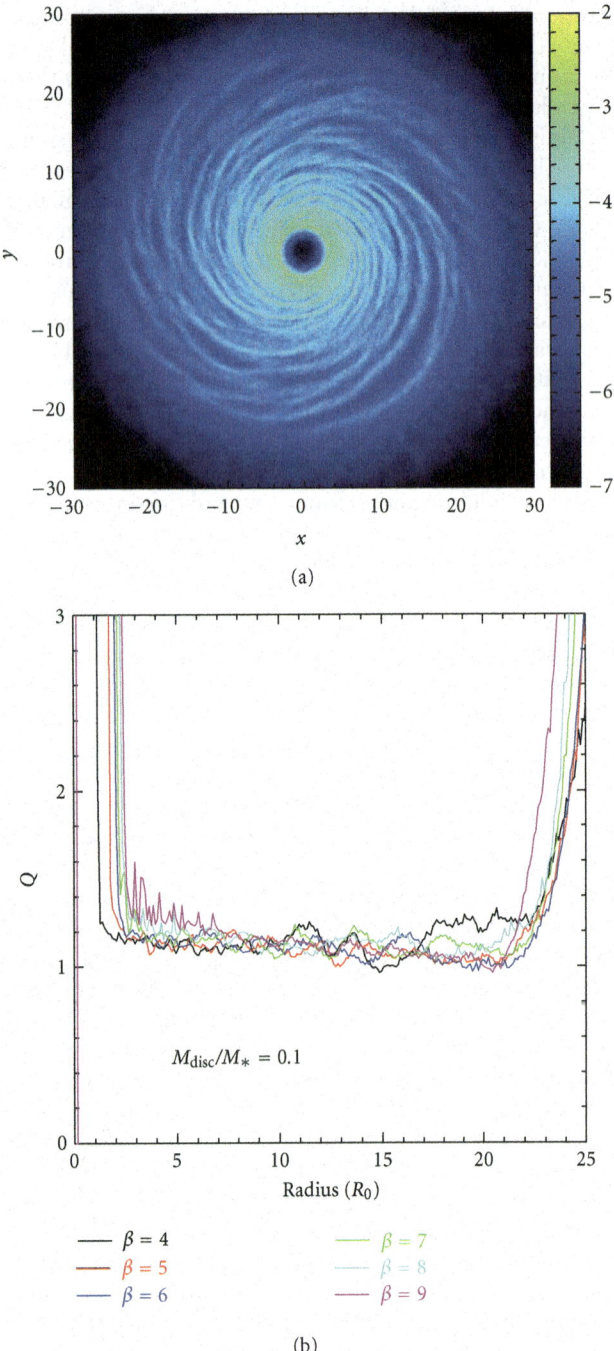

(a)

(b)

FIGURE 1: (a) Surface density of a self-gravitating disc, with $M_{disc} = 0.1\,M$ and with $\beta = 10$, where a tightly wound spiral structure is clearly seen. (b) Azimuthally averaged profiles of Q as a function of radius, for several simulations, with varying β, as indicated. From Cossins et al. [13].

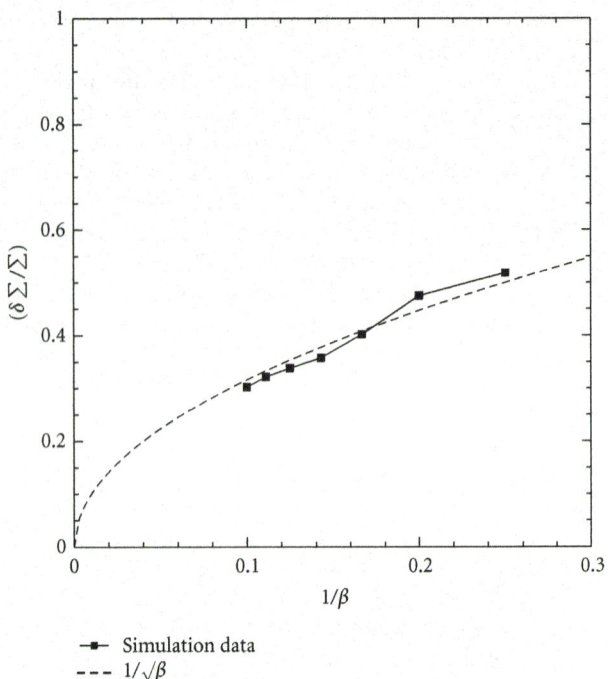

- ■ Simulation data
- - - - $1/\sqrt{\beta}$

FIGURE 2: Variation of the radially and azimuthally averaged relative surface density perturbation amplitude $\delta\Sigma/\Sigma$ with the inverse cooling parameter $1/\beta$. From Cossins et al. [13].

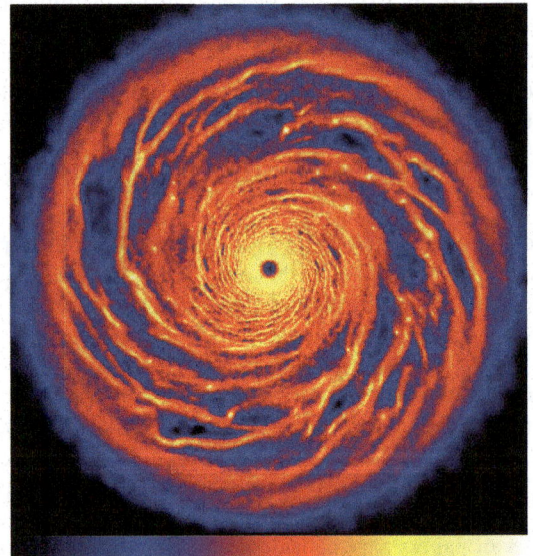

FIGURE 3: Numerical simulation of a self-gravitating disc with $M_{disc} = 0.1\,M$ and $\beta = 3$. Once unstable, the disc breaks up into numerous gravitationally bound clumps.

the disc does not reach a quasisteady self-regulated state but rather fragments into several bound objects. Figure 3 show the results of a simulation very similar to the one displayed in Figure 1, but where the cooling time is decreased to $\beta = 3$ [24]. The presence of numerous high-density clumps is clearly seen. This result can be understood in

the following way, by adopting a local approach to describe the instability. In a gravitationally unstable disc, the typical growth timescale of unstable perturbations is of the order of the dynamical timescale Ω^{-1}. The nonlinear stabilization of the perturbation only works if the heat generated by compression and shocks is not removed too efficiently from the disc through cooling. Since the perturbation grows on the dynamical timescale, if we want to avoid fragmentation, we require that cooling acts on a longer timescale. Note that the requirement that the cooling timescale be shorter than the dynamical timescale in order to result in fragmentation has been known for several years, even outside the context of disc instability [41, 42]. Note also that the existence of a critical cooling time below which the disc fragments can be easily related, through (6), to a maximum value of the perturbation amplitude that can be sustained by the disc.

The exact value of the threshold for fragmentation does depend somewhat on the specific numerical setup and ranges from $\beta = 3$ to $\beta = 6$ [20, 24, 43]. Recently, the exact value of this threshold has been the subject of intense debate, as it was discovered [44] that the threshold value appeared to increase with increasing resolution in smoothed particle hydrodynamics (SPH) simulations (see also [45, 46]). The same behaviour has also been seen in grid-based simulations [47], and it has been shown that it actually depends on the chosen initial conditions. When carefully chosen initial conditions are used, the threshold value for fragmentation does converge and it turns out to be indeed of the order of $\beta \approx 6$.

We can easily get a reasonable estimate of the mass of the fragments. Indeed, we expect $M_{frag} \approx \pi \Sigma \lambda^2$, where Σ is the local density and $\lambda \approx H$ is the typical wavelength associated with the instability. We thus obtain

$$M_{frag} \approx \pi \Sigma R^2 \left(\frac{H}{R}\right)^2 \approx \left(\frac{H}{R}\right)^3 M, \qquad (7)$$

where in the last equality we have used the fact that, for a marginally stable disc, $\pi \Sigma R^2 \approx (H/R)M$. For a typical AGN disc, where $H/R \approx 10^{-3}$–10^{-2} and, say, $M \approx 10^8 M_\odot$, the fragment mass thus corresponds to 0.1–$100 M_\odot$.

2.3. The Role of M_{disc}/M: Global versus Local Dynamics.
The issue of locality of the dynamics associated with gravitational instability is essential if one wants to construct simple viscous models for self-gravitating accretion discs [48, 49]. Indeed, it has been long realized that the spiral structure determined by the instability can efficiently transport angular momentum [50], and one may thus suppose that the instability, at the large scales where an AGN disc is unstable (and where probably the disc is too cold to support MHD instabilities, such as the magnetorotational instability, MRI), can produce the required "viscous" torque to allow the accretion of matter from ~ parsec scales down to the innermost regions where the MRI takes over and releases the accretion fuel down to the SMBH.

In the standard α-prescription for accretion disc viscosity [32], the relevant component of the viscous stress tensor $T_{R\phi}$ is simply parameterized in terms of the local pressure P, such that $T_{R\phi} \approx \alpha P$. The dimensionless parameter α is thus simply a measure of the stress tensor in units of the local pressure. One might thus be tempted to compute the stress tensor resulting from the spiral structure seen in the simulations described above and directly compute an equivalent α parameter associated to the instability. This would be obviously best done for the cases where the disc

is self-regulated and the instability saturates at a given perturbation amplitude, as discussed above. However, a fundamental problem arises in this case. This is related to the fact that the gravitational instability, is an intrinsically long-range instability and it is not clear whether the transport of energy and angular momentum associated with it can be simply expressed in terms of a local viscous process [51].

The problem is best understood in terms of a WKB analysis of the energy and angular momentum fluxes associated with the instability [13]. For a local, viscous process the torque exerted on the disc $\mathcal{L}_\alpha$ is related to the work done by viscosity $\dot{\mathcal{E}}_\alpha$ via the Keplerian rotation rate Ω, such that

$$\dot{\mathcal{E}}_\alpha = \Omega \dot{\mathcal{L}}_\alpha. \qquad (8)$$

A similar but not equal relation governs the case where potentially global effects are mediated through wave transport. In a WKB analysis, the wave angular momentum and energy densities can be obtained [30], and in turn the wave-induced torque $\dot{\mathcal{L}}_w$ and power dissipation $\dot{\mathcal{E}}_w$ are found to be related via [13]

$$\dot{\mathcal{E}}_w = \Omega_p \dot{\mathcal{L}}_w, \qquad (9)$$

where the pattern speed of the spiral perturbation is given by $\Omega_p = \omega/m$. The transport properties of gravitationally induced waves are therefore determined not by the rotation rate of the disc material (cf. (8)), but by the pattern speed of the density waves themselves. As these waves are excited or absorbed, the power exchanged with the background flow for a given stress is therefore significantly different than that dissipated by a viscous process that provides the same stress to the extent to which Ω_p is significantly different from Ω. The relative level of global versus local transport can hence be quantified via the parameter ξ, where

$$\xi = \left| \frac{\Omega - \Omega_p}{\Omega} \right|. \qquad (10)$$

The analysis of Cossins et al. [13] also allows a spectrally averaged pattern speed to be determined, and thus in turn the nonlocal transport fraction ξ can be measured from the simulations. In agreement with Lodato and Rice [21, 25], this shows that transport by gravitational waves is a predominantly local process for the systems modeled, with $\xi \approx 10\%$ for $M_{disc}/M = 0.1$ and increasing with increasing disc to star mass ratio. This is shown as a function of radius in the left-hand panels of Figure 4, where the increase in nonlocality is clearly seen with q. A corollary of this, seen from the form of (10), is that the waves remain on average close to corotation, $\Omega_p \approx \Omega$.

The right-hand panels of Figure 4 show a further interesting result obtained from the simulations of Cossins et al. [13]—the wave Mach numbers. While the heavy lines shows the values relative to an external inertial frame, the lighter lower lines give the Doppler-shifted Mach numbers $\mathcal{M}$, that is, those relative to a frame corotating with the flow. These Doppler-shifted values are almost exactly unity, implying that the density waves excited by the gravitational instability are only weakly supersonic, and furthermore this

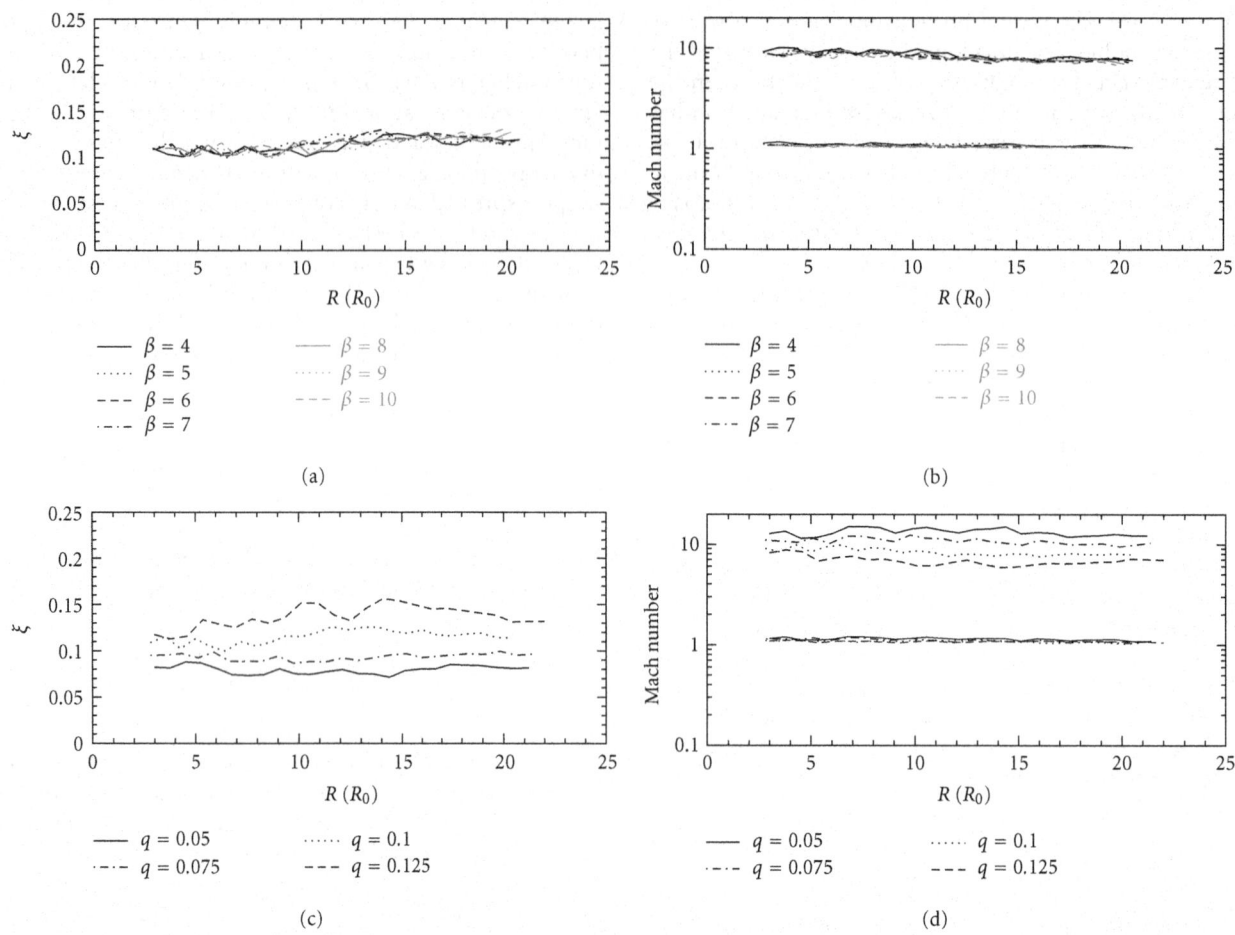

FIGURE 4: Nonlocal transport parameter ξ (a, c) and Mach Numbers (b, d) for simulations at various values of mass ratio M_{disc}/M and cooing parameter β, where β varies, $M_{\text{disc}}/M = 0.1$ and where M_{disc}/M varies, $\beta = 5$. For the Mach number plots, heavy lines denote the Mach number as measured in an inertial (static) frame, whereas light lines show the Doppler-shifted Mach number (measured in a frame corotating with the fluid). Taken from Cossins et al. [13].

result is invariant with either cooling rate or mass ratio. The gravitational instability therefore self-regulates so that not only is $Q \approx 1$, but also $\mathcal{M} \approx 1$. This result is intuitively reasonable in a quasisteady disc—subsonic waves would not impart any net heat to the disc (any compression heating is balanced by the corresponding rarefaction cooling), whereas a strong shock would be highly dissipative, leading to rapid evolution. Furthermore, the only way a fluid element can remain on a circular orbit when passing through an oblique (spiral) shock is if that shock wave has a unit Mach number.

Actually, we can also show that the above-mentioned dependence of the factor ξ on the disc-to-star mass ratio M_{disc}/M can be easily understood from the condition that these waves dissipate where they are almost sonic. In fact, using this sonic condition, we can rewrite (10) as

$$\xi = \left| \frac{\Omega - \Omega_p}{\Omega} \right| \approx \frac{c_s}{v_\phi} = \frac{H}{R} \approx \frac{M_{\text{disc}}}{M_\star}, \qquad (11)$$

where $v_\phi = \Omega R$ is the azimuthal velocity of the disc, and the last equality holds for marginally stable discs ($Q \approx 1$). This trend can actually be seen in Figure 4(c).

Thus the assumption of a local, viscous-like process for the transport associated with gravitational instabilities is only valid for light discs, where $M_{\text{disc}}/M \ll 1$. It is in such cases that one can describe the secular evolution of the disc and the associated angular momentum transport in terms of an effective viscosity, and one can even choose to measure the stress induced by the spiral structure in units of the local pressure, thus obtaining an effective α_{sg} value associated with gravitational instabilities. So, how large is the gravitationally induced α_{sg}? Lin and Pringle [6] propose the following parameterization:

$$\alpha_{\text{sg}} = \begin{cases} \eta \left(\dfrac{\overline{Q}^2}{Q^2} - 1 \right), & Q < \overline{Q}, \\ 0, & Q > \overline{Q}. \end{cases} \qquad (12)$$

Here $\overline{Q}$ is the value of Q at which the disc becomes unstable to nonaxisymmetric perturbations and η is a parameter to measure the strength of the induced torques. The above formulation is useful in practical cases, for example, when one wants to incorporate in a simple way the self-regulation mechanism in simple time-dependent models of self-gravitating discs. However, it lacks one important feature elucidated from the numerical simulations described above. In this picture, α_{sg} only depends on the local value of Q and not on the cooling timescale t_{cool}, which we have seen controls so efficiently the development of the instability. In particular, for self-regulated discs, we expect $Q \approx \overline{Q}$ and the formula above would then produce a negligibly small α_{sg}, while we know that a finite amplitude spiral structure is present in self-regulated discs and indeed it is this spiral structure that provides the heating to balance the imposed cooling rate. On the other hand, we know that the process of self-regulation and the saturation of the gravitational perturbation is fundamentally related to thermal equilibrium in the disc: the saturation amplitude of the instability is such that the power dissipated through shocks in the disc is just enough to balance the imposed cooling (hence the inverse relation described above and displayed in Figure 2). In thermal equilibrium, the value of the viscosity parameter is simply related to the cooling rate [52]:

$$\alpha_{sg} = \left| \frac{d \ln \Omega}{d \ln R} \right|^{-2} \frac{1}{\gamma(\gamma - 1)\Omega t_{cool}} = \frac{4}{9\gamma(\gamma - 1)} \frac{1}{\beta}, \quad (13)$$

where the last equality holds in the case of a Keplerian disc. Indeed, the value of the stress induced by gravitational perturbation as computed directly from simulations of self-regulated discs [21, 23] agrees very well with the value predicted by (13). Thus, in a self-regulated state, not only the fractional amplitude of the density perturbations, but also the induced stress are inversely proportional to the cooling time. Indeed, one can also interpret the fragmentation threshold in terms of α_{sg} rather than in terms of β: there is a maximum value of the stress that can be supported by the disc without fragmenting [24]. Evaluating this critical α_c from (13) using the critical value of β, one finds that $\alpha_c \approx 0.05$–0.1.

Clearly, all this applies in cases where thermal equilibrium is simply established by a balance between the viscous heating and the radiative cooling. In many interesting cases (including the outskirts of AGN discs), irradiation from the central object is going to play a major role in determining the thermal balance. In such cases, (13) should be modified, and an interesting and only rarely discussed issue is what determines fragmentation: is it the stress exceeding the critical value α_c, or is it the cooling time dropping below the critical value [53]?

What happens then for the cases where the disc mass is not much smaller than the central object mass? Here, we already know that we should expect deviations from the analysis discussed above, as transport should become significantly nonlocal. Once again, a change in behaviour has been observed in simulations [25]. The stress computed form the simulations does not agree anymore with (13), exceeding its prediction and peaking at values around unity. Furthermore, in these cases we have a situation where neither self-regulation nor fragmentation occurs. The disc simply cannot find a quasistationary nonlinear saturated, state and it keeps oscillating between periods of high spiral activity, where the stress would correspond to a local α of order unity,

to periods of low activity, characterized by a temporarily high value of Q.

A summary of the various possible behaviours of a self-gravitating disc as a function of the three main dimensionless parameters is displayed graphically in Figure 5. Such picture summarizes effectively the various results discussed up to now.

Having discussed the main features of the gravitational instability in gaseous discs, I now turn to the application of the above results to the process that relates to the formation and growth of supermassive black holes in galactic nuclei.

3. The Formation of Supermassive Black Hole Seeds

One of the most important applications of the concepts described in the previous section to the context of supermassive black hole growth is the formation of massive BH seeds from direct gas collapse at high redshift.

This issue has become particularly important due to the recent discovery of active quasars up to redshift $z \sim 6$ [54, 55] and now even to a redshift as high as $z \sim 7$ [56], which indicates that supermassive black holes, with masses up to $10^9 M_\odot$, were already in place when the Universe was only 10^9 years old and beyond. This clearly requires that the black hole growth occurred at very high rates, with an average of $1 M_\odot/\text{yr}$. Such a rapid early growth poses serious challenges to models of their formation.

Some models [57–59] assume that the seeds of supermassive black holes are the remnants of the zero-metallicity first stars (the so-called Population III stars), which are expected to be relatively massive [60, 61] and thus produce black holes with a mass of up to $100 M_\odot$. However, unless the efficiency of conversion of matter into energy through the accretion process is very low, it is impossible to grow the seeds to the required masses by $z \sim 6$-7 through Eddington-limited accretion [62]. The problem here is that when the accretion rate is large, the radiation pressure produced by the accretion luminosity can exceed the gravitational force of the black hole and thus exceeding the Eddington limit. Now, if the accretion efficiency $\epsilon = L/\dot{M}c^2$ exceeds ≈ 0.1 (where L is the accretion luminosity and c is the speed of light), the Eddington limit does not allow the large accretion rates needed to grow the seeds fast enough to become bright AGN by $z \sim 6$ [62]. Note also that the Eddington limit is linearly proportional to the black hole mass, so that the problem of accreting at very high rates is particularly important in the earliest phases of the growth, when the black hole mass is small.

The efficiency is in turn dependent on the spin of the black hole, with high spin producing very large efficiencies $\epsilon \sim 0.5$. Accretion of matter naturally tends to spin up the hole [58] and hence to increase the efficiency, thus exceeding the Eddington limit for relatively low $\dot{M}$ and preventing a fast growth of the hole. While recent calculations [63, 64] show that it is possible to keep the hole spin low if the growth occurs through several small randomly oriented accretion episodes [65], we still have to face the issue of how to produce the high infall rates required.

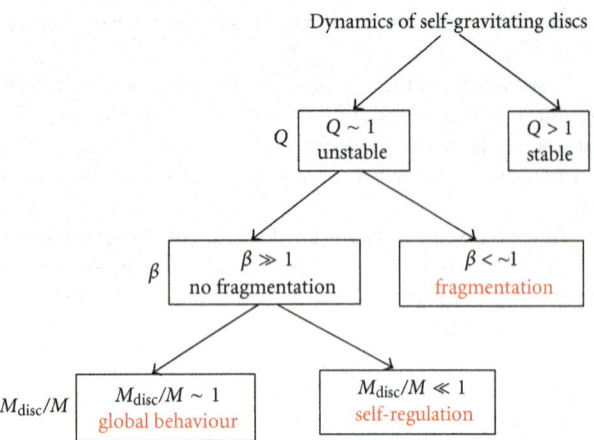

FIGURE 5: A simple diagram showing schematically the possible behaviour of a self-gravitating disc as a function of the three dimensionless parameters discussed here: Q, β, and M_{disc}/M.

Alternative models propose the direct formation of more massive seeds with masses of about $10^5 M_\odot$ directly out of the collapse of dense gas [66–72]. The key limiting factor for these models is the disposal of the angular momentum. Recently, it has been proposed [14, 71–73] that large-scale gravitational instabilities developing during the growth of pregalactic discs is the missing ingredient, able to funnel the required amount of gas into the center of the galaxy.

According to such models, the formation of the seeds of supermassive black holes occurs at a redshift $z \sim 10$–15, when the intergalactic medium had not been yet enriched by metals forming in the first stars. As a consequence, the chemical composition of the gas at this early epoch is essentially primordial, that is, the gas is mostly hydrogen and helium. The cooling properties of this gas are, therefore, relatively simple. In particular, in the absence of molecular hydrogen, the main coolant is provided by atomic hydrogen, for which the cooling timescale becomes extremely long for temperatures smaller than $\sim 10^4$ K, and we thus expect the gas to reach thermal equilibrium at a temperature T_{gas} of the order of 10^4 K.

Now, consider a dark matter halo (modeled, for simplicity, as a truncated singular isothermal sphere) of mass M_{halo} and circular velocity V_h, extending out to $r_h = GM_{\text{halo}}/V_h^2$. We also assume that the halo contains a gas mass $M_{\text{gas}} = m_d M_{\text{halo}}$, where m_d is of the order of the universal baryonic fraction, ≈ 0.1, whose angular momentum is $J_{\text{gas}} = j_d J$, where $j_d \sim m_d$. The angular momentum of the dark matter halo J is expressed in terms of its spin parameter $\lambda = J|E|^{1/2}/GM_{\text{halo}}^{5/2}$, where E is its total energy. The probability distribution of the spin parameter of dark matter halos can be obtained from cosmological N-body simulations in Warren et al. [74] and is well described by a log-normal distribution peaking at $\lambda = 0.05$.

If the virial temperature of the halo $T_{\text{vir}} \propto V_h^2$ is larger than the gas temperature T_{gas}, the gas collapses and forms a rotationally supported disc, with circular velocity V_h, determined by the gravitational field of the halo. For low values of the spin parameter λ, the resulting disc can be compact and dense. In this case, during the infall of gas onto the disc, its density rises until the stability parameter Q becomes of the order of unity. At this point, the disc starts developing a gravitational instability, which as we have seen above is able to efficiently redistribute angular momentum and allow accretion. Further infall of gas does not cause the density to rise much further, but rather it promotes an increasingly high accretion rate into the center. This process goes on until infall is over and the disc has attained a surface density low enough to be marginally gravitationally stable, that is, with $Q = \overline{Q}$. It is then possible to calculate what fraction of the infalling mass needs to be transported into the center to make the disc marginally stable, as a function of the main parameters involved. In this way, we get [14, 73]

$$M_{\text{BH}} = m_d M_{\text{halo}} \left[1 - \sqrt{\frac{8\lambda}{m_d \overline{Q}} \left(\frac{j_d}{m_d} \right) \left(\frac{T_{\text{gas}}}{T_{\text{vir}}} \right)^{1/2}} \right], \quad (14)$$

where I have suggestively called M_{BH} the accreted mass, since this mass is the total mass available for the formation of the black hole seed in the center.

However, for large halo mass, the internal torques needed to redistribute the excess baryonic mass become too large to be sustained by the disc, which might then undergo fragmentation. We have seen in the previous sections that the maximum torque that can be delivered by a quasisteady self-regulated disc is of the order of $\alpha_c \approx 0.06$. Since the infall rate of gas from the halo is proportional to $T_{\text{vir}}^{3/2}$, we expect fragmentation when the virial temperature exceeds a critical value T_{max}, given by (see [73] for details)

$$\frac{T_{\text{max}}}{T_{\text{gas}}} > \left(\frac{4\alpha_c}{m_d} \frac{1}{1 + M_{\text{BH}}/m_d M_{\text{halo}}} \right)^{2/3}. \quad (15)$$

Although it is possible, as mentioned above, that accretion proceeds even for larger values of α in a highly time-variable way when the disc mass is large, and it is also possible that accretion proceeds even in a fragmenting disc, we make here the conservative assumption that all halos that violate, (15), do fragment and do not accrete. Figure 6 illustrates the relationship between halo mass and black hole mass based on (14) for three different values of the spin parameter λ. The red line in Figure 6 corresponds to (15), so that halos on the right of the red line are expected to fragment. We can thus see that the typical mass fed into the center of such pregalactic disc is of the order of $10^3 M_\odot$ up to $10^5 M_\odot$. The typical accretion rates during this early epochs is of the order of $10^{-2} M_\odot/\text{yr}$ [73]. If such high masses are assembled as seeds of supermassive black holes at redshift 10–15, it is then easy to grow through Eddington-limited accretion to $10^9 M_\odot$ by $z = 6$, as required by observations.

Equation (14) provides a powerful link between the properties of dark matter haloes and the mass of massive seed black holes that can grow within them. As shown, the amount of mass that will be concentrated in the central regions of these pregalactic discs depends only on halo properties (such as the spin parameter λ and the fraction of baryonic mass that collapses to the disc m_d), on the ratio

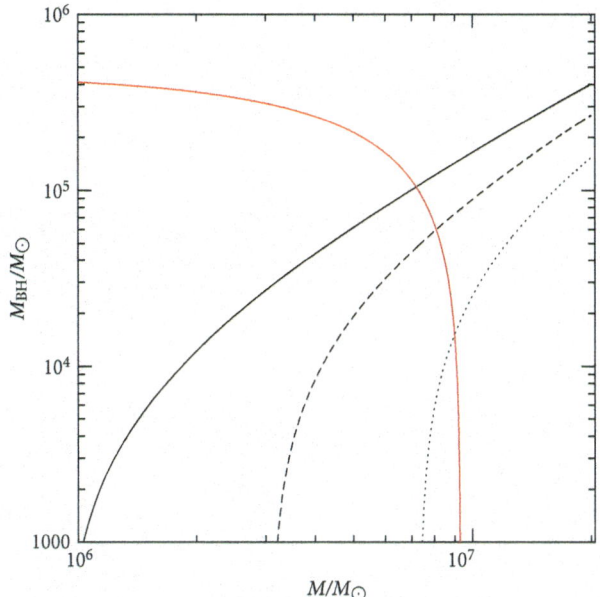

FIGURE 6: Mass available for the formation of the seed of a supermassive black hole in the center of pregalactic discs as a function of the mass of the parent dark matter halo (from [14]). The plots refer to the following choice of parameters: $\overline{Q} = 2$, $T_{gas} = 4000\,\mathrm{K}$, $m_d = j_d = 0.05$, $\lambda = 0.01$ (solid line), $\lambda = 0.015$ (dashed line), and $\lambda = 0.02$ (dotted line). The red curve shows the threshold for fragmentation from (15), with $\alpha_c = 0.06$. Halos on the right of the red line give rise to fragmenting discs.

between gas temperature and halo virial temperature, and on the threshold value of Q, which has a very small range of variation around $\overline{Q} \approx 1$. This simple model has been used to calculate several properties of the black hole population at high redshift. In particular, from the distribution of halo masses and angular momentum, it is straightforward to derive the mass function of the supermassive black hole seeds Lodato and Natarajan [14], which turns out to be strongly peaked at around $10^5 \, M_\odot$, as shown Figure 7(a). Furthermore, it is also possible to include such a simple prescription within evolutionary models that track the properties of the black hole population along cosmic time, such as merger tree models [15]. It is then interesting to see that the evolution of such a primordial seed population can naturally account for the current estimates of the density of black holes at low redshift (Figure 7(b)). In addition, an important and testable prediction of such models is that dwarf galaxies, which did not have any progenitor massive enough to seed a black hole, should not host a supermassive black hole. In particular, if the velocity dispersion of the galaxy is below ~ 50 km/sec, the probability of hosting a black hole turns out to be negligibly small [15].

A key requirement for the above model to work is that the gas in the disc is cooling very inefficiently; otherwise, it would rapidly fragment and form stars rather than accreting to the center (see Section 2 above). Indeed, we require the gas to be free of the main coolants such as metals and

molecular hydrogen. The process outlined above is thus going to be effective only before the intergalactic medium has been sufficiently enriched by metals and only where the gas is not excessively shielded by a UV background that tends to dissociate molecular hydrogen. This has led some to propose that the above mechanism only works at specific locations which satisfy the above conditions [75, 76]. Others [77] have instead proposed that even if fragmentation does occur, it would produce a compact stellar cluster whose eventual fate is still the formation of a supermassive black hole.

It is also interesting to mention that the process described analytically above has also been simulated numerically [78], and the results appear to be in substantial agreement with the analytical expectations.

Finally, note that the models described in this section only describe how can a substantial reservoir of mass be accumulated in the innermost regions of pregalactic discs. The eventual fate of this large amount of mass is not described here. Most probably, it will form a massive object at the center of the forming galaxy, such as a "quasistar" [79], where a seed black hole grows inside a large gaseous envelope which is accreting at rates which are super-Eddington for the hole, but not for the envelope itself. A detailed discussion of the physics associated with this is clearly beyond the scope of the present contribution.

4. Gravitational Instabilities, Angular Momentum Transport, and Fragmentation in AGN Discs

4.1. AGN Discs: Fragmenting or Nonfragmenting? The situation described above changes dramatically when we consider lower redshift, that is, if we now look at the outer disc in AGNs. As mentioned above, typically, the condition of marginal stability $Q \sim 1$ first occurs at a radius of the order of 0.01 pc from the central black hole. One can easily calculate the cooling rate, and the associated cooling timescale, at this radius, to find that it is typically much smaller than the dynamical time [2, 39]. Stated otherwise, the heating rate needed to keep a marginally stable ($Q \sim 1$) disc in thermal equilibrium is much larger than what can be provided by a viscous disc with reasonable values of α [1, 80]. If we now consult the results of the numerical simulations described in Section 2, we would simply conclude that the fate of such discs is to rapidly fragment into a number of bound objects. This is often interpreted as leading to intense star formation in the disc. However, it is worth noting that the dynamical time at 0.01–1 pc from a $10^8 \, M_\odot$ black hole (which is the time needed for the density perturbations to grow under the effect of the gravitational instability) (It is also easy to show that this is also the internal dynamical time of the fragments formed by instability in a $Q \sim 1$ disc.) is of the order of a few to a few thousand years. This is much shorter than the typical timescale associated with star formation in the solar neighbourhood, which is of the order of 10^6 years. Now, clearly, star formation in the Galaxy occurs under significantly different conditions, as the local molecular clouds are much less dense than the fragments

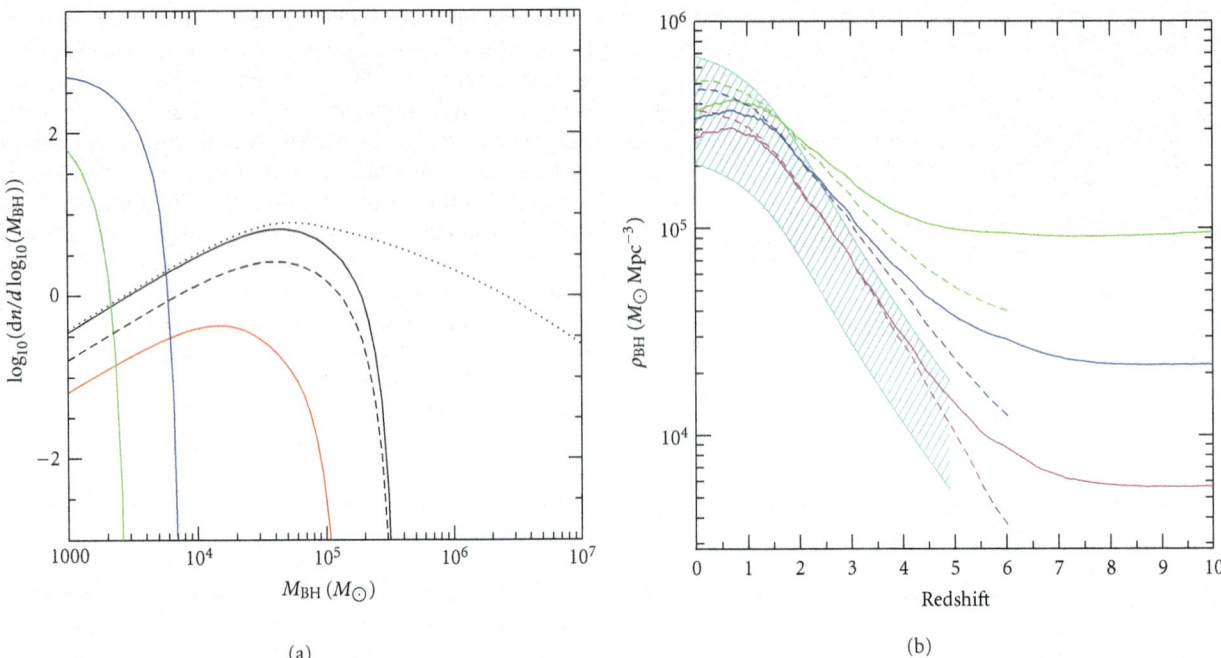

(a)   (b)

FIGURE 7: (a) Mass function of seed black holes predicted by the model based on (14) and (15). The black solid line refers to $z = 10$, while the red line refers to $z = 20$. The long-dashed line shows the effect of reducing $\overline{Q}$ from 2 to 1.5. The short dashed line shows the effect of not including the possibility of fragmentation (more details can be found in [14]). (b) The integrated density of black holes predicted from a merger tree evolution of the black hole seed population. The three solid lines refer to three different choices of the parameters (see [15] for details). The dashed area indicates the observationally permitted region from estimates of the black hole density at low redshifts. At high redshift ($z \gtrsim 6$), the density reflects the one attained after the seed formation phase, while the rise at lower redshifts indicates the growth through AGN activity.

produced in a fragmenting disc in this context. However, local star formation can also teach us something: indeed, the relatively long lifetime of molecular clouds in the Galaxy is due to the fact that rather than thermally supported, they are mostly supported by turbulent motions. The same might happen here: a fragmenting disc might produce a number of clouds whose dynamics is controlled by turbulent motions, which prevent their further collapse to form stars. In this case, the effective "cooling time," rather than the radiative timescale, would be the timescale for turbulence decay, which is of the order of the dynamical time and would thus be close to the threshold discussed in Section 2 [81].

A second thing to keep in mind is that if the disc extends to large distances, of the order of a parsec or so, its mass can become a significant fraction of the black hole mass and we might thus enter the regime where the dynamics associated with the gravitational instability is global rather than local. In this case, as already discussed, the energy balance should include some extra "global" terms, [51], arising from wave transport of energy, which might provide the required energy to prevent fragmentation in the outer disc. In this picture, a density wave might remove free rotational energy from the inner disc, but rather than dissipating it locally (as would a standard viscous process do), it might carry it a long way out along the wave and release it at large radii, where the wave is dissipated. As seen above, the evolution of such massive discs is generally highly variable, with episodes of strong

accretion and black hole feeding followed by more quiescent periods where the accretion rate is small. Such a time variable accretion model has also been sometimes proposed by Collin and Zahn [82].

From the numerical point of view, simulations of the disc dynamics in this fragmenting regime are the most challenging, as the density in the clumps rapidly rises thus slowing down significantly the simulation. As a result, we still do not know in detail what is the fate of the disc: how much mass is turned into "stars" and what fraction of the disc mass is able to accrete to the central black hole rather than onto the forming stars [83]. Clearly, if most of the disc mass ends up in stars, it would be disastrous for black hole feeding, and we would thus conclude that only gas with very low angular momentum, which would circularize within the $Q \sim 1$ radius at 0.01 pc would accrete onto the hole. This is the basic assumption behind the chaotic accretion scenario proposed recently [84, 85].

The presence of a significant stellar component within the disc can also in principle significantly affect the overall disc dynamics. Indeed, it has been proposed [86] that even if it constitutes a minor fraction of the overall disc mass, a stellar component in the disc is able to excite low-m global spiral modes, even in a relatively low mass disc, and would thus provide a significant source of angular momentum transport, thus allowing accretion even from distances of order of several parsecs [19].

In the context delineated above, an important role is played by the evidence that has been gathered in the last few years, which points to the presence of a large number of young stars very close to the supermassive black hole at the center of our own Milky Way [87, 88]. In particular, most of these stars appear to belong to two distinct stellar discs orbiting at roughly the same distance to the black hole, that is, at a distance of 0.05–0.5 pc [8, 89]. The most likely explanation for the origin of these stars in that they formed *in situ* and in particular from the fragmentation of a self-gravitating accretion disc [89, 90]. Such observations thus fit naturally in the context described above, since we know that at parsec distances an AGN accretion disc would be self-gravitating and its cooling time is expected to be short enough to induce fragmentation. The conditions in the Galactic Center might be typical of other galaxies, where a nuclear starburst can be a result of the very same mechanism [91, 92].

4.2. Hints from Maser Dynamics.

As mentioned above, there is clear observational evidence of the presence of significant mass in gas at parsec scales from the central black hole, in the form of maser emitting clumps. Such maser spots can effectively be used as a probe of the disc dynamics, as we can infer their rotation curve and hence probe the potential in the galactic nucleus. In most cases, as for example, the case of NGC 4258, the resulting rotation curve is very close to Keplerian [10], and it thus allows a very precise determination of the mass of the central BH, which for NGC 4258 is $3.6 \times 10^7 M_\odot$ (see also the recent compilation of Keplerian rotation curves obtained through maser emission by [93]).

However, in many other cases the rotation curve, while still displaying a smooth declining profile, as would be expected for a rotating disc, does not follow exactly Kepler's law. This is, for example, the case of NGC 1068 [11, 94], of the Circinus galaxy [95], and of NGC 3079 [96]. In particular, for the case of NGC 1068, the maser data are consistent with a circular velocity $v_\phi \propto r^{-0.31}$ [11]. Given the discussion above, which shows that at a scale of a fraction of a parsec, where the maser spots are detected, the disc can be self-gravitating, it is then tempting to attribute such (often small) deviation from Keplerian rotation to the contribution of the disc self-gravity.

A detailed fit to the circular velocity traced by water masers in NGC 1068 with a model which incorporates both the gravitational field of the black hole and that of the disc has been performed by Lodato and Bertin [33], by using self-regulated models of massive discs. The resulting black hole mass is $M = (8.0 \pm 0.3) 10^6 M_\odot$ and the disc mass is approximately equal to the black hole mass. From the required disc surface density, it is then possible to obtain $\dot{M} = (28.1 \pm 0.2) \alpha M_\odot/\text{yr}$. The mass accretion rate $\dot{M}$ can be estimated, for example, from the bolometric luminosity as $\dot{M} \approx 0.23 M_\odot/\text{yr}$, and we thus obtain also an estimate of $\alpha \approx 8.3 \times 10^{-3}$, which is of the right order of magnitude as would be expected from the transport induced by gravitational instabilities.

4.3. Gravitational Instabilities and the Process of Binary Black Holes Merger.

A related issue is connected to the process of black hole mergers. Black hole pairs are a natural by-product of hierarchical galaxy formation, as a consequence of the merger of two galaxies each containing a nuclear black hole. Stellar dynamical processes are able to shrink the binary down to separations of the order of 1 pc [97]. Additional gas dynamical processes can reduce the separation down to 0.1 pc or so [98, 99]. Below 0.001 pc, the emission of gravitational waves can shrink the binary further and lead to the merger of the two black holes. Such black hole mergers are indeed expected to be a primary source of gravitational radiation (and a prime target for gravitational wave detectors, such as LISA). However, an outstanding question is how to reduce the binary separation from 0.1 pc to 0.001 pc. Given the essential lack of observational evidence for sub-pc black hole binaries, we know that the process needs to be fast. It has been frequently suggested that the role of gaseous discs at sub-pc scales can provide the necessary torques to produce such fast evolution.

The problem is in several ways connected with the issues discussed above. Indeed, both in the case of mass accretion to feed a single central black hole and in the case of reducing the separation of a black hole binary, the problem is how to dispose of the large orbital angular momentum. The internal torques within an accretion disc (whether "viscous" and thus local, or globally related to gravitational instabilities) could be the natural way to remove the excess angular momentum also in the case of a binary. In reality, in the case of a binary, the angular momentum transfer process is mediated by disc tides. A secondary black hole carves an annular gap within an accretion disc. It is the gravitational force between the disc and the satellite to remove the angular momentum from the satellite and reduce the binary separation. Viscous torques within the disc are then essential in redistributing the angular momentum taken up by the gas and transport it to large radii. The circumbinary disc then evolves subject to a source of angular momentum from its inner edge, in a way that has been termed a "decretion" disc [100]. The binary evolution timescale in this case is given by [16, 101, 102]

$$t_{\text{shrink}} = \frac{M_d(a) + M_s}{M_d(a)} t_\nu, \tag{16}$$

where t_ν is the disc viscous timescale, $M_{\text{disc}}(a) = 4\pi\Sigma(a)a^2$ is a measure of the local disc mass at the binary separation a, and M_s is the mass of the secondary black hole. The dependency on t_ν indicates the fact that viscous torques in the disc are ultimately responsible for the removal of angular momentum, while the factor depending on the relative mass of the disc and of the secondary indicates that if the inertia of the secondary black hole is much larger than the disc, then the shrinking must necessarily take much longer. At 1 pc, the viscous timescale is already of the order of 10^8 years, and we thus see that if the disc mass is much less than the secondary mass, the shrinking timescale rapidly grows and can become exceedingly long for the merger to take place. Disc-assisted merger then requires large disc masses, comparable to the

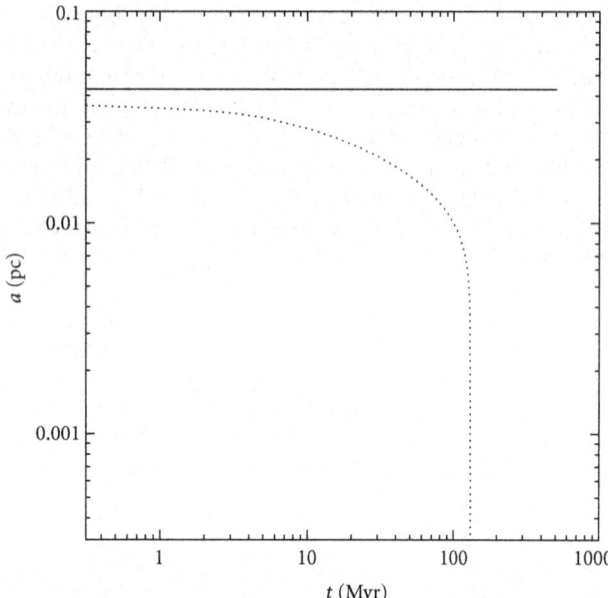

FIGURE 8: Evolution of the separation of a black hole binary with mass ratio $q = 0.1$. The initial separation here was $a_0 = 0.05\,\text{pc}$ and the disc/secondary mass ratio is $M_d/M_s = 1$. The solid curve shows the case where the disc is allowed to fragment and form stars, whereas the dotted line shows where star formation ignored. Star formation severely depletes the disc, and the remaining disc mass is not large enough to induce a black hole merger within a Hubble time. The figure taken from Lodato et al. [16].

secondary black hole, which are thus likely to be subject to gravitational instabilities.

Once again then, the dynamics of the gas disc at $\sim 0.01\,\text{pc}$ from the central black hole is essential in order to understand the evolution of the system. Lodato et al. [16] have studied the evolution of such systems. In particular, they have considered the case of a $10^8\,M_\odot$ primary black hole and of a secondary black hole with a mass ratio $q = 0.1$. They have shown that, when neglecting the possible fragmentation associated with gravitational instabilities, disc torques are able to shrink the binary and allow the merger of the two black holes within a Hubble time. On the contrary, when the disc is subject to fragmentation in the self-gravitating regime (assuming that in the self-gravitating state, enough mass is turned into stars in order to keep the disc marginally stable), the shrinking stalls and the merger does not take place. One such calculations is shown in Figure 8. Here the initial separation of the binary was 0.05 pc and the mass ratio was $q = 0.1$. The two lines refer, respectively, to the case where the disc is allowed to fragment and form stars (solid line) and where fragmentation is ignored (dotted line). The effect of fragmentation is to inhibit completely the merger process.

Fragmentation thus has a severe impact on the ability of the disc to induce a black hole merger in a reasonable time. The rapid effect of fragmentation sets an upper limit to the effective mass of a gaseous disc. To avoid the disc becoming self-gravitating, the disc must have $Q \gtrsim 1$ and so

must have a mass at most $M_d/M_p \lesssim H/R$ (see Section 2). On the other hand, we must also require that the disc mass be at least comparable to the secondary mass in order for the viscous torques to be able to remove the secondary angular momentum. With a typical aspect ratio H/R of order of a few times 10^{-3}, one can conclude that disc-assisted mergers only work for mass ratios $q \lesssim 0.001$.

5. Conclusions

AGN discs become self-gravitating at a distance of about 0.01 pc from their central black hole. The development of gravitational instabilities in the disc can be both beneficial and detrimental for the process of black hole growth. Indeed, a beneficial effect is provided by the ability of gravitational instabilities to redistribute angular momentum within the disc and thus promote accretion. A detrimental effect is instead produced by the possibility of fragmentation, which could in principle turn most of the gas mass into star and thus remove it from the accretion flow. In this contribution, I have reviewed the recent progresses that we have made in the last ten years in our understanding of the nonlinear evolution of gravitational instabilities in gaseous discs, and in particular, on the parameters (most importantly, the disc cooling rate), that determine whether the instability saturates at a finite value—thus providing a quasisteady source of angular momentum transport in a self-regulated way—or rather fragments into bound objects.

Despite the impressive progresses made from the theoretical point of view, the application to AGN discs is not straightforward. Indeed, it turns out that the cooling timescale at the radius where the disc is self gravitating is way to short to support a self-regulated state. This has led some authors to propose that AGN discs are effectively truncated at the self-gravitating radius. On the other hand, the evolution of a fragmenting disc is still not well understood, and it is not at all clear that the onset of fragmentation would totally preclude accretion. On the contrary, observational evidence such as the presence of Keplerian or quasi-Keplerian maser emitting gas at parsec scales in several AGNs, and the lack of observed black hole binaries at sub-pc scales—which in turn require the effective dynamical presence of a massive gaseous disc—hints to the importance of discs in the self-gravitating regime. Finally, it is worth noting that the presence of young stars in our own Galactic Center is indicative of the fact that star formation does effectively take place in the discs surrounding nuclear black holes.

A relatively better-understood evolution occurs at high redshifts, $z \sim 10$ or so, where the intergalactic medium was still not significantly polluted with metals and thus the cooling properties of the gas were significantly different. In these cases, fragmentation would be inhibited and the transport of angular momentum due to gravitational torques would naturally provide a way to accumulate large amounts of gas (up to $10^5\,M_\odot$) in the central regions of pregalactic discs, potentially opening the way to the rapid formation of black hole seeds by direct gas collapse.

References

[1] G. Bertin and G. Lodato, "Thermal stability of self-gravitating, optically thin accretion disks," *Astronomy and Astrophysics*, vol. 370, no. 1, pp. 342–350, 2001.

[2] J. Goodman, "Self-gravity and quasi-stellar object discs," *Monthly Notices of the Royal Astronomical Society*, vol. 339, no. 4, pp. 937–948, 2003.

[3] P. I. Kolykhalov and R. A. Sunyaev, "Disk formation through accretion of stellar wind," *Soviet Astronomy Letters*, vol. 5, pp. 180–183, 1979.

[4] S. Collin and J. P. Zahn, "Star formation and evolution in accretion disks around massive black holes: star formation and evolution in accretion disks," *Astronomy and Astrophysics*, vol. 344, no. 2, pp. 433–449, 1999.

[5] D. N. C. Lin and J. E. Pringle, "A viscosity prescription for a self-gravitating accretion disc," *Monthly Notices Royal Astronomical Society*, vol. 225, pp. 607–613, 1987.

[6] D. N. C. Lin and J. E. Pringle, "The formation and initial evolution of protostellar disks," *Astrophysical Journal*, vol. 358, no. 2, pp. 515–524, 1990.

[7] I. Shlosman, J. Frank, and M. C. Begelman, "Bars within bars: a mechanism for fuelling active galactic nuclei," *Nature*, vol. 338, no. 6210, pp. 45–47, 1989.

[8] T. Paumard, R. Genzel, F. Martins et al., "The two young star disks in the central parsec of the galaxy: properties, dynamics, and formation," *Astrophysical Journal*, vol. 643, no. 2 I, article 033, pp. 1011–1035, 2006.

[9] R. Genzel, F. Eisenhauer, and S. Gillessen, "The Galactic center massive black hole and nuclear star cluster," *Reviews of Modern Physics*, vol. 82, no. 4, pp. 3121–3195, 2010.

[10] M. Miyoshi, J. Moran, J. Herrnstein et al., "Evidence for a black hole from high rotation velocities in a sub-parsec region of NGC4258," *Nature*, vol. 373, no. 6510, pp. 127–129, 1995.

[11] L. J. Greenhill, C. R. Gwinn, R. Antonucci, and R. Barvainis, "VLBI imaging of water maser emission from the nuclear torus of NGC 1068," *Astrophysical Journal*, vol. 472, no. 1, pp. L21–L24, 1996.

[12] P. T. Kondratko, L. J. Greenhill, and J. M. Moran, "Discovery of water maser emission in five AGNs and a possible correlation between water maser and nuclear 2-10 keV luminosities," *Astrophysical Journal*, vol. 652, no. 1 I, pp. 136–145, 2006.

[13] P. Cossins, G. Lodato, and C. J. Clarke, "Characterizing the gravitational instability in cooling accretion discs," *Monthly Notices of the Royal Astronomical Society*, vol. 393, no. 4, pp. 1157–1173, 2009.

[14] G. Lodato and P. Natarajan, "The mass function of high-redshift seed black holes," *Monthly Notices of the Royal Astronomical Society*, vol. 377, no. 1, pp. L64–L68, 2007.

[15] M. Volonteri, G. Lodato, and P. Natarajan, "The evolution of massive black hole seeds," *Monthly Notices of the Royal Astronomical Society*, vol. 383, no. 3, pp. 1079–1088, 2008.

[16] G. Lodato, S. Nayakshin, A. R. King, and J. E. Pringle, "Black hole mergers: can gas discs solve the "final parsec" problem?" *Monthly Notices of the Royal Astronomical Society*, vol. 398, no. 3, pp. 1392–1402, 2009.

[17] L. Ciotti and J. P. Ostriker, "Radiative feedback from massive black holes in elliptical galaxies: agn flaring and central starburst fueled by recycled gas," *Astrophysical Journal*, vol. 665, no. 2 I, pp. 1038–1056, 2007.

[18] L. Ciotti, J. P. Ostriker, and D. Proga, "Feedback from central black holes in elliptical galaxies. III. Models with both radiative and mechanical feedback," *Astrophysical Journal*, vol. 717, no. 2, pp. 708–723, 2010.

[19] P. F. Hopkins and E. Quataert, "How do massive black holes get their gas?" *Monthly Notices of the Royal Astronomical Society*, vol. 407, no. 3, pp. 1529–1564, 2010.

[20] C. F. Gammie, "Nonlinear outcome of gravitational instability in cooling, gaseous disks," *Astrophysical Journal*, vol. 553, no. 1, pp. 174–183, 2001.

[21] G. Lodato and W. K. M. Rice, "Testing the locality of transport in self-gravitating accretion discs," *Monthly Notices of the Royal Astronomical Society*, vol. 351, no. 2, pp. 630–642, 2004.

[22] A. C. Mejia, R. H. Durisen, M. K. Pickett, and K. Cai, "The thermal regulation of gravitational instabilities in protoplanetary disks. II. Extended simulations with varied cooling rates," *Astrophysical Journal*, vol. 619, no. 2 I, pp. 1098–1113, 2005.

[23] A. C. Boley, A. C. Mejía, R. H. Durisen, K. Cai, M. K. Pickett, and P. D'Alessio, "The thermal regulation of gravitational instabilities in protoplanetary disks. III. Simulations with radiative cooling and realistic opacities," *Astrophysical Journal*, vol. 651, no. 1 I, pp. 517–534, 2006.

[24] W. K.M. Rice, G. Lodato, and P. J. Armitage, "Investigating fragmentation conditions in self-gravitating accretion discs," *Monthly Notices of the Royal Astronomical Society*, vol. 364, no. 1, pp. L56–L60, 2005.

[25] G. Lodato and W. K. M. Rice, "Testing the locality of transport in self-gravitating accretion discs - II. The massive disc case," *Monthly Notices of the Royal Astronomical Society*, vol. 358, no. 4, pp. 1489–1500, 2005.

[26] G. Lodato, "Self-gravitating accretion discs," *Nuovo Cimento Rivista Serie*, vol. 30, no. 7, pp. 293–353, 2007.

[27] R. H. Durisen, A. P. Boss, L. Mayer, A. F. Nelson, T. Quinn, and W. K. M. Rice, "Gravitational instabilities in gaseous protoplanetary disks and implications for giant planet formation," in *Protostars and Planets V*, B. Reipurth, D. Jewitt, and K. Keil, Eds., pp. 607–622, University of Arizona Press, Tucson, Ariz, USA, 2007.

[28] G. Bertin and G. Lodato, "A class of self-gravitating accretion disks," *Astronomy and Astrophysics*, vol. 350, no. 2, pp. 694–704, 1999.

[29] A. Toomre, "On the gravitational stability of a disk of stars," *The Astrophysical Journal*, vol. 139, p. 1217, 1964.

[30] G. Bertin, *Dynamics of Galaxies*, Cambridge University Press, Cambridge, UK, 2000.

[31] J. P. Ostriker and P. J. E. Peebles, "A Numerical study of the stability of flattened galaxies: or, can cold galaxies survive?" *Astrophysical Journal*, vol. 186, pp. 467–480, 1973.

[32] N. I. Shakura and R. A. Sunyaev, "Black holes in binary systems. Observational appearance," *Astronomy & Astrophysics*, vol. 24, pp. 337–355, 1973.

[33] G. Lodato and G. Bertin, "Non-Keplerian rotation in the nucleus of NGC 1068: evidence for a massive accretion disk?" *Astronomy and Astrophysics*, vol. 398, no. 2, pp. 517–524, 2003.

[34] B. Paczyński, "A model of selfgravitating accretion disk with a hot corona," *Acta Astronomica*, vol. 28, no. 3, pp. 241–251, 1978.

[35] G. Bertin, "Self-regulated accretion disks," *Astrophysical Journal*, vol. 478, no. 2, pp. L71–L74, 1997.

[36] G. Laughlin and P. Bodenheimer, "Nonaxisymmetric evolution in protostellar disks," *Astrophysical Journal*, vol. 436, no. 1, pp. 335–354, 1994.

[37] J. A. Sellwood and R. G. Carlberg, "Spiral instabilities provoked by accretion and star formation," *Astrophysical Journal*, vol. 282, pp. 61–74, 1984.

[38] L. Mayer, G. Lufkin, T. Quinn, and J. Wadsley, "Fragmentation of gravitationally unstable gaseous protoplanetary disks with radiative transfer," *Astrophysical Journal*, vol. 661, no. 1, pp. L77–L80, 2007.

[39] B. M. Johnson and C. F. Gammie, "Nonlinear outcome of gravitational instability in disks with realistic cooling," *Astrophysical Journal*, vol. 597, no. 1 I, pp. 131–141, 2003.

[40] P. Cossins, G. Lodato, and C. Clarke, "The effects of opacity on gravitational stability in protoplanetary discs," *Monthly Notices of the Royal Astronomical Society*, vol. 401, no. 4, pp. 2587–2598, 2010.

[41] M. J. Rees, "Opacity-limited hierarchical fragmentation and the masses of protostars," *Monthly Notices of the Royal Astronomical Society*, vol. 176, p. 483, 1976.

[42] J. Silk, "On the fragmentation of cosmic gas clouds. II—opacity-limited star formation," *Astrophysical Journal*, vol. 214, pp. 152–160, 1977.

[43] C. J. Clarke, E. Harper-Clark, and G. Lodato, "The response of self-gravitating protostellar discs to slow reduction in cooling time-scale: the fragmentation boundary revisited," *Monthly Notices of the Royal Astronomical Society*, vol. 381, no. 4, pp. 1543–1547, 2007.

[44] F. Meru and M. R. Bate, "Non-convergence of the critical cooling time-scale for fragmentation of self-gravitating discs," *Monthly Notices of the Royal Astronomical Society*, vol. 411, no. 1, pp. L1–L5, 2011.

[45] G. Lodato and C. J. Clarke, "Resolution requirements for smoothed particle hydrodynamics simulations of self-gravitating accretion discs," *Monthly Notices of the Royal Astronomical Society*, vol. 413, no. 4, pp. 2735–2740, 2011.

[46] G. Lodato and P. J. Cossins, "Smoothed Particle Hydrodynamics for astrophysical flows. The dynamics of protostellar discs," *The European Physical Journal Plus*, vol. 126, no. 4, 2011.

[47] S.-J. Paardekooper, C. Baruteau, and F. Meru, "Numerical convergence in self-gravitating disc simulations: initial conditions and edge effects," *Monthly Notices of the Royal Astronomical Society*, vol. 416, no. 1, pp. L65–L69, 2011.

[48] C. J. Clarke, "Pseudo-viscous modelling of self-gravitating discs and the formation of low mass ratio binaries," *Monthly Notices of the Royal Astronomical Society*, vol. 396, no. 2, pp. 1066–1074, 2009.

[49] R. R. Rafikov, "Properties of gravitoturbulent accretion disks," *Astrophysical Journal*, vol. 704, no. 1, pp. 281–291, 2009.

[50] D. Lynden-Bell and A. J. Kalnajs, "On the generating mechanism of spiral structure," *Monthly Notices of the Royal Astronomical Society*, vol. 157, p. 1, 1972.

[51] S. A. Balbus and J. C. B. Papaloizou, "On the dynamical foundations of α disks," *Astrophysical Journal*, vol. 521, no. 2, pp. 650–658, 1999.

[52] J. E. Pringle, "Accretion discs in astrophysics," *Annual Review of Astronomy and Astrophysics*, vol. 19, pp. 137–160, 1981.

[53] W. K. M. Rice, P. J. Armitage, G. Mamatsashvili, G. Lodato, and C. J. Clarke, "Stability of self-gravitating discs under irradiation," *Monthly Notices of the Royal Astronomical Society*, vol. 418, no. 2, pp. 1356–1362, 2011.

[54] X. Fan, J. F. Hennawi, G. T. Richards et al., "A survey of z > 5.7 quasars in the sloan digital sky survey. III. Discovery of five additional quasars," *Astronomical Journal*, vol. 128, no. 2, pp. 515–522, 2004.

[55] X. Fan, M. A. Strauss, G. T. Richards et al., "A survey of z > 5.7 quasars in the sloan digital sky survey. IV. Discovery of seven additional quasars," *Astronomical Journal*, vol. 131, no. 3, pp. 1203–1209, 2006.

[56] D. J. Mortlock, S. J. Warren, B. P. Venemans et al., "A luminous quasar at a redshift of z = 7.085," *Nature*, vol. 474, no. 7353, pp. 616–619, 2011.

[57] Z. Haiman and A. Loeb, "Observational signatures of the first quasars," *Astrophysical Journal*, vol. 503, no. 2, pp. 505–517, 1998.

[58] M. Volonteri, P. Madau, E. Quataert, and M. J. Rees, "The distribution and cosmic evolution of massive black hole spins," *Astrophysical Journal*, vol. 620, no. 1 I, pp. 69–77, 2005.

[59] J. S.B. Wyithe and A. Loeb, "Constraints on the process that regulates the growth of supermassive black holes based on the intrinsic scatter in the $M_{bh-\sigma sph}$ relation," *Astrophysical Journal*, vol. 634, pp. 910–920, 2005.

[60] T. Abel, G. L. Bryan, and M. L. Norman, "The formation and fragmentation of primordial molecular clouds," *Astrophysical Journal*, vol. 540, no. 1, pp. 39–44, 2000.

[61] V. Bromm, P. S. Coppi, and R. B. Larson, "The formation of the first stars. I. The primordial star-forming cloud," *Astrophysical Journal*, vol. 564, no. 1 I, pp. 23–51, 2002.

[62] M. Volonteri and M. J. Rees, "Rapid growth of high-redshift black holes," *Astrophysical Journal*, vol. 633, no. 2 I, pp. 624–629, 2005.

[63] A. R. King, S. H. Lubow, G. I. Ogilvie, and J. E. Pringle, "Aligning spinning black holes and accretion discs," *Monthly Notices of the Royal Astronomical Society*, vol. 363, no. 1, pp. 49–56, 2005.

[64] G. Lodato and J. E. Pringle, "The evolution of misaligned accretion discs and spinning black holes," *Monthly Notices of the Royal Astronomical Society*, vol. 368, no. 3, pp. 1196–1208, 2006.

[65] A. R. King and J. E. Pringle, "Growing supermassive black holes by chaotic accretion," *Monthly Notices of the Royal Astronomical Society*, vol. 373, no. 1, pp. L90–L92, 2006.

[66] M. G. Haehnelt and M. J. Rees, "The formation of nuclei in newly formed galaxies and the evolution of the quasar population," *Monthly Notices Royal Astronomical Society*, vol. 263, no. 1, pp. 168–178, 1993.

[67] M. Umemura, A. Loeb, and E. L. Turner, "Early cosmic formation of massive black holes," *Astrophysical Journal*, vol. 419, no. 2, pp. 459–468, 1993.

[68] A. Loeb and F. A. Rasio, "Collapse of primordial gas clouds and the formation of quasar black holes," *Astrophysical Journal*, vol. 432, no. 1, pp. 52–61, 1994.

[69] D. J. Eisenstein and A. Loeb, "Origin of quasar progenitors from the collapse of low-spin cosmological perturbations," *Astrophysical Journal*, vol. 443, no. 1, pp. 11–17, 1995.

[70] V. Bromm and A. Loeb, "Formation of the first supermassive black holes," *Astrophysical Journal*, vol. 596, no. 1 I, pp. 34–46, 2003.

[71] S. M. Koushiappas, J. S. Bullock, and A. Dekel, "Massive black hole seeds from low angular momentum material," *Monthly Notices of the Royal Astronomical Society*, vol. 354, no. 1, pp. 292–304, 2004.

[72] M. C. Begelman, M. Volonteri, and M. J. Rees, "Formation of supermassive black holes by direct collapse in pre-galactic haloes," *Monthly Notices of the Royal Astronomical Society*, vol. 370, no. 1, pp. 289–298, 2006.

[73] G. Lodato and P. Natarajan, "Supermassive black hole formation during the assembly of pre-galactic discs," *Monthly*

Notices of the Royal Astronomical Society, vol. 371, no. 4, pp. 1813–1823, 2006.

[74] M. S. Warren, P. J. Quinn, J. K. Salmon, and W. H. Zurek, "Dark halos formed via dissipationless collapse. I. Shapes and alignment of angular momentum," *Astrophysical Journal*, vol. 399, no. 2, pp. 405–425, 1992.

[75] M. Dijkstra, Z. Haiman, A. Mesinger, and J. S. B. Wyithe, "Fluctuations in the high-redshift Lyman-Werner background: close halo pairs as the origin of supermassive black holes," *Monthly Notices of the Royal Astronomical Society*, vol. 391, no. 4, pp. 1961–1972, 2008.

[76] C. Shang, G. L. Bryan, and Z. Haiman, "Supermassive black hole formation by direct collapse: keeping protogalactic gas H_2 free in dark matter haloes with virial temperatures T_{vir} > rsim 10^4 K," *Monthly Notices of the Royal Astronomical Society*, vol. 402, no. 2, pp. 1249–1262, 2010.

[77] B. Devecchi and M. Volonteri, "Formation of the first nuclear clusters and massive black holes at high redshift," *Astrophysical Journal*, vol. 694, no. 1, pp. 302–313, 2009.

[78] L. Mayer, S. Kazantzidis, A. Escala, and S. Callegari, "Direct formation of supermassive black holes via multi-scale gas inflows in galaxy mergers," *Nature*, vol. 466, no. 7310, pp. 1082–1083, 2010.

[79] M. C. Begelman, E. M. Rossi, and P. J. Armitage, "Quasi-stars and the cosmic evolution of massive black holes," *Monthly Notices of the Royal Astronomical Society*, vol. 387, no. 4, pp. 1649–1659, 2008.

[80] E. Sirko and J. Goodman, "Spectral energy distributions of marginally self-gravitating quasi-stellar object discs," *Monthly Notices of the Royal Astronomical Society*, vol. 341, no. 2, pp. 501–508, 2003.

[81] M. C. Begelman and I. Shlosman, "Angular momentum transfer and lack of fragmentation in self-gravitating accretion flows," *Astrophysical Journal*, vol. 702, no. 1, pp. L5–L8, 2009.

[82] S. Collin and J. P. Zahn, "Star formation in accretion discs: from the Galactic center to active galactic nuclei," *Astronomy and Astrophysics*, vol. 477, no. 2, pp. 419–435, 2008.

[83] S. Nayakshin, J. Cuadra, and V. Springel, "Simulations of star formation in a gaseous disc around Sgr A*—a failed active galactic nucleus," *Monthly Notices of the Royal Astronomical Society*, vol. 379, no. 1, pp. 21–33, 2007.

[84] A. R. King and J. E. Pringle, "Fuelling active galactic nuclei," *Monthly Notices of the Royal Astronomical Society*, vol. 377, no. 1, pp. L25–L28, 2007.

[85] A. R. King, J. E. Pringle, and J. A. Hofmann, "The evolution of black hole mass and spin in active galactic nuclei," *Monthly Notices of the Royal Astronomical Society*, vol. 385, no. 3, pp. 1621–1627, 2008.

[86] P. F. Hopkins and E. Quataert, "An analytic model of angular momentum transport by gravitational torques: from galaxies to massive black holes," *Monthly Notices of the Royal Astronomical Society*, vol. 415, no. 2, pp. 1027–1050, 2011.

[87] R. Genzel, R. Schödel, T. Ott et al., "The stellar cusp around the supermassive black hole in the Galactic center," *Astrophysical Journal*, vol. 594, no. 2 I, pp. 812–832, 2003.

[88] A. M. Ghez, S. Salim, S. D. Hornstein et al., "Stellar orbits around the galactic center black hole," *Astrophysical Journal*, vol. 620, no. 2 I, pp. 744–757, 2005.

[89] Y. Levin and A. M. Beloborodov, "Stellar disk in the Galactic center: a remnant of a dense accretion disk?" *Astrophysical Journal*, vol. 590, no. 1, pp. L33–L36, 2003.

[90] S. Nayakshin and R. Sunyaev, "The "missing" young stellar objects in the central parsec of the Galaxy: evidence for star formation in a massive accretion disc and a top-heavy initial mass function," *Monthly Notices of the Royal Astronomical Society*, vol. 364, no. 1, pp. L23–L27, 2005.

[91] S. Nayakshin, "Massive stars in subparsec rings around galactic centres," *Monthly Notices of the Royal Astronomical Society*, vol. 372, no. 1, pp. 143–150, 2006.

[92] Y. Levin, "Starbursts near supermassive black holes: young stars in the Galactic Centre, and gravitational waves in LISA band," *Monthly Notices of the Royal Astronomical Society*, vol. 374, no. 2, pp. 515–524, 2007.

[93] C. Y. Kuo, J. A. Braatz, J. J. Condon et al., "The megamaser cosmology project. III. Accurate masses of seven supermassive black holes in active galaxies with circumnuclear megamaser disks," *Astrophysical Journal*, vol. 727, no. 1, 2011.

[94] L. J. Greenhill and C. R. Gwinn, "VLBI imaging of water maser emission from a nuclear disk in NGC 1068," *Astrophysics and Space Science*, vol. 248, no. 1-2, pp. 261–267, 1997.

[95] L. J. Greenhill, P. T. Kondratko, J. E. J. Lovell et al., "The discovery of H_2O maser emission in seven active galactic nuclei and at high velocities in the Circinus galaxy," *Astrophysical Journal*, vol. 582, no. 1, pp. L11–L14, 2003.

[96] P. T. Kondratko, L. J. Greenhill, and J. M. Moran, "Evidence for a geometrically thick self-gravitating accretion disk in NGC 3079," *Astrophysical Journal*, vol. 618, no. 2 I, pp. 618–634, 2005.

[97] M. Milosavljević and D. Merritt, "Formation of galactic nuclei," *Astrophysical Journal*, vol. 563, no. 1, pp. 34–62, 2001.

[98] M. Dotti, M. Volonteri, A. Perego, M. Colpi, M. Ruszkowski, and F. Haardt, "Dual black holes in merger remnants - II. Spin evolution and gravitational recoil," *Monthly Notices of the Royal Astronomical Society*, vol. 402, no. 1, pp. 682–690, 2010.

[99] M. Dotti, M. Colpi, F. Haardt, and L. Mayer, "Supermassive black hole binaries in gaseous and stellar circumnuclear discs: orbital dynamics and gas accretion," *Monthly Notices of the Royal Astronomical Society*, vol. 379, no. 3, pp. 956–962, 2007.

[100] J. E. Pringle, "Self-induced warping of accretion discs," *Monthly Notices of the Royal Astronomical Society*, vol. 281, no. 1, pp. 357–361, 1996.

[101] D. Syer and C. J. Clarke, "Satellites in disks: regulating the accretion luminosity," *Monthly Notices of the Royal Astronomical Society*, vol. 277, p. 758, 1995.

[102] P. B. Ivanov, J. C. B. Papaloizou, and A. G. Polnarev, "The evolution of a supermassive binary caused by an accretion disc," *Monthly Notices of the Royal Astronomical Society*, vol. 307, no. 1, pp. 79–90, 1999.

8

Thirty-Year Periodicity of Cosmic Rays

Jorge Pérez-Peraza,[1] Víctor Velasco,[1] Igor Ya. Libin,[2] and K. F. Yudakhin[3]

[1] Instituto de Geofísica, Universidad Nacional Autónoma de México, C.U., 04510 Coyoacán, DF, Mexico
[2] International Academy for Appraisal and Consulting (MAOK), Moscow, Russia
[3] IZMIRAN, Academy of Sciences of Russia, Troitsk, Moscow 142092, Russia

Correspondence should be addressed to Jorge Pérez-Peraza, perperaz@yahoo.com.mx

Academic Editor: Karel Kudela

Cosmogenic isotopes have frequently been employed as proxies of ancient cosmic ray fluxes. On the basis of periodicities of the ^{10}Be time series (using data from both the South and North Poles) and the ^{14}C time series (with data from Intercal-98), we offer evidence of the existence of cosmic ray fluctuations with a periodicity of around 30 years. Results were obtained by using the wavelet transformation spectral technique, signal reconstruction by autoregressive spectral analysis (ARMA), and the Lomb-Scargle periodogram method. This 30-year periodicity seems to be significant in nature because several solar and climatic indexes exhibit the same modulation, which may indicate that the 30-year frequency of cosmic rays is probably a modulator agent for terrestrial phenomena, reflecting the control source, namely, solar activity.

1. Introduction

The importance of cosmic ray variations was pointed out long ago in a vast compendium of relevant research [1]. Here we encounter a periodicity that has not been studied within the framework of cosmic ray variations. Let us begin by emphasizing that 30-years cycles are quite common in nature: the so-called Markowitz wave (Markowitz wobble-MW) is a quasi-harmonic variation of the middle pole of the Earth with a period of 30 year and an amplitude of 0.02"-0.03" [2]. Similar results were obtained in different years by many researchers who were trying to measure the Earth's magnetic field [3–5], by evaluating the conductivity of the lower mantle and who established 60- and 30-year variations of the geomagnetic field of the lower mantle. The authors in [6], through studies of high-growth anomalies of the secular variation with foci located in South Asia and in the middle of the Indian Ocean, revealed that the increase in the current focus is the initial stage of the 30-year and 60-year variations.

Authors in [4] by evaluating the conductivity of the lower mantle have found evidence of 60 - and 30-year variations in the geomagnetic field. By conducting a qualitative analysis of timelines with data from planetary indices Ap/aa, interplanetary magnetic field intensity and sunspot numbers, as well as cosmic rays data strings taken from 1937 to 2010 [7, 8] have shown quasi-periodic three-cycle trends, which are particularly very well correlated with solar wind, polar coronal holes and the size of solar activity cycle 23 [9, 10].

Authors in [11] have found 32-year variations in temperature by analyzing the spectrum of periodical air-surface temperature fluctuations for 1423 years in Greenland ice cores, and for 1400 and 800 years in the California Arctic pine tree rings. A 30-year variation in storminess data was found by applying the "Caterpillar" method in [12]. In fact, the authors identified frequencies at 90 to 100 years, 28 to 32 years, 20 to 22 years, 9 to 13 years, and some others; it is shown that even if the accuracy of this method for determining periods is not high, the fundamental frequency obtained by the "Caterpillar" is real.

The application of wavelet analysis to the paleoclimatic proxy data [13] to large-scale atmospheric phenomena (the Atlantic Multidecadal Oscillation and Southern Oscillation Indexes), and hurricane phenomena, led to the discovery of a high coherence with periods of 30 ± 2 years, between climatic oscillations and cosmic rays (^{10}Be at the North Pole). Furthermore, some properties of hurricanes, such as their total cyclonal energy and the tropical storms appearance along the Atlantic coast of México, together with other

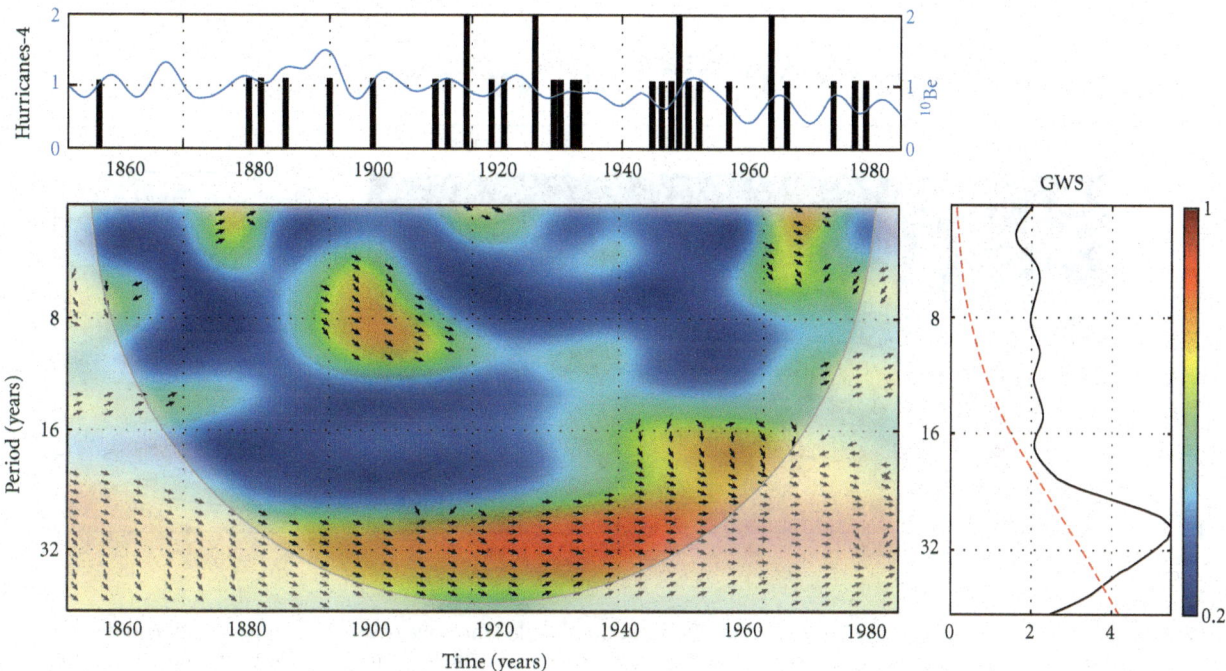

FIGURE 1: The upper panel shows time series of category-4 hurricane versus ^{10}Be. The coherence between both series appears in the middle panel, and the global wavelet spectrum (GWS) appears in the right-hand part of the figure, where the red dashed line indicates the border, also with the reliability of 95%. The scale color bar at the right indicates the level of coherence.

properties linked to hurricanes show such a 30-year cycle [14]. Figure 1 illustrates the particular case of category-4 hurricanes

Curiously, not only in nature but in several areas of contemporary human activity, such as business and commerce, 30-year cycle indexes are also found. (e.g., http://www.search.yahoo.com/search?p=30+years+cycle&ei=UTF-8&fr=moz35/). Incidentally, they are often associated with solar activity.

In the present work, we present evidence of the existence of a thirty-year periodicity for the cosmic rays: preliminary results were presented at the 2008COSPAR meeting in Montreal.

2. Data and the Spectral Wavelet Analysis

The spectral analysis of cosmic ray data from neutron monitors has been widely studied through several different methods, for instance [1, 15–18] and references included. Though most of these studies are out of the scope of this work, since they rather concern short-term periodicities, they are, however, very helpful for solar-terrestrial physics.

One of the main problems in determining significant long-term periodicities in the flux of cosmic rays is that the time series of data are relatively very short, as they have been available only for the last five to six decades, when data on cosmic rays (CR) from different stations throughout the world began to be organized and homologated. Because of this restriction, a proxy for cosmic rays has often been employed, one of which in our case allows us in a

deterministic way to provide evidence of a 30-year cycle for cosmic rays.

The cosmogenic isotopes Beryllium-10 (^{10}Be) and Carbon-14 (^{14}C) are conventionally considered to be proxies for cosmic rays, in such a way that an adequate spectral analysis may reveal important periodicities. These cosmogenic isotopes are mainly produced by galactic cosmic ray flux modulated by changes in interplanetary and geomagnetic magnetic fields. The analysis of cosmogenic isotopes stored in natural archives, such as ^{10}Be in polar ice cores and ^{14}C in tree rings, provides a means of extending our knowledge of solar variability over much longer periods (e.g., [19, 20]). In addition, the nature of climatic response to solar variability can be assessed over several time scales. It should be remembered that the analysis of the cosmogenic isotopes record is more difficult than the analysis of sunspot numbers. This is due to the fact the ^{14}C and ^{10}Be concentrations reflect the production rate, which is modulated by not only solar activity but also by atmospheric transport and deposition processes [21, 22]

Data on ^{10}Be and ^{14}C can be obtained for periods of thousands of years: we use the INTCAL 98 (http://www.depts.washington.edu/qil/) for the ^{14}C time series and the ^{10}Be time series from [21], which give the concentration found in the Dye-3 ice core (62.5 N, 43.8 W). For the South Pole we used data from [23].

The spectral techniques for analyzing periodicities of cosmophysical phenomena are very varied. The simplest technique for investigating periodicities is the Fourier Transform (FT). Although useful for stationary time series, this method is not appropriate for time series that do not fulfill

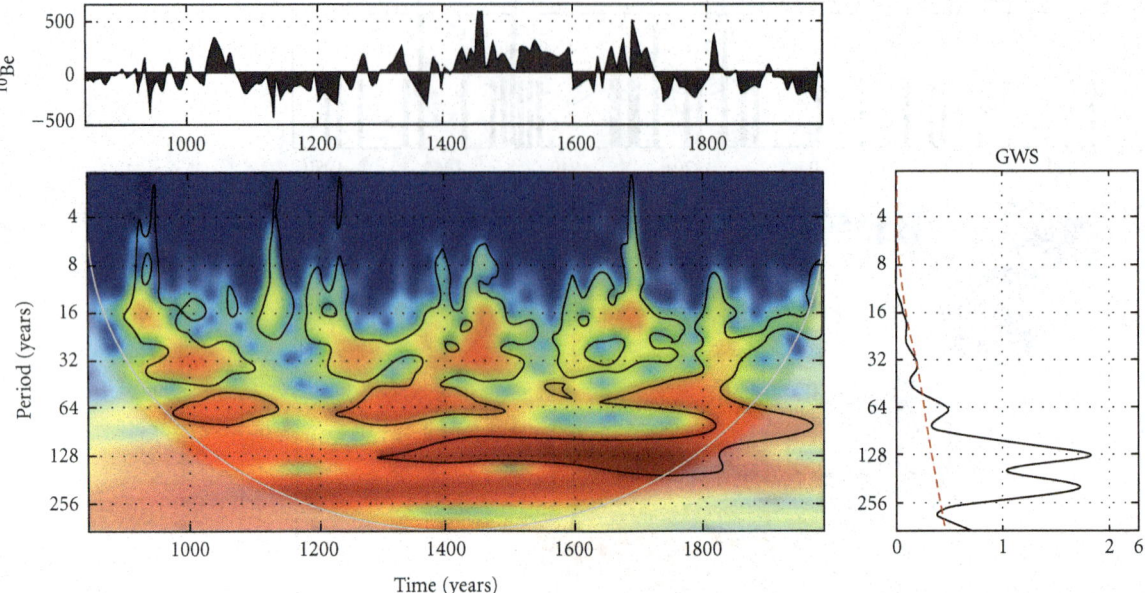

FIGURE 2: The upper panel shows the time series of the Be^{10} from the South Pole. The wavelet spectrum appears in the middle panel, and the global wavelet spectra appear in the right-most part of the figure, where the red dashed line indicates the border of 95% reliability.

the steady state condition, as is the case with cosmogenic isotopes.

In order to find the time evolution of the main frequencies of the time series, we apply the wavelet method using the Morlet mother wavelet ([24–26]. Wavelet analysis can be used for analyzing localized variations of power within a given time series at many different frequencies. However, even using this powerful tool, the reconstruction (filtering) of a signal is one of the main problems in the field, as is well known in Signals Theory and signal processing; this follows from the fact that the use of indiscriminate algorithms may lead to findings of spurious nonexistent periodicities in the time series. Furthermore, some real existing frequencies may be masked by more prominent frequencies. To avoid these problems, the Daubechies algorithm [27] has been used. This has proven to be highly efficient for the decomposition of signals in low and high frequencies, with the advantage that it does not create fictitious periodicities in the time series and, in some cases, may be more powerful than the multianalysis wavelet procedure (see Appendix A).

In Figures 2 to 7 we present the results of the wavelet analysis for the time series: the time series studied; the time series themselves are shown in the upper panel. The wavelet Morlet spectrum for the series appears in the mid panel of every figure, and the global wavelet spectrum appears in the right-hand part of the figures, where the dashed line indicates a border of 95% reliability.

It can be seen in Figure 2 that the 30-year periodicity has a confidence of 95%; however, it looks less prominent relative to the 60, 120, and 240 years frequencies, which show confidences far above 95%.

Figure 3 shows the Wavelet of ^{14}C; it can be seen that precisely at the 30-yrs. frequency there appears a small hump relative to the importance of the other periodicities. It would

then be natural to ask how to know if such periodicity really does exist with a good reliability.

This turns out to be a very complex problem, to give relevance to the periodicity of 30 years; on the other hand, it is necessary for one side to filter that signal and, on the other hand, to eliminate masking frequencies higher or lower than 30 years. It is precisely in such a situation that the *Daubechies* algorithm [27] turns out to be a very powerful tool, as shown in Figures 4 and 5 where, after the filtering process, the 30-year frequency for ^{10}Be and ^{14}C at the North Pole appears very clearly, both with high reliability, far above 95%.

The way to discern whether the periodicity of 30 years found in the ^{10}Be time series really reflects a comic ray fluctuation or is merely a local phenomenon of the cosmogenic isotope at earth level is to compare the behavior of ^{10}Be at both the North and South Poles. Since concentrations are quite different from one pole to another, it should be expected that their behavior would also be quite different if the existence of such periodicity is a local phenomenon. However, an examination of Figure 6 indicates that the wavelet coherence between the ^{10}Be at both the North and South Poles is very high, >0.9 (red color in the color bar of Figure 1)) at their common frequencies. This is an alternative way to confirm other arguments published in the literature, that this particular cosmogenic isotope is a proxy of cosmic rays.

Cosmic ray variations are mostly caused by time variations in the interplanetary magnetic field (IMF), so periodicities in cosmic ray data must be reflected in IMF data. Unfortunately, confident data from the IMF only date from the beginning of the spacecraft era, not even two 30-year cycles. However, since presumably the main source of such modulation is found in solar activity (SA); to confirm such an hypothesis we carried out a wavelet analysis of SA by

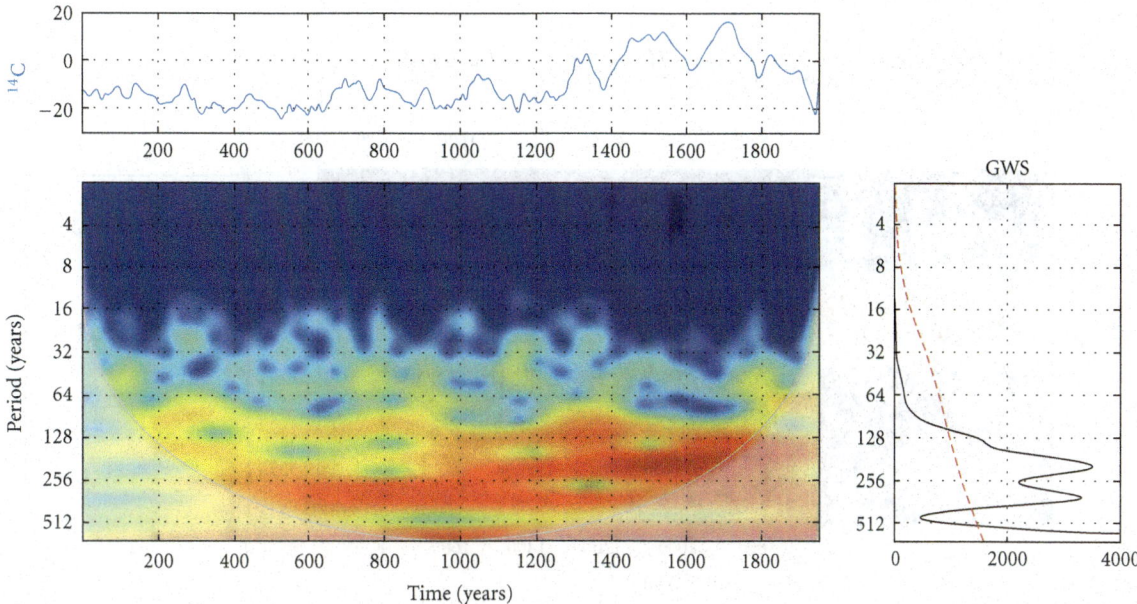

FIGURE 3: The upper panel shows the time series for the ^{14}C from the South Pole. The wavelet spectrum appears in the middle panel, and the global wavelet Spectrum appears in the right-hand part of the figure, where the red dashed line indicates a border of reliability of 95%.

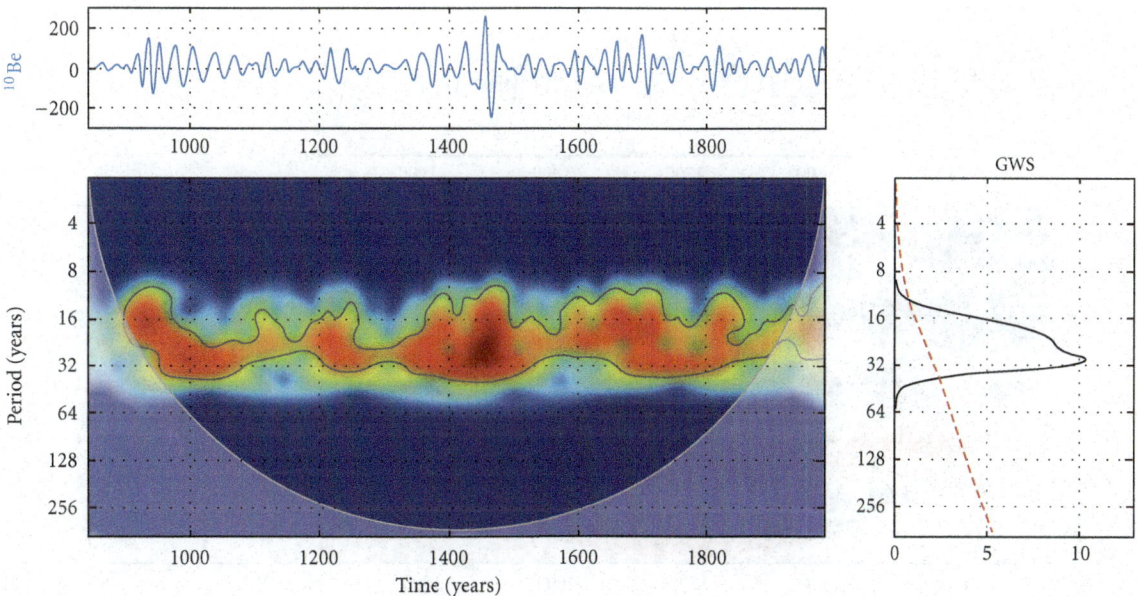

FIGURE 4: The upper panel shows the (^{10}Be) time series in the North Pole. The wavelet spectrum is a continuous one throughout all the entire time scale. The global spectrum at the right-hand panel shows that the 30-year periodicity has an extremely high confidence level.

using a time series of sunspot data (ftp://ftp.ngdc.noaa.gov/STP/SOLAR_DATA/SUNSPOT_NUMBERS/ and also http://sidc.oma.be/sunspot-data/). This analysis allows us once again to find the existence of such a 30-year frequency (Figure 7), where the results are presented without filters.

Since the 30-year periodicity appears as a small peak, (though one with higher than 95% confidence), instead of applying the Daubechies filter [27], to confirm that its existence is real, we have applied collateral methods that confirm such a periodicity for cosmic rays. In the next section the periodicity of 30 years of solar activity is clearly shown,

which in the unfiltered spectrum of Figure 7 is just barely perceptible.

3. Autoregressive Spectral Analysis ARMA

To verify the results obtained, Libin and Yudakin have calculated the mutual power spectra and coherence spectra for solar activity and temperature (Figure 8), solar activity and storminess (Figure 9), solar activity and the level of Lake Baykal (Figure 10) and, finally, solar activity and cosmic ray intensity from neutron monitor data and measurements

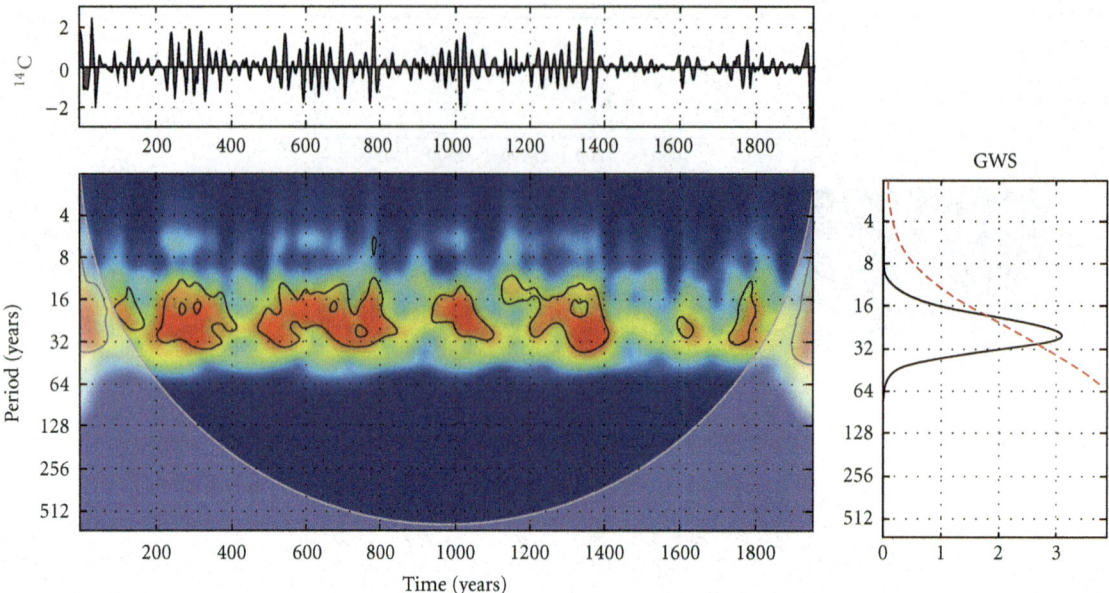

FIGURE 5: The upper panel shows the (^{14}C) time series. The wavelet spectrum is a continuous one throughout the entire time scale. The global spectrum at the right panel shows that the 30-years periodicity has a confidence level higher than 95%.

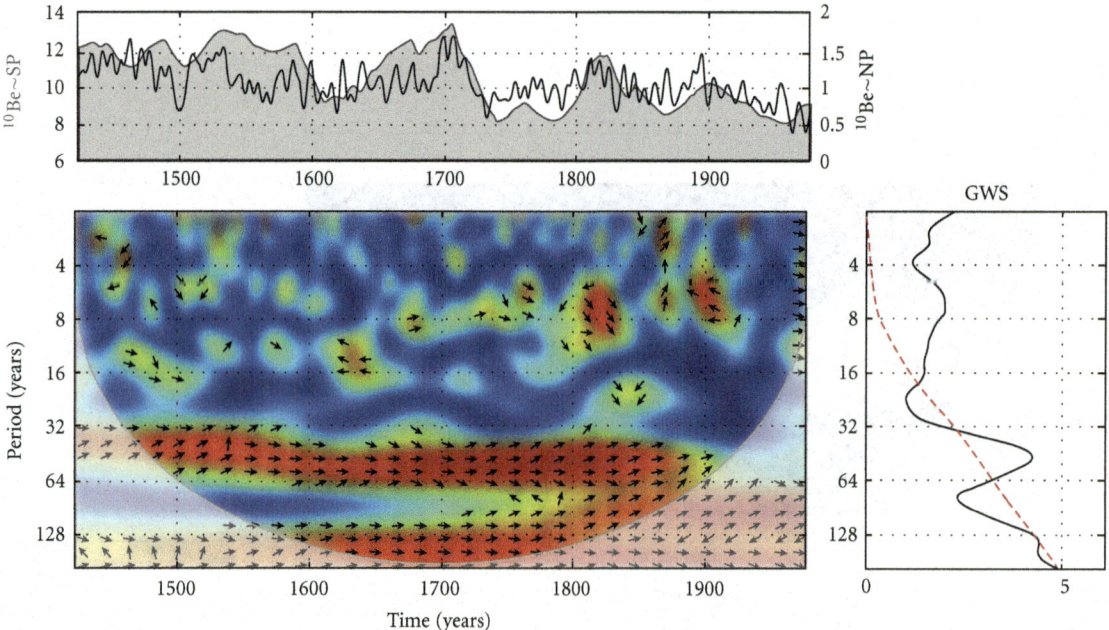

FIGURE 6: The upper panel shows the time series of ^{10}Be at the South Pole versus the ^{10}Be time series at the North Pole (black line). The middle panel shows a high coherence (>0.9) between both series, according to the color bar for coherence level in Figure 1. The right panel shows the global wavelet spectrum where the dotted line represents the border of 95% of reliability.

of ^{10}Be (Figure 11) for different measurement periods (see the data in the graphs). Calculations were made using the spectra of autoregressive spectral analysis with simultaneous ARMA (see Appendix B) and filtering the input data with the suppression of 11-year and 22-year variations.

All the above figures show the presence of stable 30-year fluctuations for almost all the processes. Although the coefficients of coherence are not always superior to a 95%

confidence interval probability the observed peaks are almost always above 90%.

4. The Standard Lomb-Scargle Periodogram Method

The author in [28] using the Lomb-Scargle technique has built periodograms based on the long-time series data:

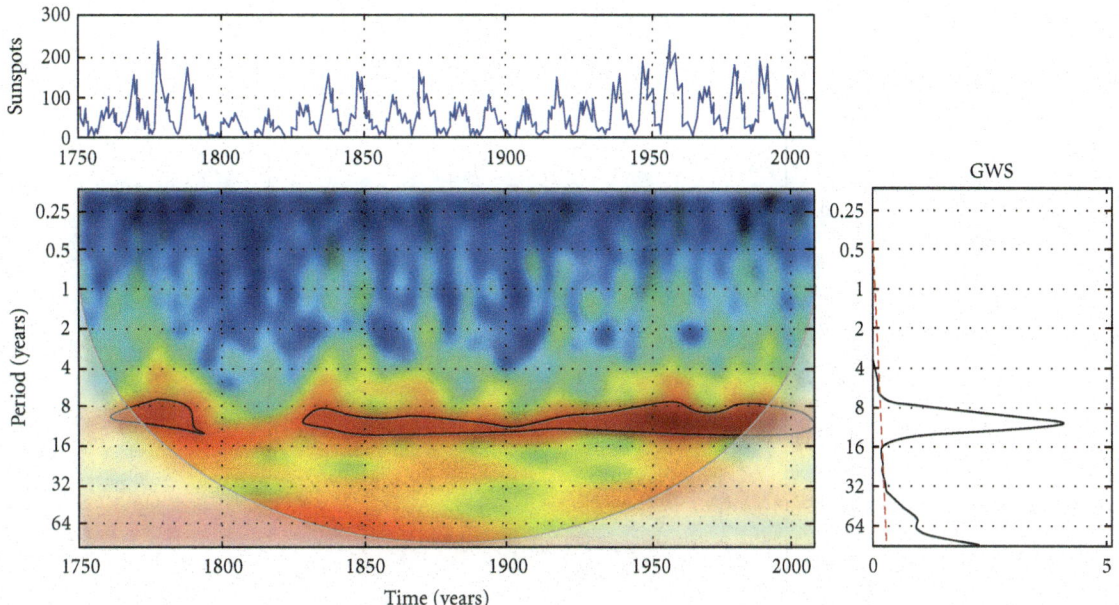

FIGURE 7: The upper panel shows the sunspot time series, the middle panel shows the wavelet spectrum, and the right panel shows the global spectrum, where the 30-year periodicity appears as a very small jump relative to the 11-year peak.

Figure 12 shows the periodogram, based on data from direct monthly stratospheric measurements of CR provided routinely over a long time period [29, 30]:

It can be seen that an approximately 34-year cycle is present in the data. A similar indication was obtained from the data of a modulation parameter of CR from [31], as shown in Figure 13.

5. Conclusions

We have shown the existence of a 30-year periodicity for cosmic rays. We can argue here that such a frequency is quite probably a modulator for terrestrial phenomena: it seems that in some way cosmic rays modulate climatic phenomena, such as the Atlantic Multidecal Oscillation (AMO) and sea-surface temperature (SST), and these, in turn, modulate hurricane development [31, 32]. Furthermore, a wavelet analysis applied to paleoclimatic proxy data for large scale atmospheric phenomena (Atlantic Multidecadal Oscillation and Southern Oscillation Indexes) has revealed coherence between climatic oscillations and cosmic rays on a 30-yrs cycle [14]. Since this periodicity is present throughout the entire interval under study, the origin of such a periodicity may be associated with the 120-year secular cycle of solar activity whose presence has been demonstrated in [17]. It corresponds to continuous periods of increasing and decreasing activity during maxima and minima of that secular cycle. The 60- and 240-year. solar activity cycles may, then, also be associatedwith the 30-year cycle. This would confirm that *Solar Activity is the source of all cosmophysical modulators of Solar-Terrestrial relationships*, through e a number of intermediaries at short, medium and long frequencies. One of them, in particular, is the 30-years periodicity found in this work, which allows for the analysis of a reasonable long-term variability for Cosmic rays.

Appendices

A.

Inverse Wavelet Transform. We use the inverse wavelet transform to obtain the decomposition of a signal which can be obtained from a time-scale filter [27]. The inverse wavelet transform is defined [25] as

$$X_n = \frac{\delta_j \delta_t^{1/2}}{C_\delta \psi_0(0)} \sum_{j=0}^{J} \frac{\text{Re}\{W_n(s_j)\}}{s_j^{1/2}}, \qquad (A.1)$$

where δ_j is the factor for scale averaging, C_δ is a constant ($\delta_j = 0.6$ and $C_\delta = 0.776$, for the Morlet wavelet), and ψ_0 removes the energy scaling. We use the inverse wavelet transform to obtain the time series.

B.

Autoregressive Moving-Average Model. The autoregressive moving-average model (ARMA) is one of the mathematical models used for the analysis and prediction of stationary time series in statistics. The ARMA model is a generalization of two simpler time series models—an autoregressive model (AR) and the moving average model (MA).

The ARMA (p, q) model, where p and q are integers that specify the order of the model is called the next generation process time series $\{X_t\}$:

$$X_t = c + \varepsilon_t + \sum_{i=1}^{p} \alpha_i X_{t-i} + \sum_{i=1}^{q} \beta_i \varepsilon_{t-i}, \qquad (B.1)$$

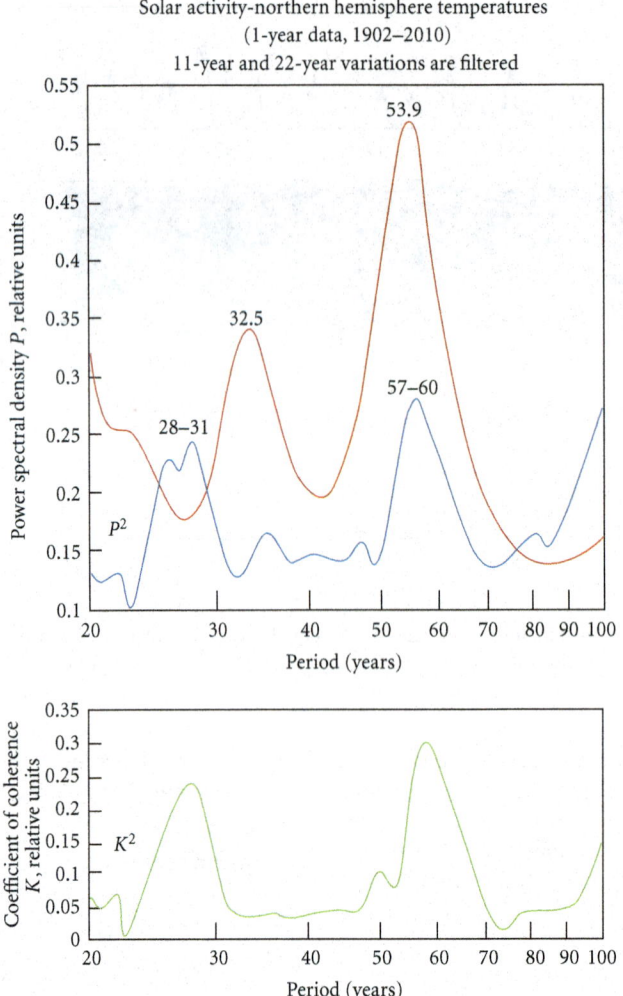

FIGURE 8: Power spectra density of solar activity and northern hemisphere temperature (1902–2010, blue) and power spectra density of greenland ice cores (red). K^2 for solar activity and temperature.

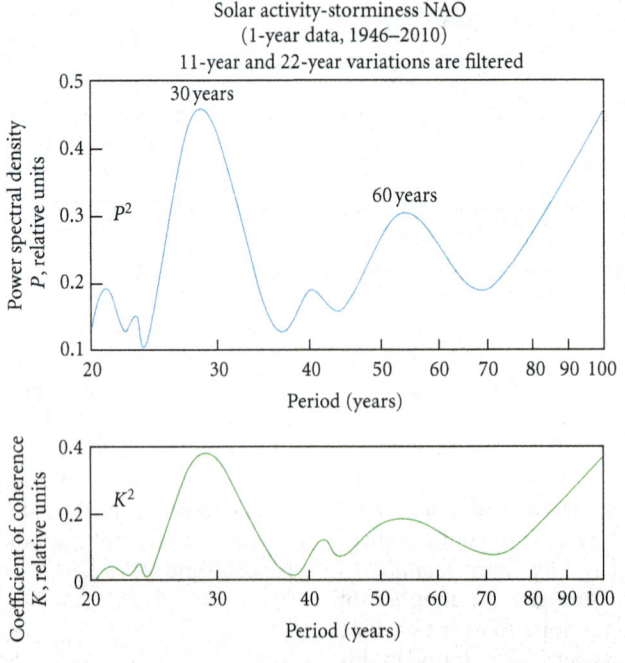

FIGURE 9: Power spectra density of solar activity and storminess (1946–2010)

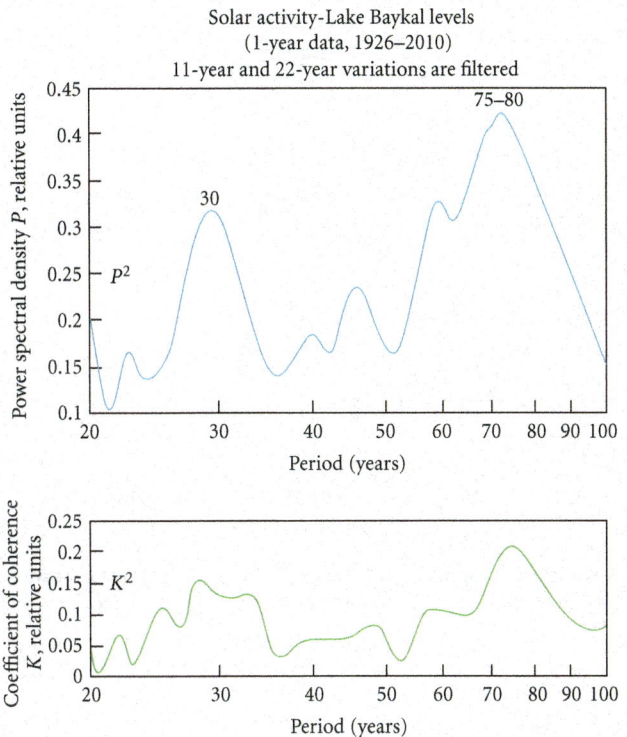

FIGURE 10: Power spectra density of solar activity and lake baykal leves l (1926–2010).

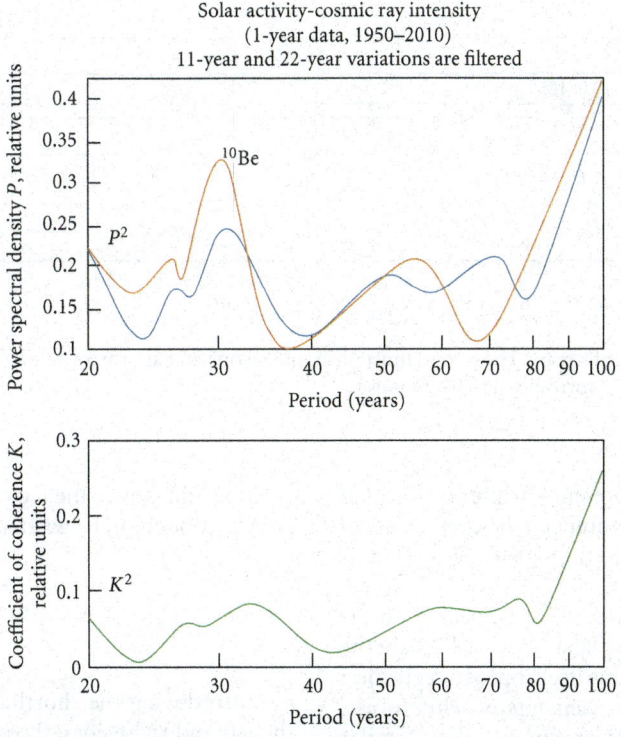

FIGURE 11: Power spectra density of solar activity and cosmic ray intensity (1950–2010, blue) and power spectra density of solar activity and ^{10}Be (1482–2010, red). K^2 for Solar Activity and Cosmic ray Intensity.

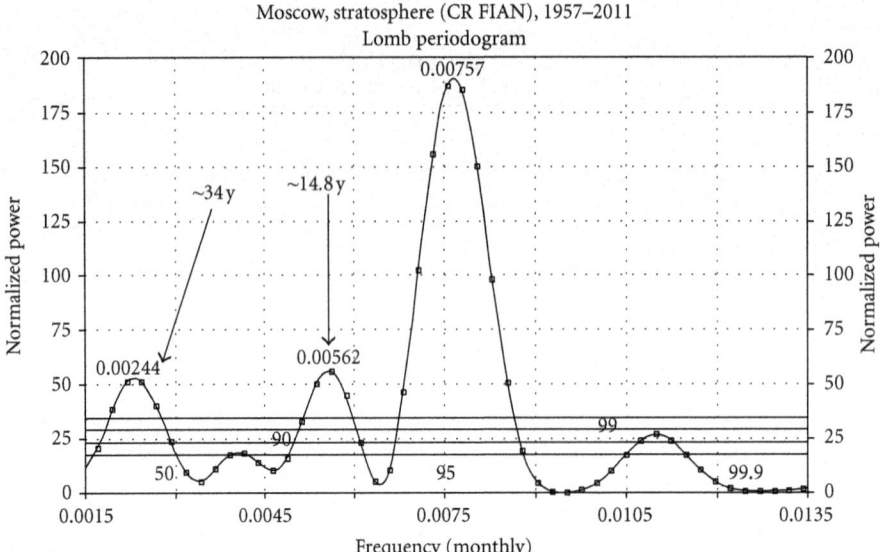

FIGURE 12: Periodogram of the monthly means of monthly stratospheric CR measurements near Moscow [29, 30]. Along with the highest peak at approximately 11 years, the approximately 34-year cycle peak and the approximately 14.8-year peak exceed a significance level of 0.99.

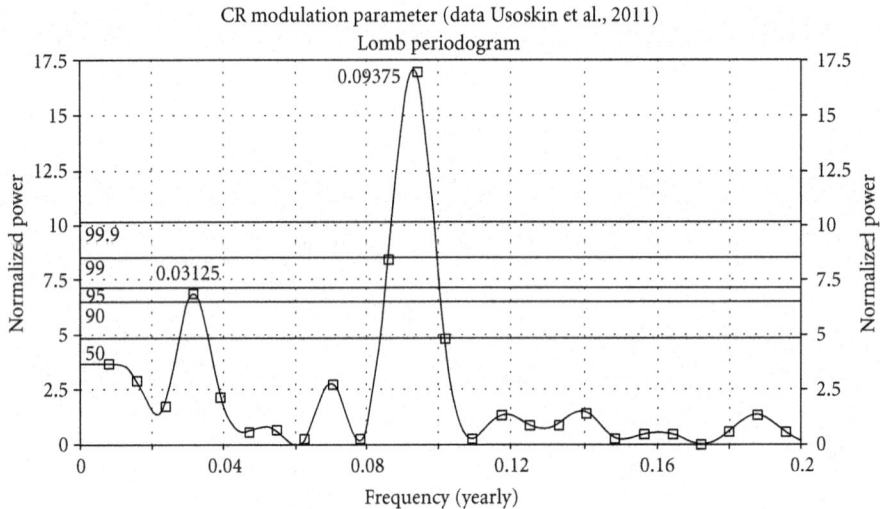

FIGURE 13: The highest maximum is at 11 years. The second most pronounced one is at approximately 0.031, corresponding to approximately 32 years. Here the levels of significance correspond to white noise.

where C is constant, $\{\varepsilon_t\}$ represents white noise, and $\alpha_1, \ldots, \alpha_p$ and $\beta_1, \ldots, \beta_q$ are real numbers, the coefficients of the autoregressive, and moving average coefficients, respectively.

Such a model can be interpreted as a linear multiple regression model, in which the explanatory variables are the past values of the dependent variable itself, but as a regression balance—moving averages of the elements of white noise. ARMA-processes are more complex compared to similar processes in a pure form; however the ARMA processes are characterized by fewer parameters, which is one of their advantages.

If we introduce the lag operator $L : Lx_t = x_{t-1}$, then the ARMA model can be written as follows:

$$\left(1 - \sum_{i=1}^{p} \alpha_i L^i\right) X_t = c + \left(1 + \sum_{i=1}^{q} \beta_i L^i\right) \varepsilon_t. \qquad (B.2)$$

Introducing the shorthand notation for polynomials of the left and right sides, the previous equation can be written as

$$\alpha(L)X_t = c + \beta(L)\varepsilon_t. \qquad (B.3)$$

For the process to be stationary, it is necessary for the roots of the characteristic polynomial of the autoregressive part $\alpha(z)$ to lie outside the unit circle in the complex plane (in modulus strictly greater than one). The stationary ARMA process can be represented as an infinite MA process:

$$X_t = \alpha^{-1}(L)c + \alpha^{-1}(L)\beta(L)\varepsilon_t = \frac{c}{a(1)} + \sum_{i=0}^{\infty} c_i \varepsilon_{t-i}. \quad (B.4)$$

For example, the process ARMA $(1,0)$ = AR (1) can be represented as an MA process of infinite order with coefficients in decreasing geometric progression:

$$X_t = \frac{c}{(1-a)} + \sum_{i=0}^{\infty} a^i \varepsilon_{t-i}. \quad (B.5)$$

Thus, the ARMA processes can be considered to be MA processes of infinite order with certain restrictions on the structure coefficients. There is a small number of parameters to describe the processes they enable rather than a complex structure. All stationary processes can be arbitrarily approximated by an ARMA model of a certain order with considerably fewer parameters than MA models use.

NonStationary (Integrated) ARMA. In the presence of unit roots of the p autoregressive polynomial, the process is nonstationary. Roots of less than unity in practice are not considered, since they are processes which exhibit explosive behavior. Accordingly, to test the stationary nature of a time series of basic tests, tests must be run for unit roots. If the tests confirm the presence of unit roots, then we need to analyze the difference between the original time series and a stationary process of the differences of one or two orders (usually the first order is sufficient and sometimes the second) of the ARMA-based model.

Such models are called ARIMA models (integrated ARMA) or Box-Jenkins models. The ARIMA model (p, d, q), where d is the order of integration (the order of differences in the original time series); p and q, the order of AR; MA the parts of the ARMA-process differences d, the order can be written in the operator form:

$$\alpha(L)\Delta^d X_t = c + \beta(L)\varepsilon_t, \quad \Delta = 1 - L. \quad (B.6)$$

The ARIMA process (p, d, q) is equivalent to the ARMA process $(p + d, q)$ with d unit roots.

To construct the ARMA model on a proxy data series of observations, it is necessary to determine the model order (numbers p and q), and then the coefficients themselves.

To determine the order of the model an investigation of these characteristics of the time series can be done, seen as its autocorrelation function and partial autocorrelation function.

To determine the coefficients the method of least squares and maximum likelihood method can be used.

ARMAX Models. In the classic ARMA model, it can add exogenous factors x. In general, the model involves not only the current values of these factors but also lagged values. Such models are usually denoted ARMAX (p, q, k), where k-lags come from the exogenous factors. In an operator form, such models can be written as follows (an exogenous factor):

$$a(L)y_t = c + b(L)\varepsilon_t + d(L)x_t, \quad (B.7)$$

where (L), $b(L)$, $d(L)$ are the order polynomial, respectively, p, q, k of the lag operator.

It should be noted that such models can be interpreted differently, for example, ADL (p, q) mode with random errors MA (q).

Acknowledgments

V. Velasco gives special thanks to the CONACyT (project 180148) for financial support to this work.

References

[1] L. I. Dorman, *Cosmic Ray Variations and Space Exploration*, North Holland, Amsterdam, 1974.

[2] V. L. Gorshkov, in *Proceedings of the 8th Congress of Astronomical Community and International Symposium 'Astronomy 2005—Status and Prospects of Development'*, MSU-GAISH, Moscow, Russia, June 2005.

[3] A. N. Pushkov and T. A. Chernova, *Features of Spatial and Temporal Structure of the Secular Variation of the Geomagnetic Field*, vol. 18, IZMIRAN, Moscow, Russia, 1972.

[4] Papitashvili, N. E. Papitashvili, N. M. Rotanova, and A. N. Pushkov, *Geomagnetism and Aeronomy*, vol. 20, pp. 711–717, 1980.

[5] N. V. Papitashvili and N. M. Rotanova, *Geomagnetism and Aeronomy*, vol. 22, no. 4, pp. 1010–1015, 1982.

[6] V. P. Golovkov, S. V. Yakovleva, and T. I. Zvereva, "Detection of fast secular variations based on the data of satellite magnetic surveys," *Geomagnetism and Aeronomy*, vol. 50, no. 2, pp. 274–277, 2010.

[7] H. S. Ahluwalia, "Three activity cycle periodicity in galactic cosmic rays," in *Proceedings of the 25th International Cosmic Ray Conference*, vol. 2, pp. 109–112, Durban, South Africa, 1997.

[8] H. S. Ahluwalia, "Meandering path to solar activity forecast for cycle 23," in *Proceedings of the 10th International Solar Wind Conference*, M. Velli, R. Bruno, and F. Malra, Eds., vol. CP679, The American Institute of Physics, 2003.

[9] H. S. Ahluwalia, "The predicted size of cycle 23 based on the inferred three-cycle quasi-periodicity of the planetary index Ap," *Journal of Geophysical Research*, vol. 103, no. A6, pp. 12, 103–12, 112, 1998.

[10] H. S. Ahluwalia, "Timelines of cosmic ray intensity, Ap, IMF, and sunspot numbers since 1937 ," *Journal of Geophysical Research*, vol. 116, Article ID A12106, 9 pages, 2011.

[11] L. B. Klyashtorin and A. A. Lyubushin, *Cyclic Climate Changes and Fish Productivity*, VNIRO Publishing, 2005.

[12] I. Libin and J. Perez-Peraza, *Helioclimatology*, MAOK, Moscow, Russia, 2009.

[13] S. T. Gray, L. J. Graumlich, J. L. Betancourt, and G. T. Pederson, "A tree-ring based reconstruction of the Atlantic Multidecadal Oscillation since 1567 A.D.," *Geophysical Research Letters*, vol. 31, no. 12, Article ID L12205, 4 pages, 2004.

[14] J. Pérez-Peraza, V. Velasco, and S. Kavlakov, "Solar, geomagnetic and cosmic ray intensity changes, preceding the cyclone

appearances around Mexico," *Advances in Space Research*, vol. 42, pp. 1601–1613, 2008.

[15] K. Kudela, H. Mavromichalaki, A. Papaioannou, and M. Gerontidou, "On mid-term periodicities in cosmic rays," *Solar Physics*, vol. 266, pp. 173–180, 2010.

[16] V. M. Velasco and B. Mendoza, "Assessing the relationship between solar activity and some large scale climatic phenomena," *Advances in Space Research*, vol. 42, no. 5, pp. 866–878, 2008.

[17] V. Velasco, J. F. Valdés-Galicia, and B. Mendoza, "The 120-year solar cycle of the cosmogenic isotopes," in *Proceedings of the 30th International Cosmic Ray Conference*, vol. 1, pp. 553–556, 2008.

[18] J. Pérez-Peraza, V. Velasco, J. Zapotitla, L. I. Miroshnichenko, E. Vashenuyk, and I. Ya Libin, "Classification of GLE's as a function of their spectral content for prognostic goals," in *Proceedings of the 32nd International Cosmic Ray Conference*, pp. 149–152, Beiging, China, 2011.

[19] M. Fligge, S. K. Solanki, and J. Beer, "Determination of solar cycle length variations using continuous wavelet transform," *Astronomy & Astrophysics*, vol. 346, pp. 313–321, 1999.

[20] M. Ossendrijver, "The solar dynamo," *The Astronomy and Astrophysics Review*, vol. 11, pp. 287–367, 2003.

[21] J. Beer, A. Blinov, G. Bonani et al., "Use of ^{10}Be in polar ice to trace the 11-year cycle of solar activity," *Nature*, vol. 347, pp. 164–166, 1990.

[22] A. M. Berggren, J. Beer, G. Possnert et al., "A 600-year annual ^{10}Be record from the NGRIP ice core, Greenland," *Geophysical Research Letters*, vol. 36, no. 11, Article ID L11801, 5 pages, 2009.

[23] E. Bard, G. Raisbeck, F. Yiou, and J. Jouzel, "Solar irradiance during the last 1200 years based on cosmogenic nuclides," *Tellus B*, vol. 52, no. 3, pp. 985–992, 2000.

[24] L. Hudgins, C. A. Friehe, and M. E. Mayer, "Wavelet transforms and atmopsheric turbulence," *Physical Review Letters*, vol. 71, pp. 3279–3282, 1993.

[25] C. Torrence and G. Compo, "A practical guide to wavelet analysis," *Bulletin of the American Meteorological Society*, vol. 79, pp. 61–78, 1998.

[26] A. Grinsted, J. C. Moore, and S. Jevrejeva, "Application of the cross wavelet transform and wavelet coherence to geophysical time series," *Nonlinear Process Geophysics*, vol. 11, pp. 561–566, 2004.

[27] I. Daubechies, *Ten Lectures on Wavelet, Society For Industrial and Applied Mathematics*, 1992, Edited by: Rulgers University and AT&T Bell Laboratories.

[28] K. Kudela, "Space weather near Earth and energetic particles: selected results," *Journal of Physics*. In press.

[29] Y. I. Stozhkov, N. S. Svirzhevsky, G. A. Bazilevskaya et al., "Fluxes of Cosmic rays in the maximum of absorption curve in the atmosphere and at the atmosphere boundary (1957–2007)," preprint FIAN, 14, 2007.

[30] Y. I. Stozhkov, N. S. Svirzhevsky, G. A. Bazilevskaya et al., "Cosmic rays in the stratosphere in 2008–2010," *Astrophysics and Space Sciences Transactions*, vol. 7, pp. 379–382, 2011.

[31] I. G. Usoskin, G. A. Bazilevskaya, and G. A. Kovaltsov, "Solar modulation parameter for cosmic rays since 1936 reconstructed from ground-based neutron monitors and ionization chambers," *Journal of Geophysical Research*, vol. 116, Article ID A02104, 9 pages, 2011.

[32] J. Pérez-Peraza, V. Velasco, and S. Kavlakov, "Wavelet coherence analysis of Atlantic hurricanes and cosmic rays," *Geofísica Internacional*, vol. 47, no. 3, pp. 231–244, 2008.

9

Testing the No-Hair Theorem with Sgr A*

Tim Johannsen

Physics Department, University of Arizona, 1118 E. 4th Street, Tucson, AZ 85721, USA

Correspondence should be addressed to Tim Johannsen, timj@physics.arizona.edu

Academic Editor: Francesco Shankar

The no-hair theorem characterizes the fundamental nature of black holes in general relativity. This theorem can be tested observationally by measuring the mass and spin of a black hole as well as its quadrupole moment, which may deviate from the expected Kerr value. Sgr A*, the supermassive black hole at the center of the Milky Way, is a prime candidate for such tests thanks to its large angular size, high brightness, and rich population of nearby stars. In this paper, I discuss a new theoretical framework for a test of the no-hair theorem that is ideal for imaging observations of Sgr A* with very long baseline interferometry (VLBI). The approach is formulated in terms of a Kerr-like spacetime that depends on a free parameter and is regular everywhere outside of the event horizon. Together with the results from astrometric and timing observations, VLBI imaging of Sgr A* may lead to a secure test of the no-hair theorem.

1. Introduction

According to the no-hair theorem, black holes are uniquely characterized by their masses and spins and are described by the Kerr metric [1–6]. Mass M and spin J are the first two multipole moments of the Kerr spacetime, and all higher-order moments can be expressed in terms of these two [7, 8]. The no-hair theorem, then, naturally leads to the expectation that all astrophysical black holes are Kerr black holes. To date, however, a definite proof for the existence of such black holes is still lacking despite a wealth of observational evidence (see discussion in, e.g., [9]).

Tests of the no-hair theorem have been suggested using observations in either the gravitational-wave [10–21] or the electromagnetic spectrum [22–31]. Both approaches are based on parametric frameworks that contain one or more free parameters in addition to mass and spin which measure potential deviations from the Kerr metric [18–20, 32–34]. If no deviation is detected, then the compact object is indeed a Kerr black hole. However, since such deviations can have a significant impact on the observed signals, the no-hair theorem may be tested in a twofold manner: if a deviation is measured to be nonzero and if general relativity is assumed, the object cannot be a black hole [18, 35]. Alternatively, if the object is otherwise known to possess an event horizon, it is

a black hole, but different from a Kerr black hole. In the latter case, the no-hair theorem would be falsified [22].

Sgr A*, the supermassive black hole at the center of the Milky Way, is a prime target for testing strong-field gravity and the no-hair theorem with electromagnetic observations (see [36] for a review). Monitoring the orbits of stars around this compact object for more than a decade has led to precise mass and distance measurements making Sgr A* the black hole with the largest angular size in the sky [37, 38]. In addition, very long baseline interferometric observations have resolved Sgr A* on event horizon scales [39]. On the theoretical side, there have been significant advances recently in the development of a framework within which the search for violations of the no-hair theorem can be carried out.

In this paper, I review this framework as well as the prospects for an observational test of the no-hair theorem with Sgr A*.

2. An Ideal Framework for Testing the No-Hair Theorem

Spacetimes of rotating stellar objects in general relativity have been studied for several decades. Due to the nonlinearity of Einstein field equations, the construction of such metrics is plagued with sometimes incredible technical challenges.

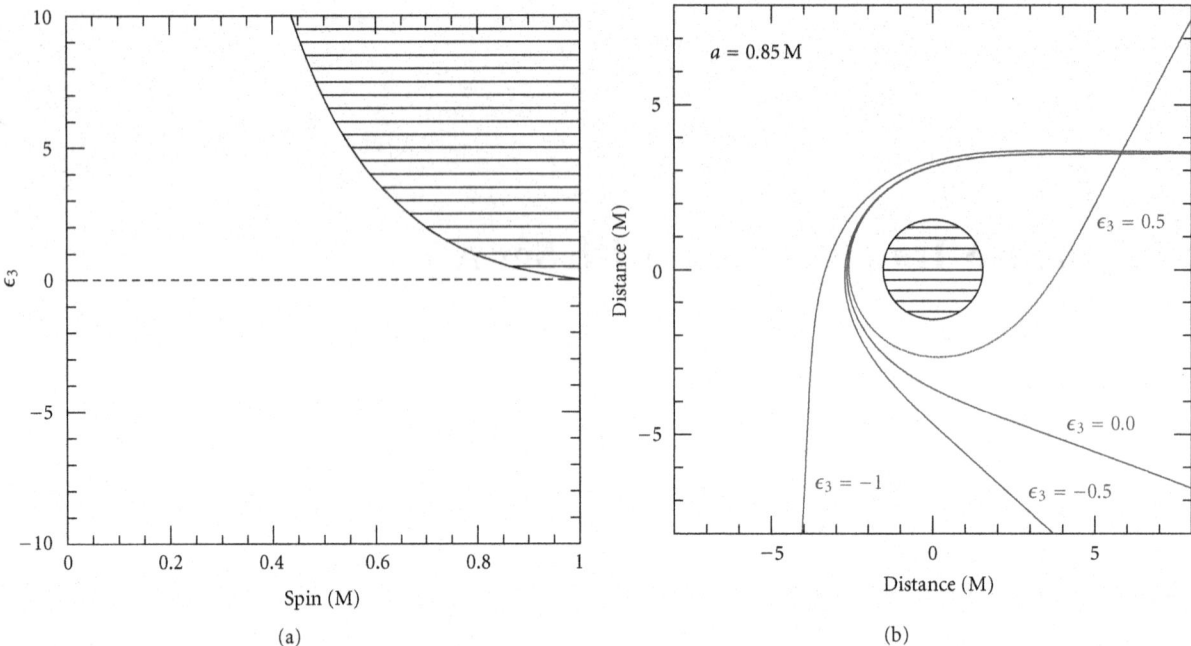

FIGURE 1: (a) Values of the parameter ϵ_3 versus the spin a, for which the central object is a black hole. The shaded region marks the excluded part of the parameter space where this object is a naked singularity. The dashed line corresponds to a Kerr black hole [34]. (b) Trajectories of photons lensed by a black hole with a (counterclockwise) spin $a = 0.85\,M$ for several values of the parameter ϵ_3. The shaded region corresponds to the event horizon of a Kerr black hole of equal spin.

Following the discovery of the Schwarzschild [40] and Kerr metrics [41] in 1916 and 1963, respectively, Hartle and Thorne [42, 43] constructed a metric for slowly rotating neutron stars that is appropriate up to the quadrupole order. Tomimatsu and Sato [44, 45] found a discrete family of spacetimes in 1972 that contain the Kerr metric as a special case. After a full decade of research, Manko and Novikov [32] found two classes of metrics in 1992 that are characterized by an arbitrary set of multipole moments. Many exact solutions of the Einstein field equations are now known [46]. Of particular interest is the subclass of stationary, axisymmetric, vacuum (SAV) solutions of the Einstein equations, and especially those metrics within this class that are also asymptotically flat. Once an explicit SAV has been found, all SAVs can in principle be generated by a series of HKX-transformations ([47, 48] and references therein), which form an infinite-dimensional Lie group [49, 50]. Each SAV is fully and uniquely specified by a set of scalar multipole moments [51, 52] and can also be generated from a given set of multipole moments [53, 54]. These solutions, however, are generally very complicated and often unphysical. For some astrophysical applications, such as the study of neutron stars, it is oftentimes more convenient to resort to a numerical solution of the field equations [55–59].

To date, there exist seven different approaches that model parametric deviations from the Kerr metric. Ryan [10–12] studied the motion of test particles in the equatorial plane of compact objects with a general expansion in Geroch-Hansen multipoles. Collins and Hughes [18], Vigeland and Hughes [19], and Vigeland et al. [33] constructed Schwarzschild and Kerr metrics with perturbations in the form of Weyl sector bumps. Glampedakis and Babak [20] designed a metric starting from the Hartle-Thorne metric [42, 43] that deviates from the Kerr metric by an independent quadrupole moment. Gair et al. [21] applied a similar technique to the Manko-Novikov metric [32] affecting the quadrupole as well as higher-order moments. Sopuerta and Yunes [60] found a metric for a slowly rotating black hole that violates parity. Vigeland et al. [33] designed parametric deviations from the Kerr metric that possess four integrals of the motion and, hence, allow for the full separability of the Hamilton-Jacobi equations. Finally, Johannsen and Psaltis constructed a metric of a rapidly rotating Kerr-like black hole [34]. Other metrics of static black holes in alternative theories of gravity have also been found (e.g., [61–64]).

Due to the no-hair theorem, the Kerr metric is the only asymptotically flat SAV in general relativity with an event horizon but no closed timelike loops [1–6]. Consequently, any parametric deviation within general relativity has to violate at least one of these prerequisites and introduces either singularities or regions with closed timelike loops outside of the event horizon, which usually occur very near to the central object at radii $r \lesssim 2\,M$ [65]. Otherwise, these metrics would render the no-hair theorem false. The relevance of this kind of pathologies depends on the astrophysical application. They play no role for tests of the no-hair theorem that only involve the orbits of objects at large distances from the horizon, as is the case for extreme mass-ratio inspirals or the motion of stars or pulsars around a black hole. They are, however, critical for the study

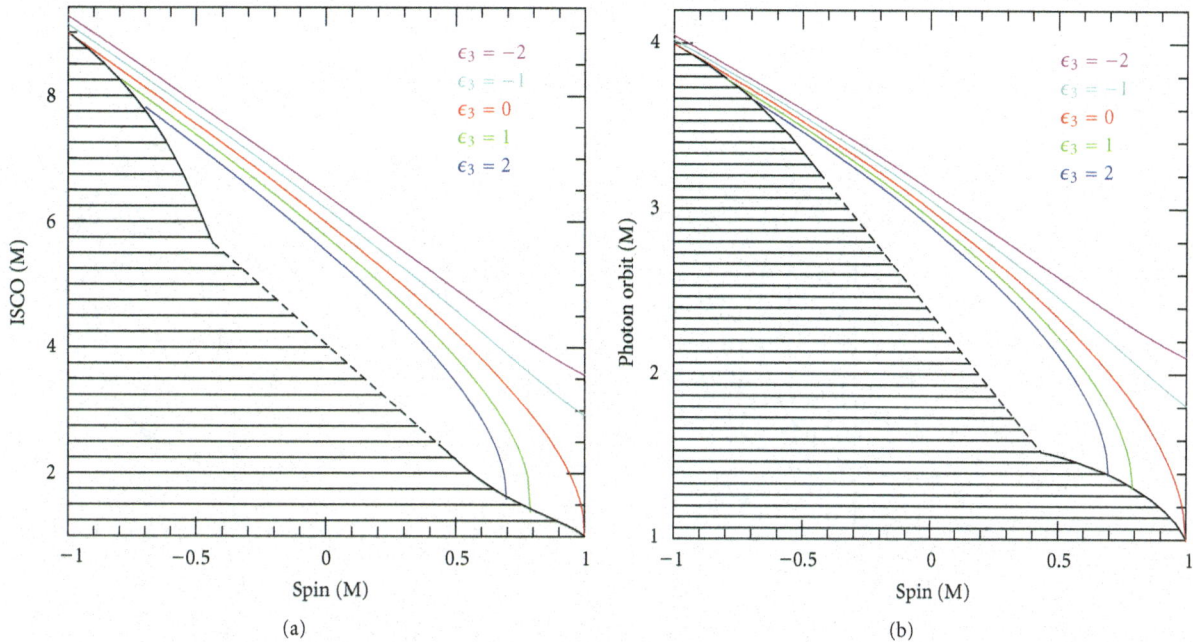

FIGURE 2: Radius of (a) the ISCO and of (b) the circular photon orbit as a function of the spin a for several values of the parameter ϵ_3. The radius of the ISCO and the circular photon orbit decrease with increasing values of the parameter ϵ_3. The shaded region marks the excluded part of the parameter space [34].

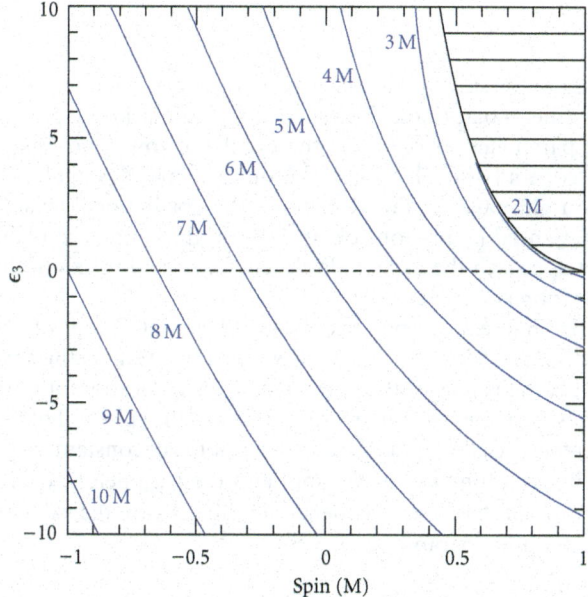

FIGURE 3: Contours of constant radius of the ISCO for values of the spin $-1 \leq a/M \leq 1$ and of the parameter $-10 \leq \epsilon_3 \leq 10$. The radius of the ISCO decreases for increasing values of the spin and the parameter ϵ_3. The shaded region marks the excluded part of the parameter space. The dashed line corresponds to the parameter space for a Kerr black hole [34].

of accretion flows around black holes [34], because the electromagnetic radiation originates predominantly from the immediate vicinity of the event horizon.

For this reason, the emission from accretion flows around black holes is most interesting for strong-field tests

of the no-hair theorem with observations across the electromagnetic spectrum ranging from X-ray observations of quasiperiodic variability, fluorescent iron lines, or continuum disk spectra [22, 24] to sub-mm imaging of supermassive black holes with VLBI [22, 23]. All of these observation techniques critically depend on the location of either the circular photon orbit or the innermost stable circular orbit (ISCO), because these orbits dominate the characteristics of the received signals.

These strong-field tests of the no-hair theorem require a very careful modeling of the inner region of the spacetime of black holes. Due to the pathologies of previously known parametric deviations, it has been necessary to impose an artificial cutoff at some radius outside of the event horizon that encloses all of the above pathologies and, thereby, shields them from the observer. Therefore, the application of parametric frameworks to such tests of the no-hair theorem in the electromagnetic spectrum has, so far, been limited to only slowly to moderately spinning black holes, for which the circular photon orbit and ISCO are still located outside of the cutoff radius [34, 65].

Recently [34], we constructed a black hole metric that is regular everywhere outside of the event horizon for all values of the spin within the allowable range and that depends on a set of free parameters in addition to mass and spin. In the case when all parameters vanish, our metric reduces smoothly to the Kerr metric. Our metric is a vacuum solution of a more general set of field equations, but otherwise fulfills all of the prerequisites of the no-hair theorem and, therefore, preserves these essential properties even if the deviation parameters from the Kerr metric are nonzero. At present, our metric constitutes the only known black hole spacetime of this kind and serves as an ideal framework for the study of

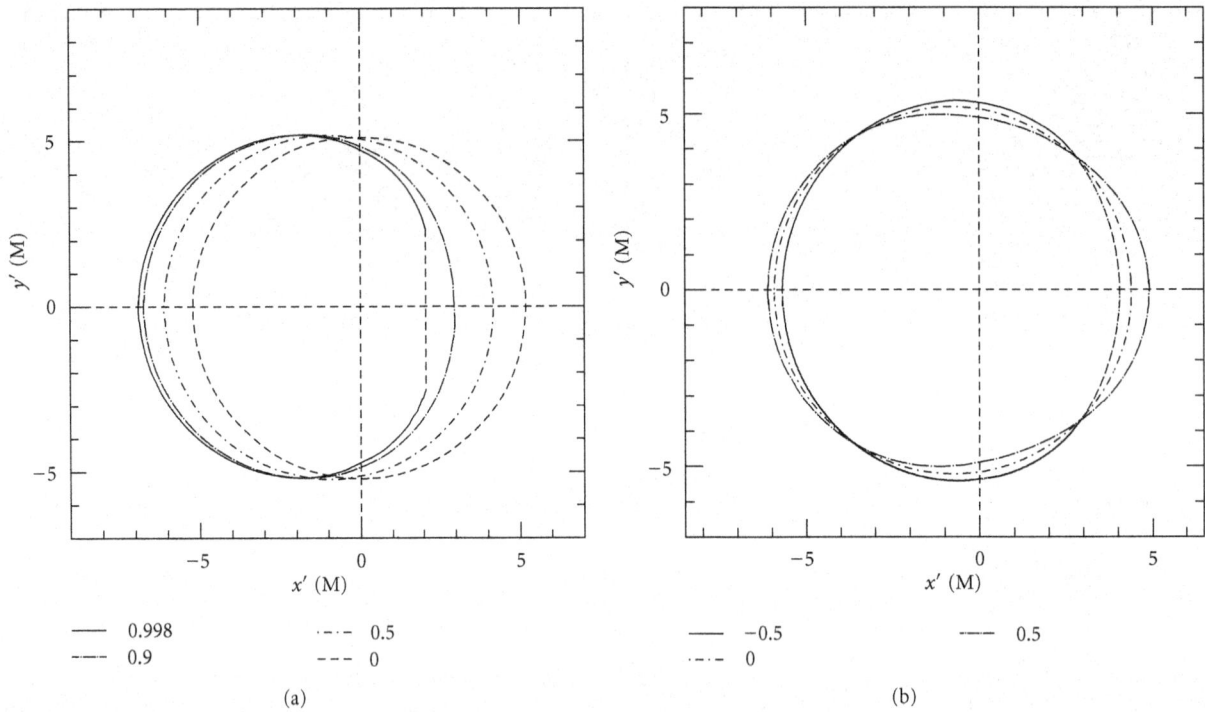

FIGURE 4: Images of rings of light of (a) a Kerr and (b) a quasi-Kerr black hole at an inclination $\cos i = 0.25$. Increasing values of the spin cause a displacement of the ring in the image plane, but the ring remains (nearly) circular for values of the spin $a \lesssim 0.9\,\mathrm{M}$ (a). Nonzero values of the deviation parameter ϵ (b) lead to an asymmetric ring image [23].

the signatures of a possible violation of the no-hair theorem from astrophysical phenomena near the event horizon of a black hole and, in particular, of Sgr A*. For the case of one additional parameter ϵ_3, our metric in Boyer-Lindquist coordinates is given by the expression [34]

$$ds^2 = -[1 + h(r,\theta)]\left(1 - \frac{2Mr}{\Sigma}\right)dt^2 - \frac{4aMr\sin^2\theta}{\Sigma}$$

$$\times [1 + h(r,\theta)]dtd\phi$$

$$+ \frac{\Sigma[1 + h(r,\theta)]}{\Delta + a^2\sin^2\theta h(r,\theta)}dr^2 + \Sigma d\theta^2$$

$$+ \left[\sin^2\theta\left(r^2 + a^2 + \frac{2a^2Mr\sin^2\theta}{\Sigma}\right)\right.$$

$$\left. + h(r,\theta)\frac{a^2(\Sigma + 2Mr)\sin^4\theta}{\Sigma}\right]d\phi^2, \qquad (1)$$

where

$$\Delta \equiv r^2 - 2Mr + a^2,$$

$$\Sigma \equiv r^2 + a^2\cos^2\theta, \qquad (2)$$

$$h(r,\theta) \equiv \epsilon_3\frac{M^3r}{\Sigma^2},$$

and $a \equiv J/M$ is the spin parameter.

In [34], we analyzed several of the key properties of our black hole metric as a function of the mass M, the spin a, and the parameter ϵ_3. The left panel in Figure 1 shows the

range of the spin and the parameter ϵ_3, for which our metric describes a black hole. The shaded region marks the part of the parameter space where the event horizon is no longer closed, and the black hole becomes a naked singularity. The right panel of Figure 1 shows the gravitational lensing experienced by photons on an orbit in the equatorial plane that approach the black hole closely for several values of the deviation parameter ϵ_3.

In Figure 2, we plot the radius of the ISCO and of the circular photon orbit, respectively, as a function of the spin for several values of the parameter ϵ_3. The location of both orbits decreases with increasing values of the spin and of the parameter ϵ_3. In Figure 3, we plot contours of constant ISCO radius as a function of the spin and the parameter ϵ_3. The location of these orbits depends significantly on the value of the deviation parameter ϵ_3.

3. Testing the No-Hair Theorem with VLBI Imaging of Sgr A*

In [23], we explored in detail the effects of a violation of the no-hair theorem for VLBI imaging using a quasi-Kerr metric [20]. This metric can be used to accurately describe Kerr-like black holes up to a spin of about $a \leq 0.4\,\mathrm{M}$. For Sgr A*, this spin range might already be sufficient ($a \leq 0.3\,\mathrm{M}$; [66, 67]).

The location of the circular photon orbit determines the size of the shadow of Sgr A* (see [68, 69]). VLBI observations are expected to image the shadow of Sgr A* and to measure the mass, spin, and inclination of this black hole (e.g., [69–73]). In addition to these parameters, the shape of the

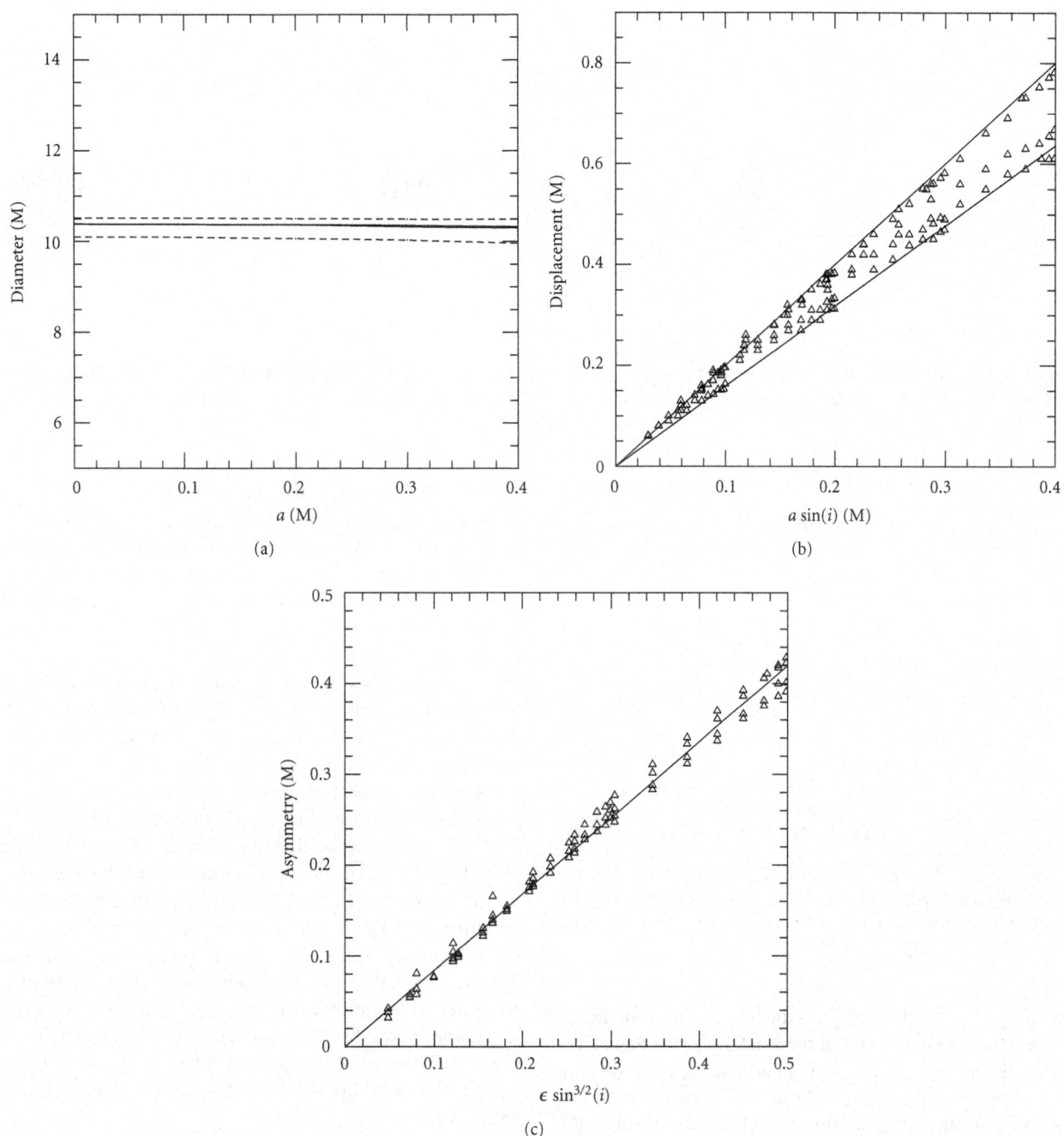

FIGURE 5: (a) The ring diameter versus spin for inclinations $17° \leq i \leq 86°$ for a Kerr black hole (solid lines) and for a quasi-Kerr black hole with a value of the deviation parameter $\epsilon = 0.5$ (dashed lines). The diameter is practically independent of the spin, inclination, and deviation parameter with a constant value of $\simeq 10.4\,M$ for a Kerr black hole. (b) The displacement of the ring of light as a function of $a \sin i$ for various values of the parameter $0 \leq \epsilon \leq 0.5$. The displacement depends only weakly on the parameter ϵ. (c) The ring asymmetry versus $\epsilon \sin^{3/2} i$ for various inclinations $17° \leq i \leq 86°$ and $0.0 \leq a/M \leq 0.4$. The asymmetry is nearly independent of the spin and hence provides a direct measure of a violation of the no-hair theorem [23].

shadow also depends uniquely on the value of the deviation parameter [23]. In practice, however, these measurements will be model dependent (e.g., [74]) and affected by finite telescope resolution (e.g., [69, 75]). Therefore, VLBI imaging may have to be complemented by additional observations such as a multiwavelength study of polarization ([73]; see also [76, 77]).

In an optically thin accretion flow such as the one around Sgr A* at sub-mm wavelengths (e.g., [66]), photons can orbit around the black hole several times before they are detected by a distant observer. This produces an image of a ring that can be significantly brighter than the underlying flow thanks to the long optical path of the contributing photons (e.g., [78]). In [23], we showed that the shape and location of this

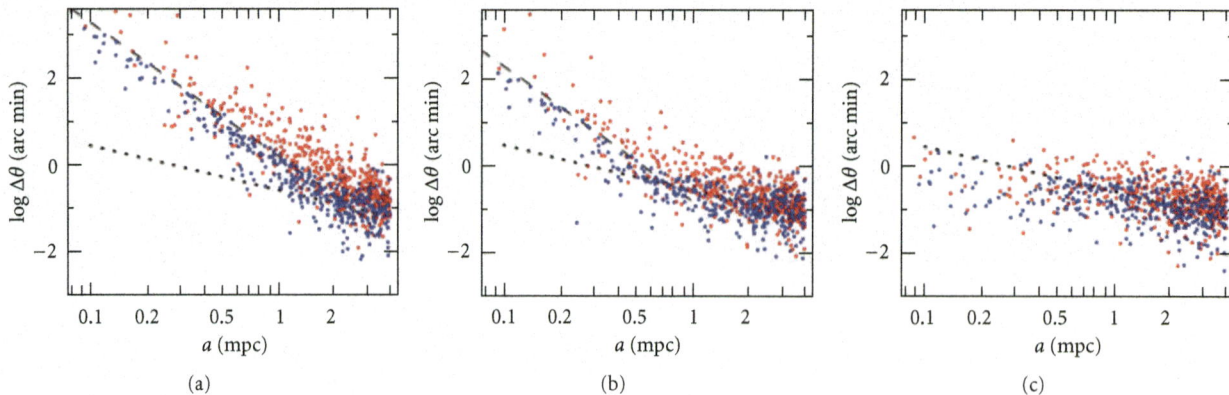

FIGURE 6: Evolution of the orbital angular momenta of stars around Sgr A* due to frame-dragging (dashed lines) and stellar perturbations (dotted lines) as measured by the angle $\Delta\theta$ as a function of the orbital semimajor axis a. The three panels correspond to a Kerr black hole with a spin (a) $a = M$, (b) $a = 0.1\,M$, and (c) $a = 0$ [30].

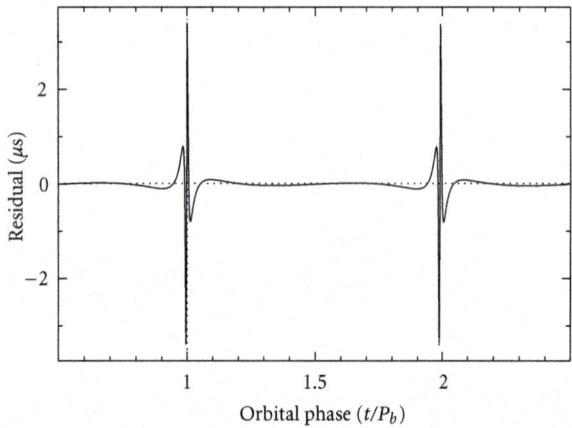

FIGURE 7: Typical timing residuals for a radio pulsar in an orbit around a black hole with a mass of $10^4\,M_\odot$ due to the presence of a nonvanishing quadrupole moment [31]. For Sgr A*, such residuals will have a similar shape but with a larger amplitude.

"ring of light" depends directly on the mass, spin, inclination, and the deviation parameter of the black hole (see Figure 4).

The diameter of the ring of light as observed by a distant observer depends predominantly on the mass of the black hole and is nearly constant for all values of the spin and disk inclination as well as for small values of the deviation parameter. For nonzero values of the spin of the black hole, the ring is displaced off center in the image plane. In all cases, the ring of a Kerr black hole remains nearly circular except for very large values of the spin $a \gtrsim 0.9\,M$. However, if Sgr A* is not a Kerr black hole, the ring becomes asymmetric in the image plane. This asymmetry is a direct measure for a violation of the no-hair theorem (see Figure 5).

4. Combining Strong-Field with Weak-Field Tests of the No-Hair Theorem

In addition to a strong-field test of the no-hair theorem with VLBI imaging of Sgr A*, there exist two other promising possibilities for performing such a test in the weak-field regime. The presence of a nonzero spin and quadrupole moment independently leads to a precession of the orbit of stars around Sgr A* at two different frequencies, which can be studied with parameterized post-Newtonian dynamics [29, 79]. Merritt et al. [30] showed that the effect of the quadrupole moment on the orbit of such stars is masked by the effect of the spin for the group of stars known to orbit Sgr A*. However, if a star can be detected within ~1000 Schwarzschild radii of Sgr A* and if it can be monitored over a sufficiently long period of time, this technique may also measure the spin (see Figure 6; [30]) and even the quadrupole moment [30, 80] together with the already obtained mass [37, 38]. Future instruments, such as GRAVITY [81], may be able to resolve the orbits of such stars providing an independent test of the no-hair theorem.

Yet another weak-field test can be performed by the observation of a radio pulsar on an orbit around Sgr A*. If present, timing observations may resolve characteristic spin-orbit residuals that are induced by the quadrupole moment and infer its magnitude (see Figure 7; [31]). Recent surveys set an upper limit for the existence of up to 90 pulsars within the central parsec of the galaxy [82] making this technique a promising third approach for testing the no-hair theorem with Sgr A*.

The fundamental properties of the black hole in the center of our galaxy can be probed with three different techniques. The combination of the results of all three approaches will lead to a secure test of the no-hair theorem with Sgr A*.

References

[1] W. Israel, "Event horizons in static vacuumspace-times," *Physical Review*, vol. 164, no. 5, pp. 1776–1779, 1967.

[2] W. Israel, "Event horizons in static electrovac space-times," *Communications in Mathematical Physics*, vol. 8, no. 3, pp. 245–260, 1968.

[3] B. Carter, "Axisymmetric black hole has only two degrees of freedom," *Physical Review Letters*, vol. 26, no. 6, pp. 331–333, 1971.

[4] S. W. Hawking, "Black holes in general relativity," *Communications in Mathematical Physics*, vol. 25, p. 152, 1972.

[5] B. Carter, *Black Holes*, Gordon and Breach, New York, NY, USA, 1973.

[6] D. C. Robinson, "Uniqueness of the Kerr black hole," *Physical Review Letters*, vol. 34, no. 14, pp. 905–906, 1975.

[7] R. Geroch, "Multipole moments. II. Curved space," *Journal of Mathematical Physics*, vol. 11, no. 8, pp. 2580–2588, 1970.

[8] R. O. Hansen, "Multipole moments of stationary space–times," *Journal of Mathematical Physics*, vol. 15, no. 1, pp. 46–53, 1974.

[9] D. Psaltis, *Compact Stellar X-Ray Sources*, Cambridge University Press, Cambridge, Mass, USA, 2006.

[10] F. D. Ryan, "Gravitational waves from the inspiral of a compact object into a massive, axisymmetric body with arbitrary multipole moments," *Physical Review D*, vol. 52, no. 10, pp. 5707–5718, 1995.

[11] F. D. Ryan, "Accuracy of estimating the multipole moments of a massive body from the gravitational waves of a binary inspiral," *Physical Review D*, vol. 56, no. 4, pp. 1845–1855, 1997.

[12] F. D. Ryan, "Scalar waves produced by a scalar charge orbiting a massive body with arbitrary multipole moments," *Physical Review D*, vol. 56, no. 12, pp. 7732–7739, 1997.

[13] L. Barack and C. Cutler, "LISA capture sources: approximate waveforms, signal-to-noise ratios, and parameter estimation accuracy," *Physical Review D*, vol. 69, no. 8, Article ID 082005, 2004.

[14] L. Barack and C. Cutler, "Using LISA extreme-mass-ratio inspiral sources to test off-Kerr deviations in the geometry of massive black holes," *Physical Review D*, vol. 75, no. 4, Article ID 042003, 2007.

[15] J. Brink, "Spacetime encodings. I. A spacetime reconstruction problem," *Physical Review D*, vol. 78, Article ID 102001, 8 pages, 2008.

[16] C. Li and G. Lovelace, "Generalization of Ryan's theorem: probing tidal coupling with gravitational waves from nearly circular, nearly equatorial, extreme-mass-ratio inspirals," *Physical Review D*, vol. 77, Article ID 064022, 10 pages, 2008.

[17] T. A. Apostolatos, G. Lukes-Gerakopoulos, and G. Contopoulos, "How to observe a non-kerr spacetime using gravitational waves," *Physical Review Letters*, vol. 103, no. 11, Article ID 111101, 2009.

[18] N. A. Collins and S. A. Hughes, "Towards a formalism for mapping the spacetimes of massive compact objects: bumpy black holes and their orbits," *Physical Review D*, vol. 69, Article ID 124022, 2004.

[19] S. J. Vigeland and S. A. Hughes, "Spacetime and orbits of bumpy black holes," *Physical Review D*, vol. 81, no. 2, Article ID 024030, 2010.

[20] K. Glampedakis and S. Babak, "Mapping spacetimes with LISA: inspiral of a test body in a "quasi-Kerr" field," *Classical and Quantum Gravity*, vol. 23, no. 12, article 013, pp. 4167–4188, 2006.

[21] J. R. Gair, C. Li, and I. Mandel, "Observable properties of orbits in exact bumpy spacetimes," *Physical Review D*, vol. 77, no. 2, Article ID 024035, 23 pages, 2008.

[22] T. Johannsen and D. Psaltis, "Testing the no-hair theorem with observations in the electromagnetic spectrum. I. Properties of a Quasi-Kerr spacetime," *Astrophysical Journal*, vol. 716, no. 1, pp. 187–197, 2010.

[23] T. Johannsen and D. Psaltis, "Testing the no-hair theorem with observations in the electromagnetic spectrum. II. Black hole images," *Astrophysical Journal*, vol. 718, no. 1, p. 446, 2010.

[24] T. Johannsen and D. Psaltis, "Testing the no-hair theorem with observations in the electromagnetic spectrum. III. Quasiperiodic variability," *Astrophysical Journal*, vol. 726, no. 1, p. 11, 2011.

[25] T. Johannsen and D. Psaltis, "Testing the no-hair theorem with observations of black holes in the electromagnetic spectrum," *Advances in Space Research*, vol. 47, p. 528, 2011.

[26] D. Psaltis and T. Johannsen, "A ray-tracing algorithm for spinning compact object spacetimes with arbitrary quadrupole moments. I. Quasi-kerr black holes," *Astrophysical Journal*. In press.

[27] C. Bambi and E. Barausse, "Constraining the quadrupole moment of stellar-mass black hole candidates with the continuum fitting method," *Astrophysical Journal*, vol. 731, p. 121, 2011.

[28] C. Bambi, "Constraint on the quadrupole moment of supermassive black hole candidates from the estimate of the mean radiative efficiency of AGN," *Physical Review D*, vol. D 83, Article ID 103003, 4 pages, 2011.

[29] C. M. Will, " Testing the general relativistic "no-hair" theorems using the galactic center black hole Sgr A*," *Astrophysical Journal*, vol. 674, p. L25, 2008.

[30] D. Merritt, T. Alexander, S. Mikkola, and C. M. Will, "Testing properties of the Galactic center black hole using stellar orbits," *Physical Review D*, vol. 81, no. 6, Article ID 062002, 2010.

[31] N. Wex and S. M. Kopeikin, "Frame dragging and other precessional effects in black hole pulsar binaries," *Astrophysical Journal*, vol. 514, no. 1, pp. 388–401, 1999.

[32] V. S. Manko and I. D. Novikov, "Generalizations of the Kerr and Kerr-Newman metrics possessing an arbitrary set of mass-multipole moments," *Classical and Quantum Gravity*, vol. 9, no. 11, article 013, pp. 2477–2487, 1992.

[33] S. J. Vigeland, N. Yunes, and L. C. Stein, "Bumpy black holes in alternative theories of gravity," *Physical Review D*, vol. 83, Article ID 104027, 16 pages, 2011.

[34] T. Johannsen and D. Psaltis, "Metric for rapidly spinning black holes suitable for strong-field tests of the no-hair theorem," *Physical Review D*, vol. 83, Article ID 124015, 16 pages, 2011.

[35] S. A. Hughes, "(Sort of) testing relativity with extreme mass ratio inspirals," in *Proceedings of the AIP Conference*, vol. 873, pp. 233–240, November 2006.

[36] D. Psaltis and T. Johannsen, "Sgr A*: the optimal testbed of strong-field gravity," *Journal of Physics*, vol. 283, Article ID 102030, 2011.

[37] A. M. Ghez, S. Salim, N. N. Weinberg et al., "Measuring distance and properties of the milky way's central supermassive black hole with stellar orbits," *Astrophysical Journal*, vol. 689, no. 2, pp. 1044–1062, 2008.

[38] S. Gillessen, F. Eisenhauer, S. Trippe et al., "Monitoring stellar orbits around the massive black hole in the galactic center," *Astrophysical Journal*, vol. 692, p. 1075, 2009.

[39] S. S. Doeleman, J. Weintroub, A. E. E. Rogers et al., "Event-horizon-scale structure in the supermassive black hole candidate at the Galactic Centre," *Nature*, vol. 455, no. 7209, pp. 78–80, 2008.

[40] K. Schwarzschild, "Über das Gravitationsfeld eines Massenpunktes nach der Einsteinschen Theorie," *Sitzungsberichte der Königlich Preussischen Akademie der Wissenschaften*, vol. 1, pp. 189–196, 1916.

[41] R. P. Kerr, "Gravitational field of a spinning mass as an example of algebraically special metrics," *Physical Review Letters*, vol. 11, no. 5, pp. 237–238, 1963.

[42] J. B. Hartle, "Slowly Rotating Relativistic Stars. I. Equations of Structure," *Astrophysical Journal*, vol. 150, p. 1005, 1967.

[43] J. B. Hartle and K. S. Thorne, "Slowly Rotating Relativistic Stars. II. Models for Neutron Stars and Supermassive Stars," *Astrophysical Journal*, vol. 153, p. 807, 1968.

[44] A. Tomimatsu and H. Sato, "New exact solution for the gravitational field of a spinning mass," *Physical Review Letters*, vol. 29, no. 19, pp. 1344–1345, 1972.

[45] A. Tomimatsu and H. Sato, "New series of exact solutions for gravitational fields of spinning masses," *Progress of Theoretical Physics*, vol. 50, no. 1, pp. 95–110, 1973.

[46] H. Stephani, D. Kramer, M. A. H. MacCallum, C. Hoenselaers, and E. Herlt, *Exact Solutions of Einstein's Field Equations*, Cambridge University Press, Cambridge, Mass, USA, 2003.

[47] I. Hauser and F. J. Ernst, "Proof of a Geroch conjecture," *Journal of Mathematical Physics*, vol. 22, no. 5, pp. 1051–1063, 1981.

[48] C. Hoenselaers, W. Kinnersley, and B. C. Xanthopoulos, "Symmetries of the stationary Einstein–Maxwell equations. VI. Transformations which generate asymptotically flat spacetimes with arbitrary multipole moments," *Journal of Mathematical Physics*, vol. 20, no. 8, p. 2530, 1979.

[49] R. Geroch, "A method for generating solutions of Einstein's equations," *Journal of Mathematical Physics*, vol. 12, no. 6, pp. 918–924, 1971.

[50] R. Geroch, "A method for generating new solutions of Einstein's equation. II," *Journal of Mathematical Physics*, vol. 13, no. 3, pp. 394–404, 1972.

[51] R. Beig and W. Simon, "Proof of a multipole conjecture due to Geroch," *Communications in Mathematical Physics*, vol. 78, p. 75, 1980.

[52] R. Beig and W. Simon, "On the multipole expansion for stationary space-times," *Proceedings of the Royal Society A*, vol. 376, pp. 333–341, 1981.

[53] N. R. Sibgatullin, *Oscillations and Waves in Strong Gravitational and Electromagnetic Fields*, Springer, Berlin, Germany, 1991.

[54] V. S. Manko and N. R. Sibgatullin, "Construction of exact solutions of the Einstein-Maxwell equations corresponding to a given behaviour of the Ernst potentials on the symmetry axis," *Classical and Quantum Gravity*, vol. 10, no. 7, article 014, pp. 1383–1404, 1993.

[55] E. M. Butterworth and J. R. Ipser, "On the structure and stability of rapidly rotating fluid bodies in general relativity. I - The numerical method for computing structure and its application to uniformly rotating homogeneous bodies," *Astrophysical Journal*, vol. 204, pp. 200–223, 1976.

[56] N. Stergioulas and J. L. Friedman, "Comparing models of rapidly rotating relativistic stars constructed by two numerical methods," *Astrophysical Journal*, vol. 444, no. 1, pp. 306–311, 1995.

[57] W. G. Laarakkers and E. Poisson, "Quadrupole moments of rotating neutron stars," *Astrophysical Journal*, vol. 512, no. 1, pp. 282–287, 1999.

[58] E. Berti, F. White, A. Maniopoulou, and M. Bruni, "Rotating neutron stars: an invariant comparison of approximate and numerical space-time models," *Monthly Notices of the Royal Astronomical Society*, vol. 358, no. 3, pp. 923–938, 2005.

[59] C. Cadeau, S. M. Morsink, D. Leaky, and S. S. Campbell, "Light curves for rapidly rotating neutron stars," *Astrophysical Journal*, vol. 654, no. 1 I, pp. 458–469, 2007.

[60] C. F. Sopuerta and N. Yunes, "Extreme- and intermediate-mass ratio inspirals in dynamical Chern-Simons modified

[61] D.-C. Dai and D. Stojkovic, "Analytic solution for a static black hole in RSII model," submitted to *General Relativity and Quantum Cosmology*.

[62] N. Yunes and L. C. Stein, "Effective gravitational wave stress-energy tensor in alternative theories of gravity," *Physical Review D*, vol. 83, p. 4002, 2011.

[63] E. Barausse, T. Jacobson, and T. P. Sotiriou, "Black holes in Einstein-aether and Horava-Lifshitz gravity," *General Relativity and Quantum Cosmology*, vol. 83, Article ID 124043, 2011.

[64] P. Figueras and T. Wiseman, "Gravity and large black holes in Randall-Sundrum II braneworlds," submitted to *High Energy Physics*.

[65] T. Johannsen et al., in preparation.

[66] A. E. Broderick, V. L. Fish, S. S. Doeleman, and A. Loeb, "Estimating the parameters of sagittarius A*'s accretion flow via millimeter vlbi," *Astrophysical Journal*, vol. 697, no. 1, pp. 45–54, 2009.

[67] A. E. Broderick, V. L. Fish, S. S. Doeleman, and A. Loeb, "Evidence for low black hole spin and physically motivated accretion models from millimeter-VLBI observations of sagittarius A*," *Astrophysical Journal*, vol. 735, p. 110, 2011.

[68] J. M. Bardeen, *Black Holes*, Gordon and Breach, New York, NY, USA, 1973.

[69] H. Falcke, F. Melia, and E. Agol, "Viewing the shadow of the black hole at the Galactic center," *Astrophysical Journal*, vol. 528, no. 1, pp. L13–L16, 2000.

[70] A. E. Broderick and A. Loeb, "Imaging bright-spots in the accretion flow near the black hole horizon of Sgr A*," *Monthly Notices of the Royal Astronomical Society*, vol. 363, no. 2, pp. 353–362, 2005.

[71] A. E. Broderick and A. Loeb, "Imaging optically-thin hotspots near the black hole horizon of Sgr A* at radio and near-infrared wavelengths," *Monthly Notices of the Royal Astronomical Society*, vol. 367, no. 3, pp. 905–916, 2006.

[72] V. L. Fish and S. S. Doeleman, *IAU Symp. 261, Relativity in Fundamental Astronomy: Dynamics, Reference Frames, and Data Analysis*, Cambridge University Press, Cambridge, UK, 2009.

[73] A. E. Broderick and A. Loeb, "Imaging optically-thin hotspots near the black hole horizon of Sgr A* at radio and near-infrared wavelengths," *Monthly Notices of the Royal Astronomical Society*, vol. 367, no. 3, pp. 905–916, 2006.

[74] A. E. Broderick and A. Loeb, "Imaging the black hole silhouette of M87: implications for jet formation and black hole spin," *Astrophysical Journal*, vol. 697, no. 2, pp. 1164–1179, 2009.

[75] R. Takahashi, "Shapes and positions of black hole shadows in accretion disks and spin parameters of black holes," *Astrophysical Journal*, vol. 611, no. 2, pp. 996–1004, 2004.

[76] J. D. Schnittman and J. H. Krolik, "X-ray polarization from accreting black holes: the thermal state," *Astrophysical Journal*, vol. 701, no. 2, pp. 1175–1187, 2009.

[77] J. D. Schnittman and J. H. Krolik, "X-ray polarization from accreting black holes: coronal emission," *Astrophysical Journal*, vol. 712, no. 2, pp. 908–924, 2010.

[78] K. Beckwith and C. Done, "Extreme gravitational lensing near rotating black holes," *Monthly Notices of the Royal Astronomical Society*, vol. 359, no. 4, pp. 1217–1228, 2005.

[79] C. M. Will, *Theory and Experiment in Gravitational Physics*, Cambridge University Press, Cambridge, Mass, USA, 1993.

[80] L. Sadeghian and C. M. Will, "Testing the black hole no-hair theorem at the galactic center: perturbing effects of stars in

the surrounding cluster," submitted to *General Relativity and Quantum Cosmology*.

[81] H. Bartko, G. Perrin, W. Brandner et al., "GRAVITY: astrometry on the galactic center and beyond," *New Astronomy Reviews*, vol. 53, no. 11-12, pp. 301–306, 2009.

[82] J.-P. MacQuart, N. Kanekar, D. A. Frail, and S. M. Ransom, "A high-frequency search for pulsars within the central parsec of Sgr A*," *Astrophysical Journal*, vol. 715, no. 2, pp. 939–946, 2010.

10

Diffusion Coefficients, Short-Term Cosmic Ray Modulation, and Convected Magnetic Structures

John J. Quenby,[1] **Tamitha Mulligan,**[2] **J. Bernard Blake,**[2] **and Diana N. A. Shaul**[1]

[1] *Blackett Laboratory, Imperial College, London SW7 2BZ, UK*
[2] *Space Sciences Department, The Aerospace Corporation, Los Angeles, CA 90009, USA*

Correspondence should be addressed to Tamitha Mulligan; tamitha.mulligan@aero.org

Academic Editor: José F. Valdés-Galicia

Three cases of large-amplitude, small spatial-scale interplanetary particle gradients observed by the anticoincidence shield (ACS) aboard the INTEGRAL spacecraft in 2006 are investigated. The high data rates provided by the INTEGRAL ACS allow an unprecedented ability to probe the fine structure of GCR propagation in the inner Heliosphere. For two of the three cases, calculating perpendicular and parallel cosmic ray diffusion coefficients based on both field and particle data results in parallel diffusion appearing to satisfy a convection gradient current balance, provided that the magnetic scattering of the particles can be described by quasi-linear theory. In the third case, perpendicular diffusion seems to dominate. The likelihood of magnetic flux rope topologies within solar ejecta affecting the local modulation is considered, and its importance in understanding the field-particle interaction for the astrophysics of nonthermal particle phenomena is discussed.

1. Introduction

A Forbush Decrease (FD) is a global transient decrease in Galactic Cosmic-Ray (GCR) intensity followed by a substantially slower recovery. Since Scott Forbush's discovery and description of these phenomena in the late 1930s, FDs have been put into context with increasing developments within heliospheric physics. In particular, detailed observations of coronal mass ejections (CMEs) and in situ observations of the solar wind and energetic particles have greatly increased understanding of the underlying physics of FDs (see review articles by [1–3]).

This investigation focuses on small amplitude and high-frequency variability in the GCR corresponding to timescales less than a few hours, much shorter than that described by the classical FD. However, small-amplitude, mHz variability in the GCR is an experimental challenge in that very large instrumental geometric factors are required in order to make statistically significant measurements of the GCR in time periods of a few minutes or less. Therefore, only a few of these investigations can be found in the literature. Among these studies, [4] establishes the existence of short-spatial-scale GCR intensity gradients of a few percent amplitude

(at >200 MeV energies) convecting with the solar wind past the Earth, coincident with the observation of an FD. This study correlates magnetic substructures within interplanetary CMEs (ICMEs) with short-scale intensity variations in the GCR. The authors in [5] investigate four simple magnetic field models for explaining short-term reductions in the GCR intensity and associated energetic particle propagation concluding that only a magnetic flux rope topology similar to that found in magnetic cloud ICMEs provides the magnetic conditions most likely to explain the overall depth of an FD.

Exploring the detailed relationship between particle intensity and magnetic field variability within the substructure of solar wind transients exhibiting large, short-period GCR fluctuations may yield new insight into energetic particle propagation within the Heliosphere. The authors of [6] originally derived energetic particle diffusion coefficients described by resonant wave scattering and field line wandering under the quasi-linear approximation. However, discrepancies of up to a factor four have been found by [7] in a strong scattering regime. In an attempt to represent observed conditions, the authors of [7] use coefficients derived empirically from solar particle propagation studies and numerical trajectory investigations in model solar wind fields. Computational models solve the transport equation describing three-dimensional long-term GCR modulation by employing empirically justified diffusion coefficients, based on the goodness of fit to the overall spatial, temporal, and energy dependence of the modulation (e.g., [8]). However, these approaches do not attempt to relate the coefficients to in situ field data. A preferred method is the more direct derivation of the radial coefficient by McDonald et al. [9], who relate the radial gradient of the long-term modulation to the convective term directly.

In Section 2 of this paper, we discuss the instrumentation and data used in the analysis. In Section 3, we discuss the observations, first focusing on the particle observations and then focusing on solar wind observations and the connection with solar wind transients. In Sections 4 and 5, we use a version of the approach in [9] to estimate the diffusion coefficients locally and then compare the result with the prediction of quasi-linear theory. In Section 6, we present our conclusions.

2. Particle and Field Data

Particle data for this investigation is obtained from the large-area ACS of the SPI spectrometer mounted on the ESA INTEGRAL gamma-ray satellite. With a 24 R_E apogee, near continuous GCR monitoring is achieved. SPI consists of an array of 19 cooled Ge detectors, hexagonal in shape and of side 3.2 cm and height 7 cm [10]. The ACS is comprised of 91 bismuth germinate blocks. Several types of signals are available for GCR monitoring. The highest energy signals include the saturated counts of the ACS (ACSSAT) and the saturated counts of the Ge detector system (GEDSAT). Here "saturated" means that the amplitude from the energy deposited in the detector is sufficient to saturate the amplifier systems. The ACSSAT threshold is ~150 MeV [11]. The GEDSAT signal has an energy threshold of 200 MeV (a consequence of the energy required to penetrate the spacecraft shielding to reach the Ge detectors and subsequently lose an additional 10 MeV to the Ge detectors) [12].

In addition to ACSSAT and GEDSAT, the system has an ACS channel that counts all triggers in the system above ~100 keV. Much higher counting rates occur in this ACS channel because of the low-energy threshold. However, the lower-energy threshold also means that there are events in the run of the ACS data due to energetic magnetospheric electrons. Luckily, ACS counts due to magnetospheric electrons have a spikey nature and time periods when they occur are easily identified and removed or ignored. The three INTEGRAL SPI channels, GEDSAT, ACSSAT, and ACS, make omnidirectional measurements with a broad and poorly known energy response. Since the GCR energy spectra peak in the energy range of hundreds of MeV per nucleon, this is not a serious issue. The compelling reason to use these INTEGRAL data is the stunning statistics provided by the relatively huge count rates that permit unprecedented temporal resolution of temporal changes in the GCR [4]. Solar wind magnetic field and plasma data are obtained from the MAG and SWEPAM instruments aboard the ACE spacecraft.

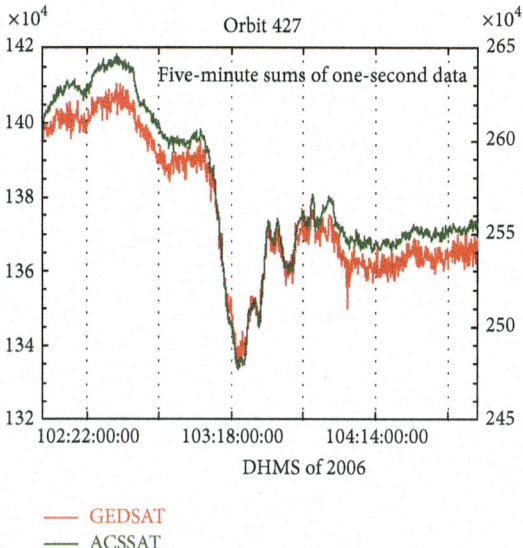

FIGURE 1: Five-minute sums of GEDSAT and ACSSAT count rates are plotted for Orbit 427.

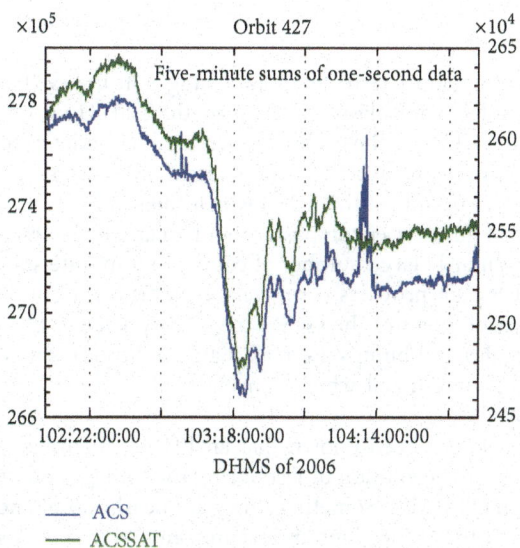

FIGURE 2: Five-minute sums of GEDSAT and ACSSAT count rates are plotted for Orbit 427.

3. Observations

3.1. GCR Study Intervals. Three periods of very rapid GCR intensity decrease were selected from 2006 INTEGRAL data, DOY 103 (orbit 427), DOY 117/118 (orbit 432), and DOY 1276/127 (orbit 435). Figure 1 compares the 5-minute sums of the GEDSAT and ACSSAT count rates for orbit 427 showing excellent agreement including the fine structure. The ACSSAT count rate is approximately twice that of the GEDSAT. Figure 2 similarly compares the ACS and ACSSAT rates. Excellent agreement is again found except for some spikes we ascribe to magnetospheric-related events. The orbits chosen for this analysis were selected based upon

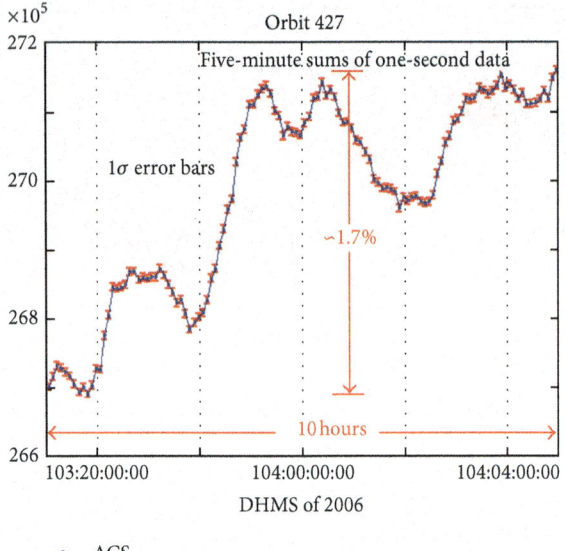

— ACS

FIGURE 3: The data plotted in Figure 2 is replotted on expanded scales. The small size of the 1σ standard error bars reveals the statistical significance of the GCR variations.

the striking short-term temporal changes in the GCR, and in the solar wind plasma and magnetic field. However the general tracking of the three channels lends confidence to the integrity of the measurement of the rapid flux variation.

The amplitude of the FD in which the sudden drop seen in both figures is encompassed is of magnitude ~3%. The extraordinarily large number of counts in 5-minute sums of the ACS data provides a marvelous statistical accuracy to the measurement of the GCR time history. Clearly many of the wiggles and bumps are real. Figure 3 illustrates this last point in more detail where conventional 1σ error bars are employed.

It has long been known that an FD has a similar time history to the evolution of Dst during the time period of the Forbush effect [13]. Sometimes they are nearly simultaneous while at other times there is a significant delay as was seen here (see Section 3.2 for more details on the Dst signature). To compensate for this effect, in Figure 4, we show the actual data for ACSSAT and Dst, while in Figure 5, Dst is shifted 14 hours earlier. The fact that such a time shift brings about a similarity in shape adds evidence for the convection model for the GCR-solar plasma structure over a time scale comparable to the duration of the rapid flux decrease studied.

Figure 6 plots the ACS data for orbit 427 along with the magnitude of the interplanetary field recorded by ACE. The field magnitude is seen to have increased abruptly at the time of the FD and remained high until the ACS recovered. This recovery did not return to pre-FD levels. (The spikes in the ACS count rate in this interval are due to leakage of magnetospheric electrons out of the radiation belts and can be safely ignored for this analysis.) To add to our confidence in the observed, rapid variability, we plot INTEGRAL in comparison with the lower statistical accuracy McMurdo neutron monitor data during the event in Figure 7.

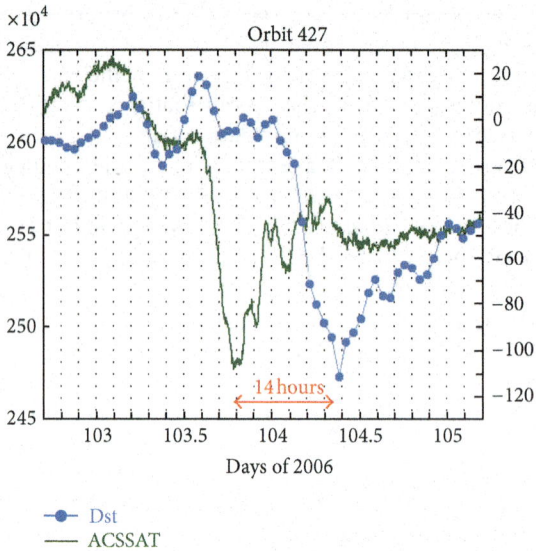

• Dst
— ACSSAT

FIGURE 4: ACSSAT for Orbit 427 plotted with the time history of Dst.

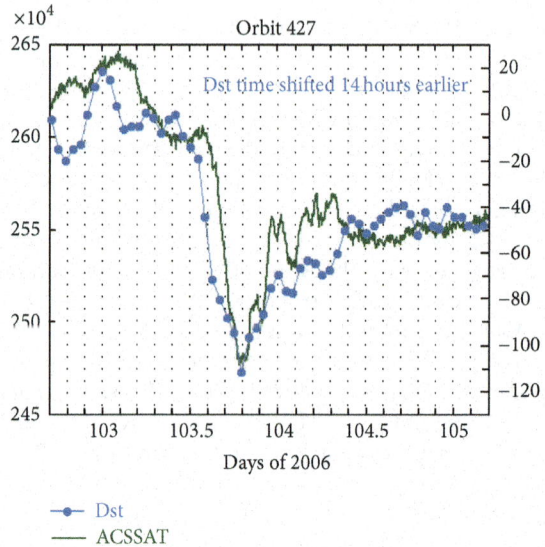

• Dst
— ACSSAT

FIGURE 5: ACSSAT for Orbit 427 plotted with the time history of Dst shifted to earlier time by fourteen hours.

3.2. Connection with Transient Solar Events. Looking at the solar wind data during this time shows the existence of an ICME. The interplanetary magnetic field and plasma conditions during a four-day period bracketing the ICME are shown in Figure 8. The bottom panel of the figure shows the ACS data. The transient spikes in the ACS data (occurring around midday of DOY 104) occur at the trailing boundary of the ICME indicated by the vertical lines. Note that the ACS data has not been time shifted to correspond to the particle signature observed at by ACE at the L1 Lagrangian point several hundred Re sunward of the Earth. The region prior to the ICME contains slow, cold plasma. The proton speed increases near the leading edge of the ICME, which causes a compression region, but no shock exists. In passing, we comment that

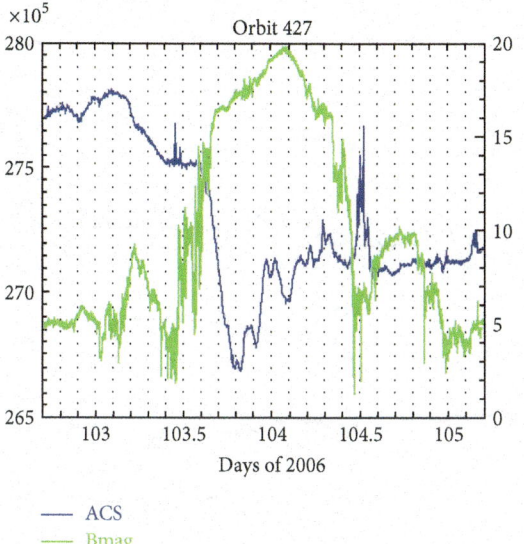

FIGURE 6: ACS is plotted for Orbit 427 along with the magnitude of the interplanetary field as seen by ACE.

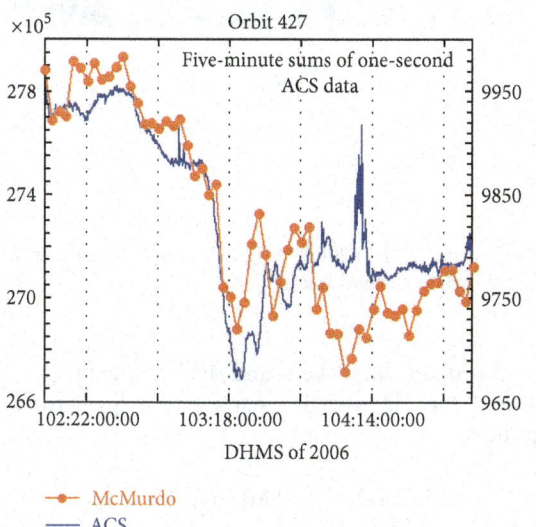

FIGURE 7: The ACS countrate is plotted with the countrate of the McMurdo neutron monitor for the time period of Orbit 427.

there is evidence that a shock front cannot be the basic cause of an FD. Close examination of the ICME reveals that the leading edge corresponds to the leading edge of the FD, but a simultaneous depression in Dst is not expected (see Figure 5) due to the presence of a strong northward component of the magnetic field (positive Bz) during the first half of the ICME. The drop in Dst is expected to occur when the field turns southward (negative Bz), which is some 12–15 hours after CME onset, consistent with the Dst drop shown in Figure 5.

A faster, warmer region is observed just downstream of the trailing ICME boundary. Taking a closer look at days 103 and 104 (04-13-06 and 04-14-06) shows some interesting details in the electron heat flux and ACS data during the period around the ICME. Inside the leading edge, sporadic

bidirectional electron heat flux persists through the end of day 103, as shown in Figure 8. During this period, the heat flux has intermittent regions of unidirectional and bidirectional streaming indicating that the spacecraft is passing through alternately open and closed magnetic field lines in rapid succession. This changing topology may correspond to the short-period oscillations in the ACS count rates. It is difficult to assess the exact correspondence without first determining the time delay from L1 (ACE) to the location of INTEGRAL.

At the beginning of day 104 (Figure 8), the electron heat flux is primarily unidirectional but then switches to counter-streaming near the trailing end of the ICME (marked with a vertical line). It is also during this time that the plasma temperature increases, although the proton speed remains nearly constant. At the ICME trailing boundary, the magnetic field reaches a minimum and the field turns northward.

Figure 9 plots the ACS count rate over the period encompassing the three events studied, together with field components and magnitude, solar wind speed, temperature and density, and low-energy electron heat flux. The two vertical bands in the figure mark periods when a simple flux rope model may be fitted. The first is in good coincidence with the day 103 event but there is no simple rope configuration fitting the day 117/118 (ICME 2) or day 126/127 (ICME 3) events. At this stage, we are only able to say that the latter two events coincide with complex transient magnetic structures likely to exist over significant regions of space.

4. Derivation of the Quasi-Linear Diffusion Coefficients

The most basic derivation of the parallel diffusion coefficient can be found in [14]. We will first outline their approach because it both provides the most accessible expression to estimate this coefficient and illustrates the problems in any representation of the scattering in a realistic field model. The work of [14] assumes that particles follow helical trajectories along a nearly uniform field line, directed in the z direction, but suffer a series of small changes in pitch angle as they encounter transverse wave packets where the spatial wavelength of the wave along the field matches the projection of the cyclotron radius along the field. A perturbing force changes the parallel component of velocity and hence via conservation of the first adiabatic invariant, the pitch angle, depending on the spatial extent of the field perturbation. This extent of the wave-particle resonance is estimated as being within a spatial wave number range $\pm k_{\parallel}/2$ where $k_{\parallel} = \omega_B/v_{\parallel}$ for gyrofrequency ω_B and particle velocity $v_{\parallel}$ along field **B**. Random pitch angle scatters summing to 90° to achieve "reflection" and evaluated for an average pitch angle lead to a parallel diffusion coefficient, which we call the Kennel-Petscheck estimate $K_{\parallel}^{K,P}$:

$$K_{\parallel}^{K,P} = \frac{vVB^2}{12\pi P(f)\,f^2},\qquad(1)$$

where the resonant frequency $f = k_{\parallel}V/2\pi$ for wind velocity V allows the particle gyration to match the spatial wavelength, and $P(f)$ is the power in waves perpendicular to **B**

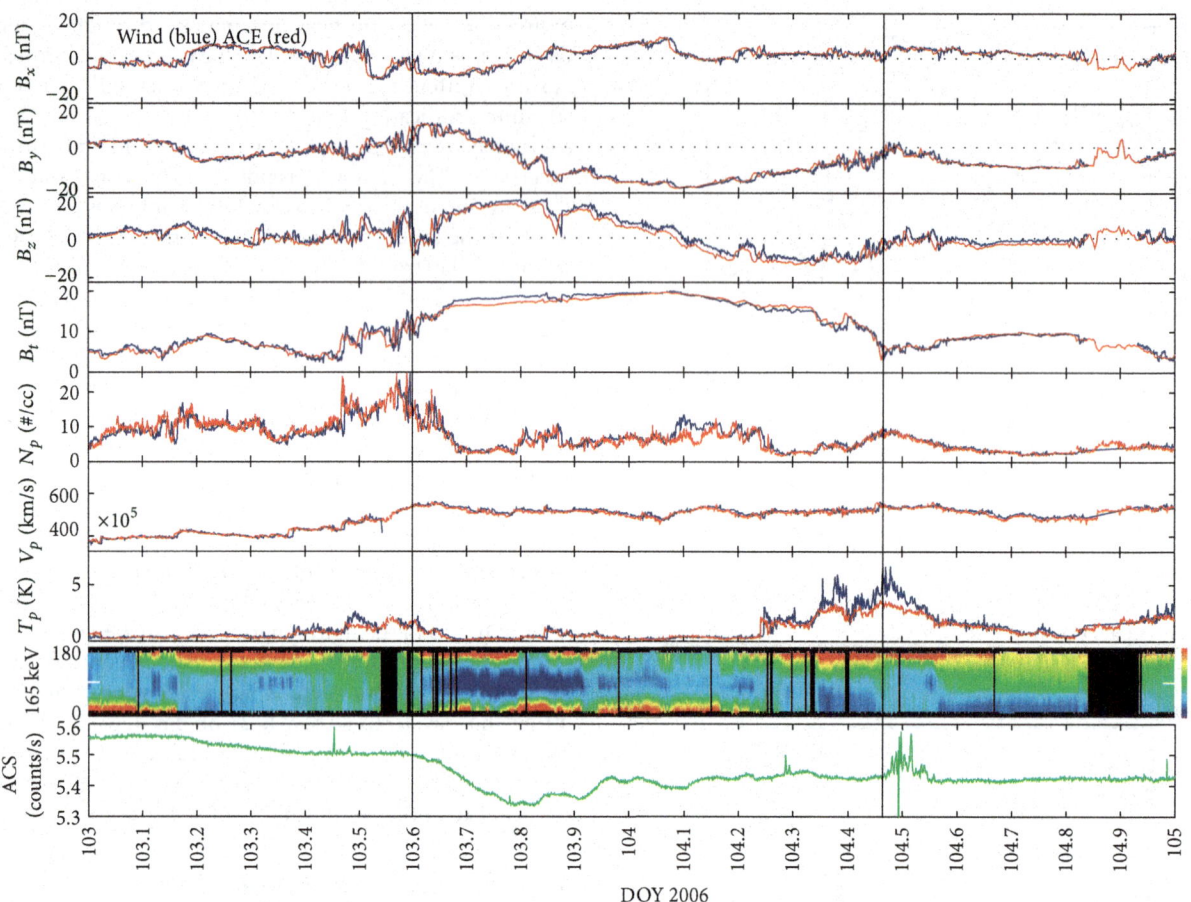

FIGURE 8: Wind and ACE IMF, plasma, and electron strahl (165 eV) data for 04-13 through 04-14 2006. INTEGRAL ACS data (unshifted) is shown in the bottom panel. The leading and trailing boundaries of the ICME are shown by vertical lines.

at f. The field model is known as a slab model where there is no dependence on x or y.

Even within the context of the assumption of transverse waves, the $K_{\|}^{K,P}$ model fails to account in detail for the differing strength of scattering power at different pitch angles, and therefore the result must depend on spectral slope. Scattering at 90° is zero on the model because wave power falls to zero at very large $k_{\|}$. A more general approach to finding the parallel diffusion coefficient, allowing the effects of waves and turbulence in three dimensions, starts with the Hall-Sturrock relation usually employed in a weakly turbulent field for the pitch angle diffusion coefficient $D_{\mu\mu}$ where μ is cosine of pitch angle:

$$D_{\mu\mu}(\mu) = Re \int_0^\infty d\epsilon \left\langle \frac{d\mu(t)}{dt} \frac{d\mu^*(t+\epsilon)}{dt} \right\rangle. \quad (2)$$

Using the equation of motion for particles with unperturbed positions x and y with the mean field in z, the quasilinear approximation which picks out the dominant term perturbing the helical orbit leads to [15]

$$\frac{d\mu}{dt} = \frac{i\omega_B}{2^{0.5}B_\circ}\left((1-\mu^2)^{0.5}\right)$$
$$\times \left[\delta B_R(x(t))e^{i\Phi} - \delta B_L(x(t))e^{-i\Phi}\right], \quad (3)$$

where Φ is phase angle. Left- and right-handed polarisation is allowed for the wavelike parts of the perturbed field components:

$$\delta B_L = \frac{1}{2^{0.5}}\left(\delta B_x + i\delta B_y\right),$$
$$\delta B_R = \frac{1}{2^{0.5}}\left(\delta B_x - i\delta B_y\right). \quad (4)$$

Various models are then chosen which define the perturbed fields via Fourier transforms representing the wave motion or turbulent, convected field structure. The integral to be performed in (2) picks out the resonance of the stationary wave and convected structure patterns with the helical orbit. Here we concentrate on the simplest or slab model which attributes the perturbations to Alfven waves propagating with k vectors parallel to the mean field which means that all change is in x or y. In [6], it was then found that

$$\frac{\langle(\Delta\mu)^2\rangle}{\Delta t} = \frac{(1-\mu^2)}{|\mu|\,v}\frac{e^2V}{m^2c^2}P_{xx}\left(f = \frac{V\omega_\circ}{2\pi\mu v}\right), \quad (5)$$

where $m = \gamma m_\circ$ and $\omega_\circ = eB/mc$, γ is the Lorenz factor, and $P_{xx} = \delta/f^n$ is here the power in one perpendicular component with f running from $-\infty$ to $+\infty$. A parallel diffusion

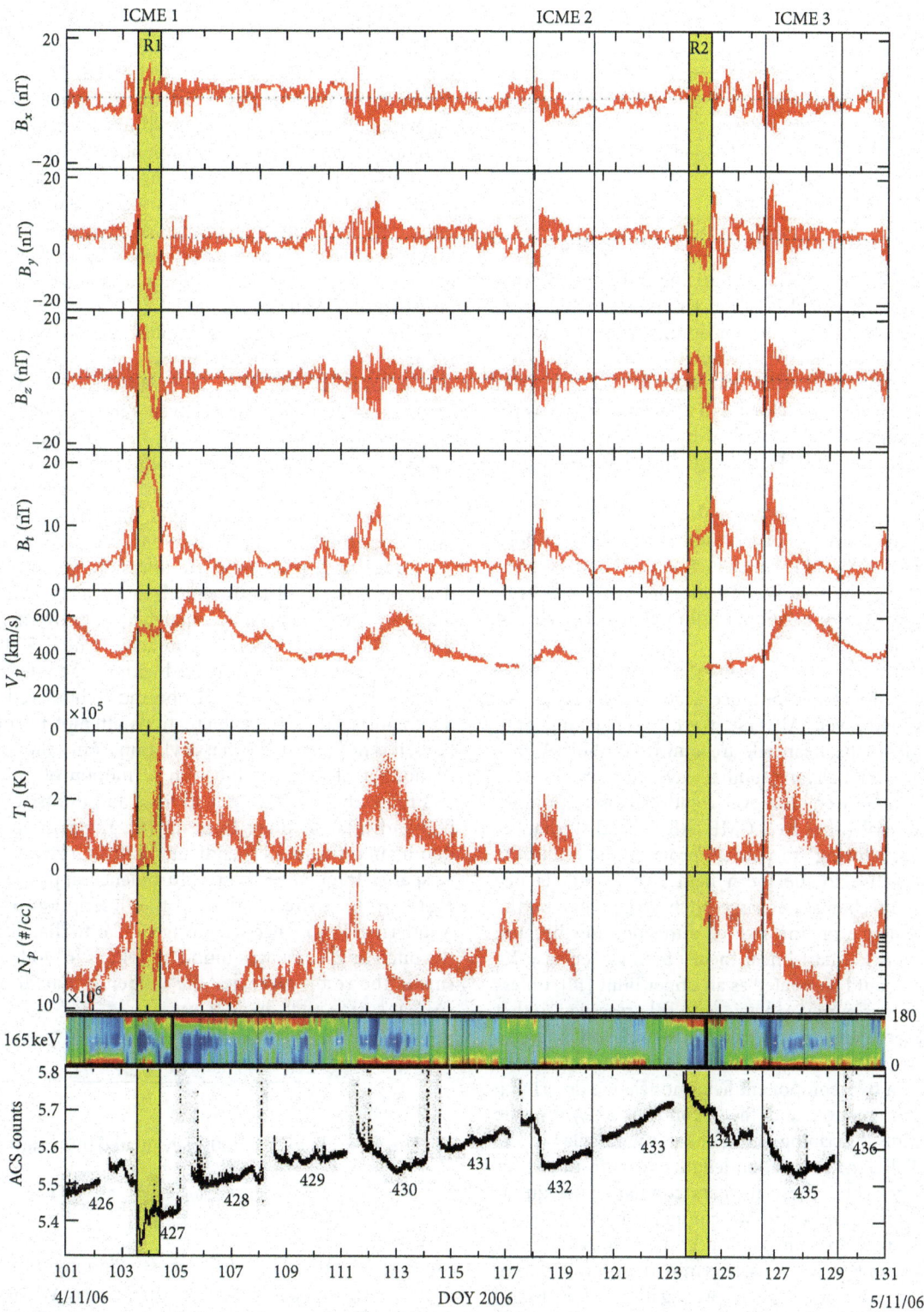

FIGURE 9: ACS count rate over entire period encompassing the three ICME events studied together with solar wind field and plasma components from ACE and electron strahl from Wind. Highlighted regions indicate presence of flux ropes.

cient is then obtained by studying the relaxation of a small anisotropy in pitch angle space and Hasselmann and Wibberenz [16] obtain

$$K_{\parallel} = \frac{v^2}{2} \int_{-1}^{1} \left[\int_{0}^{\mu'} \frac{1 - \mu^2}{\langle (\Delta\mu)^2 \rangle / \Delta t} d\mu \right] \mu' d\mu'. \qquad (6)$$

Substituting (6) into (5) yields (e.g., [17])

$$K_{\parallel}^{HW} = v^{3-n} V^{n-1} \frac{B^2}{(2\pi)^n} \frac{\omega_\circ^{n-2}}{\delta} \frac{1}{(4-n)(2-n)}. \qquad (7)$$

The symbol δ is the coefficient arising from one of three field components obtained in a Fourier transform, and it is assumed that perpendicular power dominates power in the field magnitude. Note that axial symmetry has been assumed for the perturbation spectrum and also that the cyclotron radius of the scattered particles lies in a range of spatial wave frequencies which can be described by a single power law. This assumption is satisfied by the particle and magnetic data used subsequently in this work.

Under the approximation that the power spectrum of transverse fluctuations was proportional to f^{-1}, the author's result (6) leads to a $K_{\parallel}$ value which is close to (1) and has been commonly used in the past. While values of $n \approx 1.1$ have been found during disturbed periods, a Kolmogorov-like value of 5/3 is more usual.

Subsequent work, on $K_{\parallel}$, realised that the quasi-linear theory employed did not produce scattering through 90° pitch angles. However, Alfvenic waves propagating at a variety of angles to the mean field are found in the Solar Wind, and also compressional, longitudinal waves may provide 10% of the fluctuation power. Moreover, rotational discontinuities may occur several times in a CME. All these disturbances can provide "mirroring" to particle propagation where the pitch angle relative to the mean field is near 90°. Hence the scattering mechanism at large pitch angles may be the dominant influence on the rate of diffusion, and because such corrections to quasi-linear theory tend to decrease $K_{\parallel}$ (e.g., [18]), (7) could be treated as an upper limit [18], unless most turbulence is purely Alfvenic with $|\mathbf{B}|$ preserved so that scattering at large angle is only due to a small, compressive component of the field turbulence. However, the authors of [19] introduce a two-component field model, one describing the slab approximation and the second an axisymmetric power spectrum describing variations perpendicular to the mean field. Different correlation lengths are introduced for the two components. These workers compared the quasi-linear result (QLT) and their two component model, based on average solar wind conditions at 1 AU with the "Palmer Consensus" mean free paths, as determined by a compilation of determinations from particle propagation. They found that a model with 90% of power corresponding to the slab model gave $\lambda_{\parallel}$ in the ~100 MV range at the bottom of the "Palmer" range while a model with 90% of the power concentrated in perpendicular direction fluctuations laid a factor three higher, at the top of the "Palmer" range. While the results of [19] might encourage us to believe that our use

of QLT may be only a factor two to three in error, previous numerical simulations by [7] appeared to confirm a series of investigations suggesting a larger underestimate (e.g., [17]). In [7], a field model was employed whereby each short segment of particle trajectory experienced a field obtained from actual B and V measurements as the wind flowed past a spacecraft. An equilibrium pitch angle distribution was then set up by numerically following particles within the field model after injection and until removal at boundaries. The slope of the distribution function in pitch angle space then gave $D_{\mu\mu}$. Solar particle propagation data used to deduce the spatial diffusion coefficient came from a contemporary solar particle event. The use of (7) gave an estimate a factor 8 less than that derived from the particle data and a factor 4 less than that obtained by the numerical model.

The approximation of (7) applies to an axisymmetric slab model of transverse fluctuations with no net polarisation and varying only parallel to the mean field, and we assume that corrections due to a finite Alfven speed, as implied in the equations of motion of [15], can be neglected because the particles move close to c. Our subsequent use of (7), motivated by the complexity of the subsequent approaches which are beyond the scope of this paper, must be tempered by the knowledge that it is very likely to only represent a lower limit.

Concerning perpendicular diffusion, we again start by describing the well-known and easily accessible approximation, before mentioning analytical improvements to the theory and computational approaches which will allow an assessment of the error involved with the adopted equation. Since particles of small Larmor radii compared with the dominant scale of field fluctuations attempt to follow field lines, it is not surprising that deviation from a mean direction is thought of as a combination of motion of guiding centers following wandering field lines and diffusive scattering perpendicular to these lines. In [6], it was found that the dominant cause for perpendicular diffusion lays in the power at spatial frequencies of the turbulence approaching zero, $P_{xx}(f = 0)$, corresponding to the description of field line wandering. Resonance scattering similar to that described in the pitch angle case was found by Jokipii [6] to be relatively small. The result given for the perpendicular displacement $\langle \delta x \rangle$ at cosine pitch angle μ was

$$\frac{\langle \delta x^2 \rangle}{\delta t} = \frac{\mu v V P_{xx}(f = 0)}{B_\circ^2}. \qquad (8)$$

Later, Foreman et al. [20] reevaluated this result and quote

$$K_{\perp} = 4 \times 10^{20} \left(\frac{v}{c} \right) \text{ cm}^2, \text{ sec}^{-1} \qquad (9)$$

for $P_{x,x}(f = 0) = 4 \times 10^{-6}$ Gauss^2/Hz longitudinal power, $V = 4 \times 10^7$ cm/sec, and $B = 4 \times 10^{-5}$ Gauss. We shall use (9), scaled to fit the actual field and wind parameters, according to (8).

Subsequent analytical work, especially [21], finds that the interplay between parallel and perpendicular scattering can suppress the perpendicular motion, for example, if the reversal along the field allows the gyromotion to cause the particle to sample field lines closely similar to those encountered before the scatter of parallel motion. Matthaeus et al. [21] define $K_\perp$ according to the Taylor-Green-Kubo formulation which in their case they write as

$$K_{xx} = \frac{a^2}{B_\circ} \int_0^\infty dt' \left\langle v_z(0)\, v_z(t') \right\rangle$$
$$\times \left\langle b_x[x(0),0]\, b_x[x(t'),t] \right\rangle. \qquad (10)$$

It has been assumed that the relation between the perpendicular velocity and the perpendicular field perturbation, assuming axial symmetry, is

$$v_x(t) = \frac{a v_z(t)\, b_x[x_m(t)]}{B_\circ}, \qquad (11)$$

where x_m is a mean of the gyromotion. This formula clearly describes motion following wandering field lines if $a = 1$, but in the work based on the approach outlined, a value $a = 1/3$ is taken as fitting numerical simulations corresponding to the analytical model adopted, and no theoretical justification seems to be supplied as yet for the value of the constant.

The work of [22] is the most recent and general analytical approach to finding $K_\perp$ in a medium of arbitrary turbulence, without the restrictions of, for example, the slab model or the two component model mentioned before. His nonlinear model leads to the following expression involving the power spectrum with arbitrary propagation vector $P_{xx}(\mathbf{k})$:

$$K_\perp = \frac{a^2 v^2}{3 B_\circ^2} \int d^3 k \frac{P_{xx}(\mathbf{k})}{A(\mathbf{k}) + (4/3) K_\perp k_\perp^2 + v/\lambda_\parallel}, \qquad (12)$$

where

$$A(\mathbf{k}) = \frac{(v k_\parallel)^2}{3 K_\perp k_\perp^2}. \qquad (13)$$

Suppressing the effects of parallel scattering allows this equation to tend to the field line wandering limit while restricting the power spectra to slab and axisymmetric perpendicular disturbance allows the results of [21] to be recovered. In a 3D turbulence model, the interplay of parallel and perpendicular effects is illustrated by considering the terms after the integral sign.

An alternative approach to incorporating 3D turbulence lies in computations similar to those described in [7]. The authors of [23] suppress variability in one of the three dimensions, so that $\mathbf{B}(x,y)$ is constant in z. Diffusion in z due to field fluctuations in the x-y plane was measured again by numerically tracking particles again in a steady-state experiment while drift in z was measured by following the time evolution of the distribution function. The numerical experiment had deliberately suppressed field line wandering in z. The result was a $K_\perp$ of similar magnitude to the field line wandering estimate. Net drift was consistent with the combined effects of field gradient and curvature drifts due to the x-y plane variability averaged over the one day sample used, calculated on guiding center theory. Our conclusion is

that employment of (9) could result in approximately a factor of two underestimate.

In applying the aforementioned expressions to field data, we employ ACE 4-minute component averages, using 0.71 of a day's data for convenience relating to the Matlab algorithm employed. We obtain the power in a typical perpendicular component by doing fast Fourier transforms to find the power in solar radial, tangential, and normal components as a function of frequency and then finding the sum of resolving each component in a direction perpendicular to the mean field. Power laws are used to fit these summed components. Field magnitudes are found from averages of the 4-minute total field magnitudes because the particles are actually attempting to follow the fluctuation field direction, rather than the mean field over 0.71 of a day. Solar wind data from ACE is used to determine the mean flow velocity of each selected period. Applying the aforementioned results to the field/plasma data of DOY 103, 2006 and the pairs of DOY 117, 118 and 126, 127 yields for 200 MeV protons the results of Table 1. Here ψ is the mean field-radial direction angle. It is noteworthy that the perpendicular power spectra with a negative slopes up to 1.7 are obtained, considerably steeper than for the $n = 1$ assumed in (1). Hence the scattering at high pitch angles is less effective and $K_\parallel^{K,P}$ is likely to be an underestimate.

5. Diffusion Coefficients, the GCR Gradient, and Short-Term Modulation

The maximum GCR drop in intensity during day 103 occurs between 103.6 and 103.8 Assuming that this is due to spatial convection of the feature, as is likely from previous studies of such events [4], the radial spatial gradient of the ACSSAT intensity is 3.43×10^{-14}/cm or 52 percent/AU, corresponding to 200 MeV GCR.

If we assume that a quasi-equilibrium is still maintained between convective outflow, adiabatic energy loss, and diffusive inflow,

$$CVU = -\left(K_\parallel^{cr} \cos^2\chi + K_\perp^{cr} \sin^2\chi \right) \frac{\partial U}{\partial r}, \qquad (14)$$

where C is the Compton-Getting factor appropriate to the differential number density spectrum, U, at 200 MeV and χ is the field-radial direction angle. $C = 0.81$ close to the previous solar minimum, while $\chi = 68°$ on average during the interval of the rapid GCR decrease. We will now separately investigate whether parallel or perpendicular diffusion is the most important in maintaining the assumed quasi-equilibrium with small radial streaming by successively ignoring each term on the right-hand side of the above. Putting in values yields

$$K_\parallel^{cr} = 8.4 \times 10^{21} \text{ cm}^2, \text{ sec}^{-1},$$
$$K_\perp^{cr} = 1.4 \times 10^{21} \text{ cm}^2, \text{ sec}^{-1}. \qquad (15)$$

Similar estimates are made for the other events, days 117/118 and 126/127. These cosmic-ray-derived diffusion coefficients, $K_\parallel^{cr}$ and $K_\perp^{cr}$, are also entered in Table 1. For DOY 103,

TABLE 1: Diffusion coefficients obtained from power spectra and particle gradients in cm^2, sec^{-1}.

	DOY 103	DOY 117	DOY 118	DOY 126	DOY 127
$K_{\parallel}^{K,P}$	6.3×10^{20}	9.66×10^{20}	2.41×10^{20}	1.0×10^{20}	3.40×10^{20}
$K_{\parallel}^{H,W}$	5.5×10^{21}	9.39×10^{21}	2.8×10^{21}	1.7×10^{21}	1.99×10^{21}
$K_{\perp}$	4.1×10^{20}	1.38×10^{20}	1.71×10^{20}	1.54×10^{21}	1.12×10^{21}
ψ	68	24.8	45.5	75.8	45.7
$K_{\parallel}^{cr}$	8.4×10^{21}	1.1×10^{21}	2.19×10^{21}	3.73×10^{22}	7.13×10^{21}
$K_{\perp}^{cr}$	1.4×10^{21}	5.17×10^{21}	2.15×10^{21}	2.39×10^{21}	6.79×10^{21}

and assuming the validity of the power spectral-based estimates, neither the perpendicular diffusion value nor the simple, $K_{\parallel}^{K,P}$ parallel diffusion case can satisfy the quasi-equilibrium. However, because the field-based $K_{\parallel}^{H,W}$ and gradient-based, $K_{\parallel}^{cr}$ estimates are close, parallel diffusion seems to satisfy quasi-equilibrium. For the event of days 117/118 parallel diffusion again seems to give a better agreement between power spectrum calculated and gradient-derived diffusion coefficients, especially as the major part of the observed sharp decrease best corresponds to day 118 power spectral analysis. However for the event of days 126/127, perpendicular diffusion seems to give a little better accord, especially as the major drop lies around the day 126/127 boundary. Note the high ψ angle on day 126 which might indicate the reason for the dominance of perpendicular diffusion. On the other hand, previously noted, marked discrepancy between quasi-linear theory and diffusion computations based on actual field values may be in play here.

Inspection of Figure 8, Section 3.2, shows a rapid field rotation at the time of the rapid decrease with radial component going through zero and with the dominant field component normal to the ecliptic in the event of day 103. It is possible to imagine that nearer the sun the GCR intensity was significantly reduced by a field configuration unfavourable to radial diffusion. This allowed a new quasi-equilibrium to be set up with inflow from a large angle to the ecliptic plane and radial direction from some entry point at the boundary of the traveling large magnetic disturbance. What we see during the three events is the quasi-equilibrium expressed by (14) as it is convected past the Earth. The short-term modulation under consideration here is thus essentially due to a barrier mechanism, because true radial diffusion is largely inhibited during the events.

All three field components exhibit large changes but only one event clearly fits a flux rope model. A bidirectional heat flux persisting through to the end of day 103 has already been noted. We have already discussed the flux rope as a suitable GCR barrier [4, 5], although the detail of the proposed mechanism is different here.

6. Conclusions

We have set out to demonstrate the likely validity of a model for significant short-term reduction of the GCR intensity and the use of this model to obtain information on energetic particle diffusion. Approximate theoretical values of the parallel and perpendicular diffusion coefficients have been obtained appropriate to the actual field turbulence present at the time of passage of three selected events. Reference to previous analytical and Monte Carlo numerical studies taking into account 2- or 3-dimensional field representations, not present in the slab approximation of the adopted calculation of the coefficients, allowed an assessment of errors in our procedure to be made. A sufficiently satisfactory explanation of the three events was provided by the idea of a quasi-equilibrium field/particle structure being convected past the Earth with a single, dominant mode of diffusive propagation, either parallel or perpendicular to the mean field of the event. Reasonable self-consistency between diffusion coefficients derived by particle gradient and analytic theory was achieved. Nevertheless, there is every reason to urge the application of more comprehensive modeling, for example, by the methods of [7, 19, 22], to better check the extent of this agreement.

Three-dimensional models of the overall heliospheric modulation assume diffusive scattering varying smoothly over large spatial scales with at most a dependence on angle with respect to the solar equatorial plane. The relative importance of this, near isotropic and homogeneous turbulence model, compared with barrier effects discussed in this work must depend on the abundance of flux rope magnetic topologies within the solar wind.

In terms of the wider, astrophysical significance, the method proposed here of using local particle conditions to check diffusion coefficients with locally derived plasma parameters seems unprecedented as a way of validating the ability to model cosmic wave-particle interaction in a collisionless regime. One only has to point to the extensive use made of quasi-linear theory for the interaction in nonthermal astrophysical shock modeling to appreciate the worth of the investigation.

Acknowledgments

D. N. A. Shaul acknowledges Science & Technology Facilities Council (STFC) support. T. Mulligan and J. B. Blake acknowledge the support for this work under NASA Grant CG189307NGA. The authors would like to thank the INTEGRAL team for supplying the INTEGRAL data, the ACE MAG and SWEPAM teams for making their data available on the ACE Science Center website (http://www.srl.caltech.edu/ACE/ASC/), and the Wind 3DP team for making their data available on the WIND 3-D Plasma and Energetic Particle Investigation Web site (http://sprg.ssl.berkeley.edu/wind3dp/esahome.html).

References

[1] J. A. Lockwood, "Forbush decreases in the cosmic radiation," *Space Science Reviews*, vol. 12, no. 5, pp. 658–715, 1971.

[2] H. V. Cane, "Coronal mass ejections and forbush decreases," *Space Science Reviews*, vol. 93, no. 1-2, pp. 55–77, 2000.

[3] I. G. Richardson, "Energetic particles and corotating interaction regions in the solar wind," *Space Science Reviews*, vol. 111, no. 3-4, pp. 267–376, 2004.

[4] T. Mulligan, J. B. Blake, D. Shaul et al., "Short-period variability in the galactic cosmic ray intensity: high statistical resolution observations and interpretation around the time of a Forbush decrease in August 2006," *Journal of Geophysical Research A*, vol. 114, no. 7, Article ID A07105, 2009.

[5] J. J. Quenby, T. Mulligan, J. B. Blake, J. E. Mazur, and D. Shaul, "Local and nonlocal geometry of interplanetary coronal mass ejections: galactic cosmic ray (GCR) short-period variations and magnetic field modeling," *Journal of Geophysical Research*, vol. 113, no. A10, 2008.

[6] J. R. Jokipii, "Cosmic-ray propagation. I. Charged particles in a random magnetic field," *The Astrophysical Journal*, vol. 146, p. 480, 1966.

[7] J. F. Valdés-Galicia, G. Wibberenz, J. J. Quenby, X. Moussas, G. Green, and F. M. Neubauer, "Pitch angle scattering of solar particles: comparison of "particle" and "field" approach—I. Strong scattering," *Solar Physics*, vol. 117, no. 1, pp. 135–156, 1988.

[8] U. D. Langer, M. S. Potgieter, and W. R. Webber, "Modelling of "barrier" modulation for cosmic ray protons in the outer heliosphere," *Advances in Space Research*, vol. 34, no. 1, pp. 138–143, 2004.

[9] F. McDonald, Z. Fujii, P. Ferrando et al., "The cosmic ray radial and latitudinal intensity gradients in the inner and outer heliosphere 1996-2001.3," in *Proceedings of the 27th International Cosmic Ray Conference (ICRC'01)*, vol. 10, p. 3906, Hamburg, Germany, August 2001.

[10] G. Vedrenne, J. P. Roques, V. Schönfelder et al., "SPI: the spectrometer aboard INTEGRAL," *Astronomy and Astrophysics*, vol. 411, no. 1, pp. L63–L70, 2003.

[11] P. Jean, G. Vedrenne, J. P. Roques et al., "SPI instrumental background characteristics," *Astronomy and Astrophysics*, vol. 411, no. 1, pp. L107–L112, 2003.

[12] B. J. Teegarden, J. Naya, H. Seifert et al., "SPI: a high resolution imaging spectrometer for INTEGRAL," in *Proceedings of the 4th Compton Symposium*, C. D. Dermer, M. S. Strickman, and J. D. Kurfess, Eds., vol. 410 of *AIP*, 1997.

[13] K. Kudela and R. Brenkus, "Cosmic ray decreases and geomagnetic activity: list of events 1982–2002," *Journal of Atmospheric and Solar-Terrestrial Physics*, vol. 66, no. 13-14, pp. 1121–1126, 2004.

[14] C. F. Kennel, H. Petscheck, and H. E, "Limit on stably trapped particle fluxes," *Journal of Geophysical Research*, vol. 71, no. 1, pp. 1–28, 1966.

[15] A. Teufel and R. Schlickeiser, "Analytic calculation of the parallel mean free path of heliospheric cosmic rays: I. Dynamical magnetic slab turbulence and random sweeping slab turbulence," *Astronomy and Astrophysics*, vol. 393, no. 2, pp. 703–715, 2002.

[16] K. Hasselmann and G. Wibberenz, "A note on the parallel diffusion coefficient," *The Astrophysical Journal*, vol. 162, p. 1049, 1970.

[17] J. J. Quenby, G. E. Morfill, and A. C. Durney, "The solar proton diffusion mean free path and the anisotropic particle event of November 18, 1968," *Journal of Geophysical Research*, vol. 79, no. 1, pp. 9–16, 1974.

[18] M. L. Goldstein, "The mean free path of low-rigidity cosmic rays," *Journal of Geophysical Research*, vol. 85, no. A6, pp. 3033–3036, 1980.

[19] A. Shalchi, J. W. Bieber, W. H. Matthaeus, and R. Schlickeiser, "Parallel and perpendicular transport of heliospheric cosmic rays in an improved dynamical turbulence model," *Astrophysical Journal*, vol. 642, no. 1, pp. 230–243, 2006.

[20] M. A. Foreman, J. R. Jokippi, and A. J. Owens, "Cosmic-ray streaming perpendicular to the mean magnetic field," *The Astrophysical Journal*, vol. 192, pp. 535–540, 1974.

[21] W. H. Matthaeus, G. Qin, J. W. Bieber, and G. P. Zank, "Nonlinear collisionless perpendicular diffusion of charged particles," *Astrophysical Journal Letters*, vol. 590, no. 1, pp. L53–L56, 2003.

[22] A. Shalchi, "A unified particle diffusion theory for cross-field scattering: subdiffusion, recovery of diffusion, and diffusion in three-dimensional turbulence," *Astrophysical Journal Letters*, vol. 720, no. 2, pp. L127–L130, 2010.

[23] X. Moussas, J. J. Quenby, and J. H. Valdes-Galicia, "Drift motion and perpendicular diffusion of energetic particles in interplanetary space based on spacecraft data," *Astrophysics and Space Science*, vol. 86, no. 1, pp. 197–207, 1982.

11

M94 as a Unique Testbed for Black Hole Mass Estimates and AGN Activity at Low Luminosities

Anca Constantin[1] and Anil C. Seth[2]

[1] Department of Physics and Astronomy, James Madison University, Harrisonburg, VA 22807, USA
[2] Department of Physics and Astronomy, University of Utah, Salt Lake City, UT 84112, USA

Correspondence should be addressed to Anca Constantin, constaax@jmu.edu

Academic Editor: Antonis Georgakakis

We discuss the peculiar nature of the nucleus of M94 (NGC 4736) in the context of new measurements of the broad Hα emission from *HST*-STIS observations. We show that this component is unambiguously associated with the high-resolution X-ray, radio, and variable UV sources detected at the optical nucleus of this galaxy. These multiwavelength observations suggest that NGC 4736 is one of the least luminous broad-line (type 1) LINERs, with $L_{bol} = 2.5 \times 10^{40}$ erg s^{-1}. This LINER galaxy has also possibly the least luminous broad-line region known ($L_{Hα} = 2.2 \times 10^{37}$ erg s^{-1}). We compare black hole mass estimates of this system to the recently measured $\sim 7 \times 10^{6} M_{\odot}$ dynamical black hole mass measurement. The fundamental plane and M-σ^* relationship roughly agree with the measured black hole mass, while other accretion-based estimates (the M-FWHM(Hα) relation, empirical correlation of BH mass with high-ionization mid-IR emission lines, and the X-ray excess variance) provide much lower estimates ($\sim 10^{3} M_{\odot}$). An energy budget test shows that the AGN in this system may be deficient in ionizing radiation relative to the observed emission-line activity. This deficiency may result from source variability or the superposition of multiple sources including supernovae.

1. Introduction: LINERs and M94

Most high-mass galaxies are known to host massive black holes (BH), some passively lurking in their centers while others are actively accreting surrounding material (e.g., [1] and references therein). The mechanism that causes a BH's activity to turn on and off is still largely unknown. Understanding the structure of the active galactic nuclei (AGN) at their lowest luminosities is crucial to determining the physical and possibly evolutionary links between the most luminous galaxy centers and the passive ones. However, at low luminosities it is difficult to disentangle the various emission mechanisms that could be concurrently present in galaxy centers. As a consequence, the dominant power source of a large majority of actively line-emitting galaxies remains ambiguous (e.g., [2] for a review).

Diagnostic diagrams (e.g., [3–6]) are relatively successful in separating out bona fide accretion sources (Seyferts) from nuclei whose emission-line activity is mainly powered by young, hot stars (H II galaxies), based on emission line ratios.

At least 50% of the strong line emitters fall easily onto the H II locus; however, only less than 10% are of the Seyfert type [5, 7, 8]. A large fraction of the objects "in between" these two categories, that exhibit relatively low levels of ionization (i.e., low values of [O III]/Hβ), maintain reasonably strong forbidden line activity (i.e., high values of [N II]/Hα) and are classified as low-ionization nuclear emission regions (LINERs). The reminder are usually called transition objects (Ts). Whether Ts and Ls are powered, at least partly, by accreting BHs, and thus could be called AGN, is a matter of continuous debate [2].

There are some typical emission characteristics that are considered to be particularly good indications that accretion onto a massive BH is an important, if not the dominant source of ionization in some Ls, and possibly Ts. The detection of broad Hα emission, regardless of its strength or luminosity, is generally considered to be *the* clue to AGN emission. Some LINERs (and maybe Ts as well) exhibit these features; however, the majority of these systems show only narrow emission, which could be generated by shocks, poststarbursts, or other processes unrelated with accretion. Observations outside of the optical wavelengths often reveal AGN signatures in ambiguous and even starburst galaxy nuclei. X-rays are particularly good tracers of accretion; however, they are not efficient in distinguishing AGN at $L_X \lesssim 10^{42}$ erg s^{-1}, where contamination by X-ray binaries can be significant. X-rays are also unlikely to detect heavily absorbed AGN (i.e., Compton thick; $N_H > 1.5 \times 10^{24}$ cm^{-2}). On the other hand, mid-IR high-ionization emission lines like [Ne V] $\lambda 14.32 \mu m$, $24.32 \mu m$ (97.1 eV) appear to be a trustworthy indicator of AGN activity [9–11] due to the extreme conditions (i.e., very hard ionizing radiation) required to produce them; because these features have considerably lower optical depth, their detection can also reveal Compton-thick AGN ([12], e.g., NGC 1068). This technique has now been applied to reveal new and large numbers of optically unidentified AGN [13, 14], providing thus sensitive improvements on previous AGN censuses. These studies, along with X-ray and radio studies of nearby galaxies (e.g., [15–17]). suggest that a majority of LINERs and a large fraction of transition galaxies might in fact host accreting black holes. The presence of an accreting black hole does not however guarantee that the accretion power is the dominant source of ionization of those galaxy nuclei. A more recent assessment of the energy budget of LINERs by Eracleous et al. [18] argues that in 85% of LINERs the AGN ionizing photons are not sufficient for producing the observed nebular emission, and thus other power sources are likely to dominate.

Some new potentially powerful insights into the excitation mechanism of the low-luminosity AGN (LLAGN), and in particular the ambiguous sources, come from recent studies of large statistical samples of nearby galaxy nuclei, which reveal a potential H II → S → T → LINER → *Passive Galaxies* evolutionary sequence in the process of BH growth within galaxies [19, 20]. This sequence traces trends in (1) increasing host halo mass, (2) increasing environmental density, (3) increasing central BH mass and host stellar mass, (4) decreasing BH accretion rate, (5) aging of the stellar population associated with their nuclei, and (6) decreasing in the amount of dust obscuration, which might translate into a decrease in the amount of material available for star forming or accretion. In this picture, Seyferts and Ts are transition phases between the initial onset of accretion, usually swamped by the star-forming gas and associated dust, which is seen optically as an H II system, and the final phase of accretion observed as LINERs of already massive BHs. While this idea is supported by various other independent observational studies of low-luminosity AGN and starburst galaxies [21, 22], along with state-of-the-art hydrodynamical models for the life cycles of the most luminous AGN (e.g., [23, 24]) it is very probable that not *all* sources fit into this scenario. It is also important that this evolutionary sequence idea is tested on samples that span a narrow distribution in Hubble types; because most of the H IIs and Ts are relatively late-type disk galaxies that, likely, never experienced a (recent) major merger, it is possible that the trigger of such a sequence is different from the merger that initiates a similar life cycle at high luminosities. Nevertheless,

investigating the challenges certain objects bring to this idea is useful for identifying and quantifying the caveats associated with this sequence. These objects may also be at interesting stages in their galaxy evolution.

NGC 4736 (or M94, UGC 7996) is a captivating example of an ambiguous galaxy nucleus, which poses challenges to the general understanding of AGN phenomena, including the above-mentioned sequence. This object is one of the closest ($d = 4.3$ Mpc; [25]) nearly face-on spiral, with a SAab Hubble Type. Its proximity enables study of details that would be unobservable in more distant systems. Its nucleus has a low-luminosity LINER spectrum, but has been also included in catalogs of transition objects [26] or Seyfert 2s (e.g., [27]). The AGN nature of this object has been constantly debated, an aging starburst being a compelling alternative [28–31]. The galaxy presents a ring of H II regions at a radius of ~50″, red arcs at ~ 15″, a high-surface-brightness nuclear region, and high far-infrared bulge emission [32, 33]. Its intricate structure of off-nuclear compact source detections in X-ray [30], radio [31], and UV [29], that do not necessarily match with each other, certainly increases the ambiguity associated with the nature of the main nuclear ionization mechanism. A common implied scenario in all of these studies is that this system is probably in the final stages of a merger.

We reexamine here NGC 4736 in the context of additional evidence for its AGN nature, which is the detection of a broad Hα component in its nuclear spectrum, as observed by the *Hubble Space Telescope* (*HST*) with the Space Telescope Imaging Spectrograph (STIS). We also gather multiwavelength data and show that source of the broad H-alpha emission line is coincident with a compact X-ray and radio source. These observations suggest that NGC 4736 hosts a broad-line region of significantly low luminosity, which makes this object one of the least luminous LINERs with strong evidence for BH accretion. Interestingly, BH mass indicators calibrated on rapidly accreting Seyfert galaxies give highly discrepant mass estimates in this more quiescent system. The $\sim 10^7 M_{\odot}$ value given by the M-σ^* is two orders of magnitude higher than the values obtained via estimators based on the observed emission (X-ray variability, scaling relations, mid-IR emission) which, although physically independent of each other, give a consistent result of $\sim 10^5 M_{\odot}$. We are thus facing the following conundrum: either (a) the standard AGN BH mass indicators do not necessarily apply to sources emitting in this low-luminosity regime as, probably, the emission mechanism is fundamentally different from that associated with higher-luminosity AGN, or (b) the emission signatures do not trace accretion onto the central BH. We discuss possible resolutions of this discrepancy in Section 5, where we propose some rather exotic scenarios.

2. Data Compilation and Analysis

NGC 4736 has been quite extensively observed across the whole electromagnetic spectrum. We present in this section the multiwavelength observations of this galaxy nucleus, in connection to new measurements of the broad Hα emission detected with *HST*-STIS.

2.1. The Broad Hα Emission Observed with HST-STIS. High-resolution optical spectra of the NGC 4736 nucleus were obtained with *HST*-STIS on July 2002. Data are publicly available (Prop ID 8591) but have not been published. The observations were obtained with the $52'' \times 0.''1$ aperture oriented at PA = 49.65°, with the slit centered along the major axis of the starlight distribution; two cosmic-ray split exposures were obtained, one being slightly shifted in the slit direction. The total combined exposure time is ~4000 s. The G750M grating was set at 6581 Å, with a scale of $0.''05$/pixel, with no binning. We reduced the spectra using IRAF (IRAF is distributed by NOAO, which is operated by AURA Inc., under contract with the National Science Foundation) and the STIS reduction pipeline maintained by the Space Telescope Science Institute [34]. This reduction included image combination and cosmic-ray rejection, flux calibration, and correction of the wavelength to the heliocentric frame. To measure the nuclear nebular line-emission properties we extracted the 1-dimensional aperture spectrum five pixels wide ($0.''25$) centered on the continuum peak. The extracted spectrum thus consists of the central emission convolved with the STIS spatial point-spread function (PSF) and sampled over a rectangular aperture of $0.''25 \times 0.''1$. The measurements span 6295–6867 Å with a resolution of 0.87 Å ($\sigma_{inst} = 17 \, \text{km s}^{-1}$).

Figure 1 shows the resulting spectrum together with the best matching continuum model, spectral fits of the pure emission-line component, and the associated residuals. The continuum model of the underlying stellar population is obtained via a χ^2 minimization of a nonnegative least-squares fit between the observed spectrum and a sum of discrete star bursts of different ages, adopted from the Bruzual and Charlot [35] stellar population synthesis templates, together with dust attenuation modeled as an additional free parameter. The continuum fitting is performed using an adaptation to our data of Christy Tremonti's code [36]. The pure emission-line spectrum, obtained by subtracting the modeled continuum from the observed spectrum, is fit by a combination of linear continuum and Gaussian components. In addition to narrow emission, the Hα + [N II] feature shows clear evidence for a broader feature. A flux ratio of 1 : 3 was assumed for the [N II] doublet, as dictated by the branching ratio [37]; the [O I] feature and the [N II] and [S II] doublets were assumed to share common velocity (red) shifts and widths. The best-fitting Gaussian parameters were derived via an interactive χ^2 minimization, using SPECFIT [38]. Because of the generally low signal to noise of the 2-d spectrum we are not able to test whether our measurements of the broad Hα feature could be corrupted by a possible rotating disk as in the case of M84 (e.g., [39]); the match in the widths of the [S II] and [N II] narrow features (as long as that of the narrow Hα) argue however against such a significant effect.

Measurements related to the broad Hα feature from the STIS measurements are listed in Table 1. The fractional contribution of this broad component to the total flux of the Hα + [N II] blend is 80%. Compared to other galaxies with broad-line emission in the Palomar survey [40], the width of the broad Hα is typical (1570 km s^{-1}). However,

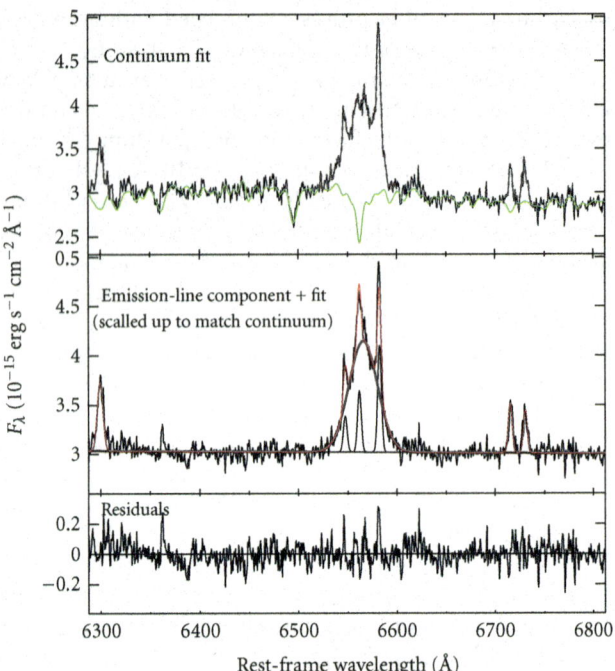

FIGURE 1: Spectral fits to the *HST*-STIS spectrum of NGC 4736, based on a 5-pixel extraction aperture, which corresponds to a total emission area of 0.025 arcsec². *Upper panel*: The thin continuous line shows the observed data, while the green line shows the best-fit stellar population model obtained by clipping out the hashed emission-line region. *Middle panel*: The continuum subtracted spectrum, that is, the emission-line component, together with the corresponding spectral fit, that have been shifted by an additive constant to match the initial continuum level, for illustration purposes. The thin continuous lines show the individual Gaussian components (1 per narrow line), the thick grey continuous line represents the broad Hα feature, and the red line is the final fit to the observed spectrum. *Lower panel*: The residuals after the subtraction of the model fit.

TABLE 1: Broad Hα emission measurements.

Quantity	Value	Notes
f_{blend}	0.80	Fraction of broad Hα to Hα + [N II] blend.
$f_{Hα}$	0.93	Fraction of broad Hα to total Hα
FWHM(Hα^{broad})	1570	±110; in km s^{-1}
Δv	140	±20; in km s^{-1}; broad relative to narrow Hα
$\log F(Hα^{broad})$	−13.38	Observed flux (erg s^{-1}cm^{-2})
$\log L(Hα^{broad})$	37.96	Observed luminosity (erg s^{-1})[a]

[a]assuming distance d = 4.3 Mpc.

at just 9.1×10^{37} ergs s^{-1}, the Hα luminosity is lower than any known broad-line sources in the Palomar sample or the Sloan Digital Sky Survey [40, 41]. This includes the famous low luminosity Seyfert 1 galaxy, NGC 4395, which has a luminosity of 1.2×10^{38} ergs s^{-1} [42] and is at a similar distance to NGC 4736. Note also that the broad Hα feature is redshifted by ~140 ± 20 km s^{-1} relative to the narrow emission, that is locked at the systemic velocity of the object; velocity offsets between the broad Hα and narrow lines are not unusual [43].

Evidence for broad Hα emission has also been presented in a recent PCA tomography study applied to this nucleus [44]. The observations reported in this case come from the Gemini Multi Object Spectrograph (GMOS)-IFU data cube and have been obtained 4 years after the *HST*-STIS spectrum had been acquired. Thus, the broad Hα line associated with the nucleus of M94 appears to be persistent for at least 4 years. They derive a broad Hα luminosity of 6×10^{38} ergs s^{-1}, $\gtrsim 6$ times brighter in the GMOS data than the one we detect in the STIS spectrum. The difference could result from true variability in the broad-line luminosity or could be due to differences in aperture size or placement. We note that Steiner et al. [44] suggest the broad-line region is offset from the photocenter of the galaxy by 0.″15, while our spectral 0.″25 × 0.″1 aperture coincides with the photocenter. Even with their higher Hα luminosity, this source is among the least luminous known broad-line AGN.

2.2. Ground-Based Optical Spectroscopy and the Spectral Classification.

Ground-based optical spectra of the NGC 4736 nucleus are available from the Palomar spectroscopic survey by Ho et al. [43] and the integrated spectrophotometric survey of Moustakas and Kennicutt [45] conducted with the 2.3 m Bok telescope, as well as from new data we acquired on February 2008 at the MMT Observatory. The MMT spectrum is obtained with the Blue Chanel Spectrograph, the 500 grooves/mm grating used in first order with the 1″ slit, and covers λλ3800–7000 with 3.6 Å resolution. None of these data show indications of broad Hα emission. This outcome is not surprising given the feature's flux in the *HST* spectrum, which would be very difficult to discern in the 1″ or 2″ apertures employed in these ground-based observations, which are at least one order of magnitude larger than that used in the *HST*-STIS observations. Through a 2″ aperture (~40 pc), NGC 4736's emission complex Hα + [N II] is generally heavily swamped by the host stellar light.

Probably as expected, measurements of NGC 4736's narrow-line emission are consistent with this system being powered at least partially by a nonstellar source, regardless of the resolution of its observation. Figure 2 shows the location of this nucleus in the 3-dimensional diagnostic diagram usually employed in classifying emission-line galaxies (e.g., [3, 4, 6]), for different sets of data mapping ~10 to ~100 pc. As with the *HST*-STIS observations, the ground-based measurements are performed after the host stellar contribution has been subtracted from the observed spectrum by means of absorption galaxy template fits. Our MMT spectrum offers only an upper limit for the [O III]/Hβ while the Palomar line flux measurements are at least 50% uncertain. We list the measurements of the nebular emission of this nucleus as observed by Palomar, Bok, MMTO, and *HST*-STIS in Table 2. Because measurements of the [O III]/Hβ ratio are not available in the *HST* spectrum, we show the high-resolution measurements in Figure 2 using the [O III]/Hβ from the Palomar catalog; it is readily apparent that the high-resolution data are consistent with a LINER or a Seyfert

TABLE 2: Emission line measurements[a].

line name	*HST*-STIS	Palomar[b]	Bok[c]	MMTO[d]
[O I] λ6300	5.03 ± 0.31	6.03	2.8 ± 0.9	7.0 ± 1.0
Hα (narrow)	3.00 ± 0.32	> 25.12	13.5 ± 1.9	22.0 ± 2.4
[N II] λ6583	5.25 ± 0.25	54.01	21.5 ± 1.5	39.3 ± 2.9
[S II] λ6716	2.48 ± 0.18	18.59	8.3 ± 1.1	15.5 ± 0.3
[S II] λ6731	2.16 ± 0.18	16.33	7.2 ± 1.1	9.8 ± 0.3

[a]All fluxes are in units of 10^{-15} erg s^{-1} cm^{-2} and represent the observed values, not corrected for reddening.
[b]nonphotometric conditions; line ratios are at least 50% uncertain.
[c]2.″5 slit, 2.3 m telescope; photometric conditions [45].
[d]1″ slit; fluxes are in units of 10^{-15} erg s^{-1} cm^{-2}.

classification for this object and reveal its type 1 (broad-lined) character.

2.3. The Radio, UV, and X-Ray Observations and the Astrometric Coincidence with the Broad Hα Emission.

Across the electromagnetic spectrum, the center of this galaxy exhibits a complex morphology. There is a plethora of bright radio, UV, and X-ray sources detected in the center of NGC 4736. There is a nuclear compact (15 GHz) radio source measured by Nagar et al. [46] that appears to be associated with that of the brightest of the two close (8.49 GHz) radio sources revealed by Körding et al. [31], and with that of the brightest of the two ultraviolet point sources, which varies on a 10-year time scale [29]. *Chandra* observations [30] reveal numerous discrete X-ray sources in the inner galaxy; the second brightest X-ray source, X2 ($L_{X,2-10\,keV} = 5.9 \times 10^{38}$ erg s^{-1}), coincides within the errors with the nucleus position. The off-nuclear radio, UV, and X-ray sources are apparently unrelated to each other.

The relation to possible optical counterparts of these observations is not well constrained in the literature. We thus investigated this issue, with a particular interest in the degree to which our newly detected broad Hα coincides in position with the multi-λ detections. We used for this purpose a variety of archival images from the *HST*: WFPC2/PC data in the F555W filter data (PID: 5741 and 10402), HRC data in the F250W and F330W filters (PID: 9454), and NICMOS/NIC3 data in the F160W filter (PID: 9360). All data were downloaded from the *HST* archive, and images were drizzled together when required. Absolute astrometry was performed on these data as well as the STIS observations (taken with the slit out) using a ground-based V-band image obtained from the *Spitzer Infrared Nearby Galaxies Survey* (SINGS) ancillary data by Kennicutt et al. [48]. These SINGS observations were aligned to the USNO-B system using ~80 stars, and then astrometry of the *HST* data was obtained by degrading the resolution of the F555W image to match the SINGS image. All other *HST* images were then matched with the astrometry-corrected F555W frame to an accuracy of < 0.″05. The absolute error on this astrometry is about 0.″2 and is dominated by scatter of stars in the SINGS image relative to the USNO-B positions.

Figure 3 illustrates the result of this data compilation and the corresponding radio, UV, and X-ray source matches. *It is*

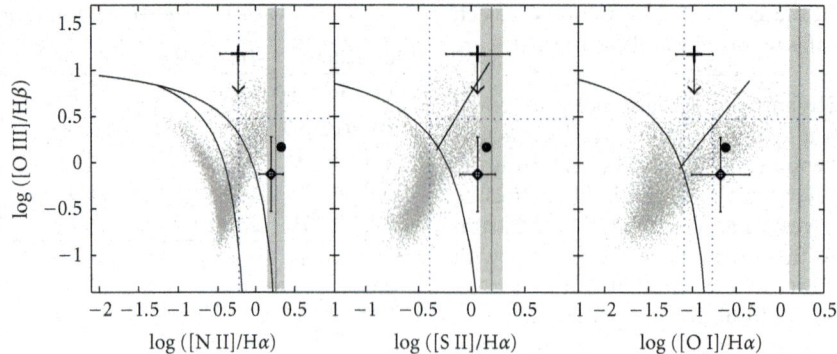

FIGURE 2: NGC 4736 within emission-line diagnostic diagrams. Filled circles reflect measurements based on the Palomar observations. Diamonds represent measurements from Moustakas and Kennicutt Jr. [45], while crosses show measurements from our MMT spectrum. The vertical bands indicate the *HST*-STIS measurements and the associated errors (with no correction for reddening; [O III] and Hβ are not available in the *HST* spectra). The solid (black) curves indicate the Kewley et al. [6] classification, while the dotted (blue) lines indicate criteria used by Ho et al. [43]. The background grey points correspond to measurements of SDSS nearby galaxies from Constantin et al. [19].

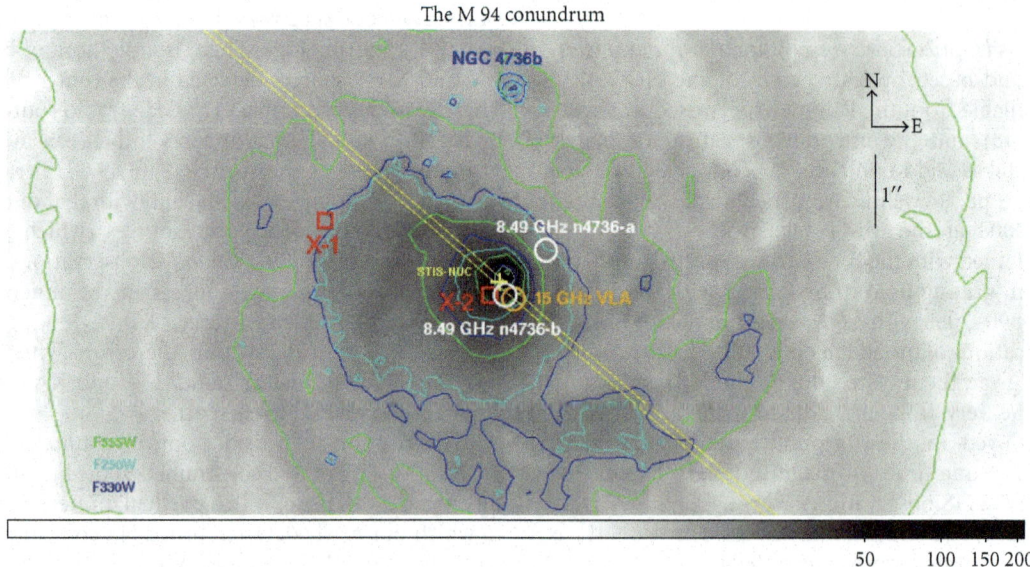

FIGURE 3: The mosaic-ed F555W (V-band; combination of 5 WFPC2 images taken at two different epochs) image of the NGC4736 nucleus, with intensity contours overlaid (green). The position of the STIS-nucleus is indicated (by the yellow slit) along with the locations of the 2 most luminous and closest to the nucleus hard X-ray sources, X1 and X2, as detected with Chandra (red squares), along with the position of the 15 GHz VLA radio core ([46], orange circle), and those of the two radio compact sources detected by Körding et al. [31] (white circles). The blue and cyan contours correspond to the F330W and F250W observations from Maoz et al. [29]; note the presence of NGC 4736b, $\approx 2.''5$ to the north of the nucleus. The absolute astrometry is good to within $0.''2$ (pixel size is $0.''05$ for F555W, and $0.''025$ for F250W and F330W).

clear that, within $< 0.''2$ *(4 pc), there is an obvious astrometric match in the nuclear X-ray, UV, optical, radio compact sources, and the newly detected broad Hα emission line.* The nucleus position as observed by STIS has RA: 12 h 50 m 53 s.20, DEC: 41°07′13.″40. The off-nuclear X1, that is the brightest compact X-ray source detected in this galaxy nucleus, has no counterpart at other wavelengths. Same is true for the off-nuclear 8.49 GHz detection, that is only 1″ (~20 pc) away from the nucleus. The off-nuclear UV source is the only one detected in optical light.

3. NGC 4736's Nuclear Emission across the Electromagnetic Spectrum and L_{Bol}

With the wealth of data available for this galaxy nucleus, we are able to build its least contaminated X-ray to radio nuclear spectral energy distribution (SED). The multiwavelength observations of the sources detected in the very central regions of NGC 4736, plotted in $\nu L\nu$ units, are displayed in Figure 4. The X-ray detection X2 is represented here by a power law, estimated based on its photon indices $\Gamma = 1.6$,

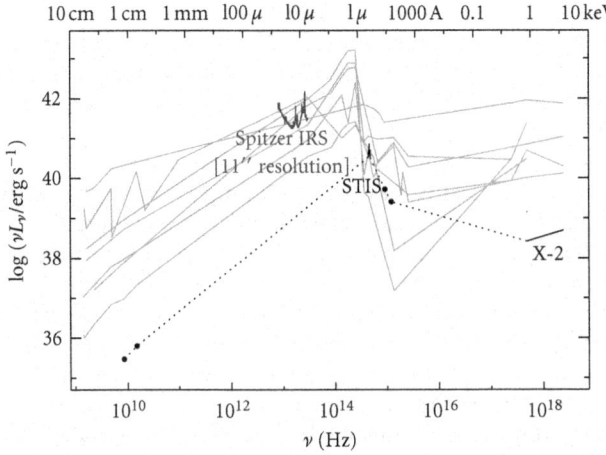

TABLE 3: Nuclear SED data.

log(ν/Hz)	log(νL_ν/erg/s)	Resolution	Obs. date	Instr.
18.38	38.69	0.″15	5/2000	Chandra-ACIS
17.68	38.42	0.″15	5/2000	Chandra-ACIS
15.08	39.41	0.″5	6/2003	ACS F250W
14.96	39.72	0.″5	6/2003	ACS F330W
14.68	39.91	0.″1	6/2002	HST-STIS
14.64	39.93	0.″1	6/2002	HST-STIS
13.48	40.58	4.″7	5/2004	IRS-SH
13.19	40.33	11.″1	5/2004	IRS-SH/LH
12.91	40.87	11.″1	5/2004	IRS-LH
10.18	35.79	0.″15	1/2001	VLA, A config.
9.93	35.47	0.″24	6–10/2003	VLA, A config.

FIGURE 4: Radio to X-ray nuclear SED of NGC 4736 superposed on SEDs of LLAGN from Ho [47]. The optical STIS spectrum stands out as an amorphous blob featuring the strong Hα emission feature. The 2–10 keV Chandra X-ray detection X2 is depicted as a power-law corresponding to Γ = 1.6. The low spatial resolution Spitzer IRS observations are shown for comparison.

where the absorbing column density is fixed at the Galactic value. Both F250W and F330W UV observations are plotted, only for the nuclear detections. The optical data are represented by the HST-STIS spectrum, featuring the strong Hα emission line. Measurements corresponding to all the radio observations discussed above are indicated. For the sake of completeness, we also include in this plot lower spatial-resolution observations from SINGS IRS [48] as well. We show all of these measurements superposed on the SEDs of LLAGN from Ho [47]. No artificial normalization has been performed, and there is no correction for absorption in either the Galactic extinction (except for the X-ray data) or intrinsic to NGC 4736.

With X2 and n4736-b as the X-ray and the 8.5 GHz counterparts of the broad Hα detection, respectively and assuming that the continuum could be described as simple power laws between the points we present data for, this object's nuclear SED corresponds to $L_{bol} \approx 2.5 \times 10^{40}$ erg s^{-1}. Spitzer data is not included in estimating L_{bol} because they do not reflect the IR emission of the uncontaminated nucleus; the aperture used in these observations is very large (see Table 3), including the entire field of view shown in Figure 3. This luminosity makes NGC 4736 one of the least luminous LINERs with strong evidence of BH accretion. This source is thus a critical signpost of BH accretion at extremely low levels. Note that this object's weak emission is most likely not caused by obscuration; Ho et al. [43] list a Balmer decrement of Hα/Hβ = 3.1 (albeit highly uncertain, with a probable error of ± 100%), and Eracleous et al. [30] provide an upper limit for the neutral hydrogen column density of $N_H < 3.3 \times 10^{20}$ cm^{-2}.

3.1. Comparison with Other Observed LLAGN. Figure 4 shows that NGC 4736's SED is very similar to those of other LLAGN [47]. This suggests that, in spite of the significant

difference in L_{bol}, there is no fundamental transition in the accretion mode in this source compared to other LLAGN.

The similarity of the SED means that previously proposed bolometric corrections for LLAGN appear to work quite well for NGC 4736. The value estimated based on the correction to the [O III] line luminosity that Heckman et al. [49] proposed to work well for Seyferts is $L_{bol} \approx 3500 \times L_{[O iii]} = 4.5 \times 10^{40}$ erg s^{-1} (here $L_{[O III]}$ is measured from the ground within the Palomar survey and is not corrected for intrinsic reddening). If the more recent assessment of the $L_{bol} = 600 \times L_{[O III]}$ bolometric correction of kauffmann and Heckman [50] is used, with $L_{[O III]}$ corrected for reddening based on the Balmer decrement listed above and a $\tau \propto \lambda^{-0.7}$ attenuation law [51], then $L_{bol} = 0.93 \times 10^{40}$ erg s^{-1}, which compares satisfactorily with the measured L_{bol}. The average bolometric correction to the observed 2–10 keV X-ray luminosity proposed by Ho [2], with $L/L_X = 16$, results in a somewhat lower value, 0.94×10^{40} erg s^{-1}; with the more recent $L/L_X = 50$ average bolometric correction to the 2–10 keV luminosity of Eracleous et al. [52], $L_{bol} = 1.8 \times 10^{40}$ erg s^{-1} and, thus, very close to our integrated value. Within a typical uncertainty of ~50%, all of these estimates are consistent with the measured L_{bol} value. Other, more uncertain bolometric indicators are also consistent with our integration: the correlation between the mid-IR [Ne V] emission-line luminosity and bolometric luminosity derived by Satyapal et al. [53] for a small sample of much brighter nearby AGN, that has a large scatter (~1 order of magnitude), gives a $L_{bol} = 1.6 \times 10^{40}$ erg s^{-1}, which is very close to our measured value.

This good match among quite a variety of bolometric estimators derived independently from the multiwavelength properties of AGN supports the AGN interpretation for this system. However, there are two details of its emission spectrum which are unlike typical broad-lined LLAGN.

First, despite the fact that the broad Hα emission of NGC 4736 is one of the weakest measured among type 1 AGN, the total Hα emission is peculiarly strong relative to the X-ray counterpart. The ratio $L_X(2–10 \text{ keV})/L_{H\alpha}$ is ~6, and thus lower than the median of ~15 exhibited by the type 1 AGN (and low-z quasars) included in, for example, Ho et al. [54] study that revealed a relatively tight correlation between the two types of emission. With the higher broad Hα flux from Steiner et al. [44], the ratio is even lower, $L_X(2\text{--}10\,\text{keV})/L_{\text{H}\alpha} \approx 1$. This finding is surprising given that the low-luminosity nearby galaxies that deviate from the $L_X - L_{\text{H}\alpha}$ linear scaling toward lower values of the $L_X/L_{\text{H}\alpha}$ ratios are the type 2 sources, mostly transition objects, where the ionization mechanism is not necessarily dominated by an AGN type of source. Following, for example, Ho et al. [54] arguments, the unusually low $L_X/L_{\text{H}\alpha}$ ratio measured in the nucleus of NGC 4736 suggests that either: (1) the optical line emission is not powered exclusively by a central AGN, or (2) the X-ray emission in this system arises, at least partially, from a non-AGN source, for example, an X-ray binary. A final possibility is that the source is highly variable (as suggested by the UV, X-ray, and Hα observations), and thus, the unusual $L_X/L_{\text{H}\alpha}$ is the result of measurements made at different times.

Second, NGC 4736's nebular emission shows a number of peculiarities. To start with, the electron density in NGC 4736's nebular emission is very low, and it does not show the typical gradient exhibited by AGN, that is, increase toward the more nuclear regions (Constantin et al. 2012, in prep.). In both the Palomar and HST-STIS observations, the ratio [S II] $\lambda\lambda$ 6716/6731 is ~1.26 (Table 2), implying an electron density of $n_e \sim 10^2\,\text{cm}^{-3}$, which lies at the lowest end in the distribution of particle densities measured in the Palomar objects [55].

Given the lack of a density gradient in the line-emitting region it is then rather peculiar to observe a significant increase (by a factor of ~6) in the [O I]/Hα line flux ratio in the HST spectrum relative to the large aperture observations, which would usually be interpreted as an indication for a more pronounced AGN-like ionization in the more central regions. There are two reasons for this: (1) [O I] requires a significantly hard radiation field, that is, that of an AGN, to sustain a sufficiently extensive partially ionized zone in clouds optically thick to Lyman continuum and thus to produce such a strong feature. Since the ionization potential of [O I] matches that of H very well, large differences in the [O I]/Hα ratios are expected between accretion and nonaccretion sources. (2) given the lack of a density stratification in this nucleus, the degree of ionization of the emitting gas is expected to diminish with radius as the ionizing radiation emerging from a nuclear source falls off in density as r^{-2} and thus produce strong gradients in the [O I]/Hα ratio.

In summary, NGC 4736 may be a broad-lined AGN, with an SED similar to other LLAGN, but it is atypical in several aspects whose physical origin remains unclear. Given the above-listed possible explanations for the peculiar $L_X/L_{\text{H}\alpha}$ ratio, we evaluate in the next subsection whether photoionization by the weak AGN in this system is sufficiently powerful to balance the emission cooling in this system.

3.2. Comparison of Ionizing and Emission-Line Power. The multiwavelength observations of this system allow for a relatively rigorous assessment of whether the photoionization by this system's AGN can power the measured emission-line

luminosities and in particular that of the broad Hα component. Following Eracleous et al. [18], we can run an energy budget test via a direct comparison of the Hα luminosity and count rate with the ionizing luminosity $L_i = L_{1\,\text{Ry}-100\,\text{keV}}$ and the ionizing photon rate $Q_i = Q_{1\,\text{Ry}-100\,\text{keV}}$.

It is important to treat separately the broad and narrow emission line features as there is strong evidence that they originate from regions of significantly different physical conditions. The broad Hα comes from a much more compact and much denser emitting gas than the narrow Balmer and forbidden lines. The difference in density is at least 3 orders of magnitude; the critical density for collisional excitation of the [O I] (~ $2 \times 10^6\,\text{cm}^{-3}$) can be used to estimate the gas density of the broad-line region, as [O I] does not exhibit a broad component in this nucleus' emission. The difference in size is expected to be ~2 orders of magnitude (e.g., reverberation mapping of nearby Seyfert galaxies measured to be <1 week, [56]). Thus, the mechanisms that can operate in these two regions are expected to be qualitatively different. The photon and energy balance conditions should reflect these differences and thus should differ as well; in particular, potential contributions to the Hα emission via collisional excitation should be minimal for the narrow-line-emitting region, but important for the broad component.

Based on photoionization models (i.e., Cloudy, v94.0; see [57]) computed by Lewis et al. [58] for a wide range of ionization parameters, densities, and metallicities, Eracleous et al. [18] find that energy balance in a line-emitting nebula requires that $L_i > 18(\pm 2)L_{\text{H}\alpha}/f_c$, where f_c is the covering factor or the fraction of the ionizing luminosity of the AGN that is absorbed by the line-emitting gas. These models are covering electron densities that appear to encompass both the broad- and the narrow-line-emitting regions in NGC 4736; thus, with only a fraction of 10% of ionizing photons being absorbed by the line-emitting gas, the minimum energy balance condition for AGN ionization is given by $L_i/L_{\text{H}\alpha} > 180$, for both narrow and broad emission features. In the same time, however, a minimum requirement for photon balance can be quite different for the two emission regions: for the narrow-line component $Q_i > 2.2Q_{\text{H}\alpha}$, corresponding to the case B recombination (i.e., one Hα photon is emitted for every 2.2 recombinations); for the denser broad-line-emitting region, the number of H-alpha photons that can be produced for each ionization can be at least 7-8 times higher than the standard case B estimate or $Q_i \gtrsim 0.25Q_{\text{H}\alpha}$ (e.g., [37]).

With the L_i and Q_i values already calculated by Eracleous et al. [18] by integrating M94's nuclear SED assuming that pairs of points could be connected by a power law, and with the Hα measurements from all HST and ground-based observations presented above, we can proceed with the comparison. Table 4 lists the $L_{\text{H}\alpha}$, $Q_{\text{H}\alpha}$, and associated ratios $L_i/L_{\text{H}\alpha}$, $Q_i/Q_{\text{H}\alpha}$ for all of these optical spectroscopic observations, where the narrow and broad Hα measurements are shown separately.

It is readily apparent that while $L_i/L_{\text{H}\alpha}$ is in general >18 ($f_c = 1$), it is almost never >180 ($f_c = 0.1$); the only exception is the narrow Hα emission measured in the HST aperture. Thus, a dominant AGN ionization of the narrow-line

TABLE 4: Ionizing and emission-line power.

	$\log L_{H\alpha}$[a]	$\log Q_{H\alpha}$[a]	$L_i = L_{H\alpha}$[b]	$Q_i = Q_{H\alpha}$[c]
Palomar	>37.7	>49.3	<93	<2.75
Bok	37.5	49.0	165.9	4.9
MMTO	37.5	49.2	104.7	3.1
HST-STIS, narrow	36.8	48.4	759	22.4
HST-STIS, broad	37.9	49.5	57.5	1.6
PCA tomography	38.8	50.3	8.7	0.25

[a]the luminosities are measured in $\mathrm{erg\,s^{-1}}$ and the photon rates in $\mathrm{s^{-1}}$.

[b]the minimum energy balance condition for AGN ionization is given by $L_i/L_{H\alpha} \gtrsim 180$ (when a fraction of 10% of ionizing photons photons is absorbed by the line-emitting gas).

[c]the minimum photon balance condition for AGN ionization is given by $Q_i/Q_{H\alpha} > 2.2$ for the narrow line regions, and $Q_i/Q_{H\alpha} \gtrsim 0.25$ for the broad component.

region in this nucleus is definitely possible in the HST aperture, but only for $f_c \gtrsim 18\%$ at larger radial distances. For the HST broad Hα feature, an AGN ionization is possible only if $f_c \gtrsim 30\%$. The PCA tomography measurement of the broad Hα argues, however, against a balanced energy budget originating entirely in an AGN-like power mechanism, even when a maximum covering factor is considered; the AGN-produced energy falls short of the required amount by at least 50%. Interestingly, the measured $Q_i/Q_{H\alpha}$ ratio is well within the required photon balance corresponding to an AGN excitation for both the narrow and the broad emission features. Thus, with one clear exception, the AGN in NGC 4736 appears to be capable of providing enough photons to explain the observed Hα luminosity but only for relatively high f_c values.

Simply because the AGN-like SED of NGC 4736 can explain the majority of the ionization energy and photon rate does not imply that the actual mechanism is an AGN, alternative excitation mechanisms must be explored. Possible options are (i) we are missing ionizing photons from accretion onto the central BH, or (ii) there are other power sources that could make up the power deficit in this system, particularly for producing the broad Hα feature.

The first alternative could be possible if we were observing an "echo" of a previous epoch of more violent accretion, a few hundred years ago. This idea has been explored by Eracleous et al. [59] who showed that the reverberation of an ionizing flare in the nebula can produce LINER-like emission-line ratios. In this scenario, it is expected that the central UV source and the [O III] line would also follow the decay of the ionizing continuum; while the UV observations are not providing clear evidence for such a decay over the course of one decade, the multiple optical spectroscopic observations that we present in Figure 2 are consistent with a possible decrease in time (a few years) in the [O III] flux (in this model, Hβ, [S II], [O I] are expected to decay very slowly, in 60–250 years). Nevertheless, this duty-cycle hypothesis also requires that the broad-line emission fades immediately if the ionizing continuum declines, and the PCA tomography observations show that this is not happening in this source. The echo of such an ionizing continuum flare should also be

detectable in a narrow-band [O III] λ5007 image in the form of a ring, which is not readily observed in M94.

The second alternative of power sources other than accretion has been often proposed in explaining the emission-line spectra of LINERs, with mixed success (see Section 1). The most probable alternative sources appear to be the mechanical power delivered by compact radio jets, along with photoionization by young or post-AGB stars from old or intermediate-age stellar populations. Shock models are highly unlikely to produce broad emission features and, thus, are not favored in this case. Recent star-formation activity remains, however, a viable option, particularly in light of relatively new discoveries of peculiar supernovae with broad Hα features that do not appear to fade in time. This idea is discussed in more detail in Section 5.

4. The Black Hole Mass

Given the unusual energetics of M94's nucleus, it is important to investigate whether the BH mass estimators derived for rapidly accreting Seyferts, which appear to be widely used for AGN, also work in the low-luminosity regime flagged by this particular system.

There are a variety of indirect methods that can be used to estimate the mass of BHs in galaxy centers. The available multiwavelength measurements of the nuclear emission for NGC 4736 allow the calculation of M_{BH} based on four different techniques, as well as with the M-σ^* relation. In this section, we explore and compare the results of these five methods, along with their consequences for this object's energetics, and a comparison with a recent dynamical measurement of $6.68(5.14$–$8.22) \times 10^6\, M_\odot$ for this object, which is listed in Kormendy et al. [60] as obtained from (Gebhardt et al. 2012, in prep.). We first present the methods and associated M_{BH} calculations and then discuss the shortcomings of each measurement.

(1) Using the M-σ^* relation, established for quiescent nearby galaxies, including both ellipticals and spirals with classical bulges, as quantified by Gültekin et al. [1], for $\sigma^* = 110 \pm 5\,\mathrm{km\,s^{-1}}$ [61], $M_{BH} = 1.05 \pm 0.64 \times 10^7\,M_\odot$; within errors, this value agrees well with Gebhardt et al. dynamical measurement. With this value, $L_{bol}/L_{Edd} \approx 2 \times 10^{-5}$, which is consistent with the range of values within LINERs are expected to (e.g., [2, 62]).

(2) Using the HST measurements of the FWHM and the luminosity of the broad Hα component (see Table 1) within the scaling relation based solely on observations of this broad emission feature, as derived by Greene and Ho [63], we obtain $M_{BH} \approx 3 \times 10^4\,M_\odot$; the fractional uncertainty associated with this measurement is $\sim 30\%$, and includes both the scatter in the scaling relation and the errors in the line measurements. With the broad Hα line luminosity measured in the PCA tomography study [44], the BH mass would increase (by a factor of 6.5) to $M_{BH} = 1.9 \times 10^5\,M_\odot$. The difference in these two values could be considered the most conservative uncertainty associated with this BH mass estimate. The corresponding Eddington ratios for these BH masses are $L_{bol}/L_{Edd} \approx 1$–$7 \times 10^{-3}$.

(3) Mid IR detection and measurements of the [Ne V] (14.32 μm) and [O IV] (25.89 μm) emission lines [64] give, via the empirical correlations between the MIR line luminosities and reverberation mapping-based M_{BH} values presented by Dasyra et al. [65], a black hole mass of $2.3\pm0.4\times 10^5\,M_\odot$ and $1.7\pm0.5\times10^5\,M_\odot$, respectively. The corresponding Eddington ratio is in this case $L_{bol}/L_{Edd} \approx 1.0\times10^{-3}$. Note that the scatter adopted for these relations is only a lower limit of the real value; thus, the uncertainty may be larger.

(4) The normalized X-ray excess variance method, as described in Papadakis et al. [66], applied to Chandra observations [30] gives $M_{BH} = 2.5 \pm 1.7 \times 10^5\,M_\odot$. To be specific, we used for this calculation the excess variance $\sigma = 0.06 \pm 0.04$, $L_{bol} = 2.5 \times 10^{40}$ erg s^{-1} as derived in Section 3, and $\nu_{lf} = 1/T = 1/14$ h^{-1} (where T is the length of the light curve), to estimate the break frequency ν_{bf} and then M_{BH} via equations 4 and 6, respectively, of Papadakis et al. [66].

(5) The M_{BH} of this system can also be obtained via the "fundamental plane of black hole activity" that relates black hole mass to the emitted compact radio $L_R = \nu L_\nu (5\,\text{GHz})$ and hard X-ray luminosities $L_X (2\text{--}10\,\text{keV})$ and spans nine orders of magnitude in black hole mass. We calculate $L_R = 1.7 \times 10^{35}$ erg s^{-1} using the two 8.5 GHz and 15 GHz radio measurements presented in this paper (Figure 3, Table 3), with a flux modeled by a powerlaw $S_\nu \propto \nu^{-\alpha}$. Using the empirical fits of Merloni et al. [67] for the fundamental plane relation, $\log(M_{BH}/M_\odot) = 5.9 \pm 1.1$ for the black hole in this system; note that this measurement and its associated errors embrace all of the estimates presented above and, thus, do not provide any additional constraint to the M_{BH}. Interestingly, the latest derivation of the fundamental plane relation [68], applied to very low nuclear galactic luminosities, provides a $\log(M_{BH}/M_\odot) = 7.2 \pm 0.4$ for this system, which remains consistent only with the M-σ^* value, as it departs considerably from those provided by the relations employing AGN emission.

There is a significant inconsistency between the value given by the M-σ^* relation, which is supported by Gebhardt's dynamical measurement, and those based on the AGN emission. The M-σ^* relation suggests a BH mass in NGC 4736 of $1 \times 10^7\,M_\odot$ and a correspondingly low L_{bol}/L_{Edd} of $\sim 2 \times 10^{-5}$. Three of the estimates based on the AGN emission converge to black hole masses of $\sim 10^5\,M_\odot$ (and $L_{bol}/L_{Edd} \sim 10^{-3}$), showing a surprisingly consistent departure of two orders of magnitude from the M-σ^* estimate. The fundamental plane relation provides a M_{BH} value right in between these two different situations, however, with no real additional constraint, due to its associated large uncertainty. These differences are somewhat puzzling given that the BH mass estimates based on nuclear emission properties are all calibrated to follow the M-σ^* for high-mass black holes ($M_{BH} \gtrsim 10^6\,M_\odot$). It is however true that the calibrators are biased toward nearby Seyfert galaxies with much higher Eddington ratios than that of NGC 4736. We briefly discuss in the following subsections more specific weaknesses of each measurement.

4.1. Caveats of the M-σ^* Relation. The M-σ^* relation is expected to provide a reliable estimate of the BH mass as it

is based on the strong correlation between dynamical mass measurements of supermassive BHs and their host properties [1, 69–71]. This relationship is derived primarily from ellipticals and spirals with classical bulges (formed during major mergers). Recent observations suggest it may not be valid for samples of later-type spirals which more commonly host pseudobulges (formed via secular disk processes) (e.g., [72]). Because the distinction between classical and pseudobulges is based on formation, it does not simply correlate with observable properties [73, 74]. In general pseudobulges are less luminous, have lower bulge-to-total ratios, and have ongoing star formation and lower Sérsic indices than classical bulges. The possibility that the NGC 4736 is in fact a pseudobulge provides a solution to the apparent conflict in BH mass estimates.

Nevertheless, pseudobulges are difficult to identify, and there is not yet a consensus as to what defines them. To complicate things further, classical and pseudobulges can exist within the same galaxy (e.g., NGC 2787; [75]). The classification of the NGC 4736 bulge is ambiguous: Fisher and Drory [74] classifies it as a pseudobulge based on its nuclear spiral and bar and low Sérsic index ($n = 1.3$). However, they also find it has a low star formation rate, more typical of classical bulges, and thus classify it as an "Inactive Pseudobulge;" if NGC 4736 hosts a pseudobulge, it is an atypical one. We have created a surface brightness profile from NICMOS and 2MASS H-band data from the Large Galaxy Atlas [76] and found results that conflict with the fits of Fisher and Drory [74]. Specifically, we find Sérsic indices for the bulge of $n = 2.3$–3.0 depending on the radial range and type of fit (single versus double Sérsic), which are consistent with a classical one (as shown in [74]).

Our surface brightness profile fits also reveal the presence of a nuclear star cluster within the central $\sim0.6''$ (~12 pc), with an H-band magnitude of ~12.5. Such nuclear star clusters are common in early-type spiral galaxies [77]. The luminosity and mass of nuclear star clusters are known to scale with bulge luminosity and mass [78–80], and the NGC 4736 nuclear star cluster has a luminosity that is 0.1% of its bulge, typical for nuclear star clusters [81]. Nuclear star clusters commonly coexist with black holes, but there are a very limited number of cases where masses for both can be estimated [82]. In these cases, including the Milky Way, the BH mass is similar to the mass of the nuclear star cluster within an order of magnitude. There is also some evidence that the ratio of BH mass to nuclear star cluster mass increases with spheroid mass [83]. For NGC 4736, assuming an old population with an H-band $M/L \sim 0.7$, the nuclear cluster would have a mass of $\sim 2 \times 10^7\,M_\odot$. This mass is quite similar to the M-σ^* BH mass estimate, suggesting thus the presence of a similar-sized BH.

4.2. Caveats of the BLR Scaling Relation. The BLR scaling relations were derived using high-luminosity systems (i.e., Seyferts, not LINERs) and have been scaled to match the dynamical black hole detections of BHs with masses $>10^6\,M_\odot$. Thus, their applicability to NGC 4736, where they yield an estimate of $\sim 10^5\,M_\odot$, is a (perhaps unwise) extrapolation. These relations assume that the BLR is virialized due to proximity to the BH [84] and thus the 5100 Å continuum luminosity correlates with the emissivity-weighted radius of the BLR and thus with the BH mass [85]. The overall errors associated with BH mass measurements based on these relations do not exceed 0.5 dex [86–88], and there appears to be good consistency with BH masses obtained via the M-σ^* relation for relatively bright AGN ($L > 10^{42}$ erg s^{-1}), with $M_{BH} \gtrsim 10^6 M_\odot$ [89]. More recent calibrations of the radius-luminosity relationship on which these techniques are based infer that BH masses have been overestimated, however, only by up to a factor of ~3 [90]. This latter study also indicates a trend toward larger uncertainties and larger amount of overestimation in the BH with decreasing luminosity; however, no conclusive results are yet available for this regime.

4.3. Caveats of the MIR Line Correlation. The Dasyra et al. [65] empirical relation between the MIR line emission properties and the BH mass is derived using reverberation mapping BH masses and, thus, as with the previous method, may not apply to systems with $M_{BH} < 10^6 M_\odot$. The relation holds for systems with $L_{bol}/L_{Edd} > 0.003$, but not necessarily beyond this range. The $M_{BH} \approx 10^5 M_\odot$ given by this method places this object at the low end of this L_{bol}/L_{Edd} range; on the other hand, a more massive and thus quiescent BH would correspond to a [Ne V] luminosity of a few orders of magnitude lower than that measured. The scatter associated with this relation remains in average 0.5 dex and thus cannot account for the ~2-3 orders of magnitude difference between this value and that obtained via M-σ^*.

4.4. Issues with Estimating the BH Mass from X-Rays. The BH mass derivation based on X-ray variability, or more precisely on the relation between the excess variance and M_{BH}, relies on the hypothesis of a universal power spectral density function (PSD) shape and amplitude in AGN, which is based on the idea that the X-ray variability mechanism and the accretion efficiency are the same for all AGN, at all redshifts. These assumptions appear to hold for the objects involved in deriving and testing this relation [66]; that sample is, however, small and rather biased toward luminous (X-ray) sources, with log $L_X/(\text{erg s}^{-1}) > 41.5$. None of the 2–10 keV sources detected in the nucleus of NGC 4736 are in this luminosity range. Nevertheless, Galactic BHs ($M_{bh} < 10^2 M_\odot$) in their hard states show variability properties that match well those of AGN, both of these types of sources falling on the same projection of the T_B-M_{bh}-L_{bol}/L_{Edd} plane, where $T_B = 1/\nu_{bf}$, suggesting that an extrapolation to the intermediate mass BH (or lower L) level is practicable.

4.5. Problems with the Fundamental Plane. Finally, it appears that the main problem with the fundamental plane is that the relation is not sharpened enough to provide strong constraints on BH masses. Because the relation spans nine orders of magnitude in BH mass, it is expected to equally apply to any value of M_{bh} in the range we are interested in. It is also the case that a wide variety of BH accretion models (e.g., with efficient and inefficient flows for the X-ray emission or associating X-ray flux with synchrotron emission

near the base of a jet) are consistent with this relation (e.g., [91, 92]), suggesting that a large diversity of accretion modes or rates are accommodated.

Nevertheless, the fundamental plane remains best constrained only for systems with $M_{BH} \gtrsim 10^6 M_\odot$ and for very low nuclear luminosities, that is, with negligible or zero AGN contribution that allow a dynamical measurement of their BH mass [68]. In this regime, the latest derivation of the fundamental plane relation provides for M94 a M_{BH} value consistent with the M-σ^* estimate and the dynamical measurement, while it would not reliably constrain a $M_{BH} \lesssim 10^6 M_\odot$ value. The scatter in the fundamental plane relation increases for lower BH masses, with higher Eddington ratios [68, 93].

5. Discussion: Alternative Power Generation Mechanisms

The detection of broad Hα emission, combined with the spatial coincidence of this emission with the detection of X-ray, UV, and radio compact sources, provides strong evidence that this galaxy hosts an accreting massive black hole. The SED of this nucleus makes this object one of the lowest luminosity LINER with a distinct contribution to the total emission by black hole accretion. In this scenario, the nucleus is an AGN, and the presence of off-nuclear sources, particularly the radio and the UV detections, may result from remnant jet activity emerging from the nucleus. It is definitely exciting to detect AGN activity at energy levels equal to that of several young supernova remnants of the Cas A variety [94], an OB association which hosts a high-mass X-ray binary [30] or simply a group of five late O supergiants [28].

Nevertheless, the presence of a number of unusual off-nuclear sources, coupled with the apparent deficit in the photoionizing photons, and with the fact that the BH mass estimates based on AGN emission appear to fail for this object, encourages us to explore alternative scenarios for the NGC 4736 nucleus. In particular, it is very likely that the BH mass estimates that exploit the multiwavelength AGN characteristics do not work for this system because at least some of this emission is not the result of BH accretion.

There are certain (peculiar) kinds of core-collapse SNe that present multiwavelength observations and in particular broad Hα emission components with characteristics that are very similar to those we measure in the nucleus of M94. The SN 2005ip, presented by Smith et al. [95], is a very good example. The so-called intermediate Hα component associated with this SN ejecta presents the same FWHM and brightness level as the broad Hα detected in the STIS aperture, and it does not show any sign of diminishing its strength over more than 3-year period (see their Figure 7; a *very* broad component with FWHM $\gtrsim 10000$ km s^{-1} like the one exhibited by SN 2005ip would not be measurable in the galaxy spectrum as it would be completely swamped in the continuum stellar light, even in the *HST*-STIS observations).

SNe as luminous as X2, with intrinsically hard X-ray spectra (photon index $\Gamma \lesssim 1$), have certainly been encountered, for cases observed few years after the explosion (e.g., ATe #1023; [96]); for these cases, L_X/L_{bol} ratios are high relative to more standard SN cases [97]. Another particular example of a SN that matches well the measurements of the nuclear emission in NGC 4736 is SDSS J09529.56 + 214313.3 [98], which is believed (but not confirmed) to be a SN type IIn. For this system, the broad Hα component stays strong for at least three years, and its $L_X(2-10\,\text{keV})/L_{H\alpha} \approx 1$, matching thus very well the surprisingly low value measured for NGC 4736.

The peculiar nebular, radio, and UV characteristics of the nucleus of NGC 4736 appear to also compare well the SN phenomenology. The electron densities (or [S II] line flux ratios) measured in this galaxy center are matching exactly the ones measured in the environments of the extraordinary type IIn SNe we compare here with, for example, SDSS J09529.56 + 214313.3. The two different band radio observations of this nucleus are consistent with emission from extragalactic SNe, in both intensity and decline rate in flux density [99–101]. Moreover, the compact radio off-nuclear detection, only $1''$ (20 pc) away from the nuclear one [31], could be interpreted as the result of shock emission associated with a nuclear core-collapse SN. Measurements of the brightness of the off-nuclear UV detection in F250W and F330W place this object into the O star of late (5-ish) type spectral category, fitting thus well into the idea that this nucleus could simply be a star-forming site and that NGC 4736a's emission includes significant contributions from a SN which exploded close to the weakly active (and massive, $M_{BH} \sim 10^{(6-7)} M_\odot$) central BH.

This SN contribution scenario may have its own drawbacks. If the broad Hα emission has actually gotten brighter by a factor of 6, as suggested by the comparison between the HST-STIS and GMOS observations, the SN interpretation becomes problematic; variability of the broad-line region originating in the AGN could, in principle, account for this effect. Also, the observed intraday (hour-scale) X-ray variability measured for X2 remains yet to be detected in a supernova and conflicts with the physical scale over which X-ray emission is expected in SN remnants; thus, this particular behavior may remain strictly associated with the AGN. Additional observations would be necessary to fully confirm or rule out the SN scenario. Specifically, new high-resolution observations in the optical, UV, or X-rays would be able to confirm if there is a fading in the light curve, as expected from a SN, and would also allow accurate localization of the source. High S/N optical spectra would much better resolve the emission-line profile and a possible temporal evolution.

6. Conclusion

We have presented here an exhaustive multiwavelength analysis of the nuclear emission properties of NGC 4736, prompted by new measurements of a broad Hα emission component detected in its high-resolution HST-STIS optical spectrum. This broad Hα component, with a luminosity of $9 \times 10^{37}\,\text{ergs s}^{-1}$, is one of the lowest luminosity broad line known. This broad Hα is coincident with a compact bright X-ray and radio source. Our measurements of this object's spectral energy distribution reveal a bolometric luminosity

of $L_{bol} \approx 2.5 \times 10^{40}\,\text{erg s}^{-1}$, that categorizes NGC 4736 as one of the least luminous LINERs with strong evidence for BH accretion. Our comparison of five independent BH mass estimates reveals a discrepancy of two orders of magnitude between the value $\sim 10^7 M_\odot$ predicted by the M-σ^* relation and the value $\sim 10^5 M_\odot$ toward which methods based on AGN emission activity in optical, mid-IR, and X-ray, seem to converge; the fifth method is provided by the fundamental plane relation, which, however, due to its large associated uncertainties, does not offer any additional constraint to this comparison.

We conclude that this system's BH mass cannot be reliably estimated via standard AGN BH mass indicators because the nuclear emission in this system is not entirely tracing the accretion onto the central BH. Our assessment of the energy budgets of the ionizing and emission-line power suggests a possible deficit in the AGN ionization and production of a broad Hα emission feature which can be made up by a peculiar kind of Type IIn SN that matches well the nuclear emission of NGC 4736 over the whole electromagnetic spectrum and supports this galaxy nucleus' general aging starburst-like appearance.

Acknowledgments

The authors thank the anonymous referee for constructive comments that helped them improved the paper. Support for this work was provided by NASA through Grant no. HST-AR-11749.01-A from the STScI, which is operated by the Association of Universities for Research in Astronomy, Inc., under NASA Contract NAS5-26555, based on observations made with the NASA/ESA *Hubble Space Telescope*, obtained from the data archive at the Space Telescope Science Institute. STScI is operated by the Association of Universities for Research in Astronomy, Inc., under NASA Contract NAS 5-26555. Some observations reported here were obtained at the MMT Observatory, a joint facility of the Smithsonian Institution and the University of Arizona.

References

[1] K. Gültekin, D. O. Richstone, K. Gebhardt et al., "the m-σ and m-l relations in galactic bulges, and determinations of their intrinsic scatter," *Astrophysical Journal*, vol. 698, no. 1, p. 198, 2009.

[2] L. C. Ho, "Nuclear activity in nearby galaxies," *Annual Review of Astronomy and Astrophysics*, vol. 46, pp. 475–539, 2008.

[3] J. A. Baldwin, M. M. Phillips, and R. Terlevich, "Classification parameters for the emission-line spectra of extragalactic objects," *Astronomical Society of the Pacific*, vol. 93, pp. 5–19, 1981.

[4] S. Veilleux and D. E. Osterbrock, "Spectral classification of emission-line galaxies," *Astrophysical Journal, Supplement Series*, vol. 63, pp. 295–301, 1987.

[5] G. Kauffmann, T. M. Heckman, C. Tremonti et al., "The host galaxies of active galactic nuclei," *Monthly Notices of the Royal Astronomical Society*, vol. 346, no. 4, pp. 1055–1077, 2003.

[6] L. J. Kewley, B. Groves, G. Kauffmann, and T. Heckman, "The host galaxies and classification of active galactic nuclei," *Monthly Notices of the Royal Astronomical Society*, vol. 372, no. 3, pp. 961–976, 2006.

[7] L. C. Ho, A. V. Filippenko, and W. L. W. Sargent, "A search for "dwarf" seyfert nuclei. V. Demographics of nuclear activity in nearby galaxies," *Astrophysical Journal*, vol. 487, no. 2, pp. 568–578, 1997.

[8] A. Constantin and M. S. Vogeley, "The clustering of low-luminosity active galactic nuclei," *Astrophysical Journal*, vol. 650, no. 2 I, pp. 727–748, 2006.

[9] D. W. Weedman, L. Hao, S. J. U. Higdon et al., "Mid-infrared spectra of classical AGNs observed with the spitzer space telescope," *Astrophysical Journal*, vol. 633, no. 2 I, pp. 706–716, 2005.

[10] L. Armus, J. Bernard-Salas, H. W. W. Spoon et al., "Detection of the buried active galactic nucleus in NGC 6240 with the infrared spectrograph on the Spitzer Space Telescope," *Astrophysical Journal*, vol. 640, no. 1 I, pp. 204–210, 2006.

[11] N. P. Abel and S. Satyapal, "[Ne v] Emission in optically classified starbursts," *Astrophysical Journal*, vol. 678, no. 2, pp. 686–692, 2008.

[12] E. Sturm, D. Lutz, A. Verma et al., "Mid-infrared line diagnostics of active galaxies: a spectroscopic AGN survey with ISO-SWS," *Astronomy and Astrophysics*, vol. 393, no. 3, pp. 821–841, 2002.

[13] S. Satyapal, D. Vega, R. P. Dudik, N. P. Abel, and T. Heckman, "Spitzer uncovers active galactic nuclei missed by optical surveys in seven late-type galaxies," *Astrophysical Journal*, vol. 677, no. 2, pp. 926–942, 2008.

[14] A. D. Goulding and D. M. Alexander, "Towards a complete census of AGN in nearby Galaxies: a large population of optically unidentified AGN," *Monthly Notices of the Royal Astronomical Society*, vol. 398, no. 3, pp. 1165–1193, 2009.

[15] N. M. Nagar, H. Falcke, and A. S. Wilson, "Radio sources in low-luminosity active galactic nuclei IV. Radio luminosity function, importance of jet power, and radio properties of the complete Palomar sample," *Astronomy and Astrophysics*, vol. 435, no. 2, pp. 521–543, 2005.

[16] W. M. Zhang, R. Soria, S. N. Zhang, D. A. Swartz, and J. Liu, "A census of X-ray nuclear activity in nearby galaxies," *Astrophysical Journal*, vol. 699, no. 1, pp. 281–297, 2009.

[17] L.-B. Desroches, J. E. Greene, and L. C. Ho, "X-ray properties of intermediate-mass black holes in active galaxies. II. X-ray-bright accretion and possible evidence for slim disks," *Astrophysical Journal*, vol. 698, p. 1515, 2009.

[18] M. Eracleous, J. A. Hwang, and H. M. L. G. Flohic, "An assessment of the energy budgets of low-ionization nuclear emission regions," *Astrophysical Journal*, vol. 711, no. 2, pp. 796–810, 2010.

[19] A. Constantin, F. Hoyle, and M. S. Vogeley, "Active galactic nuclei in void regions," *Astrophysical Journal*, vol. 673, no. 2, pp. 715–729, 2008.

[20] A. Constantin, P. Green, T. Aldcroft et al., "Probing the balance of AGN and star-forming activity in the local universe with ChaMP," *Astrophysical Journal*, vol. 705, no. 2, pp. 1336–1355, 2009.

[21] K. Schawinski, D. Thomas, M. Sarzi et al., "Observational evidence for AGN feedback in early-type galaxies," *Monthly Notices of the Royal Astronomical Society*, vol. 382, no. 4, pp. 1415–1431, 2007.

[22] K. Schawinski, C. M. Urry, S. Virani et al., "Galaxy zoo: the fundamentally different co-evolution of supermassive black holes and their early- and late-type host galaxies," *Astrophysical Journal*, vol. 711, no. 1, pp. 284–302, 2010.

[23] T. Di Matteo, V. Springel, and L. Ilernquist, "Energy input from quasars regulates the growth and activity of black holes

and their host galaxies," *Nature*, vol. 433, no. 7026, pp. 604–607, 2005.

[24] P. F. Hopkins, L. Hernquist, T. J. Cox, T. Di Matteo, B. Robertson, and V. Springel, "A unified, merger-driven model of the origin of starbursts, quasars, the cosmic X-ray background, supermassive black holes, and galaxy spheroids," *Astrophysical Journal, Supplement Series*, vol. 163, no. 1, pp. 1–49, 2006.

[25] R. B. Tully and J. R. Fisher, *Catalog of Nearby Galaxies*, Cambridge University Press, Cambridge, UK, 1988.

[26] A. V. Filippenko and W. L. W. Sargent, "A search for "Dwarf" seyfert 1 nuclei. I. The initial data and results," *Astrophysical Journal, Supplement Series*, vol. 57, pp. 503–522, 1985.

[27] P. F. Spinelli, T. Storchi-Bergmann, C. H. Brandt, and D. Calzetti, "An atlas of Hubble space telescope stis spectra of Seyfert galaxies," *Astrophysical Journal, Supplement Series*, vol. 166, no. 2, pp. 498–504, 2006.

[28] D. Maoz, A. V. Filippenko, L. C. Ho et al., "Detection of compact ultraviolet nuclear emission in liner galaxies," *Astrophysical Journal*, vol. 440, no. 1, pp. 91–99, 1995.

[29] D. Maoz, N. M. Nagar, H. Falcke, and A. S. Wilson, "The murmur of the sleeping black hole: detection of nuclear ultraviolet variability in LINER galaxies," *Astrophysical Journal*, vol. 625, no. 2 I, pp. 699–715, 2005.

[30] M. Eracleous, J. C. Shields, G. Chartas, and E. C. Moran, "Three LINERs under the Chandra X-ray microscope," *Astrophysical Journal*, vol. 565, no. 1 I, pp. 108–124, 2002.

[31] E. Körding, E. Colbert, and H. Falcke, "A radio monitoring survey of ultra-luminous X-ray sources," *Astronomy and Astrophysics*, vol. 436, no. 2, pp. 427–436, 2005.

[32] A. L. Kinney, R. C. Bohlin, D. Calzetti, N. Panagia, and R. F. G. Wyse, "An atlas of ultraviolet spectra of star-forming galaxies," *Astrophysical Journal, Supplement Series*, vol. 86, no. 1, pp. 5–93, 1993.

[33] B. J. Smith, P. M. Harvey, C. Colomé, C. Y. Zhang, J. DiFrancesco, and R. W. Pogge, "Far-infrared emission from the bulges of early-type spirals: kao observations of NGC 4736 (M94) and NGC 3627 (M66)," *Astrophysical Journal*, vol. 425, no. 1, pp. 91–102, 1994.

[34] L. Dressel, P. Hodge, and P. Barrett, "wx2d: a PyRAF routine to resample spectral images," *Instrument Science Report STIS 2007-04*, 2007.

[35] G. Bruzual and S. Charlot, "Stellar population synthesis at the resolution of 2003," *Monthly Notices of the Royal Astronomical Society*, vol. 344, no. 4, pp. 1000–1028, 2003.

[36] C. A. Tremonti, T. M. Heckman, G. Kauffmann et al., "The origin of the mass-metallicity relation: insights from 53,000 star-forming galaxies in the sloan digital sky survey," *Astrophysical Journal*, vol. 613, no. 2 I, pp. 898–913, 2004.

[37] D. E. Osterbrock, *Astrophysics of Gaseus Nebulae and Active Galactic Nuclei*, University Science Books, 1989.

[38] G. Kriss, "Fitting models to UV and optical spectral data," in *Astronomical Data Analysis Software and Systems III*, D. R. Crabtree, R. J. Hanisch, and J. Barnes, Eds., vol. 61 of *Astronomical Society of the Pacific Conference Series*, p. 437, 1994.

[39] J. L. Walsh, A. J. Barth, and M. Sarzi, "The supermassive black hole in M84 revisited," *Astrophysical Journal*, vol. 721, no. 1, pp. 762–776, 2010.

[40] L. C. Ho, A. V. Filippenko, W. L. W. Sargent, and C. Y. Peng, "A search for "dwarf" seyfert nuclei. IV. Nuclei with broad Hα emission," *Astrophysical Journal, Supplement Series*, vol. 112, no. 2, pp. 391–414, 1997.

[41] J. E. Greene and L. C. Ho, "A new sample of low-mass black holes in active galaxies," *Astrophysical Journal Letters*, vol. 670, no. 1, pp. 92–104, 2007.

[42] A. V. Filippenko and W. L. W. Sargent, "Discovery of an extremely low luminosity Seyfert 1 nucleus in the dwarf galaxy NGC 4395," *The Astrophysical Journal*, vol. 342, pp. 11–14, 1989.

[43] L. C. Ho, A. V. Filippenko, and W. L. W. Sargent, "A search for "dwarf" seyfert nuclei. III. Spectroscopic parameters and properties of the host galaxies," *Astrophysical Journal, Supplement Series*, vol. 112, no. 2, pp. 315–390, 1997.

[44] J. E. Steiner, R. B. Menezes, T. V. Ricci, and A. S. Oliveira, "PCA Tomography: how to extract information from data cubes," *Monthly Notices of the Royal Astronomical Society*, vol. 395, no. 1, pp. 64–75, 2009.

[45] J. Moustakas and R. C. Kennicutt Jr., "An integrated spectrophotometric survey of nearby star-forming galaxies," *Astrophysical Journal, Supplement Series*, vol. 164, no. 1, pp. 81–98, 2006.

[46] N. M. Nagar, H. Falcke, A. S. Wilson, and J. S. Ulvesta, "Radio sources in low-luminosity active galactic nuclei. III. "AGNs" in a distance-limited sample of "LLAGNs"," *Astronomy and Astrophysics*, vol. 392, no. 1, pp. 53–82, 2002.

[47] L. C. Ho, "The spectral energy distributions of low-luminosity active galactic nuclei," *Astrophysical Journal*, vol. 516, no. 2, pp. 672–682, 1999.

[48] R. C. Kennicutt, L. Armus, G. Bendo et al., "SINGS: the SIRTF nearby galaxies survey," *Publications of the Astronomical Society of the Pacific*, vol. 115, no. 810, pp. 928–952, 2003.

[49] T. M. Heckman, G. Kauffmann, J. Brinchmann, S. Charlot, C. Tremonti, and S. D. M. White, "Present-day growth of black holes and bulges: the sloan digital sky survey perspective," *Astrophysical Journal*, vol. 613, no. 1 I, pp. 109–118, 2004.

[50] G. Kauffmann and T. M. Heckman, "Feast and Famine: regulation of black hole growth in low-redshift galaxies," *Monthly Notices of the Royal Astronomical Society*, vol. 397, no. 1, pp. 135–147, 2009.

[51] S. Charlot and S. M. Fall, "A simple model for the absorption of starlight by dust in galaxies," *Astrophysical Journal*, vol. 539, no. 2, pp. 718–731, 2000.

[52] M. Eracleous, J. A. Hwang, and H. M. L. G. Flohic, "Spectral energy distributions of weak active galactic nuclei associated with low-ionization nuclear emission regions," *Astrophysical Journal, Supplement Series*, vol. 187, no. 1, pp. 135–148, 2010.

[53] S. Satyapal, D. Vega, T. Heckman, B. O'Halloran, and R. Dudik, "The discovery of an active galactic nucleus in the late-type galaxy NGC 3621: spitzer spectroscopic observations," *Astrophysical Journal*, vol. 663, no. 1, pp. L9–L12, 2007.

[54] L. C. Ho, E. D. Feigelson, L. K. Townsley et al., "Detection of nuclear X-ray sources in nearby galaxies with chandra," *Astrophysical Journal*, vol. 549, no. 1, pp. L51–L54, 2001.

[55] L. C. Ho, A. V. Filippenko, and W. L. W. Sargent, "A search for "dwarf" Seyfert nuclei. VI. Properties of emission-line nuclei in nearby galaxies," *Astrophysical Journal*, vol. 583, no. 1 I, pp. 159–177, 2003.

[56] K. D. Denney, B. M. Peterson, R. W. Pogge et al., "Reverberation mapping measurements of black hole masses in six local Seyfert galaxies," *Astrophysical Journal*, vol. 721, no. 1, pp. 715–737, 2010.

[57] G. J. Ferland, K. T. Korista, D. A. Verner, J. W. Ferguson, J. B. Kingdon, and E. M. Verner, "CLOUDY 90: Numerical simulation of plasmas and their spectra," *Publications of the Astronomical Society of the Pacific*, vol. 110, no. 749, pp. 761–778, 1998.

[58] K. T. Lewis, M. Eracleous, and R. M. Sambruna, "Emission-line diagnostics of the central engines of weak-line radio galaxies," *Astrophysical Journal*, vol. 593, no. 1 I, pp. 115–126, 2003.

[59] M. Eracleous, M. Livio, and L. Binette, "A duty cycle hypothesis for the central engines of LINERs," *Astrophysical Journal Letters*, vol. 445, p. L1, 1995.

[60] J. Kormendy, R. Bender, and M. E. Cornell, "Supermassive black holes do not correlate with galaxy disks or pseudobulges," *Nature*, vol. 469, no. 7330, pp. 374–376, 2011.

[61] A. J. Barth, L. C. Ho, and W. L. W. Sargent, "A study of the direct fitting method for measurement of galaxy velocity dispersions," *Astronomical Journal*, vol. 124, no. 5, pp. 2607–2614, 2002.

[62] L. C. Ho, "Black hole demography from nearby active galactic nuclei," in *Coevolution of Black Holes and Galaxies*, L. C. Ho, Ed., the Carnegie Observatories Astrophysics Series, p. 293, Cambridge University Press, 2004.

[63] J. E. Greene and L. C. Ho, "Estimating black hole masses in active galaxies using the Hα emission line," *Astrophysical Journal*, vol. 630, no. 1 I, pp. 122–129, 2005.

[64] R. P. Dudik, S. Satyapal, and D. Marcu, "A *spitzer* spectroscopic survey of low-ionization nuclear emission-line regions: characterization of the central source," *Astrophysical Journal*, vol. 691, p. 1501, 2009.

[65] K. M. Dasyra, L. C. Ho, L. Armus et al., "High-ionization mid-infrared lines as black hole mass and bolometric luminosity indicators in active galactic nuclei," *Astrophysical Journal*, vol. 674, no. 1, pp. L9–L12, 2008.

[66] I. E. Papadakis, E. Chatzopoulos, D. Athanasiadis, A. Markowitz, and I. Georgantopoulos, "The long-term X-ray variability properties of AGNs in the Lockman Hole region," *Astronomy and Astrophysics*, vol. 487, no. 2, pp. 475–483, 2008.

[67] A. Merloni, S. Heinz, and T. Di Matteo, "A fundamental plane of black hole activity," *Monthly Notices of the Royal Astronomical Society*, vol. 345, no. 4, pp. 1057–1076, 2003.

[68] K. Gültekin, E. M. Cackett, J. M. Miller et al., "the fundamental plane of accretion onto black holes with dynamical masses," *The Astrophysical Journal*, vol. 706, no. 1, p. 404, 2009.

[69] L. Ferrarese and D. Merritt, "A fundamental relation between supermassive black holes and their host galaxies," *Astrophysical Journal*, vol. 539, no. 1, pp. L9–L12, 2000.

[70] K. Gebhardt, R. Bender, G. Bower et al., "A relationship between nuclear black hole mass and galaxy velocity dispersion," *Astrophysical Journal*, vol. 539, no. 1, pp. L13–L16, 2000.

[71] S. Tremaine, K. Gebhardt, R. Bender et al., "The slope of the black hole mass versus velocity dispersion correlation," *Astrophysical Journal*, vol. 574, no. 2 I, pp. 740–753, 2002.

[72] J. E. Greene, C. Y. Peng, M. Kim et al., "Precise black hole masses from megamaser disks: black hole-bulge relations at low mass," *Astrophysical Journal*, vol. 721, no. 1, pp. 26–45, 2010.

[73] J. Kormendy and R. C. Kennicutt, "Secular evolution and the formation of pseudobulges in disk galaxies," *Annual Review of Astronomy and Astrophysics*, vol. 42, pp. 603–683, 2004.

[74] D. B. Fisher and N. Drory, "Bulges of nearby galaxies with Spitzer: scaling relations in pseudobulges and classical bulges," *Astrophysical Journal*, vol. 716, no. 2, pp. 942–969, 2010.

[75] P. Erwin, J. C. Vega Beltrán, A. W. Graham, and J. E. Beckman, "When is a bulge not a bulge? Inner disks masquerading as bulges in NGC 2787 and NGC 3945," *Astrophysical Journal Letters*, vol. 597, no. 2 I, pp. 929–947, 2003.

[76] T. H. Jarrett, T. Chester, R. Cutri, S. E. Schneider, and J. P. Huchra, "The 2mass large galaxy atlas," *Astronomical Journal*, vol. 125, no. 2, pp. 525–554, 2003.

[77] C. M. Carollo, M. Stiavelli, M. Seigar, T. P. De Zeeuw, and H. Dejonghe, "Spiral galaxies with HST/NICMOS. I. Nuclear morphologies, color maps, and distinct nuclei," *Astronomical Journal*, vol. 123, no. 1, pp. 159–183, 2002.

[78] M. Balcells, A. W. Graham, L. Domínguez-Palmero, and R. F. Peletier, "Galactic bulges from Hubble Space Telescope near-infrared camera multi-object spectrometer observations: the lack of r1/4 bulges," *Astrophysical Journal*, vol. 582, no. 2, pp. L79–L82, 2003.

[79] L. Ferrarese, P. Côté, E. D. Bontà et al., "A fundamental relation between compact stellar nuclei, supermassive black holes, and their host galaxies," *Astrophysical Journal*, vol. 644, pp. 21–24, 2006.

[80] J. Rossa, R. P. Van Der Marel, T. Böker et al., "Hubble space telescope stis spectra of nuclear star clusters in spiral galaxies: dependence of age and mass on hubble type," *Astronomical Journal*, vol. 132, no. 3, pp. 1074–1099, 2006.

[81] P. Côté, S. Piatek, L. Ferrarese et al., "The ACS Virgo Cluster Survey. VIII. The nuclei of early-type galaxies," *Astrophysical Journal, Supplement Series*, vol. 165, no. 1, pp. 57–94, 2006.

[82] A. Seth, M. Agüeros, D. Lee, and A. Basu-Zych, "The coincidence of nuclear star clusters and active galactic nuclei," *Astrophysical Journal*, vol. 678, no. 1, pp. 116–130, 2008.

[83] A. W. Graham and L. R. Spitler, "Quantifying the coexistence of massive black holes and dense nuclear star clusters," *Monthly Notices of the Royal Astronomical Society*, vol. 397, no. 4, pp. 2148–2162, 2009.

[84] B. M. Peterson and A. Wandel, "Keplerian motion of broad-line region gas as evidence for supermassive black holes in active galactic nuclei," *Astrophysical Journal*, vol. 521, no. 2, pp. L95–L98, 1999.

[85] S. Kaspi, P. S. Smith, H. Netzer, D. Maoz, B. T. Jannuzi, and U. Giveon, "Reverberation measurements for 17 quasars and the size-mass-luminosity relations in active galactic nuclei," *Astrophysical Journal*, vol. 533, no. 2, pp. 631–649, 2000.

[86] M. Vestergaard, "Determining central black hole masses in distant active galaxies," *Astrophysical Journal*, vol. 571, no. 2 I, pp. 733–752, 2002.

[87] C. H. Nelson, R. F. Green, G. Bower, K. Gebhardt, and D. Weistrop, "The relationship between black hole mass and velocity dispersion in Seyfert 1 galaxies," *Astrophysical Journal*, vol. 615, no. 2 I, pp. 652–661, 2004.

[88] C. A. Onken, L. Ferrarese, D. Merritt et al., "Supermassive black holes in active galactic nuclei. II. Calibration of the black hole mass-velocity dispersion relationship for active galactic nuclei," *Astrophysical Journal*, vol. 615, no. 2 I, pp. 645–651, 2004.

[89] A. J. Barth, J. E. Greene, and C. H. O. Luis, "Dwarf Seyfert 1 nuclei and the low-mass end of the MBH-σ relation," *Astrophysical Journal*, vol. 619, no. 2, pp. L151–L154, 2005.

[90] M. C. Bentz, B. M. Peterson, H. Netzer, R. W. Pogge, and M. Vestergaard, "The radius-luminosity relationship for active galactic nuclei: the effect of host-galaxy starlight on luminosity measurements. II. the full sample of reverberation-mapped agns," *Astrophysical Journal*, vol. 697, no. 1, pp. 160–181, 2009.

[91] H. Falcke and P. L. Biermann, "The jet-disk symbiosis. I. Radio to X-ray emission models for quasars," *Astronomy and Astrophysics*, vol. 293, pp. 665–682, 1995.

[92] S. Heinz and R. A. Sunyaev, "The non-linear dependence of flux on black hole mass and accretion rate in core-dominated jets," *Monthly Notices of the Royal Astronomical Society*, vol. 343, no. 3, pp. L59–L64, 2003.

[93] E. Körding, H. Falcke, and S. Corbel, "Refining the fundamental plane of accreting black holes," *Astronomy and Astrophysics*, vol. 456, no. 2, pp. 439–450, 2006.

[94] J. L. Turner and P. T. P. Ho, "Bright radio continuum emission from star formation in the cores of nearby spiral galaxies," *Astrophysical Journal*, vol. 421, no. 1, pp. 122–139, 1994.

[95] N. Smith, J. M. Silverman, R. Chornock et al., "Coronal lines and dust formation in SN 2005ip: not the brightest, but the hottest type IIn supernova," *Astrophysical Journal*, vol. 695, no. 2, pp. 1334–1350, 2009.

[96] D. Pooley, S. Immler, and A. V. Filippenko, "Chandra observation of SN 2005kd: very luminous and hard X-ray emission," *The Astronomer's Telegram*, vol. 1023, p. 1, 2007.

[97] S. Immler, P. J. Brown, P. Milne et al., "X-ray, UV, and optical observations of supernova 2006bp with Swift: detection of early X-ray emission," *Astrophysical Journal*, vol. 664, no. 1 I, pp. 435–442, 2007.

[98] S. Komossa, H. Zhou, A. Rau et al., "NTT, spitzer, and chandra spectroscopy of SDSSJ095209.56+214313.3: the most luminous coronal-line supernova ever observed, or a stellar tidal disruption event?" *Astrophysical Journal*, vol. 701, no. 1, pp. 105–121, 2009.

[99] S. D. Van Dyk, K. W. Weiler, R. A. Sramek, and N. Panagia, "SN 1988Z: the most distant radio supernova," *Astrophysical Journal*, vol. 419, no. 2, pp. L69–L72, 1993.

[100] C. L. Williams, N. Panagia, S. D. Van Dyk, C. K. Lacey, K. W. Weiler, and R. A. Sramek, "Radio emission from SN 1988Z and very massive star evolution," *Astrophysical Journal*, vol. 581, no. 1 I, pp. 396–403, 2002.

[101] P. Chandra, V. V. Dwarkadas, A. Ray, S. Immler, and D. Pooley, "X-rays from the explosion site: 15 years of light curves of SN 1993J," *Astrophysical Journal*, vol. 699, no. 1, pp. 388–399, 2009.

12

Retrograde versus Prograde Models of Accreting Black Holes

David Garofalo

Department of Physics, Columbia University, New York, NY 10027, USA

Correspondence should be addressed to David Garofalo; david.a.garofalo@gmail.com

Academic Editor: Alberto J. Castro-Tirado

There is a general consensus that magnetic fields, accretion disks, and rotating black holes are instrumental in the generation of the most powerful sources of energy in the known universe. Nonetheless, because magnetized accretion onto rotating black holes involves both the complications of nonlinear magnetohydrodynamics that currently cannot fully be treated numerically, and uncertainties about the origin of magnetic fields that at present are part of the input, the space of possible solutions remains less constrained. Consequently, the literature still bears witness to the proliferation of rather different black hole engine models. But the accumulated wealth of observational data is now sufficient to meaningfully distinguish between them. It is in this light that this critical paper compares the recent retrograde framework with standard "spin paradigm" prograde models.

1. Introduction

When Roy Kerr presented his solution at the Texas Symposium almost five decades ago, the astronomical community, ironically, was too busy with the recent discovery of quasars to pay attention. But the importance of black holes is now grounded in observations pointing to a breadth and depth of a black hole impact that is likely still underestimated. In galaxies, black holes appear to participate in triggering and in quenching star formation, in heating and expanding gas, and in altering the mode of accretion [1–7]. They are connected to bulges and stellar dispersions at the spatial extremes of galaxies [8, 9], and in many cases the impact of black holes appears on cluster environments as well [1, 10–13]. But black hole influence is also observed on smaller scales, with stellar-mass black holes producing a rich panoply of observational signatures [14].

Black holes are both spatially and gravitationally irrelevant to galaxies as a whole, so the influence they exert is thought to occur during active phases when large amounts of energy spew from their centers in both kinetic form and radiation spanning the entire electromagnetic spectrum. Because the black hole scaling relations are ubiquitous, perhaps all galaxies experience an active phase during which black holes reveal their presence to the larger galaxy. And this active phase would involve accretion onto supermassive black holes [15, 16], the formation of powerful winds [17], and in some cases jet formation [18, 19], but the details of how jets are related to accretion remain elusive.

The earliest analytic models involve accretion in either thin-disk or advection-dominated form [15, 16, 20, 21], and/or spin energy extraction from black holes ([18]; henceforth BZ). Because of the nature of the relativistic gravitational potential, the thin-disk accretion model [16] is characterized by the existence of stable circular orbits with an inner boundary inwards of which gas plunges rapidly onto the black hole [22]. This innermost stable circular orbit (ISCO) depends on the spin of the black hole and the relative orientation of the black hole angular momentum relative to the angular momentum of the accretion flow. When the two angular momenta are aligned and the magnitude of the black hole angular momentum (or spin) is large, the accretion disk lives closer to the black hole which translates into a greater accretion efficiency (up to $0.42\,\dot{M}c^2$, with $\dot{M}$ as the accretion rate onto the black hole). This is due to the presence of accretion material close in to the black hole where it can tap into the strong gravitational potential and reprocess that energy further into the disk. As the black hole spin value drops toward zero, the accretion efficiency drops as well, reaching $0.06\,\dot{M}c^2$ at zero spin, due to an ISCO located further away from the black hole, so less energy is reprocessed into the disk. In the retrograde regime, where the black hole's angular momentum is opposite that of the accretion flow (i.e., they rotate in opposite directions), the accretion efficiency is even lower than that of the zero spin. Unsurprisingly, this is due to an ISCO that moves even further outwards.

The efficiency of the jet production in the BZ model depends on the spin value of the black hole and the strength of the magnetic field surrounding the black hole via

$$L_{BZ} \propto B^2 a^2, \tag{1}$$

where B is the magnetic field strength on the black hole and a is the spin parameter which is a dimensionless number (varying in magnitude from 0 to 1) characterizing the angular momentum of the black hole. Therefore, the disk and jet efficiency scale directly with prograde black hole spin (i.e., they both increase or decrease together). This framework is referred to in the literature as the "spin paradigm" [23–26] and, as we shall see when confronted with observation, is fraught with a host of difficulties, most of them emerging in the last few years.

Due to jet efficiency issues in the spin paradigm, hybrid spin paradigm models have been developed that in addition to jets from black holes include the contribution of jet outflow from the disk ([19]; henceforth BP) and the ability of the ergosphere to enhance the strength of the black hole-driven jet [27–29]. Hybrid spin paradigms explore the ramifications of high/low black hole spin surrounded by thin- disk or advection-dominated accretion in an attempt to enhance the jet efficiency. These ideas have been in part a response to arguments against powerful jets in black hole systems due to diffusive effects [30–32], suggesting that the black hole-threading magnetic field would be too weak for the BZ mechanism to explain the observations. This has also been a concern of general relativistic magnetohydrodynamic simulations (GRMHD) which we will discuss in some detail [33–36].

While the most radical departure from the spin paradigm, referred to as the "gap paradigm" [37], is also constructed from the basic building blocks of the other models (including BZ, BP, and thin/thick disks), it also, and crucially, hinges on the ability of retrograde accretion to produce the most powerful jets. This framework, therefore, argues that in addition to high/low black hole spin and thin-disk/advection dominated accretion, retrograde/prograde directions between disk and black hole also matter.

This paper is not an introduction to black hole engines (for an exhaustive review and detailed treatment see [38]). Its purpose is to highlight the differences between standard prograde models of accreting black holes and the recent retrograde framework. And because the track record on the discussion of compatibility between theory and observation has been poor, that is the focus of this paper.

2. On Supermassive Black Hole Formation

The underlying assumption of the black hole paradigm for AGN that supermassive black holes are produced in galactic centers is not a completed story because the mechanism for producing 10^9 solar mass black holes at early times remains unknown. Specifically, if the high redshift FRII radio quasars are modeled as the result of cold gas accretion onto black holes [39] with nonzero spin, constraints are imposed on the black hole formation scenarios. While models of spiral galaxies suggest that central black hole accretion is produced by secular processes [40–43] onto smaller-sized black holes (i.e., 10^5–10^7 solar masses) that originate from primordial seeding mechanisms that are still debated, the supermassive black holes in the centers of elliptical galaxies are presumed to be the product of galaxy mergers, where two already massive black holes come together. However, merging black holes suffer the so-called "last parsec problem" [44]. Because the two black holes are unlikely to fall straight into each other but find themselves orbiting one another, a successful merger comes about only if the binary angular momentum can be extracted. While the conventional wisdom has been that the excess angular momentum ends up in stars and gases that are close to the center of the newly formed galaxy, a problem emerges. Stars and gases may well acquire the angular momentum, but they are then pushed outwards or away from the black holes, leaving the two black holes isolated, without a complete repository for shedding the remaining angular momentum required to merge the black holes. As a result, the merger process stalls at a characteristic distance of about 1 pc, well short of the 0.01 pc necessary for gravitational waves to complete the job [44]. While signatures of binary black holes are observed [45], it appears to be only a small fraction (about 10%) in the local universe but very small (about 0.3%) for quasars at all redshifts [46]. For the largest black holes, the merger process generally appears to operate to completion. How does this happen?

While a number of possible mechanisms have been proposed, recent numerical simulations of retrograde accretion onto binary black holes show that the efficiency of this process in extracting both the energy and angular momentum of the binary is greater than that of a prograde disk [47]. Here, the angular momentum of the accretion flow is antiparallel to that of the black hole binary, that is, the accretion flow is rotating in the opposite direction compared to the direction in which black holes rotate about one another. Because angular momentum acquired by the retrograde flow causes the accretion flow to move further inward toward the binary, unlike in the prograde case, a repository for the angular momentum of the binary continues to be present throughout the merging process, and the black holes successfully shed their binary angular momentum as they approach the subparsec regime.

Interestingly, retrograde accretion onto a massive black hole is the starting assumption of the gap paradigm [37]. However, simulations suggest that the postmerger black hole will tend to have low retrograde spin [47], which imposes hitherto unclear constraints on the gap paradigm. Given that the strengths of magnetic fields threading black holes are unknown, the extent to which highest spinning retrograde black holes (as opposed to intermediate retrograde spin values) are needed is weakly constrained. Once magnetic field strengths in the very central regions are better estimated observationally, the model predictions can be squared with the observed redshift distribution of radio loud quasars. In more detail, once the simulations specify the fraction of postmerger systems producing rapid retrograde spin, using as input both the merger function versus redshift and magnetic field values in the inner accretion flow, we can determine the number of retrograde systems predicted by the model that will satisfy the observed requirements of the powerful FRII radio quasars, allowing the model to be tested in this respect. Spin and hybrid spin paradigm models are built on the assumption that FRII quasars have high prograde black hole spin. This difference, as we shall see in the next section, is crucial.

3. On Flat Spectrum Radio Quasars

If spin and hybrid spin paradigm models postulate rapid spinning black holes surrounded by a thin-disk accretion generated in postmerger ellipticals with powerful jets in a prograde configuration, the ISCO is closer to the horizon. Interestingly, the evidence in Flat-Spectrum Radio Quasars (FSRQs) points to ISCO values that are not close to the black hole but compatible with retrograde accretion [48–50]. This issue is addressed in spin paradigm and hybrid spin paradigm models via the added assumption that FSRQs are rapidly accreting systems, which may produce radiatively inefficient, geometrically thick disks [20, 31, 51, 52]. The purpose of this additional assumption is twofold. First, there is a need to explain the absence of strong X-ray reprocessing features from the inner regions, which would exist under the assumption of a radiatively efficient, geometrically thin disk plus hot corona models [53] within the context of high prograde spin. Second, numerical simulations have suggested that strong jets are produced only in geometrically thick accretion systems, but there is tension here. Radiatively inefficient accretion will tend to wash out or be incompatible with prominent signatures of thermal accretion such as broad optical lines and big blue continuum bumps, which means that as you consider thicker disk geometry, you tend to weaken the thermal disk signatures, but the absence of evidence for ISCOs close to the black hole come from the possibility of modeling the system via radiatively efficient thin disk accretion. In fact, the thermal spectra in such objects produce big blue bumps which are strong signatures of radiatively efficient thin-disk accretion with maximum disk temperatures lower than expected [54]. That tension is avoided in the gap paradigm since the most powerful jets in that framework are produced in thin-disk systems with largest gap regions where the ISCO is further out from the black hole, and the less reprocessed energy through the disk is compatible with a lower peak frequency for the big blue bump compared to radio quiet quasars [55]. In high prograde spin models, the need for both advection-dominated accretion as well as thin disks to model the observations suggests an additional way out, namely disk truncation. Here, the radiatively efficient nature of thin disks abruptly ceases to operate at some location in the disk, giving rise to radiatively inefficient thick-disk geometry inwards of that location [56]. However, this picture produces tension with the assumption

of scale invariance (discussed in Section 4) because X-ray binaries in either soft states or transitory burst states appear not to be the small-scale versions of FRII quasars. Finally, we should point out that recent work suggests the absence of jet power dependence on disk thickness [57], which removes the necessity to associate powerful jets with radiatively inefficient accretion.

4. On FRII Radio Quasars and Ballistic Microquasar Jets

Although from a theoretical perspective the detailed physics such as accretion disk temperature and density and outflow power do depend on black hole mass, the mechanism for producing that outflow does not. In other words, the nature of the mechanism producing the outflow is the same regardless of the size of the accreting black hole. This so-called "scale invariance" produces additional constraints on the modeling of the most powerful sources with jets—FRII radio quasars— and their time evolution. To see this, we must compare with their small-scale counterparts.

Stellar mass black hole accretion systems—X-ray binaries [58]—undergo transitions between states with jets (with relativistic gamma factors whose value is less than about 2), dominated by a hard X-ray spectrum and modeled as radiatively inefficient advection dominated accretion, and states without jets but with softer X-rays that can be well-fit by optically thick, geometrically thin disk models. And the evolutionary cycle involves a progression from hard state jets to states without jets but soft X-ray spectra and back, with a transitory but more powerful and collimated jet outflow (with a relativistic gamma factor whose value is greater than 2) produced in the hard to soft transition [59]. While scale invariance does not require that AGN jets follow this specific time evolution, it does imply large-scale analogs among AGN. In particular, the brightest hard state jets observed in X-ray binaries may be the small-scale counterpart to the FRI radio galaxies; the lower-luminosity hard state X-ray binaries may be the small-scale version of the low-luminosity AGN, while the jet-less soft state in X-ray binaries may be the small-scale counterpart to quasars and radio quiet AGN [60–62].

Explanations of the nature of the powerful transitory jets in X-ray binaries have been explored in terms of an unstable transition between hard state and soft state [20, 63]. And the interpretation of these states has emerged from the suggestion that lifetimes of FRII radio quasars are as short compared to lifetimes of FRI radio galaxies to the same degree, as transitory jets are to hard state jets in X-ray binaries. While the underlying physical model for the generation of these transitory jets is not in itself problematic, its straightforward time evolution scale-free extension to AGN violates the observations. The cyclical nature of X-ray binary state transitions, in fact, would imply that on average there would be no straightforward redshift dependence in the density of FRI versus FRII objects. Observationally, however, FRII quasar density peaks at higher redshifts (about $z = 2$), while the density of FRI radio galaxies and black hole masses increases toward lower redshifts [64, 65].

Alternatively, one could choose to explore a scale-invariant violating framework. Although this would circumvent the constraints discussed above, no successful scale-free violating models currently exist. In fact, recently discovered powerful jets in spiral galaxies (gamma ray loud Narrow Line Seyfert 1s) with lower mass black holes [66] further strain the tension between the scale-free spin paradigm and observations by forcing us to revisit models in which jets avoid spirals. It is worth pausing at this point to emphasize the degree to which the spin paradigm is grounded in the notion that spiral galaxies have low black hole spin and cannot thus produce powerful jets [26, 67, 68]. The scale-invariant gap paradigm predicts that jets in spiral galaxies exist in lower prograde spin black hole systems in radiatively efficient accretion states [37]. Hence, it would be model- constraining to apply the broad iron line fluorescence method to such systems [69].

5. On the Efficiency Requirements in Powerful Radio Galaxies

With the exception of the early and isolated pioneering work of Wilson [70], GRMHD simulations have been exploring black hole accretion for the past decade. While early numerical work supported the basic analytic BZ framework by showing that energy extraction from black holes is possible [33, 35, 71, 72], GRMHD simulations have also recently begun addressing the problems discussed above concerning jet efficiency. The first GRMHD simulations were disappointing in this respect because they produced jet powers considerably less than $\dot{M}c^2$ (by jet efficiency we mean the ratio of jet power to accretion power). Recent observations, in fact, showed that in many powerful radio galaxies in hot cluster environments, the jet power is at least an order of magnitude greater than its accretion power [73–75], in some cases about 25 times the accretion power [76]. In an attempt to address this problem, recent numerical work has focused on initial conditions for the magnetic fields that produce a greater advection of magnetic flux on the black hole [77, 78]. These "flooded" magnetospheres have enhanced efficiencies that increase with prograde values of spin up to about $3\,\dot{M}c^2$ at the highest spin values [79]. While this is an improvement, we are still a factor of 10 or more shy of explaining the powerful radio galaxies (e.g., the galaxy cluster RBS 797—[80]). In addition, more mild efficiency requirements in systems such as M87 also appear difficult to explain in GRMHD (even in the context of the ad hoc assumption of "flux flooding") as VLBI imaging of the central region of M87 indicates the possibility of black hole spin values around 0.6 [81]. In fact, jet efficiency in GRMHD is a steep function of black hole spin and drops down to about 30% for intermediate values of prograde spin [79], making it difficult to explain the measured jet efficiency in M87 around 5% at the Bondi radius and therefore possibly making it orders of magnitude larger near the black hole due to outflows [82]. The gap paradigm, on the other hand, produces less stringent accretion requirements for intermediate prograde spin values [37, 83]. It is unfortunate that Doeleman et al. [81] give the false impression that the possible prograde nature of the black hole in M87 constrains current models. In fact, none of their cited references suggest that M87's supermassive black hole should be retrograde. On the other hand, detailed modeling of the behavior of low angular momentum accretion onto M87 suggests caution on black hole spin inference in this source [84]. More generally, the proximity of M87's jet provides a wonderful opportunity for studying the shock-producing interaction of the jet with its environment in a classic FRI radio galaxy in detail [85, 86].

6. On Scale Invariance and FRI Radio Galaxies

FRI radio galaxy jets are generally weaker and less collimated on kpc scales than their powerful FRII counterparts with an observed distribution captured by the so-called Owen-Ledlow diagram. We mentioned that FRI radio galaxies are thought to satisfy scale invariance as the large-scale analogs of bright hard state X-ray binaries. When this assumption is coupled with recent observations, constraints are produced. Among these are the observed absence of any clear black hole spin dependence in X-ray binaries in their hard states [87] and the claim of a spin dependence in the transient ballistic jet [88].

It is worth emphasizing that the observed lack of difference in jet power between systems with different black hole spin in hard state X-ray binaries [87] does not constitute evidence that black hole spin energy extraction is not occurring—although it is compatible with that notion. It is evident that whichever process influences jet power, it produces a black hole spin dependence that is roughly flat. The possibility of a flat spin dependence is problematic for GRMHD which has consistently produced steep spin dependencies of jet power of the form $a^2 - a^6$ [89]. While it is clear that spin cannot be the only factor in determining the presence of a jet [90], the details of the jet-disk connection remain unresolved.

What about the apparent spin dependence observed during the transition state [88]? What mechanism operates to reveal the value of black hole spin as the accretion state transitions from advection dominated to radiatively efficient and thin? Narayan & McClintock suggest a previously proposed explanation, which is that as the system moves toward the soft state and the inner edge of the disk moves inward toward the ISCO shocks may occur leading to the formation of pc scale jets [59], which would be the small-scale counterpart of the quasar jets. However, this notion breaks scale invariance because—as discussed above—it implies an absence of FRII versus FRI density dependence on redshift, which appears directly incompatible with the redshift distribution of the AGN sources.

The gap paradigm, on the other hand, is explicitly scale invariant in this respect because the nature of the radio-loud to radio-quiet transition hinges on the observational results of Neilsen & Lee [91] and is further strengthened by Ponti et al. [92], which come from X-ray binaries. And the picture goes like this: as the disk transits toward a soft state, and the inner edge moves toward the ISCO, the radiative wind component of the disk increases which serves to collimate the black hole or BZ jet. As Neilsen and Lee point out, the effect is transitory since the disk wind eventually quenches the inner jet, but these considerations have implications for a spin dependence in the jet-quenching ability. This is because the wind outflow power is proportional to the energy generated at that location in the disk, which depends on the black hole spin (the larger the prograde spin value, the greater the wind efficiency). One must conclude, therefore, that the jet-quenching ability of the disk wind will be greater for larger prograde spin values. The consequence of adopting this scale-free framework, therefore, is the existence of black hole X-ray binaries at lower prograde spin whose jets are never fully quenched. The prediction in the gap paradigm is that some X-ray binaries will have jets that are more quenched in high states (corresponding to higher values of prograde spins), and others will have jets that are less fully quenched in high states (with lower values of prograde spins). The implication of this for AGN is the evolution of systems, that if persistently radiatively efficient in their accretion state, they evolve into radio-quiet quasars or radio-quiet AGN from FRII quasar states [37]. In addition, this picture suggests an explanation for the small fraction of observed FRI quasar objects in that they occupy a small range on the black hole spin spectrum (above zero but in the lower range of the prograde values).

Therefore, while the analysis of Narayan and McClintock suggesting black hole spin evidence in X-ray binaries is compatible with the gap paradigm, I would argue that the conditions producing ballistic jets in X-ray binaries are not relevant to those in high redshift AGN jets, so they cannot be the small-scale analogs of the FRII quasar jets. In spin and hybrid spin paradigm models, no problem-free, scale-invariant analog exists for the ballistic jets. In the gap paradigm, instead, FRII quasar jets are modeled as retrograde systems, and because time evolution (discussed later) does not produce thin radiatively efficient accretion following ADAF states, the large-scale analog of the ballistic jet is not observed in nature.

7. On the Radio-Loud/Radio-Quiet Dichotomy

In addition to the jet-disk connection, another major unresolved issue in astrophysics is the nature of the radio-loud/radio-quiet dichotomy. Observationally, we find that only 15–20% of active galaxies are radio-loud. A scale-free extension of the observations of X-ray binary state transitions does not explain the quantitative nature of the dichotomy. But if we consider retrograde accretion as a model for radio-loud quasars, we find a space for addressing the issue. Regardless of what assumption we begin with in terms of the fraction of postmerger objects that are retrograde compared with the number of prograde ones, the time evolution will take some of the radio-loud objects and naturally evolve them into radio-quiet ones. Time evolution, in fact, will spin the retrograde black hole down to zero spin in less than 10^7 years at the Eddington rate and after a few more 10^7 years, the black hole spin will be intermediate prograde. For the fraction of systems that persist in their radiatively efficient mode of accretion, the gap paradigm prescribes that they will turn into radio-quiet

objects. However, a fraction of these originally retrograde systems will not remain radiatively efficient in their accretion states but become ADAFs. These become FRI radio galaxies so it is only a fraction of a fraction of the original radio-loud retrograde objects that become radio-quiet quasars/AGN. In short, there is a natural mechanism in the paradigm for taking originally radio loud quasars and turning a fraction of them into radio-quiet quasars/AGN. In addition, the less massive black holes in retrograde configurations may flip to prograde configurations (discussed later), thereby taking originally radio-loud systems and turning them into radio-quiet ones. This means that although the quantitative fraction of radio-loud to radio-quiet objects depends on the specific initial conditions, a qualitative explanation for having more radio-quiet quasars/AGN compared to radio-loud ones emerges naturally. Once we acquire quantitative statements about the fraction of retrograde versus prograde configurations forming in mergers, the energetics required to turn FRII objects into FRIs, and the expected fraction of spin flips, we will be able to constrain the gap paradigm in relation to the observed 15–20% value of the radio-loud/radio-quiet dichotomy.

8. On Gamma-Ray Loud NLS1s and Jets in Spiral Galaxies

Until recently, spiral galaxies have generally entered into the AGN classification as radio-quiet objects which has prompted spin paradigm models to assume they have low-spinning black holes in their centers [26, 67, 68]. These ideas are grounded in the possibility of chaotic accretion whose purpose is to produce low-spinning black holes on average in such systems [93], but evidence is growing that the fraction of high-spinning prograde black holes is large in Seyferts [94, 95]. In addition, powerful jets have recently been discovered in spirals [66, 96–98]. Within the context of the spin paradigm, spiral AGN with powerful jets must contain high-spinning black holes. However, the Eddington ratio luminosities in these objects are difficult to reconcile with this notion because they appear to live in a regime of Eddington ratio luminosity that is intermediate between FSRQs and PG quasars [26, 98]. More precisely, FSRQs are observed to have lower Eddington ratios compared to gamma-ray loud NLS1s. In spin paradigm models, FSRQs and gamma-ray loud NLS1s should occupy the same region of the Eddington ratio luminosity because they would naturally be scale-free equivalents. In other words, if we model these two classes of AGN as prescribed in the spin paradigm, that is, as high prograde spinning black holes surrounded by truncated inner accretion disks, their powerful jets differ only because the black hole masses are larger in FSRQs. Therefore, their Eddington ratios would be expected to be equal. However, what we find is that on average gamma-ray loud NLS1s have larger Eddington ratios. This difference requires an explanation. In the gap paradigm, on the other hand, spiral AGN with jets are lower-spinning prograde systems so their disk efficiencies are naturally sandwiched between FSRQs (which are modeled as retrograde systems with smallest disk efficiencies) and radio-quiet AGN/quasars (which are modeled as the highest-spinning prograde systems, and thus with the largest disk efficiency). The notion that the overwhelming fraction of observed radio-quiet AGN as high-spinning prograde systems being due to a selection effect [95] stands in contrast to these intermediate-Eddington luminosities in gamma NLS1s.

9. On GRMHD

Numerical simulations of magnetized black hole accretion involve the general relativistic version of the Maxwell equations, which amounts to an advection and diffusive terms competing in the time evolution equation for the magnetic field, but this physics is not fully implemented in general relativistic simulations [99, 100]. Due to complications with the numerics, the diffusive term is absent. Hence, there is no physical diffusion in GRMHD (see Bucciantini and Del Zanna [101]) for recent issues and progress on resistive GRMHD). This fact has received far less emphasis than it should. What is the impact of this absence of physics? Nonrelativistic simulations of protostellar disks including resistive MHD [102, 103] show that the so-called "non-ideal" terms—that is, the terms that capture diffusive effects—are as important to the outcome as the ideal ones. Given that general relativity involves more intense dynamical interaction between the gas and magnetic field, especially close to the black hole in the ergosphere [104, 105], diffusive terms should be at least as important there as in the Newtonian regime, and most likely more so by some order(s) of magnitude. Unfortunately, this issue is no longer mentioned in the GRMHD literature. In addition, recent work in plasma theory, suggests that even if GRMHD simulations did include the standard diffusive terms, they would still produce results that are off by orders of magnitude due to the so-called "stochastic flux-freezing" [106].

The other issue confronting numerical simulations of magnetized black hole accretion involves the nature and geometry of the magnetic field [107]. In the absence of some naturally expected large-scale magnetic field configuration in the nuclei of galaxies, the geometry must be put in by hand according to some ad hoc prescription. In fact, due to computational limits, the geometry in GRMHD simulations tends to be specified and restricted to the inner regions of the accretion disks, usually in the form of internal loops of magnetic field that produce a zero net flux, a highly unnatural state of affairs. This would not be much of an issue were it not for the fact that different initial magnetic field configurations produce sufficiently different black hole-threading magnetic fields and therefore different jet powers [108, 109]. The combination of these two issues—choice for magnetic field geometry and absence of physical diffusion—can yield unrealistic detailed dependencies such as on the black hole spin [83]. While certain dynamical aspects of actual astrophysical magnetized black hole accretion should be reproduced in GRMHD, the arguments suggest that detailed dependencies cannot yet be properly captured (see

[107] Section 4), and it is within this context that analytic and semianalytic models remain essential.

10. On Time Evolution

The observed redshift distribution of radio galaxies and quasars provides a space for the most stringent constraints on theory [110, 111]. The gap paradigm postulates the presence of retrograde accretion in the most massive black hole mergers but not in spiral galaxies and black hole X-ray binaries, where the difficulty in forming a retrograde system constrains them to a prograde accretion regime. While the inverse relation between accretion and jet efficiency in the gap paradigm model has been the primary focus so far, the time evolution and consequent unification are arguably its most attractive features. This unification connects radio-loud objects (or sources with jets) to one another and to radio-quiet ones (or sources without jets) and in doing so produces an explanation for their redshift distribution. Despite claims in the recent literature [26, 67, 68], the spin paradigm continues to be at odds with observations in this respect, as discussed below.

The postmerger retrograde systems, that is, the starting assumption of the gap paradigm involve cold, radiatively efficient, and Shakura and Sunyaev type accretion with powerful jets, with cold accretion being the result of post-merger funneling of cold gas into the nuclear region [112]. The subsequent time evolution does not require additional assumptions. This point requires emphasis. Accretion simply spins the black hole down and then up again in the prograde regime, so the gap region decreases in size with time. On the foundation of that simple picture, the model proposes an explanation for the evolution of radio-loud quasars into radio-quiet quasars and/or radio galaxies. Accordingly, the fraction of FRII objects is larger at higher redshifts and gives way to a dominance of FRI objects at lower redshifts. An explanation is also found in this picture for the absence of merger signatures associated with FRI radio galaxies [113] in that they are the late-state evolution of originally retrograde black hole accretion systems that in turn were triggered by galaxy mergers. It is worth noting that on purely accretion timescales, the Eddington limited accretion will spin down to zero spin a maximally retrograde black hole in just under 10 million years. This timescale is beautifully and otherwise coincidentally compatible with estimates of FRII lifetimes [114]. Also, this framework provides a means for understanding that scaling relations should differ between radio-loud AGN and radio-quiet AGN, the former being determined by a combination of different types of jets (FRII/FRI) and accretion (radiatively efficient to radiatively inefficient), the latter being governed by disk winds in radiatively efficient accretion.

Spin paradigm models, on the other hand, struggle at the outset in modeling the powerful FRII quasars because they are forced to high spin and prograde accretion which maximizes both the jet and accretion efficiency. The already mentioned tension that such a combination of requirements produces is further compounded by its difficulty to explain the inverse relationship between accretion outflow power in radio-loud quasars versus radio-quiet quasars. In other words, accretion outflow power should be greater in sources that have high prograde spin compared to those with low spin, yet the sources with the most powerful jets invariably show distributions spanning a range of weaker disk outflows [115, 116]. Martinez-Sansigre & Rawlings [67] argue that it is the low-number statistics that creates this distribution and that once the number of sources increases, the wind power in radio-loud quasars will be shown to dominate over the radio-quiet population. But what would the observations have to be to support an inverse relationship between accretion wind power and radio-loud AGN? In other words, what would the observations have to be to support the gap paradigm? They would be exactly what they are. In addition, FRII quasars have on average larger black hole masses compared to radio-quiet quasars which means that observations of faster winds in radio-quiet quasars is even more statistically significant. Hence, the conclusion of Martinez-Sansigre & Rawlings is a hope, one that spin paradigm and hybrid spin paradigm models share explicitly or implicitly, but one that unfortunately is the hallmark of a nonfalsifiable framework.

The observations of low-luminosity FRII radio galaxies at intermediate redshifts between FRII quasars and low-luminosity FRI radio galaxies is a natural outcome of time evolution in the gap paradigm (as described in detail in [37]). The redshift distribution of low-luminosity FRII radio galaxies in the spin paradigm, on the other hand, finds no natural explanation in terms of time evolution, and certainly not in a scale-invariant sense. And, as pointed out above, the jet efficiency requirements [73, 75, 76] in these objects are currently unresolved within spin paradigm and hybrid spin paradigm models regardless of explanations that deal with their redshift distributions. The gap paradigm, again, produces much less stringent requirements on the accretion rate [80].

Because retrograde systems may be unstable [117], and naturally tend to prograde systems, a possible explanation to the observation that radio-loud AGN host the most massive black holes ([118] and references therein) emerges in the gap paradigm. In fact, the larger the black holes, the more stable their retrograde accretion phase, while smaller black holes would tend to be less stable. The picture in the gap paradigm would go as follows: any high-spinning retrograde accretion system with a less massive black hole may flip to a prograde configuration, thereby becoming a radio-quiet quasar or AGN. The transition from the retrograde to the prograde regime is also an attractive place to explore explanations for X-shaped radio galaxies and the missing objects of the so-called "blazar envelope" [119].

11. Conclusions

Despite a common foundation grounded in accretion and black hole spin, the gap paradigm and spin and hybrid spin paradigms (SHSPs) differ in substantial ways. They are listed below. The emphasis of this paper has been on the observations and the extent to which theoretical differences can currently be constrained.

(1) In the gap paradigm, powerful FRII quasars involve retrograde accretion onto rotating black holes; while in SHSPs, FRII quasars involve prograde accretion onto rotating black holes.

(2) In the gap paradigm, FRII quasars evolve into FRII radio galaxies or FRI quasars and eventually either into FRI radio galaxies or radio-quiet quasars; while in SHSPs, FRII quasar evolution has no clear or model-constrained scale-invariant progression.

(3) In the gap paradigm, FRI quasars live in a range of lower prograde black hole spin values; whereas in SHSPs, they are high-spinning prograde systems.

(4) Whereas in the gap paradigm radio-quiet quasars are high-spin prograde systems surrounded by cold, thin-disk accretion, in SHSPs, they are low-spin systems surrounded by cold, thin-disk accretion.

(5) Whereas in the gap paradigm FRII objects will have a greater density at higher redshifts compared to FRI objects, in SHSPs, the natural scale invariant approach provides no redshift-dependent difference between FRI and FRIIs.

(6) Whereas in the gap paradigm spiral galaxies generate smaller-scale equivalents of FRI quasars, in SHSPs, spiral galaxies generally do not produce jets because they tend to include the assumption of secular processes and chaotic accretion, but when they do have jets, the spin must be prograde and high.

(7) Whereas in the gap paradigm the distribution of the black hole spin in the local universe should be prograde because such is the natural outcome of prolonged accretion, in SHSPs, the distribution of the black hole spin in the local universe should be around zero.

(8) Whereas in the gap paradigm X-shaped radio galaxies may find an explanation at the transition between retrograde and prograde systems, in SHSPs, there are no natural explanations for their presence along the boundary of the Owen-Ledlow diagram.

(9) Whereas in the gap paradigm FSRQs should have lower-Eddington ratios compared to gamma NLS1s, in SHSPs, their Eddington ratios should be equivalent.

Acknowledgment

The author thanks Rob Fender for setting him straight on a number of issues.

References

[1] B. R. McNamara and P. E. J. Nulsen, "Heating hot atmospheres with active galactic nuclei," *Annual Review of Astronomy and Astrophysics*, vol. 45, pp. 117–175, 2007.

[2] N. P. H. Nesvadba, M. D. Lehnert, C. De Breuck, A. M. Gilbert, and W. Van Breugel, "Evidence for powerful AGN winds at high redshift: Dynamics of galactic outflows in radio galaxies during the 'quasar Era," *Astronomy and Astrophysics*, vol. 491, no. 2, pp. 407–424, 2008.

[3] A. Mangalam, Gopal-Krishna, and P. J. Wiita, "The changing interstellar medium of massive elliptical galaxies and cosmic evolution of radio galaxies and quasars," *Monthly Notices of the Royal Astronomical Society*, vol. 397, no. 4, pp. 2216–2224, 2009.

[4] A. Cattaneo, S. M. Faber, J. Binney et al., "The role of black holes in galaxy formation and evolution," *Nature*, vol. 460, pp. 213–219, 2009.

[5] V. Antonuccio-Delogu and J. Silk, "Active galactic nuclei activity: self-regulation from backflow," *Monthly Notices of the Royal Astronomical Society*, vol. 405, no. 2, pp. 1303–1314, 2010.

[6] F. van de Voort, J. Schaye, C. M. Booth, and C. D. Vecchia, "The drop in the cosmic star formation rate below redshift 2 is caused by a change in the mode of gas accretion and by active galactic nucleus feedback," *Monthly Notices of the Royal Astronomical Society*, vol. 415, no. 3, pp. 2782–2789, 2011.

[7] W. Ishibashi and A. C. Fabian, "Active galactic nucleus feedback and triggering of star formation in galaxies," *Monthly Notices of the Royal Astronomical Society*, vol. 427, no. 4, pp. 2998–3005, 2012.

[8] J. Magorrian, S. Tremaine, D. Richstone et al., "The demography of massive dark objects in galaxy centers," *Astronomical Journal*, vol. 115, no. 6, pp. 2285–2305, 1998.

[9] K. Gebhardt, "A relationship between nuclear black hole mass and galaxy velocity dispersion," *The Astrophysical Journal Letters*, vol. 539, no. 1, article L13, 2000.

[10] L. Bîrzan, D. A. Rafferty, B. R. McNamara, M. W. Wise, and P. E. J. Nulsen, "A systematic study of radio-induced X-ray cavities in clusters, groups, and galaxies," *The Astrophysical Journal*, vol. 607, no. 2, article 800, 2004.

[11] D. A. Rafferty, "the feedback-regulated growth of black holes and bulges through gas accretion and starbursts in cluster central dominant galaxies," *The Astrophysical Journal*, vol. 652, no. 1, article 216, 2006.

[12] B. R. McNamara and P. E. J. Nulsen, "Mechanical feedback from active galactic nuclei in galaxies, groups and clusters," *New Journal of Physics*, vol. 14, Article ID 055023, 2012.

[13] M. Gitti, F. Brighenti, and B. R. McNamara, "Evidence for AGN feedback in galaxy clusters and groups," *Advances in Astronomy*, vol. 2012, Article ID 950641, 24 pages, 2012.

[14] R. A. Remillard and J. E. McClintock, "X-ray properties of black-hole binaries," *Annual Review of Astronomy and Astrophysics*, vol. 44, pp. 49–92, 2006.

[15] N. L. Shakura and R. A. Sunyaev, "Black holes in binary systems. Observational appearance," *Astronomy and Astrophysics*, vol. 24, pp. 337–355, 1973.

[16] I. Novikov and K. Thorne, "Astrophysics of black holes," in *Black Holes (Les Astres Occlus)*, C. DeWitt and B. S. DeWitt, Eds., pp. 343–450, Gordon Breach, New York, NY, USA, 1973.

[17] J. Silk and M. J. Rees, "Quasars and galaxy formation," *Astronomy & Astrophysics*, vol. 331, pp. L1–L4, 1998.

[18] R. D. Blandford and R. L. Znajek, "Electromagnetic extraction of energy from Kerr black holes," *Monthly Notices of the Royal Astronomical Society*, vol. 179, pp. 433–456, 1977.

[19] R. D. Blandford and D. G. Payne, "Hydromagnetic flows from accretion discs and the production of radio jets," *Monthly Notices of the Royal Astronomical Society*, vol. 199, pp. 883–903, 1982.

[20] D. L. Meier, "The association of jet production with geometrically thick accretion flows and black hole rotation," *The Astrophysical Journal Letters*, vol. 548, no. 1, article L9, 2001.

[21] R. Narayan and I. Yi, "Advection-dominated accretion: self-similarity and bipolar outflows," *Astrophysical Journal*, vol. 444, no. 1, part 1, pp. 231–243, 1995.

[22] J. M. Bardeen, W. H. Press, and S. A. Teukolsky, "Rotating black holes: locally nonrotating frames, energy extraction, and scalar synchrotron radiation," *Astrophysical Journal*, vol. 178, pp. 347–370, 1972.

[23] R. D. Blandford, "Physical processes in active galactic nuclei," in *Active Galactic Nuclei*, T. J. L. Courvoisier and M. Mayor, Eds., pp. 161–275, Springer, Berlin, Germany, 1990.

[24] A. S. Wilson and E. J. M. Colbert, "The difference between radio-loud and radio-quiet active galaxies," *Astrophysical Journal*, vol. 438, no. 1, part 1, p. 62.

[25] R. Moderski, M. Sikora, and J. P. Lasota, "On the spin paradigm and the radio dichotomy of quasars," *Monthly Notices of the Royal Astronomical Society*, vol. 301, no. 1, pp. 142–148, 1998.

[26] M. Sikora, Ł. Stawarz, and J. P. Lasota, "Radio loudness of active galactic nuclei: observational facts and theoretical implications," *Astrophysical Journal*, vol. 658, no. 2, pp. 815–828, 2007.

[27] B. Punsly and F. V. Coroniti, "Ergosphere-driven winds," *Astrophysical Journal Letters*, vol. 354, no. 2, pp. 583–615, 1990.

[28] D. L. Meier, "A magnetically switched, rotating black hole model for the production of extragalactic radio jets and the fanaroff and riley class division," *The Astrophysical Journal*, vol. 522, no. 2, article 753, 1999.

[29] R. S. Nemmen, R. G. Bower, A. Babul, and T. Storchi-Bergmann, "Models for jet power in elliptical galaxies: a case for rapidly spinning black holes," *Monthly Notices of the Royal Astronomical Society*, vol. 377, no. 4, pp. 1652–1662, 2007.

[30] S. H. Lubow, J. C. B. Papaloizou, and J. E. Pringle, "On the stability of magnetic Wind-Driven accretion discs," *Monthly Notices of the Royal Astronomical Society*, vol. 268, no. 4, pp. 1010–1014, 1994.

[31] M. Livio, "extracting energy from black holes: the relative importance of the Blandford-Znajek mechanism," *The Astrophysical Journal*, vol. 512, no. 1, article 100, 1999.

[32] P. Ghosh and M. A. Abramowicz, "Electromagnetic extraction of rotational energy from disc-fed black holes: the strength of the Blandford-Znajek process," *Monthly Notices of the Royal Astronomical Society*, vol. 292, no. 4, pp. 887–895, 1997.

[33] J.-P. De Villiers, J. F. Hawley, and J. H. Krolik, "Magnetically driven accretion flows in the kerr metric. I. models and overall structure," *The Astrophysical Journal*, vol. 599, no. 2, article 1238, 2003.

[34] J.-P. De Villiers, J. F. Hawley, J. H. Krolik, and S. Hirose, "Magnetically driven accretion in the Kerr metric. III. Unbound outflows," *The Astrophysical Journal*, vol. 620, no. 2, article 878, 2005.

[35] J. C. McKinney and C. F. Gammie, "A measurement of the electromagnetic luminosity of a Kerr black hole," *Astrophysical Journal*, vol. 611, no. 2 I, pp. 977–995, 2004.

[36] J. C. McKinney, "General relativistic magnetohydrodynamic simulations of the jet formation and large-scale propagation from black hole accretion systems," *Monthly Notices of the Royal Astronomical Society*, vol. 368, no. 4, pp. 1561–1582, 2006.

[37] D. Garofalo, D. A. Evans, and R. M. Sambruna, "The evolution of radio-loud active galactic nuclei as a function of black hole spin," *Monthly Notices of the Royal Astronomical Society*, vol. 406, no. 2, pp. 975–986, 2010.

[38] D. L. Meier, *Black Hole Astrophysics: the Engine Paradigm*, Springer, Berlin, Germany, 2012.

[39] M. J. Hardcastle, D. A. Evans, and J. H. Croston, "Hot and cold gas accretion and feedback in radio-loud active galaxies," *Monthly Notices of the Royal Astronomical Society*, vol. 376, no. 4, pp. 1849–1856, 2007.

[40] D. M. Crenshaw, S. B. Kraemer, and J. R. Gabel, "The host galaxies of narrow-line seyfert 1 galaxies: evidence for bar-driven fueling," *The Astronomical Journal*, vol. 126, no. 4, article 1690, 2003.

[41] R. I. Davies, "a close look at star formation around active galactic nuclei," *The Astrophysical Journal*, vol. 671, no. 2, article 1388, 2007.

[42] Y.-M. Chen, J.-M. Wang, C.-S. Yan, C. Hu, and S. Zhang, "The starburst-active galactic nucleus connection: the role of young stellar populations in fueling supermassive black holes," *The Astrophysical Journal Letters*, vol. 695, no. 2, article L130, 2009.

[43] G. O. de Xivry, R. Davies, M. Schartmann et al., "The role of secular evolution in the black hole growth of narrow-line Seyfert 1 galaxies," *Monthly Notices of the Royal Astronomical Society*, vol. 417, no. 4, pp. 2721–2736, 2011.

[44] M. Milosavljevic and D. Merritt, "Formation of galactic nuclei," *The Astrophysical Journal*, vol. 563, no. 1, article 34, 2001.

[45] J. M. Comerford, B. F. Gerke, D. Stern et al., "Kiloparsec-scale spatial offsets in double-peaked narrow-line active galactic nuclei. I. Markers for selection of compelling dual active galactic nucleus candidates," *Astrophysical Journal*, vol. 753, no. 1, article 42, 2012.

[46] M. Volonteri, F. Haardt, and P. Madau, "The assembly and merging history of supermassive black holes in hierarchical models of galaxy formation," *Astrophysical Journal Letters*, vol. 582, no. 2 I, pp. 559–573, 2003.

[47] C. J. Nixon, P. J. Cossins, A. R. King, and J. E. Pringle, "Retrograde accretion and merging supermassive black holes," *Monthly Notices of the Royal Astronomical Society*, vol. 412, no. 3, pp. 1591–1598, 2011.

[48] J. Kataoka, J. N. Reeves, K. Iwasawa et al., "Probing the disk-jet connection of the radio galaxy 3C 120 observed with Suzaku," *Publications of the Astronomical Society of Japan*, vol. 59, no. 2, pp. 279–297, 2007.

[49] R. M. Sambruna, J. N. Reeves, V. Braito et al., "Structure of the accretion flow in broad-line radio galaxies: the case of 3C 390.3," *Astrophysical Journal Letters*, vol. 700, no. 2, pp. 1473–1487, 2009.

[50] R. M. Sambruna, F. Tombesi, J. N. Reeves et al., "The *SUZAKU* view of 3C 382," *The Astrophysical Journal*, vol. 734, no. 2, article 105, 2011.

[51] D. Lynden-Bell, "Gravity power," *Physica Scripta*, vol. 17, no. 3, article 185, p. 185, 1978.

[52] L. Nobili, E. Turolla, and M. Calvani, "Are thick accretion disks the «central engine» for astrophysical jets?" *Lettere Al Nuovo Cimento Series 2*, vol. 35, no. 10, pp. 335–340, 1982.

[53] F. Haardt and L. Maraschi, "A two-phase model for the X-ray emission from Seyfert galaxies," *Astrophysical Journal Letters*, vol. 380, no. 2, pp. L51–L54, 1991.

[54] E. W. Bonning, G. A. Shields, A. C. Stevens, and S. Salviander, "Accretion disk temperatures of QSOs: constraints from the emission lines," http://arxiv.org/abs/1210.6997.

[55] L. Ballo, V. Braito, J. N. Reeves, R. M. Sambruna, and F. Tombesi, "The high-energy view of the broad-line radio galaxy 3C 111," *Monthly Notices of the Royal Astronomical Society*, vol. 418, no. 4, pp. 2367–2380, 2011.

[56] A. A. Esin, J. E. McClintock, and R. Narayan, "Advection-dominated accretion and the spectral states of black hole X-ray binaries: application to nova muscae 1991," *The Astrophysical Journal*, vol. 489, no. 2, article 865, 1997.

[57] P. C. Fragile, "No correlation between disc scale height and jet power in GRMHD simulations," *Monthly Notices of the Royal Astronomical Society*, vol. 424, no. 1, pp. 524–531.

[58] W. H. G. Lewin and M. van der Klis, Eds., *Compact Stellar X-ray Sources*, Cambridge Astrophysics Series, no. 39, Cambridge University Press, Cambridge, UK, 2006.

[59] R. P. Fender, T. M. Belloni, and E. Gallo, "Towards a unified model for black hole X-ray binary jets," *Monthly Notices of the Royal Astronomical Society*, vol. 355, no. 4, pp. 1105–1118, 2004.

[60] T. J. Maccarone, E. Gallo, and R. Fender, "The connection between radio-quiet active galactic nuclei and the high/soft state of X-ray binaries," *Monthly Notices of the Royal Astronomical Society*, vol. 345, pp. L19–L24, 2003.

[61] E. G. Koerding, S. Jester, and R. Fender, "Accretion states and radio loudness in active galactic nuclei: analogies with X-ray binaries," *Monthly Notices of the Royal Astronomical Society*, vol. 372, no. 3, pp. 1366–1378, 2006.

[62] S. Markoff, "From multiwavelength to mass scaling: accretion and ejection in microquasars and AGN," in *The Jet Paradigm*, Lecture Notes in Physics, pp. 143–172, Springer, Berlin, Germany, 2010.

[63] E. Szuszkiewicz and J. C. Miller, "limit-cycle behaviour of thermally unstable accretion flows on to black holes," *Monthly Notices of the Royal Astronomical Society*, vol. 298, no. 3, pp. 888–896, 1998.

[64] V. Smolcic, "the radio AGN population dichotomy: green valley seyferts versus red sequence low-excitation active galactic nuclei," *The Astrophysical Journal Letters*, vol. 699, no. 1, article L43, p. L43, 2009.

[65] V. Smolčić, G. Zamorani, E. Schinnerer et al., "Cosmic evolution of radio selected active galactic nuclei in the cosmos field," *Astrophysical Journal Letters*, vol. 696, no. 1, pp. 24–39, 2009.

[66] L. Foschini, "Evidence of powerful relativistic jets in narrow-line Seyfert 1 galaxies," in *Proceedings of the Workshop Narrow-Line Seyfert 1 Galaxies and Their Place in the Universe*, vol. NLS1, p. 024, 2011.

[67] A. Martinez-Sansigre and S. Rawlings, "Observational constraints on the spin of the most massive black holes from radio observations," *Monthly Notices of the Royal Astronomical Society*, vol. 414, no. 3, pp. 1937–1964, 2011.

[68] N. Fanidakis, C. M. Baugh, A. J. Benson et al., "Grand unification of AGN activity in the ΛCDM cosmology," *Monthly Notices of the Royal Astronomical Society*, vol. 410, no. 1, pp. 53–74, 2011.

[69] L. W. Brenneman and C. S. Reynolds, "Relativistic broadening of iron emission lines in a sample of active galactic nuclei," *Astrophysical Journal Letters*, vol. 702, no. 2, pp. 1367–1386, 2009.

[70] J. R. Wilson, "Numerical integration of the equations of relativistic hydrodynamics," in *Proceedings of the Marcel Grossman Meeting*, R. Ruffini, Ed., p. 393, North Holland, 1977.

[71] S. S. Komissarov, "Direct numerical simulations of the Blandford-Znajek effect ," *Monthly Notices of the Royal Astronomical Society*, vol. 326, no. 3, pp. L41–L44, 2001.

[72] S. Hirose, J. H. Krolik, J.-P. De Villiers, and J. F. Hawley, "Magnetically driven accretion flows in the Kerr metric. II. structure of the magnetic field," *The Astrophysical Journal*, vol. 606, no. 2, article 1083, 2004.

[73] B. R. McNamara, "An energetic agn outburst powered by a rapidly spinning supermassive black hole or an accreting ultramassive black hole," *The Astrophysical Journal*, vol. 698, no. 1, article 594, 2009.

[74] B. R. McNamara, M. Rohanizadegan, and P. E. J. Nulsen, "Are radio active galactic nuclei powered by accretion or black hole spin?" *Astrophysical Journal Letters*, vol. 727, no. 1, 2011.

[75] C. A. C. Fernandes, "Evidence for a maximum jet efficiency for the most powerful radio galaxies," *Monthly Notices of the Royal Astronomical Society*, vol. 411, no. 3, pp. 1909–1916, 2011.

[76] B. Punsly, "high jet efficiency and simulations of black hole magnetospheres," *The Astrophysical Journal Letters*, vol. 728, no. 1, article L17, 2011.

[77] A. Tchekhovskoy, R. Narayan, and J. C. McKinney, "Efficient generation of jets from magnetically arrested accretion on a rapidly spinning black hole," *Monthly Notices of the Royal Astronomical Society Letters*, vol. 418, no. 1, pp. L79–L83, 2011.

[78] J. C. McKinney, A. Tchekhovskoy, and R. D. Blandford, "General relativistic magnetohydrodynamic simulations of magnetically choked accretion flows around black holes," *Monthly Notices of the Royal Astronomical Society*, vol. 423, no. 4, pp. 3083–3117, 2012.

[79] A. Tchekhovskoy, J. C. McKinney, and R. Narayan, "General relativistic modeling of magnetized jets from accreting black holes," *Journal of Physics: Conference Series*, vol. 372, no. 1, Article ID 012040, 2012.

[80] K. W. Cavagnolo, B. R. McNamara, M. W. Wise et al., "A powerful AGN outburst in RBS 797," *The Astrophysical Journal*, vol. 732, no. 2, article 71, 2011.

[81] S. S. Doeleman, V. L. Fish, D. E. Schenck et al., "Jet-launching structure resolved near the supermassive black hole in M87," *Science*, vol. 338, no. 6105, pp. 355–358, 2012.

[82] Allen, "The relation between accretion rate and jet power in X-ray luminous elliptical galaxies," *Monthly Notices of the Royal Astronomical Society*, vol. 372, no. 1, pp. 21–30, 2006.

[83] D. Garofalo, "The spin dependence of the blandford-znajek effect," *Astrophysical Journal Letters*, vol. 699, no. 1, pp. 400–408, 2009.

[84] T. K. Das, S. Nag, S. Hegde et al., "Behaviour of low angular momentum relativistic accretion close to the event horizon," http://arxiv.org/abs/1211.6952.

[85] M. Nakamura, D. Garofalo, and D. L. Meier, "A magnetohydrodynamic model of the M87 jet. i. superluminal knot ejections from HST-1 as trails of quad relativistic MHD shocks," *The Astrophysical Journal*, vol. 721, no. 2, article 1783, 2010.

[86] P. Polko, D. L. Meier, and S. Markoff, "Determining the optimal locations for shock acceleration in magnetohydrodynamical jets," *Astrophysical Journal Letters*, vol. 723, no. 2, pp. 1343–1350, 2010.

[87] R. P. Fender, E. Gallo, and D. Russell, "No evidence for black hole spin powering of jets in X-ray binaries," *Monthly Notices of the Royal Astronomical Society*, vol. 406, no. 3, pp. 1425–1434, 2010.

[88] R. Narayan and J. E. McClintock, "Observational evidence for a correlation between jet power and black hole spin," *Monthly Notices of the Royal Astronomical Society Letters*, vol. 419, no. 1, pp. L69–L73, 2012.

[89] A. Tchekhovskoy, R. Narayan, and J. C. McKinney, "Black hole spin and the radio loud/quiet dichotomy of active galactic nuclei," *Astrophysical Journal Letters*, vol. 711, no. 1, pp. 50–63, 2010.

[90] J. E. McClintock, R. Narayan, S. W. Davis et al., "Measuring the spins of accreting black holes," *Classical and Quantum Gravity*, vol. 28, no. 11, article 114009, 2011.

[91] J. Neilsen and J. C. Lee, "Accretion disk winds as the jet suppression mechanism in the microquasar GRS 1915+105," *Nature*, vol. 458, pp. 481–484, 2009.

[92] G. Ponti, R. P. Fender, M. C. Begelman, R. J. H. Dunn, J. Neilsen, and M. Coriat, "Ubiquitous equatorial accretion disc winds in black hole soft states," *Monthly Notices of the Royal Astronomical Society Letters*, vol. 422, no. 1, pp. L11–L15, 2012.

[93] A. R. King, J. E. Pringle, and J. A. Hoffman, "The evolution of black hole mass and spin in active galactic nuclei," *Monthly Notices of the Royal Astronomical Society*, vol. 385, no. 3, pp. 1621–1627, 2008.

[94] A. Zoghbi, A. C. Fabian, P. Uttley et al., "Broad iron L line and X-ray reverberation in 1H0707-495," *Monthly Notices of the Royal Astronomical Society*, vol. 401, no. 4, pp. 2419–2432, 2010.

[95] L. W. Brenneman, C. S. Reynolds, M. A. Nowak et al., "THE spin of the supermassive black hole in NGC 3783," *The Astrophysical Journal*, vol. 736, no. 2, article 103, 2011.

[96] S. Komossa, W. Voges, D. Xu et al., "Radio-loud narrow-line type 1 Quasars," *Astronomical Journal*, vol. 132, no. 2, pp. 531–545, 2006.

[97] L. Foschini, "Gamma-ray emission from Narrow-Line Seyfert 1 galaxies and implications on the jets unification," *AIP Conference Proceedings*, vol. 1505, pp. 574–577, 2012.

[98] L. Foschini, "Powerful relativistic jets in spiral galaxies," *International Journal of Modern Physics*, vol. 8, pp. 172–177, 2012.

[99] D. L. Meier, "Ohm's law in the fast lane: general relativistic charge dynamics," *The Astrophysical Journal*, vol. 605, no. 1, article 340, 2004.

[100] J. A. Font, "Numerical hydrodynamics and magnetohydrodynamics in general relativity," *Living Reviews in Relativity*, vol. 11, article 7, 2008.

[101] N. Bucciantini and L. Del Zanna, "A fully covariant mean-field dynamo closure for numerical 3 + 1 resistive GRMHD," *Monthly Notices of the Royal Astronomical Society*, vol. 428, no. 1, pp. 71–85, 2013.

[102] R. Krasnopolsky, Z. Y. Li, and H. Shang, "Disk formation enabled by enhanced resistivity," *Astrophysical Journal Letters*, vol. 716, no. 2, pp. 1541–1550, 2010.

[103] R. Krasnopolsky, Z.-Y. Li, and H. Shang, "Disk formation in magnetized clouds enabled by the hall effect," *The Astrophysical Journal*, vol. 733, no. 1, article 54, 2011.

[104] S. Koide and K. Arai, "Energy extraction from a rotating black hole by magnetic reconnection in the ergosphere," *Astrophysical Journal Letters*, vol. 682, no. 2, pp. 1124–1133, 2008.

[105] S. Koide, "Generalized general relativistic magnetohydrodynamic equations and distinctive plasma dynamics around rotating black holes," *The Astrophysical Journal*, vol. 798, no. 2, article 1459, 2010.

[106] G. L. Eyink, "Stochastic flux freezing and magnetic dynamo," *Physical Review E*, vol. 83, no. 5, article 056405, 25 pages, 2011.

[107] E. G. Blackman, "Accretion disks and dynamos: toward a unified mean field theory ," *Physica Scripta*, vol. 86, no. 5, Article ID 058202, 2012.

[108] K. Beckwith, J. F. Hawley, and J. H. Krolik, "The influence of magnetic field geometry on the evolution of black hole accretion flows: similar disks, drastically different jets," *The Astrophysical Journal*, vol. 678, no. 2, 1180, 2008.

[109] J. C. McKinney and R. D. Blandford, "Stability of relativistic jets from rotating, accreting black holes via fully three-dimensional magnetohydrodynamic simulations," *Monthly Notices of the Royal Astronomical Society Letters*, vol. 394, no. 1, pp. L126–L130, 2009.

[110] M. Balokovic, "Disclosing the radio loudness distribution dichotomy in quasars: an unbiased Monte Carlo approach

applied to the SDSS-first quasar sample," *The Astrophysical Journal*, vol. 750, no. 1, article 30, 2012.

[111] L. Ballo, F. J. H. Heras, X. Barcons, and F. J. Carrera, "Exploring X-ray and radio emission of type 1 AGN up to $z \sim 2.3$," *Astronomy & Astrophysics*, vol. 545, article 66, 15 pages, 2012.

[112] J. E. Barnes and L. E. Hernquist, "Fueling starburst galaxies with gas-rich mergers," *Astrophysical Journal Letters*, vol. 370, no. 2, pp. L65–L68, 1991.

[113] B. H. C. Emonts, R. Morganti, C. Struve et al., "Large-scale H i in nearby radio galaxies-II. The nature of classical low-power radio sources," *Monthly Notices of the Royal Astronomical Society*, vol. 406, no. 2, pp. 987–1006, 2010.

[114] C. P. O'Dea, R. A. Daly, P. Kharb, K. A. Freeman, and S. A. Baum, "Physical properties of very powerful FRII radio galaxies," *Astronomy & Astrophysics*, vol. 494, no. 2, article 488, 2009.

[115] F. Tombesi, R. M. Sambruna, J. N. Reeves et al., "Discovery of ultra-fast outflows in a sample of broad-line radio galaxies observed with *SsUZAKU*," *The Astrophysical Journal*, vol. 719, no. 1, article 700, 2010.

[116] F. Tombesi, M. Cappi, J. N. Reeves et al., "Evidence for ultra-fast outflows in radio-quiet AGNs," *Astronomy & Astrophysics*, vol. 521, article A57, 35 pages, 2010.

[117] A. Perego, M. Dotti, M. Colpi, and M. Volonteri, "Mass and spin co-evolution during the alignment of a black hole in a warped accretion disc," *Monthly Notices of the Royal Astronomical Society*, vol. 399, no. 4, pp. 2249–2263, 2009.

[118] D. J. E. Floyd, J. S. Dunlop, M. J. Kukula et al., "Star formation in luminous quasar host galaxies at $z = 1 - 2$," *Monthly Notices of the Royal Astronomical Society*, vol. 429, no. 1, pp. 2–19, 2013.

[119] E. T. Meyer, "From the blazar sequence to the blazar envelope: revisiting the relativistic jet dichotomy in radio-loud active galactic nuclei," *The Astrophysical Journal*, vol. 740, no. 2, article 98, 2011.

13

Spinning Dust Emission from Wobbling Grains: Important Physical Effects and Implications

Thiem Hoang[1] and A. Lazarian[2]

[1]Canadian Institute for Theoretical Astrophysics, University of Toronto, 60 St. George Street, Toronto, ON, Canada M5S 3H8
[2]Department of Astronomy, University of Wisconsin-Madison, Madison, WI 53705, USA

Correspondence should be addressed to Thiem Hoang, hoang@cita.utoronto.ca

Academic Editor: Laurent Verstraete

We review the major progress on the modeling of electric dipole emission from rapidly spinning tiny dust grains, including polycyclic aromatic hydrocarbons (PAHs). We begin by summarizing the original model of spinning dust proposed by Draine and Lazarian and recent theoretical results improving the Draine and Lazarian model. The paper is focused on important physical effects that were disregarded in earlier studies for the sake of simplicity and recently accounted for by us, including grain wobbling due to internal relaxation, impulsive excitation by single-ion collisions, the triaxiality of grain shape, charge fluctuations, and the turbulent nature of astrophysical environments. Implications of the spinning dust for constraining the physical properties of ultrasmall dust grains and environmental conditions are discussed. We discuss the alignment of tiny dust grains and the possibility of polarized spinning dust emission. Suggestions for constraining the alignment of tiny grains and polarization of spinning dust are also discussed.

1. Introduction

Diffuse Galactic microwave emission carries important information on the fundamental properties of the interstellar medium, but it also interferes with cosmic microwave background (CMB) experiments (see Bouchet et al. [1] and Tegmark et al. [2]). Precision cosmology with *Wilkinson Microwave Anisotropy Probe* (WMAP) and Planck satellite requires a good model of the microwave foreground emission to allow for reliable subtraction of Galactic contamination from the CMB radiation.

The discovery of an anomalous microwave emission (hereafter AME) in the range from 10–100 GHz illustrates well the treacherous nature of dust. Until very recently, it has been thought that there are three major components of the diffuse Galactic foreground: synchrotron emission, free-free radiation from plasma (thermal bremsstrahlung), and thermal emission from dust. In the microwave range, the latter is subdominant, leaving essentially two components. However, it is exactly in this range that an anomalous

emission component was reported (Kogut et al. [3, 4]). In the paper by De Oliveira-Costa et al. [5], this emission was nicknamed "Foreground X," which properly reflects its mysterious nature. This component is spatially correlated with $100\,\mu$m thermal dust emission, but its intensity is much higher than one would expect by directly extrapolating the thermal dust emission spectrum to the microwave range.

An early explanation for AME was proposed by Draine and Lazarian model [6, 7] (hereafter DL98 model), where it was identified as electric dipole emission from very small grains (mostly containing polycyclic aromatic hydrocarbons—PAHs) that spin rapidly due to several processes, including gas-grain interactions and dust infrared emission. Although spinning dust emission had been discussed previously (see Erickson [8] and Ferrara and Dettmar [9]), Draine and Lazarian were the first to include the variety of excitation and damping processes that are relevant for very small grains.

While the DL98 model appears to be in general agreement with observations (see [10, 11]), it did not account for some important effects, namely, the nonsphericity of grain shapes, the internal relaxation within grain, and the transient spinup due to ion collisions.

This induced more recent work in order to improve the original DL98 model. The recent papers include Ali-Haïmoud et al. [12], Hoang et al. [13], Ysard and Verstraete [14], Hoang et al. [15], Silsbee et al. [16]. In this paper, we review both the original DL98 model and the ways that it has been improved recently. We focus on the improvement of the dynamics of PAHs and the important physical effects associated with these ultrasmall grains. Recent reviews of the subject include Draine and Lazarian [17], Lazarian and Prunet [18], and Lazarian and Finkbeiner [10].

In Section 2, we briefly present the history of AME and discuss the original DL98 model including their basic assumptions. Section 3 presents our principal results improving the DL98 model from Hoang et al. [13, 15]. From Section 4 to Section 6, we review the grain rotational dynamics and discuss our general approach to calculate power spectrum of spinning dust emission, grain angular momentum distribution, and emissivity for PAHs of arbitrary shapes. In Section 7, we discuss the implications of spinning dust for constraining physical parameters of PAHs as well as environmental conditions. The possibility of polarization of spinning dust and its constraint is discussed in Section 8. A summary of the present paper is given in Section 9.

2. The Original DL98 Model

2.1. Anomalous Microwave Emission and PAHs.
The emission spectrum of diffuse interstellar dust was mostly obtained by the *Infrared Astronomy Satellite* (IRAS) and infrared spectrometers on the *Cosmic Background Explorer* (COBE) and on the *Infrared Telescope in Space* (IRTS). The emission at short wavelength ($\lambda < 50\,\mu m$) arises from transiently heated ultrasmall grains (e.g., PAHs). These grains have such a small heat capacity that the absorption of a single ultraviolet (UV) starlight photon ($\sim 6\,eV$) raises their temperature to $T_{vib} > 200\,K$. Typically, these grains have less than 300 atoms and can be viewed as large molecules rather than dust particles. They are, however, sufficiently numerous to account for most of the prominent 2175 Å absorption features and for $\sim 35\%$ of the total starlight absorption (see, e.g., Li and Draine [19]).

The thermal (vibrational) emissivity of these grains is thought to be negligible at low frequency, because they spend most of their time cold and only emit most of their energy when they are hot. These ultrasmall grains (PAHs) are invoked in the DL98 model to account for the anomalous microwave emission (AME) that was measured in observations.

The first detection of anomalous dust-correlated emission by COBE (Kogut et al. [3, 4]) was quickly followed by detections in the data sets from Saskatoon (de Oliveira-Costa et al. [20]), OVRO (Leitch et al. [21]), the 19 GHz survey (De Oliveira-Costa et al. [22]), de Oliveira-Costa et al. ([23]). Initially, AME was identified as thermal bremsstrahlung from ionized gas correlated with dust (Kogut et al. [3])

and was presumably produced by photoionized cloud rims (McCullough et al. [24]). This idea was scrutinized in Draine and Lazarian [6] and criticized on energetic grounds. Poor correlation of Hα with 100 μm emission also argued against the free-free explanation (McCullough et al. [24]). These arguments are summarized in [17]. Later, [25] used Wisconsin H-Alpha Mapper (WHAM) survey data and established that the free-free emission "is about an order of magnitude below Foreground X over the entire range of frequencies and latitudes where it is detected." The authors concluded that the Foreground X cannot be explained as the free-free emission. Additional evidence supporting this conclusion has come from a study at 5, 8, and 10 GHz by Finkbeiner et al. [26] of several dark clouds and HII regions, two of which show a significantly rising spectrum from 5 to 10 GHz.

The recent Wilkinson Microwave Anisotropy Probe (WMAP) data were used to claim a lower limit of 5% for the spinning dust fraction at 23 GHz (Bennett et al. [27]). However, other models of spinning dust are not ruled out by the WMAP data and in fact fit reasonably well. Finkbeiner [11] performed a fit to WMAP data using a CMB template, a free-free template (based on Hα-correlated emission plus hot gas emission near the Galactic center), a soft synchrotron template traced by the 408 MHz map, a thermal dust extrapolation (Finkbeiner et al. [28]), and a spinning dust template consisting of dust column density times T_d^3. This fit results in excellent χ^2/dof values of 1.6, 1.09, 1.08, 1.05, and 1.08 at 23, 33, 41, 61, and 94 GHz and a reasonable spectral shape for the average spinning dust spectrum.

This WMAP analysis alone does not rule out the Bennett et al. [27] hypothesis of hard synchrotron emission, but when it is combined with the Green Bank Galactic Plane survey data (Langston et al. [29]) at 8 and 14 GHz, spinning dust appears to provide a much better fit than hard synchrotron (Finkbeiner et al. [30]).

Spinning dust emission has recently been reported in a wide range of astrophysical environments, including general ISM (Gold et al. [31, 32] and Collaboration et al. [33]), star-forming regions in the nearby galaxy NGC 6946 (Scaife et al. [34, 35]), and Perseus and Ophiuchus clouds (Casassus et al. [36] and Tibbs et al. [37]). Early Planck results have been interpreted as showing a microwave emission excess from the spinning dust in the Magellanic Clouds (Bot et al. [38]; Collaboration et al. [33]).

2.2. Basic Assumptions

(i) The smallest PAH particles of a few Angstroms are expected to be planar. The grain size a is defined as the radius of an equivalent sphere of the same mass. PAHs are assumed to be planar, disklike with height L and radius R for $a < a_2$, and spherical for $a \geq a_2$. The value $a_2 = 6$ Å is adopted.

(ii) PAHs usually have electric dipole moment $\boldsymbol{\mu}$ arising from asymmetric polar molecules or substructures (*intrinsic dipole moment*) and from the asymmetric distribution of grain charge. The latter is shown to be less important.

(iii) The grain spins around its symmetry axis a_1 with angular momentum J parallel to a_1, and J is isotropically oriented in space.

(iv) For a fixed angular momentum, the spinning grain emits electric dipole radiation at a *unique* frequency mode ν, which is equal to the rotational frequency, that is, $\nu = \omega/2\pi$.

(v) A grain in the gas experiences collisions with neutral atoms and ions, interacts with passing ions (plasma-grain interactions), emits infrared photons following UV absorption, and emits electric dipole radiation. All these processes result in the damping and excitation of grain rotation, that is, they change grain angular momentum J and velocity ω.

(vi) Due to the excitation of various aforementioned processes, the grain angular velocity randomly fluctuates and its distribution can be approximated as the Maxwellian distribution function $f_{Mw}(\omega)$.

(vii) The total emissivity per H atom of the electric dipole radiation from spinning dust at the frequency ν is given by

$$\frac{j_\nu}{n_H} = \frac{1}{4\pi} \frac{1}{n_H} \int_{a_{min}}^{a_{max}} da \frac{dn}{da} 4\pi \omega^2 f_{Mw}(\omega) 2\pi \left(\frac{2\mu_\perp^2 \omega^4}{3c^3} \right), \quad (1)$$

where n_H is the density of H nuclei, $\mu_\perp$ is the electric dipole moment perpendicular to the rotation axis, and dn/da is the grain size distribution function with a in the range from a_{min} to a_{max}.

3. Improved Model of Spinning Dust Emission

Ali-Haïmoud et al. [12] revisited the spinning dust model and presented an analytic solution of the Fokker-Planck (FP) equation that describes the rotational excitation of a spherical grain if the discrete nature of impulses from single-ion collisions can be neglected.

Hoang et al. [13] (hereafter HDL10) improved the DL98 model by accounting for a number of physical effects. The main modifications in their improved model of spinning dust emission are as follows.

(i) Disk-like grains rotate with their grain symmetry axis a_1 that is not perfectly aligned with angular momentum J. The disaligned rotation of J with a_1 causes the wobbling of the grain principal axes with respect to J due to internal thermal fluctuations.

(ii) The power spectrum of a freely spinning grain is obtained using Fourier transform.

(iii) Distribution function of grain angular momentum, J, and velocity, ω, is obtained exactly using the Langevin equation (LE) for the evolution of J in an inertial coordinate system.

(iv) The limiting cases of fast internal relaxation and no internal relaxation are both considered for calculations of the angular momentum distribution and emissivity of spinning dust.

(v) Infrequent collisions of single ions which deposit an angular momentum larger than the grain angular momentum prior to the collision are treated as Poisson-distributed events.

The wobbling disk-like grain has anisotropic rotational damping and excitation. Such an anisotropy can increase the peak emissivity by a factor ~ 2 and increases the peak frequency by a factor 1.4–1.8, compared to the results from the DL98 model.

The effects of grain wobbling on electric dipole emission were independently studied in Silsbee et al. [16] using the FP equation approach, but they disregarded the transient spinup by infrequent single-ion collisions and considered two limiting cases of dust grain temperature, $T_d \to 0$ and $T_d \to \infty$.

Further improvements of the DL98 model were performed in Hoang et al. [15], where a couple of additional effects were taken into account:

(i) emission from very small grains of triaxial ellipsoid (*irregular*) shape with the principal moments of inertia $I_1 \geq I_2 \geq I_3$,

(ii) effects of the orientation of dipole moment μ within grain body for different regimes of internal thermal fluctuations,

(iii) effects of compressible turbulence on the spinning dust emission.

The work found that a freely rotating irregular grain with a given angular momentum radiates at multiple frequency modes. The resulting spinning dust spectrum has peak frequency and emissivity increasing with the degree of grain shape irregularity, which is defined by $I_1 : I_2 : I_3$. Considering the transient heating of grains by UV photons, the study found that the spinning dust emissivity in the case of strong thermal fluctuations is less sensitive to the orientation of μ than in the case of weak thermal fluctuations. In addition, the emission in a turbulent medium increases by a factor from 1.2–1.4 relative to that in a uniform medium, as sonic Mach number M_s increases from 2–7. The latter Mach numbers are relevant to cold phases of the ISM (see Hoang et al. [15] for more details).

4. Grain Rotational Configuration and Power Spectrum

A discussion of the basic physical processes involved in spinning dust can be found in the review by Yacine Ali-Haïmoud, which can be found in the same volume. There, the use of Fokker-Planck equation for describing grain dynamics is discussed. Here, we discuss our numerical approach based on Fourier transform and the Langevin equation, which exhibits a number of advantages to the FP equation when numerical studies of grain dynamics are performed and arbitrary shape of PAHs is considered. We summarize a general approach to find the spinning dust emissivity from grains of triaxial ellipsoid shape with $I_1 > I_2 > I_3$ subject to fast internal relaxation.

4.1. Torque-Free Motion and Internal Relaxation.

The dynamics of a triaxial (*irregular*) grain is more complicated than that of a disk-like grain with $I_2 = I_3$. Indeed, in addition to the precession of the axis of major inertia $\mathbf{a}_1$ around $\mathbf{J}$ as in the disk-like grain, the axis $\mathbf{a}_1$ wobbles rapidly, resulting in the variation of the angle θ between $\mathbf{a}_1$ and $\mathbf{J}$ (see Figure 1).

To describe the torque-free motion of an irregular grain having a rotational energy E_{rot}, the conserved quantities are taken, including the angular momentum $\mathbf{J}$, and a dimensionless parameter that characterizes the deviation of the grain rotational energy from its minimum value

$$q = \frac{2I_1 E_{rot}}{J^2}. \tag{2}$$

The orientation of the triaxial grain in the lab system is completely described by three Euler angles ψ, ϕ, and θ (see, e.g., Hoang et al. [15]). Following [39], we define the total number of states s in phase space for q ranging from 1 to q as

$$s \equiv 1 - \frac{2}{\pi} \int_0^{\psi_1} d\psi \left[\frac{I_3(I_1 - I_2 q) + I_1(I_2 - I_3)\cos^2\psi}{I_3(I_1 - I_2) + I_1(I_2 - I_3)\cos^2\psi} \right]^{1/2}, \tag{3}$$

where

$$\psi_1 = \cos^{-1} \left[\frac{I_3(I_2 q - I_1)}{I_1(I_2 - I_3)} \right]^{1/2}, \tag{4}$$

for $q > q_{sp}$ and $\psi_1 = \pi/2$ for $q \le q_{sp}$, with $q_{sp} \equiv I_1/I_2$ being the separatrix between the two regimes.

The intramolecular vibrational-rotational energy transfer process (IVRET) due to imperfect elasticity occurs on a timescale 10^{-2} s, for a grain of a few angstroms (Purcell [40]), which is shorter than the IR emission time. So, when the vibrational energy decreases due to IR emission, as long as the vibrational-rotational (V-R) energy exchange exists, interactions between vibrational and rotational systems maintain a thermal equilibrium, that is, $T_{rot} \approx T_{vib}$. As a result, the LTE distribution function of rotational energy reads (hereafter VRE regime; see Lazarian and Roberge [41])

$$f_{VRE}(s, J) \propto \exp\left(-\frac{E_{rot}}{k_B T_{rot}}\right) \approx \exp\left(-\frac{E_{rot}}{k_B T_{vib}}\right). \tag{5}$$

Substituting E_{rot} as a function of J and q from (2) into (5), the distribution function for the rotational energy becomes

$$f_{VRE}(s, J) = A \exp\left(-\frac{q(s)J^2}{2I_1 k_B T_{vib}}\right), \tag{6}$$

where A is a normalization constant such that $\int_0^1 f_{VRE}(s, J)ds = 1$.

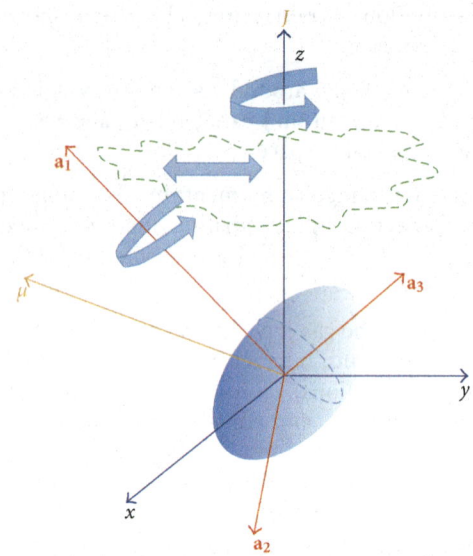

FIGURE 1: Rotational configuration of a triaxial ellipsoid characterized by three principal axes: $\mathbf{a}_1$, $\mathbf{a}_2$, and $\mathbf{a}_3$ in the inertial coordinate system xyz. Grain angular momentum $\mathbf{J}$ is conserved in the absence of external torques and directed along z-axis. The torque-free motion of the triaxial grain comprises the rotation around the axis of major inertia $\mathbf{a}_1$, the precession of $\mathbf{a}_1$ around $\mathbf{J}$, and the wobbling of $\mathbf{a}_1$ with respect to $\mathbf{J}$. The dipole moment $\boldsymbol{\mu}$, which is fixed to grain body, moves together with the grain and thus radiates electric dipole emission.

4.2. Power Spectrum of a Freely Spinning Grain.

Consider a grain with a dipole moment $\boldsymbol{\mu}$ fixed in the grain body rotating with an angular momentum $\mathbf{J}$. If the grain only spins around its symmetry axis, then the rotating dipole moment emits radiation at a unique frequency ν equal to the rotational frequency, that is, $\nu = \omega/2\pi$ (see DL98). The power spectrum for this case is simply a delta function $\delta(\nu - \omega/2\pi)$ with a unique frequency mode.

For an irregular grain of triaxial ellipsoid shape, the grain rotational dynamics is more complicated. In general, one can also obtain analytical expressions for power spectrum, but it is rather tedious. To find the power spectrum of a freely rotating irregular grain, Hoang et al. [13, 15] have employed a more simple brute force approach based on the Fourier transform approach. First, they represent the dipole moment $\boldsymbol{\mu}$ in an inertial coordinate system, and then they compute its second derivative. We obtain

$$\ddot{\boldsymbol{\mu}} = \sum_{i=1}^{3} \mu_i \ddot{\mathbf{a}}_i, \tag{7}$$

where μ_i are components of $\boldsymbol{\mu}$ along principal axes $\mathbf{a}_i$, $\ddot{\mathbf{a}}_i$ are second derivatives of $\mathbf{a}_i$ with respect to time, and $i = 1, 2$, and 3.

The instantaneous emission power by the rotating dipole moment is equal to

$$P_{ed}(J, q, t) = \frac{2}{3c^3} \ddot{\boldsymbol{\mu}}^2. \tag{8}$$

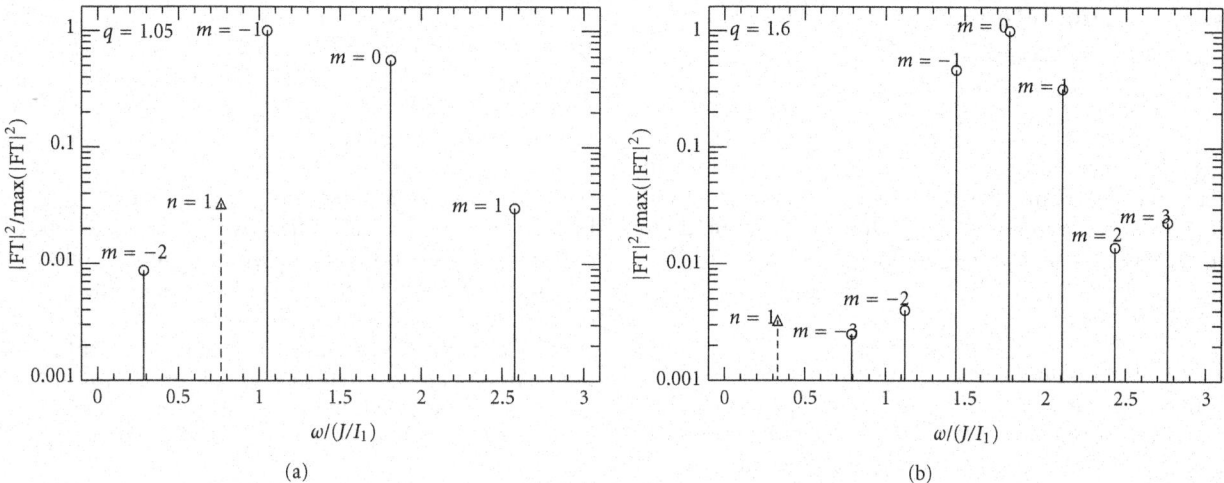

FIGURE 2: Normalized power spectrum of a torque-free rotating irregular grain with $I_1 : I_2 : I_3 = 1 : 0.6 : 0.5$ for the different values of $q = 1.05, 1.60$ (i.e., $q < q_{sp} \equiv I_1/I_2$) and $q = 1.81 > q_{sp}$. The components of $|FT(\ddot{\mu}_x)|^2/\max(|FT(\ddot{\mu}_x)|^2)$ (or $|FT(\ddot{\mu}_y)|^2/\max(|FT(\ddot{\mu}_x)|^2)$) are indicated by circles, while the components of $|FT(\ddot{\mu}_z)|^2/\max(|FT(\ddot{\mu}_x)|^2)$ are indicated by triangles. Orders of in-plane modes m and out-of-plane modes n are indicated, and case 1 ($\mu_1 = \mu/\sqrt{3}$) of μ orientation is assumed. The figure is reproduced from Hoang et al. [15].

The power spectrum is then obtained from the Fourier transform (FT) for the components of $\ddot{\mu}$. For example, the amplitude of $\ddot{\mu}_x$ at the frequency ν_k is defined as

$$\ddot{\mu}_{x,k} = \int_{-\infty}^{+\infty} \ddot{\mu}_x(t) \exp(-i2\pi\nu_k t) dt, \qquad (9)$$

where k denotes the frequency mode. The emission power at the positive frequency ν_k is given by

$$P_{ed,k}(J,q) = \frac{4}{3c^3}\left(\ddot{\mu}_{x,k}^2 + \ddot{\mu}_{z,k}^2 + \ddot{\mu}_{z,k}^2\right), \qquad (10)$$

where the factor 2 arises from the positive/negative frequency symmetry of the Fourier spectrum. To reduce the spectral leakage in the FT, we convolve the time-dependent function $\ddot{\mu}$ with the Blackman-Harris window function (see Harris, 1978). The power spectrum then needs to be corrected for the power loss due to the window function.

The total emission power from all frequency modes for a given J and q then becomes

$$P_{ed}(J,q) = \sum_k P_{ed,k}(J,q) \equiv \frac{1}{T}\int_0^T dt\left(\frac{2}{3c^3}\ddot{\mu}^2\right), \qquad (11)$$

where T is the integration time. (this is the result of Parseval's theorem).

Figure 2 presents normalized power spectra (squared amplitude of Fourier transforms), $|FT(\mu_{x,y})|^2/\max(|FT(\mu_x)|^2)$ and $|FT(\mu_z)|^2/\max(|FT(\mu_x)|^2)$, for the components $\ddot{\mu}_x$ (or $\ddot{\mu}_y$) and $\ddot{\mu}_z$ for a freely rotating irregular grain having the ratio of moments of inertia $I_1 : I_2 : I_3 = 1 : 0.6 : 0.5$ and for various q. Circles and triangles indicated with m and n denote peaks of the power spectrum for oscillating components of $\ddot{\mu}_x$ (or $\ddot{\mu}_y$) and $\ddot{\mu}_z$, respectively. The horizontal axis is the angular frequency of emission modes normalized over the frequency of emission when the grain spins around its shortest axis.

Multiple frequency modes are observed in the power spectra of the irregular grain, but in Figure 2, we show only the modes with power no less than 10^{-3} the maximum value. One can see that in the case with large $q = 1.6$, the modes with $\omega/(J/I_1) > 1$ have increasing power, while the modes with $\omega/(J/I_1) < 1$ have decreasing power. It indicates that if grain rotational energy is increased so that the grain spends a significant fraction of time rotating with large q, then the grain should radiate larger rotational emission.

Although one should not expect the analytical expression of power spectrum for the triaxial grain, the frequency modes can be approximately found. Indeed, for $q < I_1/I_2$, we found that power spectra for $\ddot{\mu}_x$ (or $\ddot{\mu}_y$) have angular frequency modes

$$\omega_m \approx \langle\dot{\phi}\rangle + m\langle|\dot{\psi}|\rangle, \qquad (12)$$

where the bracket denotes the averaging value over time, and $m = 0, \pm1, \pm2\ldots$ denote the order of the mode. The frequency modes for $\ddot{\mu}_z$ are given by

$$\omega_n = n\langle|\dot{\psi}|\rangle, \qquad (13)$$

where n is integer and $n \geq 1$.

In the following, the emission modes induced by the oscillation of μ_x or μ_y, which lie in the $\hat{\mathbf{x}}\hat{\mathbf{y}}$ plane, perpendicular to $\mathbf{J}$, are called in-plane modes, and those induced by the oscillation of μ_z in the direction perpendicular to the $\hat{\mathbf{x}}\hat{\mathbf{y}}$ plane, are called out-of-plane modes. The order of mode is denoted by m and n, respectively. Figure 2 also shows that the emission power for out-of-plane modes ω_n is negligible compared to the power emitted by in-plane modes ω_m.

Emission power spectra are numerically calculated to find ω_k and $P_{ed,k}$, as functions of J and q, for the various ratio of moments of inertia $I_1 : I_2 : I_3$. The obtained data will be used later to compute spinning dust emissivity.

5. Grain Angular Momentum Distribution: Langevin Equation

5.1. Langevin Equation. To find the exact distribution function for grain angular momentum **J**, Hoang et al. [13] and Hoang et al. [15] proposed a numerical approach based on the Langevin equation. Basically, they numerically solved the Langevin equation describing the evolution of three components of **J** in an inertial coordinate system. They read

$$dJ_i = A_i dt + \sqrt{B_{ii}} dq_i, \quad \text{for } i = x, y, z, \quad (14)$$

where dq_i are random Gaussian variables with $\langle dq_i^2 \rangle = dt$, and $A_i = \langle \Delta J_i / \Delta t \rangle$ and $B_{ii} = \langle (\Delta J_i)^2 / \Delta t \rangle$ are damping and diffusion coefficients defined in the inertial coordinate system. Detailed expressions of these coefficients can be found in Hoang et al. [13] and Hoang et al. [15].

For an irregular grain, and to simplify calculations, we adopt the A_i and B_{ii} for a disk-like grain obtained in HDL10. Following DL98b and HDL10, the disk-like grain has the radius R and thickness $L = 3.35$ Å, and the ratio of moments of inertia is along and perpendicular to the grain symmetry axis $h = I_\parallel / I_\perp$. Thereby, the effect of nonaxisymmetry on A_i and B_{ii} is ignored, and we only examine the effect of grain wobbling resulting from the grain triaxiality.

In dimensionless units, $\mathbf{J}' \equiv \mathbf{J}/I_\parallel \omega_{T,\parallel}$ with $\omega_{T,\parallel} \equiv (2k_B T_{gas}/I_\parallel)^{1/2}$ being the thermal angular velocity of the grain along the grain symmetry axis and $t' \equiv t/\tau_{H,\parallel}$; (14) becomes

$$dJ_i' = A_i' dt' + \sqrt{B_{ii}'} dq_i', \quad (15)$$

where $\langle dq_i'^2 \rangle = dt'$,

$$A_i' = -\frac{J_i'}{\tau_{gas,eff}'} - \frac{2}{3} \frac{J_i'^3}{\tau_{ed,eff}'}, \quad (16)$$

$$B_{ii}' = \frac{B_{ii}}{2I_\parallel k_B T_{gas}} \tau_{H,\parallel},$$

where

$$\tau_{gas,eff}' = \frac{\tau_{gas,eff}}{\tau_{H,\parallel}} = \frac{F_{tot,\parallel}^{-1}}{\cos^2\theta + \gamma_H \sin^2\theta},$$

$$\gamma_H = \frac{F_{tot,\perp} \tau_{H,\parallel}}{F_{tot,\parallel} \tau_{H,\perp}}, \quad \tau_{ed,eff}' = \frac{\tau_{ed,eff}}{\tau_{H,\parallel}}, \quad (17)$$

where $\tau_{H,\parallel}$ and $\tau_{H,\perp}$ are rotational damping times due to gas of purely hydrogen atom for rotation along parallel and perpendicular direction to the grain symmetry axis $\mathbf{a}_1$, $\tau_{ed,eff}$ is the effective damping time due to electric dipole emission (see HDL10, HLD11), θ is the angle between $\mathbf{a}_1$ and **J**, and $F_{tot,\parallel}$ and $F_{tot,\perp}$ are total damping coefficients parallel and perpendicular to $\mathbf{a}_1$ (see HDL10). In the case of fast internal relaxation, the diffusion coefficients A and B are averaged over the distribution function f_{VRE}.

The Langevin equation (15) is solved using the numerical integration with a constant timestep. At each timestep, the angular momentum J_i obtained from LEs is recorded and later used to find the distribution function f_J with normalization $\int_0^\infty f_J dJ = 1$.

5.2. Advantages of the Langevin Equation Approach. There are two apparent advantages of the LE approach. First, it allows us to treat the spinning dust emission from grains with an arbitrary grain vibrational temperature. Second, the impulsive excitation by single-ion collisions, which can deposit an amount of angular momentum greater than the grain angular momentum prior the collision, is easily included in (14) (see [15]). Next, we briefly discuss the effect of impulsive excitations arising from single-ion collisions.

DL98b showed that for grains smaller than 7 Å, the angular impulse due to an individual ion-grain collision may be comparable to the grain angular momentum prior the collision. Thus, infrequent hits of ions can result in the transient rotational excitation for very small grains.

Let τ_{icoll}^{-1} be the mean rate of ion collisions with the grain given by

$$\tau_{icoll}^{-1} = f(Z_g = 0) n_i \pi a^2 \left(\frac{8k_B T_{gas}}{m_i \pi} \right)^{1/2} \left[1 + \frac{\sqrt{\pi}}{2} \Phi \right]$$

$$+ \sum_{Z_g \neq 0} f(Z_g) n_1 \pi a^2 \left(\frac{8k_B T_{gas}}{m_i \pi} \right)^{1/2} g\left(\frac{Z_g Z_i e^2}{a k_B T_{gas}} \right), \quad (18)$$

where $\Phi = (2Z_i^2 e^2 / a k_B T_{gas})^{1/2}$, $g(x) = 1 - x$ for $x < 0$ and $g(x) = e^{-x}$ for $x > 0$, and $f(Z_g)$ is the grain charge distribution function. The probability of the next collision occurring in $[t, t + dt]$ is

$$dP = \tau_{icoll}^{-1} \exp\left(\frac{-t}{\tau_{icoll}} \right) dt. \quad (19)$$

The rms angular momentum per ion collision $\langle \delta J^2 \rangle$ is inferred by dividing the total rms angular momentum by the collision rate, and its final formula is given in Hoang et al. [13].

Provided that the random moment of a single-ion collision is obtained from (19), the angular momentum that the grain acquires through each single-ion collision can easily be incorporated into the Langevin equation (14). Hoang et al. [13] found that the impulsive excitations of ions extend the distribution of grain angular momentum to the region of high angular momentum (see next section for its effect on spinning dust emission).

6. Spinning Dust Emissivity

6.1. Spinning Grain of Triaxial Ellipsoid Shape. An irregular grain rotating with a given angular momentum J radiates at frequency modes $\omega_k \equiv \omega_m$ with $m = 0, \pm 1, \pm 2 \ldots$ and $\omega_k \equiv \omega_n$ with $n = 1, 2, 3 \ldots$ (see (12) and (13)). For simplicity, let ω_{m_i} denote the former and ω_{n_i} denote the latter where i indicates the value for m and n. These frequency modes depend on the parameter $q(s)$, which is determined by the internal thermal fluctuations within the grain.

To find the spinning dust emissivity by a grain at an observational frequency ν, first we need to know how much emission that is contributed by each mode ω_k.

Consider an irregular grain rotating with the angular momentum J; the probability of finding the emission at the angular frequency ω depends on the probability of finding the value ω such that

$$pdf(\omega \mid J)d\omega = f_{\text{VRE}}(s,J)ds = A \exp\left(-\frac{q(s)J^2}{2I_1 k_B T_{\text{vib}}}\right)ds, \tag{20}$$

where we assumed the VRE regime with f_{VRE} given by (6).

For the mode $\omega \equiv \omega_k(s)$, from (20) we can derive

$$pdf_k(\omega \mid J) = \left(\frac{\partial \omega_k}{\partial s}\right)^{-1} f_{\text{VRE}}(s,J). \tag{21}$$

The emissivity from the mode k is calculated as

$$j_{\nu,k}^a = \frac{1}{4\pi} \int_{J_l}^{J_u} P_{\text{ed},k}(J,q_{\leq}) f_J(J) pdf_k(\omega \mid J) 2\pi dJ$$
$$+ \frac{1}{4\pi} \int_{J_l}^{J_u} P_{\text{ed},k}(J,q_{>}) f_J(J) pdf_k(\omega \mid J) 2\pi dJ, \tag{22}$$

where $q_{\leq}$ and $q_{>}$ denote $q \leq q_{\text{sp}}$ and $q > q_{\text{sp}}$, respectively, J_l and J_u are lower and upper limits for J corresponding to a given angular frequency $\omega_k(J,q) = \omega$, and 2π appears due to the change of variable from ν to ω.

Emissivity by a grain of size a at the observation frequency ν arising from all emission modes is then

$$j_\nu^a \equiv \sum_k j_{\nu,k}^a. \tag{23}$$

Consider, for example, the emission mode $k \equiv m_0$. For the case I_2 which is slightly larger than I_3, this mode has the angular frequency $\omega_{m_0} = \langle \dot{\phi} \rangle = (J/I_1)q_0$ with q_0 obtained from calculation of ω_{m_0}, which is independent of q for $q < q_{\text{sp}}$ (q_0 approaches I_1/I_2 as $I_3 \to I_2$, i.e., when irregular shape becomes spheroid). As a result,

$$pdf_{m_0}(\omega \mid J) = \delta\left(\omega - \left(\frac{J}{I_1}\right)q_0\right). \tag{24}$$

Thus, the first term of (22), denoted by $j_{\nu,m_0,\leq}^a$, is rewritten as

$$j_{\nu,m_0,\leq}^a = \frac{1}{2} \int_{J_l}^{J_u} P_{\text{ed},m_0}(J,q_{\leq}) f_J(J) \delta\left(\omega - \left(\frac{J}{I_1}\right)q_0\right) dJ$$
$$= \frac{1}{2} \frac{I_1 f_J(J_0)}{q_0} P_{\text{ed},m_0}(J_0,q(s)), \tag{25}$$

where $J_0 = I_1\omega/q_0$, and the value of $q(s)$ remains to be determined.

For $q > q_{\text{sp}}$, $\langle \dot{\phi} \rangle$ is a function of q. Hence, the emissivity (22) for the mode $k \equiv m_0$ becomes

$$j_{\nu,m_0}^a = \frac{1}{2} \frac{I_1 f_J(J_0)}{q_0} \int_0^{s_{\text{sp}}} ds P_{\text{ed},m_0}(J_0,q(s)) f_{\text{VRE}}(J_0,s)$$
$$+ \frac{1}{2} \int_{J_l}^{J_u} P_{\text{ed},m_0}(J,q_{>}) f_J(J) pdf_{m_0}(\omega \mid J) dJ, \tag{26}$$

where s_{sp} is the value of s corresponding to $q = q_{\text{sp}}$, and the term $P_{\text{ed},m_0}(J_0,q(s))$ in (25) has been replaced by its average value over the internal thermal distribution f_{VRE}.

The emissivity per H is obtained by integrating j_ν^a over the grain size distribution

$$\frac{j_\nu}{n_H} = \frac{1}{n_H} \int_{a_{\min}}^{a_{\max}} da \frac{dn}{da} j_\nu^a, \tag{27}$$

where j_ν^a is given by (23).

6.2. A Degenerate Case: Grains of Disk-Like Shape. The spinning dust emissivity from disk-like grains (e.g., $I_2 = I_3$) is a degenerate case of triaxial grains. Basically, a disk-like grain with an angular momentum $\mathbf{J}$ radiates at four frequency modes as follows:

$$\omega_{m_i} \equiv \dot{\phi} + i\dot{\psi} = \frac{J}{I_{\|}}[h + i(1-h)\cos\theta],$$
$$\omega_{n_1} \equiv \dot{\psi} = \frac{J}{I_{\|}}(1-h)\cos\theta, \tag{28}$$

where $i = 0$ and ± 1 (see HDL10 and [12]).

The emission power of these modes are given by the following analytical forms (HDL10 and [16]):

$$P_{\omega_{m_0}} = \frac{2\mu_{\|}^2}{3c^3} \omega_{m_0}^4 \sin^2\theta,$$
$$P_{\omega_{m_{\pm 1}}} = \frac{\mu_{\perp}^2}{6c^3} \omega_{m_{\pm 1}}^4 (1 \pm \cos\theta)^2, \tag{29}$$
$$P_{\omega_{n_1}} = \frac{2\mu_{\perp}^2}{3c^3} \omega_{n_1}^4 \sin^2\theta.$$

For the disk-like grain, from (3), the number of states in phase space s for q spanning from $1 - q$ becomes

$$s = 1 - \left(\frac{h-q}{h-1}\right)^{1/2} = 1 - \cos\theta, \tag{30}$$

where $q = 1 + (h-1)\sin^2\theta$ has been used. Thus, for an arbitrary mode with frequency ω_k, we obtain

$$pdf_k(\omega \mid J)d\omega = f_{\text{VRE}}(s,J)ds = f_{\text{VRE}}(\theta,J)\sin\theta \, d\theta. \tag{31}$$

Taking use of $\omega = \omega_k(J,\theta)$, we derive

$$pdf_k(\omega \mid J) = f_{\text{VRE}}(\theta,J)\left(\frac{\partial \omega_k}{\partial \theta}\right)^{-1}\sin\theta. \tag{32}$$

Therefore, by substituting (29) in (22), the emissivity at the observation frequency $\nu = \omega/(2\pi)$ from a disk-like grain of size a is now given by

$$j_\nu^a \equiv \frac{1}{2} \frac{f_J(I_{\|}\omega/h)}{h} \frac{2\mu_{\|}^2}{3c^3} \omega^4 \langle \sin^2\theta \rangle$$
$$+ \frac{1}{2} \frac{\mu_{\perp}^2}{6c^3} \omega^4 \int_{J_l}^{J_u} pdf_{m_1}(\omega \mid J) f_J(J) \, dJ$$
$$+ \frac{1}{2} \frac{\mu_{\perp}^2}{6c^3} \omega^4 \int_{J_l}^{J_u} pdf_{m_{-1}}(\omega \mid J) f_J(J) \, dJ$$
$$+ \frac{1}{2} \frac{\mu_{\perp}^2}{3c^3} \omega^4 \int_{J_l}^{J_u} pdf_{n_1}(\omega \mid J) f_J(J) \, dJ, \tag{33}$$

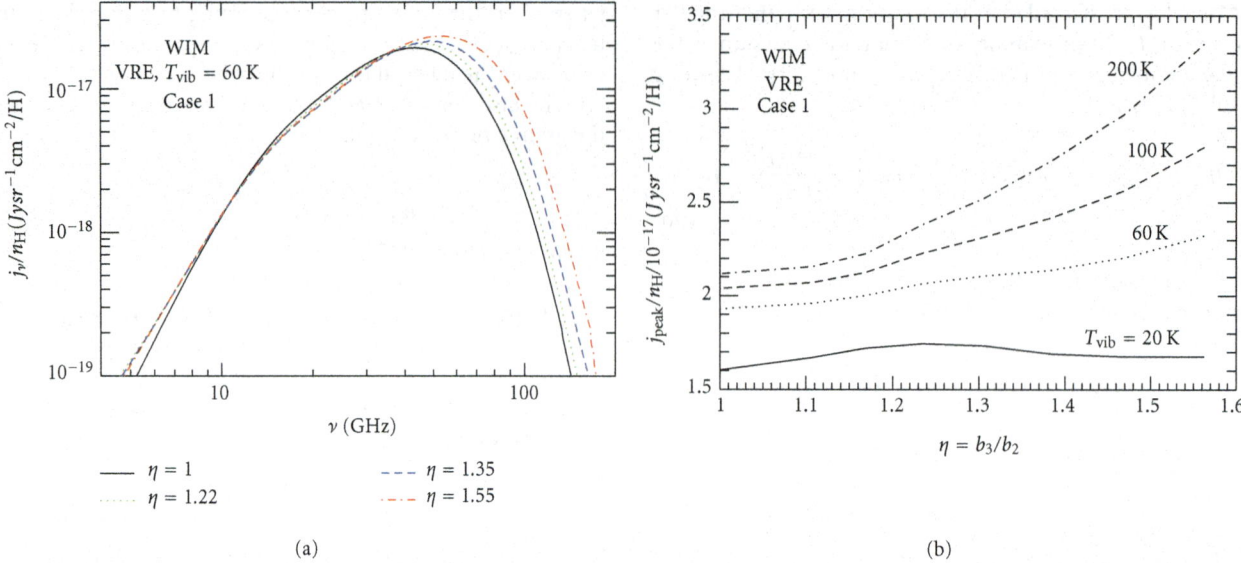

FIGURE 3: Emissivity per H from irregular grains of different degrees of irregularity $\eta = b_3/b_2$ with $T_{\rm vib} = 60$ and 200 K in the WIM for the case in which the electric dipole moment is isotropically oriented in the grain body (i.e., case 1 with $\mu_1 = \mu/\sqrt{3}$). The emission spectrum shifts to higher frequency as η decreases (i.e., grain becomes more irregular). Here, the grain mass is held fixed as η changes. The figure is reproduced from Hoang et al. [15].

where $pdf_{m_{\pm 1}}$ and pdf_{n_1} are easily derived by using (32) for $\omega_{m_{\pm 1}}$ and ω_{n_1}, and $J_l = I_\| \omega/(2h - 1)$ and $J_u = I_\| \omega$ for $m_{\pm 1}$ mode, $J_l = I_\| \omega/(h - 1)$ and $J_u = \infty$ for n_1 mode.

6.3. Emissivity.

Hoang et al. [15] assumed that the smallest grains of size $a \leq a_2 = 6$ Å have irregular shape and larger grains are spherical. To compare the emissivity from an irregular grain with that from a disk-like grain, they considered the simplest case of the irregular shape in which the circular cross section of the disk-like grain is adjusted to the elliptical cross section. The emission by two grains of different shapes with the same mass M and thickness L is under interest; therefore, the semiaxes of the elliptical disk are constrained by the grain mass

$$M = \pi R^2 L = \pi b_2 b_3 L, \qquad (34)$$

where $R = (4a^3/3L)^{1/2}$ is the radius of the disk-like grain, b_2 and b_3 are the length of semiaxes $\mathbf{a}_2$ and $\mathbf{a}_3$, and $b_1 = L$ is kept constant. Assuming that the circular disk is compressed by a factor $\alpha \leq 1$ along $\mathbf{a}_2$, then (34) yields

$$b_2 = \alpha R, \qquad b_3 = \alpha^{-1} R. \qquad (35)$$

Denote the parameter by $\eta \equiv b_3/b_2 = \alpha^{-2}$, then the degree of grain shape irregularity is completely characterized by η.

For each grain size a, the parameter η is increased from $\eta = 1$ to $\eta = \eta_{\max}$. However, $\eta_{\max}$ is constrained by the fact that the shortest axis $\mathbf{a}_2$ should not be shorter than the grain thickness L. The value $\eta_{\max} \sim 3/2$ is conservatively chosen.

Although the irregular grain can radiate at a large number of frequency modes, only the modes with the order $|m| \leq 2$ are important. The higher-order modes contribute

less than $\sim 0.5\%$ to the total emission, and thus they are neglected. Hoang et al. [15] assumed that grains smaller than a_2 have a fixed vibrational temperature $T_{\rm vib}$ (see Hoang et al. [15] for the detailed treatment of $T_{\rm vib}$ distribution), and that for the instantaneous value of J, the rotational energy has a probability distribution $f_{\rm VRE}$ (i.e., VRE regime, see (6)).

The grain size distribution dn/da from Draine and Li [42] is adopted with the total to selective extinction $R_V = 3.1$ and the total carbon abundance per hydrogen nucleus $b_C = 5.5 \times 10^{-5}$ in carbonaceous grains with $a_{\min} = 3.55$ Å and $a_{\max} = 100$ Å.

The spinning dust emissivity is calculated for a so-called model A (similar to DL98b; HDL10), in which 25% of grains have the electric dipole moment parameter $\beta = 2\beta_0$, 50% have $\beta = \beta_0$, and 25% have $\beta = 0.5\beta_0$ with $\beta_0 = 0.4$ D. In the rest of the paper, the notation model A is omitted, unless stated otherwise.

The left panel in Figure 3 shows the spinning dust emissivity for different degrees of irregularity η and with a dust temperature $T_{\rm vib} = 60$ K in the WIM. The emission spectrum for a given $T_{\rm vib}$ shifts to higher frequency as η decreases (i.e., the degree of grain irregularity increases), but their spectral profiles remain similar. The right panel shows the increase of peak emissivity $J_{\rm peak}$ with increasing η.

One particular feature in Figure 3(b) is that for axisymmetric grains ($\eta = 1$), the emissivity increases by a factor of 1.3 with $T_{\rm vib}$ increasing from 20 to 200 K. However, for the irregular grain with high triaxiality $\eta = 1.5$, the emissivity increases by a factor of 2. The peak frequency is increased by a factor of 1.4.

This feature is easy to understand because the irregular grain radiate at more frequency modes than the axisymmetric grain. As a result, for the grain temperature to increase to

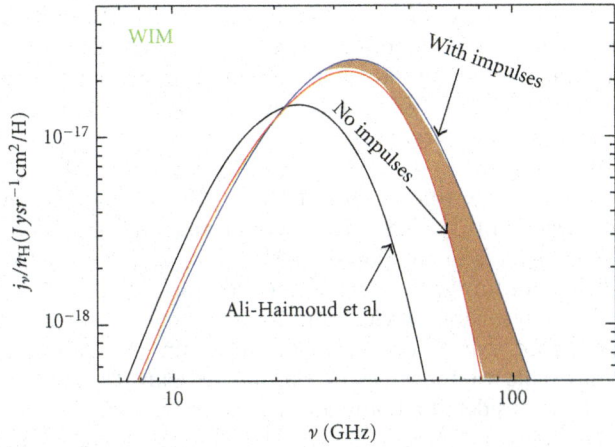

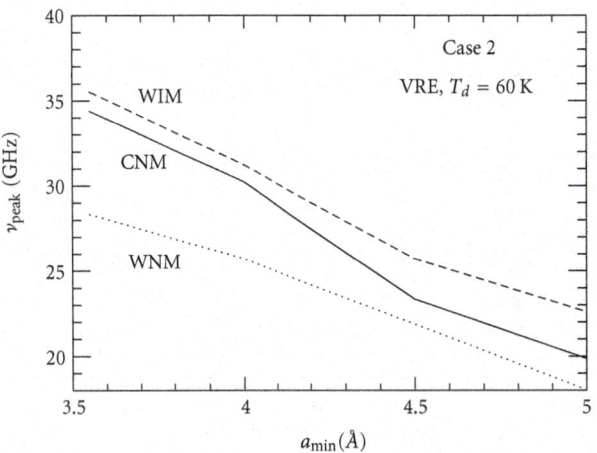

FIGURE 4: Emissivity per H obtained for WIM without ionic impulses using the Fokker-Planck equation from Ali-Haïmoud et al. [12] and with impulses using our LE simulations for grain wobbling. The spectra are efficiently broadened as a result of impulses (see blue line). The figure is reproduced from Hoang et al. [13].

FIGURE 5: Decrease of the peak frequency ν_{peak} of spinning dust spectrum with the lower cutoff of grain size distribution a_{min} for various environmental conditions. The figure is reproduced from Hoang et al. [15].

a sufficiently high value, it results in the uniform distribution of the angle between grain symmetry axis and angular momentum, so that the spinning dust emissivity becomes saturated. On the other hand, for the triaxial grain, as T_{vib} increases, it allows the grain to rotate about its axis of minimum inertia (smallest moment of inertia). As a result, the grain radiates at frequency modes with higher frequency and power.

In the case of efficient IVRET, vibrational energy is converted to rotational emission, which results in the increase of both emissivity and peak frequency. As shown, the energy transfer is more efficient for the more irregular grain. The reason for this is that the more irregular grain allows the grain to spend a larger fraction of time rotating along the axis of minor inertia.

The effect of impulsive excitations by single-ion collisions is shown in Figure 4. One can see that the impulses from ions can increase the emissivity by ~23% and slightly increase the peak frequency (see Figure 4). The tail of high frequency part is obviously extended due to the contribution from ionic impulses with large angular momentum.

7. Constraining Spinning Dust Parameters and Implications

Spinning dust emission involves a number of parameters, including grain physical parameters and environmental parameters. Among them, the grain dipole moment and gas density are two most important parameters, but they can be constrained using theoretical modeling combined with observation data (see, e.g., Dobler et al. [43] and Hoang et al. [15]). In the following, we discuss a number of parameters, which are shown to be important but more difficult to constrain through observation.

7.1. Lower Cutoff of Grain Size Distribution a_{min}.
The spinning dust emission spectrum is sensitive to the population of tiny dust grains, and its peak frequency is mostly determined by the smallest PAHs. Let a_{min} be the size of the smallest PAHs. When a_{min} is increased, the peak frequency ν_{peak} decreases accordingly.

Figure 5 shows the variation of ν_{peak} as a function of a_{min} for various environments for the case in which the grain dipole moment lies in the grain plane (Case 2) with $\mu_1 = 0$ and with the VRE regime ($T_d = 60$ K). As expected, ν_{peak} decreases generically with a_{min} increasing. Thus, in addition to grain dipole moment, the lower cutoff of grain size also plays an important role.

7.2. Constraining the Shape of Very Small Grains.
Very small grains and PAHs are expected to be nonspherical. However, constraining grain triaxiality using spinning dust appears rather challenging. In the simplest case where the grain shape can be approximated as a triaxial ellipsoid, the possibility is still low because there are many parameters involved in the spinning dust.

7.3. Can Compressible Turbulence Be Observed through Spinning Dust Emission?
The discussion of interstellar conditions adopted in DL98 and other works on spinning dust was limited by idealized interstellar phases. It is now recognized that turbulence plays an important role in shaping the interstellar medium.

For spinning dust, the turbulence can increase the emissivity due to its nonlinear dependence on material density. Indeed, in a medium with density fluctuations, the effective emissivity is

$$\langle j_\nu \rangle = \int_0^1 f(x) j_\nu(x\langle \rho \rangle) dx, \qquad (36)$$

where $f(x)dx$ is the fraction of the mass with $\rho/\langle \rho \rangle \in (x, x + dx)$. We use compression distributions $f(x)$ obtained from

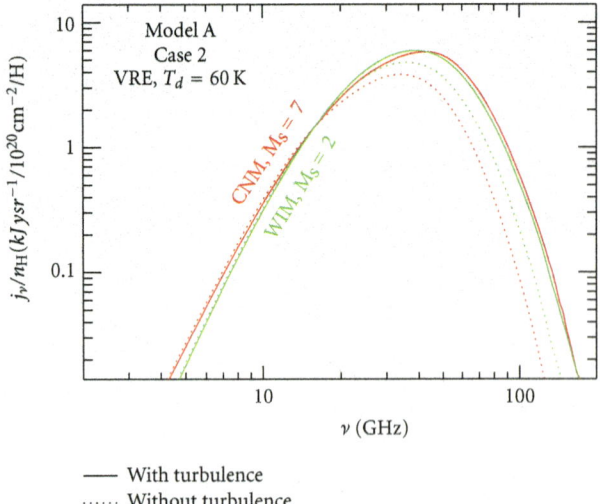

— With turbulence
······ Without turbulence

FIGURE 6: Spinning dust emissivity per H in the presence of compressible turbulence with sonic Mach number $M_s = 2$ and 7, compared to that from uniform medium with $n_H = \overline{n}_H$ for the CNM (red) and WIM (green). The peak emissivity is increased, and the spectrum is shifted to higher frequency due to compressible turbulence. Case 2 ($\mu_1 = 0$) of μ orientation is considered. Figure reproduced from Hoang et al. [15].

MHD simulations for $M_s = 2$ and 7 to evaluate $\langle j_\nu \rangle$ for the WIM and CNM, respectively.

We assume the case 2 ($\mu_1 = 0$) of μ orientation. The resulting effective emissivity is compared with the emissivity from the uniform medium in Figure 6. It can be seen that the turbulent compression increases the emissivity and shifts the peak to higher ν_{peak}. The increase of emissivity is significant for strong turbulent medium.

The distribution of phases, for instance, CNM and WNM of the ISM at high latitudes, can be obtained from absorption lines. Similarly, by studying fluctuations of emission, it is possible to constrain parameters of turbulence. In an idealized case of a single-phase medium with fluctuations of density with a given characteristic size, one can estimate the value of the 3D fluctuation by studying the 2D fluctuations of column density. More sophisticated techniques for obtaining sonic Mach numbers have been developed recently (see Kowal et al. [44], Esquivel and Lazarian [45] and Burkhart et al. [46]). (It may be seen that Alfven Mach numbers have subdominant effect on the distribution of densities (see Kowal et al. [44]). Thus, in our study we did not vary the Alfven Mach number.) In particular, Burkhart et al. [47], using just column density fluctuations of the SMC, obtained a distribution of Mach numbers corresponding to the independent measurements obtained using Doppler shifts and absorption data. With such an input, it is feasible to quantify the effect of turbulence in actual observational studies of spinning dust emission.

7.4. Effect of Dust Acceleration on Spinning Dust Emission. Collisions of ultrasmall grains with ions and neutrals in plasma appear to be a dominant mechanism of rotational

excitations for spinning dust emission, particularly, in dark clouds where UV photons are blocked out. Current spinning dust models assume Brownian motion of grains relative to gas, but it is known that grains may move with suprathermal velocities due to acceleration by turbulence (see, e.g., [48], Yan and Lazarian [49, 50], and Hoang et al. [51]) and random charge fluctuations (Ivlev et al. [52] and Hoang and Lazarian [53]). The latter mechanism, namely, random charge fluctuations-induced acceleration, is found to be efficient for tiny grains (Hoang and Lazarian [53]).

The resonant acceleration by fast modes of MHD turbulence, which occurs when the grain gyroradius is comparable to the scale of turbulence eddy (i.e., $r_g \sim k^{-1}$), is considered a dominant mechanism for large grains ($>10^{-5}$ cm), whereas it is negligible for ultrasmall grains because the grain gyroradius falls below the cutoff scale of the turbulence due to viscous damping (see Yan et al. [50] and Hoang et al. [51]).

In highly ionized media (e.g., WIM, HII regions), the resonant acceleration by MHD turbulence may become important for ultrasmall grains because the damping cutoff of MHD turbulence is suppressed due to the decrease of viscous neutral damping. We also note that recent observations by Paladini et al. [54]) revealed that PAHs and ultrasmall grains may be present in HII regions, as shown through their 8 μm and 24 μm emission features, respectively.

Thus, assuming that grain rotational kinetic energy is equal to its translational energy, the acceleration by these aforementioned processes is expected to increase the spinning dust emission. Further studies should take this issue into account.

8. Polarization of Spinning Dust Emission and Alignment of Ultrasmall Grains

8.1. Polarization of Anomalous Microwave Emission. Spinning dust emission is an important foreground component that contaminates with the CMB radiation in the frequency 10–90 GHz. An understanding of how much is this emission component polarized is becoming a pressing question for future CMB B-mode missions.

Recent observational studies (Dickinson et al. [55], López-Caraballo et al. [56], and Macellari et al. [57]) showed that the average polarization of AME is between 2 and 5%. In the last years, significant progress has been made in understanding spinning dust emission, both in theory and observation, but the principal mechanism of alignment of ultrasmall grains is not well understood.

8.2. Alignment of Ultrasmall Dust Grains. Grain alignment is an exciting problem (see Lazarian [58] for a review). The most promising mechanism for the grain alignment is based on radiative torques. Proposed originally by Dolginov and Mytrophanov [59], it is related to the interaction of unpolarized radiation with irregular grains. The numerical studies in Draine and Weingartner [60, 61] showed the efficiency and promise of the radiative torques (which later were termed RATs). The physical picture of the RAT alignment and a detailed study of important relevant effects are presented in Lazarian and Hoang [62, 63] and Hoang and Lazarian [64–66]. However, the efficiency of RATs plummets as the size of grains gets much smaller than the radiation wavelength. Therefore, this mechanism, which seems to provide a good correspondence with the optical and infrared data (see Lazarian [58] and Whittet et al. [67]), cannot be applicable to ultrasmall spinning dust.

Microwave emission from spinning grains is expected to be polarized if grains are aligned. Alignment of ultrasmall grains (essentially PAHs) is likely to be different from alignment of large (i.e., $a > 10^{-6}$ cm) grains as discussed previously. One of the mechanisms that might produce the alignment of the ultrasmall grains is the paramagnetic dissipation mechanism proposed by Davis and Greenstein [68]. The Davis-Greenstein alignment mechanism (Davis and Greenstein [68] and Roberge and Lazarian [69]) is straightforward; for a spinning grain, the component of the interstellar magnetic field perpendicular to the grain angular velocity varies in grain coordinates, resulting in time-dependent magnetization, associated energy dissipation, and a torque acting on the grain [68]. As a result, grains tend to rotate with angular momenta parallel to the interstellar magnetic field.

Lazarian and Draine [70] (henceforth LD00) found that the traditional picture of paramagnetic relaxation is incomplete, since it disregards the so-called "Barnett magnetization" (Landau and Lifshitz [71]). The Barnett effect, the inverse of the Einstein-de Haas effect, consists of the spontaneous magnetization of a paramagnetic body rotating in field-free space. This effect can be understood in terms of the lattice-sharing part of its angular momentum with the spin system. Therefore, the implicit assumption in Davis and Greenstein [68] that the magnetization within a *rotating grain* in a *static* magnetic field is equivalent to the magnetization within a *stationary grain* in a *rotating* magnetic field is clearly not exact.

LD00 accounted for the "Barnett magnetization" and termed the effect of enhanced paramagnetic relaxation arising from grain magnetization "resonance paramagnetic relaxation." It is clear from Figure 7 that resonance paramagnetic relaxation persists at the frequencies when the Davis-Greenstein relaxation vanishes. However, the polarization is marginal for $\nu > 35$ GHz anyhow. The discontinuity at ~20 GHz is due to the assumption that smaller grains are planar and larger grains are spherical. The microwave emission will be polarized in the plane perpendicular to the magnetic field because the angular momentum is partially aligned with the magnetic field.

8.3. Constraining the Alignment of Ultrasmall Grains

8.3.1. Can We Constrain the Alignment of Ultrasmall Grains through Polarization of Midinfrared (2–12 μm) Emission Features? The answer to this question is "probably not." Indeed, as discussed earlier, midinfrared emission from ultrasmall grains takes place as they absorb UV photons. These photons raise grain vibrational temperature, randomizing grain axes

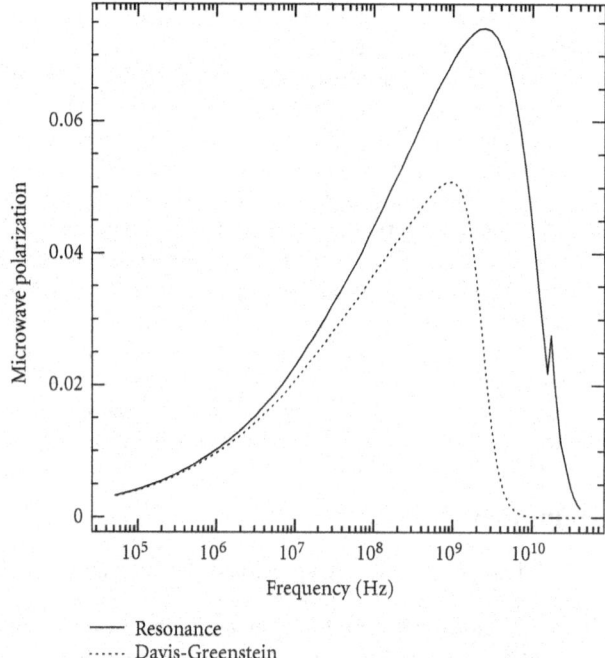

FIGURE 7: Polarization for both resonance paramagnetic relaxation and Davis-Greenstein relaxation for grains in the cold interstellar medium as a function of frequency (from LD00). For resonance relaxation the saturation effects are neglected, which means that the upper curves correspond to the *maximal* values allowed by the resonant paramagnetic mechanism.

in relation to its angular momentum (see Lazarian and Roberge [72]). Taking values for Barnett relaxation from Lazarian and Draine [73], we estimate the randomization time of the 10^{-7} cm grain to be 2×10^{-6} s, which is less than the grain cooling time due to IR emission. As a result, the emanating infrared emission will be polarized very marginally. If, however, Barnett relaxation is suppressed, the randomization time will be determined by inelastic relaxation (Lazarian and Efroimsky [74]) and will be ~ 0.1 s, which would entail a partial polarization of infrared emission.

8.3.2. Can We Constrain the Alignment of Ultrasmall Grains via the Ultraviolet Polarization? PAHs and ultrasmall grains that produce spinning dust emission are likely the same particles that produce the prominent UV absorption feature at 2175 Å (see, e.g., Draine and Li [42]). The lack of polarization excess at 2175 Å is consistent with the expectation that the PAHs are poorly aligned. However, the small degree of polarization (see Wolff et al. [75]) indicates that there must be some residual alignment of ultrasmall grains. The constraint for such a residual alignment can be obtained by fitting the theoretical model with the UV polarization of starlight (Martin [76]). When the residual alignment is available, one can predict the polarization level of spinning dust.

Apart from the emission from spinning dust, another new type of emission from dust is possible. Draine and

Lazarian [77] noticed that the strongly magnetized material is capable of producing much more microwave thermal emission compared with nonmagnetic grains. They suggested this as a possible alternative to spinning dust emission, which can be responsible for a part or even most of the anomalous microwave emission. Such an emission can be strongly polarized, making anomalous emission an important contaminant in terms of CMB polarization studies.

Further research showed that at the frequencies 20–90 GHz, the spinning dust dominates. However, Draine and Hensley [78] performed new calculations of microwave response of strongly magnetic grains. At higher frequencies, this new extensive study of evaluating microwave emissivity of strongly magnetic grains showed that magnetic dipole response of interstellar dust may be extremely important.

9. Summary

The principal points discussed previously are as follows.

(i) The model of spinning dust emission proposed by DL98 proved to be capable of explaining anomalous microwave emission, and its predictions were confirmed by numerous observations since the introduction of the model.

(ii) The DL98 spinning dust model has been improved recently by including the effects of thermal fluctuations within dust grains, impulsive excitations with single ions, transient heating by UV photons, triaxiality of grain shape, and compressible turbulence, which made the spinning dust model more realistic.

(iii) Spinning dust emission involves a number of grain physical parameters and environmental parameters. With the latest progress on theoretical modeling and observations, the possibility of using spinning dust as a diagnostic tool for physical parameters of ultrasmall dust is open.

(iv) The spinning dust emission is expected to be partially polarized, but further studies on alignment of ultrasmall grains and modeling of spinning dust polarization are vitally required.

Acknowledgments

A. Lazarian acknowledges the support of the Center for Magnetic Self-Organization and the NASA Grant no. NNX11AD32G. T. Hoang is grateful for fruitful discussions with Peter Martin on constraining the alignment of ultrasmall grains and polarization of anomalous microwave emission.

References

[1] F. R. Bouchet, S. Prunet, and S. K. Sethi, "Multifrequency Wiener filtering of cosmic microwave background data with polarization," *Monthly Notices of the Royal Astronomical Society*, vol. 302, no. 4, pp. 663–676, 1999.

[2] M. Tegmark, D. J. Eisenstein, W. Hu, and A. de Oliveira-Costa, "Foregrounds and forecasts for the cosmic microwave background," *The Astrophysical Journal*, vol. 530, no. 1, pp. 133–165, 2000.

[3] A. Kogut, A. J. Banday, C. L. Bennett, K. M. Górski, G. Hinshaw, and W. T. Reach, "High-latitude galactic emission in the cobe differential microwave radiometer 2 year sky maps," *The Astrophysical Journal*, vol. 460, no. 1, pp. 1–9, 1996.

[4] A. Kogut, A. J. Banday, C. L. Bennett et al., "Microwave emission at high galactic latitudes in the four-year DMR sky maps," *The Astrophysical Journal*, vol. 464, no. 1, pp. L5–L9, 1996.

[5] A. De Oliveira-Costa, M. Tegmark, D. P. Finkbeiner et al., "A new spin on galactic dust," *The Astrophysical Journal*, vol. 567, no. 1, pp. 363–369, 2002.

[6] B. T. Draine and A. Lazarian, "Electric dipole radiation from spinning dust grains," *The Astrophysical Journal*, vol. 508, no. 1, pp. 157–179, 1998.

[7] B. T. Draine and A. Lazarian, "Diffuse galactic emission from spinning dust grains," *The Astrophysical Journal*, vol. 494, no. 1, pp. L19–L22, 1998.

[8] W. C. Erickson, "A mechanism of non-thermal radio-noise origin," *The Astrophysical Journal*, vol. 126, p. 480, 1957.

[9] A. Ferrara and R. J. Dettmar, "Radio-emitting dust in the free electron layer of spiral galaxies: testing the disk/halo interface," *The Astrophysical Journal*, vol. 427, no. 1, pp. 155–159, 1994.

[10] A. Lazarian and D. Finkbeiner, "Microwave emission from aligned dust," *New Astronomy Reviews*, vol. 47, no. 11-12, pp. 1107–1116, 2003.

[11] D. P. Finkbeiner, "Microwave interstellar medium emission observed by the Wilkinson Microwave Anisotropy Probe," *The Astrophysical Journal*, vol. 614, no. 1, pp. 186–193, 2004.

[12] Y. Ali-Haïmoud, C. M. Hirata, and C. Dickinson, "A refined model for spinning dust radiation," *Monthly Notices of the Royal Astronomical Society*, vol. 395, no. 2, pp. 1055–1078, 2009.

[13] T. Hoang, B. T. Draine, and A. Lazarian, "Improving the model of emission from spinning dust: effects of grain wobbling and transient spin-up," *The Astrophysical Journal*, vol. 715, no. 2, pp. 1462–1485, 2010.

[14] N. Ysard and L. Verstraete, "The long-wavelength emission of interstellar PAHs: characterizing the spinning dust contribution," *Astronomy and Astrophysics*, vol. 509, article A12, 2010.

[15] T. Hoang, A. Lazarian, and B. T. Draine, "Spinning dust emission: effects of irregular grain shape, transient heating, and comparison with wilkinson microwave anisotropy probe results," *The Astrophysical Journal*, vol. 741, p. 87, 2011.

[16] K. Silsbee, Y. Ali-Haïmoud, and C. M. Hirata, "Spinning dust emission: the effect of rotation around a non-principal axis," *Monthly Notices of the Royal Astronomical Society*, vol. 411, no. 4, pp. 2750–2769, 2011.

[17] B. T. Draine and A. Lazarian, "Microwave emission from galactic dust grains," *Microwave Foregrounds*, vol. 181, p. 133, 1999.

[18] A. Lazarian and S. Prunet, "Polarized microwave emission from dust," in *Proceedings of the Workshop on Astrophysical Polarized Backgrounds*, vol. 609, pp. 32–43, March 2002.

[19] A. Li and B. T. Draine, "Infrared emission from interstellar dust. II. The diffuse interstellar medium," *The Astrophysical Journal*, vol. 554, no. 2, pp. 778–802, 2001.

[20] A. de Oliveira-Costa, A. Kogut, M. J. Deviln, C. Barth Netterfield, L. A. Page, and E. J. Wollack, "Galactic microwave emission at degree angular scales," *The Astrophysical Journal*, vol. 482, no. 1, pp. L17–L20, 1997.

[21] E. M. Leitch, A. C. S. Readhead, T. J. Pearson, and S. T. Myers, "An anomalous component of galactic emission," *The Astrophysical Journal*, vol. 486, no. 1, pp. L23–L26, 1997.

[22] A. De Olivera-Costa, M. Tegmark, L. A. Page, and S. P. Boughn, "Galactic emission at 19 GHz," *The Astrophysical Journal*, vol. 509, no. 1, pp. L9–L12, 1998.

[23] A. de Oliveira-Costa, M. Tegmark, C. M. Gutiérrez et al., "Cross-correlation of tenerife data with galactic templates-evidence for spinning dust?" *The Astrophysical Journal*, vol. 527, no. 1, pp. L9–L12, 1999.

[24] P. R. McCullough, J. E. Gaustad, W. Rosing, and D. van Buren, "Implications of H observations for studies of the CMB," *Microwave Foregrounds*, vol. 181, p. 253, 1999.

[25] A. de Oliveira-Costa, M. Tegmark, M. J. Devlin et al., "Galactic contamination in the qmap experiment," *The Astrophysical Journal*, vol. 542, no. 1, pp. L5–L8, 2000.

[26] D. P. Finkbeiner, D. J. Schlegel, C. Frank, and C. Heiles, "Tentative detection of electric dipole emission from rapidly rotating dust grains," *The Astrophysical Journal*, vol. 566, no. 2, pp. 898–904, 2002.

[27] C. L. Bennett, R. S. Hill, G. Hinshaw et al., "First-year Wilkinson Microwave Anisotropy Probe (WMAP) observations: foreground emission," *The Astrophysical Journal*, vol. 148, no. 1, pp. 97–117, 2003.

[28] D. P. Finkbeiner, M. Davis, and D. J. Schlegel, "Extrapolation of galactic dust emission at 100 microns to cosmic microwave background radiation frequencies using FIRAS," *The Astrophysical Journal*, vol. 524, no. 2, pp. 867–886, 1999.

[29] G. Langston, A. Minter, L. D'Addario, K. Eberhardt, K. Koski, and J. Zuber, "The first Galactic plane survey at 8.35 and 14.35 GHz," *The Astronomical Journal*, vol. 119, no. 6, pp. 2801–2827, 2000.

[30] D. P. Finkbeiner, G. I. Langston, and A. H. Minter, "Microwave interstellar medium emission in the green bank galactic plane survey: evidence for spinning dust," *The Astrophysical Journal*, vol. 617, no. 1, pp. 350–359, 2004.

[31] B. Gold, C. L. Bennett, R. S. Hill et al., "Five-year wilkinson microwave anisotropy probe observations: galactic foreground emission," *The Astrophysical Journal*, vol. 180, no. 2, pp. 265–282, 2009.

[32] B. Gold, N. Odegard, J. L. Weiland et al., "Seven-year wilkinson microwave anisotropy probe (WMAP∗) observations: galactic foreground emission," *The Astrophysical Journal*, vol. 192, no. 2, article 15, 2011.

[33] P. Collaboration, P. A. R. Ade, N. Aghanim et al., "Planck early results. xx. new light on anomalous microwave emission from spinning dust grains," *Astronomy and Astrophysics*, vol. 536, article A20, 2011.

[34] A. M. M. Scaife, B. Nikolic, D. A. Green et al., "Microwave observations of spinning dust emission in NGC 6946," *Monthly Notices of the Royal Astronomical Society*, vol. 406, no. 1, pp. L45–L49, 2010.

[35] C. T. Tibbs, R. Paladini, M. Compiègne et al., "A multi-wavelength investigation of RCW175: an H II region harboring spinning dust emission," *The Astrophysical Journal*, vol. 754, no. 2, p. 94, 2012.

[36] S. Casassus, C. Dickinson, K. Cleary et al., "Centimetre-wave continuum radiation from the ρ Ophiuchi molecular cloud," *Monthly Notices of the Royal Astronomical Society*, vol. 391, no. 3, pp. 1075–1090, 2008.

[37] C. T. Tibbs, N. Flagey, R. Paladini et al., "Spitzer characterization of dust in an anomalous emission region: the Perseus cloud," *Monthly Notices of the Royal Astronomical Society*, vol. 418, no. 3, pp. 1889–1900, 2011.

[38] C. Bot, N. Ysard, D. Paradis et al., "Submillimeter to centimeter excess emission from the magellanic clouds. II. on the nature of the excess," *Astronomy and Astrophysics*, vol. 523, p. 20, 2010.

[39] J. C. Weingartner and B. T. Draine, "Radiative torques on interstellar grains. III. Dynamics with thermal relaxation," *The Astrophysical Journal*, vol. 589, no. 1, pp. 289–318, 2003.

[40] E. M. Purcell, "Suprathermal rotation of interstellar grains," *The Astrophysical Journal*, vol. 231, pp. 404–416, 1979.

[41] A. Lazarian and W. G. Roberge, "Barnett relaxation in thermally rotating grains," *The Astrophysical Journal*, vol. 484, no. 1, pp. 230–237, 1997.

[42] B. T. Draine and A. Li, "Infrared emission from interstellar dust. IV. The silicate-graphite-PAH model in the post-Spitzer ERA," *The Astrophysical Journal*, vol. 657, no. 2, pp. 810–837, 2007.

[43] G. Dobler, B. Draine, and D. P. Finkbeiner, "Constraining spinning dust parameters with the wmap five-year data," *The Astrophysical Journal*, vol. 699, no. 2, pp. 1374–1388, 2009.

[44] G. Kowal, A. Lazarian, and A. Beresnyak, "Density fluctuations in MHD turbulence: spectra, intermittency, and topology," *The Astrophysical Journal*, vol. 658, no. 1, pp. 423–445, 2007.

[45] A. Esquivel and A. Lazarian, "Tsallis statistics as a tool for studying interstellar turbulence," *The Astrophysical Journal*, vol. 710, no. 1, pp. 125–132, 2010.

[46] B. Burkhart, D. Falceta-Gonçalves, G. Kowal, and A. Lazarian, "Density studies of mhd interstellar turbulence: statistical moments, correlations and bispectrum," *The Astrophysical Journal*, vol. 693, p. 250, 2009.

[47] B. Burkhart, S. Stanimirović, A. Lazarian, and G. Kowal, "Characterizing magnetohydrodynamic turbulence in the small magellanic cloud," *The Astrophysical Journal*, vol. 708, no. 2, pp. 1204–1220, 2010.

[48] B. T. Draine, "Grain evolution in dark clouds," in *Protostars and Planets II*, pp. 621–640, University of Arizona Press, Tucson, Ariz, USA, 1985.

[49] H. Yan and A. Lazarian, "Grain acceleration by magneto-hydrodynamic turbulence: gyroresonance mechanism," *The Astrophysical Journal*, vol. 592, no. 1, pp. L33–L36, 2003.

[50] T. Hoang and A. Lazarian, "Acceleration of very small dust grains due to random charge fluctuations," *The Astrophysical Journal*, vol. 761, no. 2, p. 96, 2012.

[51] T. Hoang, A. Lazarian, and R. Schlickeiser, "Revisiting acceleration of charged grains in magnetohydrodynamic turbulence," *The Astrophysical Journal*, vol. 747, no. 1, p. 54, 2012.

[52] A. V. Ivlev, A. Lazarian, V. N. Tsytovich, U. de Angelis, T. Hoang, and G. E. Morfill, "Acceleration of small astrophysical grains due to charge fluctuations," *The Astrophysical Journal*, vol. 723, no. 1, pp. 612–619, 2010.

[53] T. Hoang and A. Lazarian, "Acceleration of small dust grains due to charge fluctuations," arXiv, 1112.3409, 2011.

[54] R. Paladini, G. Umana, M. Veneziani et al., "Spitzer and Herschel multiwavelength characterization of the dust content of evolved HII regions," *The Astrophysical Journal*, vol. 760, no. 2, 2012.

[55] C. Dickinson, M. Peel, and M. Vidal, "New constraints on the polarization of anomalous microwave emission in nearby molecular clouds," *Monthly Notices of the Royal Astronomical Society*, vol. 418, no. 1, pp. L35–L39, 2011.

[56] C. H. López-Caraballo, J. A. Rubīo-Martín, R. Rebolo, and R. Génova-Santos, "Constraints on the polarization of the anomalous microwave emission in the perseus molecular complex from seven-year wmap data," *The Astrophysical Journal*, vol. 729, no. 1, article 25, 2011.

[57] N. Macellari, E. Pierpaoli, C. Dickinson, and J. E. Vaillancourt, "Galactic foreground contributions to the 5-year Wilkinson Microwave Anisotropy Probe maps," *Monthly Notices of the Royal Astronomical Society*, vol. 418, no. 2, pp. 888–905, 2011.

[58] A. Lazarian, "Tracing magnetic fields with aligned grains," *Journal of Quantitative Spectroscopy and Radiative Transfer*, vol. 106, no. 1–3, pp. 225–256, 2007.

[59] A. Z. Dolginov and I. G. Mytrophanov, "Orientation of cosmic dust grains," *Astrophysics and Space Science*, vol. 43, no. 2, pp. 291–317, 1976.

[60] B. T. Draine and J. C. Weingartner, "Radiative torques on interstellar grains. I. Superthermal spin-up," *The Astrophysical Journal*, vol. 470, no. 1, pp. 551–565, 1996.

[61] B. T. Draine and J. C. Weingartner, "Radiative torques on interstellar grains. II. Grain alignment," *The Astrophysical Journal*, vol. 480, no. 2, pp. 633–646, 1997.

[62] A. Lazarian and T. Hoang, "Radiative torques: analytical model and basic properties," *Monthly Notices of the Royal Astronomical Society*, vol. 378, no. 3, pp. 910–946, 2007.

[63] A. Lazarian and T. Hoang, "Alignment of dust with magnetic inclusions: radiative torques and superparamagnetic barnett and nuclear relaxation," *The Astrophysical Journal*, vol. 676, no. 1, pp. L25–L28, 2008.

[64] T. Hoang and A. Lazarian, "Radiative torque alignment: essential physical processes," *Monthly Notices of the Royal Astronomical Society*, vol. 388, no. 1, pp. 117–143, 2008.

[65] T. Hoang and A. Lazarian, "Grain alignment induced by radiative torques: effects of internal relaxation of energy and complex radiation field," *The Astrophysical Journal*, vol. 697, no. 2, pp. 1316–1333, 2009.

[66] T. Hoang and A. Lazarian, "Radiative torques alignment in the presence of pinwheel torques," *The Astrophysical Journal*, vol. 695, pp. 1457–1476, 2009.

[67] D. C. B. Whittet, J. H. Hough, A. Lazarian, and T. Hoang, "The efficiency of grain alignment in dense interstellar clouds: a reassessment of constraints from near-infrared polarization," *The Astrophysical Journal*, vol. 674, no. 1, pp. 304–315, 2008.

[68] L. Davis and J. L. Greenstein, "The polarization of starlight by aligned dust grains," *The Astrophysical Journal*, vol. 114, p. 206, 1951.

[69] W. G. Roberge and A. Lazarian, "Davis-Greenstein alignment of oblate spheroidal grains," *Monthly Notices of the Royal Astronomical Society*, vol. 305, pp. 615–630, 1999.

[70] A. Lazarian and B. T. Draine, "Resonance paramagnetic relaxation and alignmentof amall grains," *The Astrophysical Journal*, vol. 536, no. 1, pp. L15–L18, 2000.

[71] L. D. Landau and E. M. Lifshitz, *Electrodynamics of Continuous Media*, Addison-Wesley, Reading, Mass, USA, 1960.

[72] A. Lazarian and W. G. Roberge, "Cosmic rays and grain alignment," *Monthly Notices of the Royal Astronomical Society*, vol. 287, no. 4, pp. 941–946, 1997.

[73] A. Lazarian and B. T. Draine, "Nuclear spin relaxation within interstellar grains," *The Astrophysical Journal*, vol. 520, no. 1, pp. L67–L70, 1999.

[74] A. Lazarian and M. Efroimsky, "Inelastic dissipation in a freely rotating body: application to cosmic dust alignment," *Monthly Notices of the Royal Astronomical Society*, vol. 303, no. 4, pp. 673–684, 1999.

[75] M. J. Wolff, G. C. Clayton, S. H. Kim, P. G. Martin, and C. M. Anderson, "Ultraviolet interstellar linear polarization. III. Features," *The Astrophysical Journal*, vol. 478, no. 1, pp. 395–402, 1997.

[76] P. G. Martin, "On predicting the polarization of low frequency emission by diffuse interstellar terstellar dust," in *Sky Polarisation at Far-infrared to Radio Wavelengths: The Galactic Screen Before the Cosmic Microwave Background*, M. -A. Miville-Deschênes and F. Boulanger, Eds., vol. 23 of *EAS Publications Series*, pp. 165–188, 2007.

[77] B. T. Draine and A. Lazarian, "Magnetic dipole microwave emission from dust grains," *The Astrophysical Journal*, vol. 512, no. 2, pp. 740–754, 1999.

[78] B. T. Draine and B. Hensley, "Magnetic nanoparticles in the interstella medium: emission spectrum and polarization," arXiv, astro-ph.GA, 2012.

Permissions

The contributors of this book come from diverse backgrounds, making this book a truly international effort. This book will bring forth new frontiers with its revolutionizing research information and detailed analysis of the nascent developments around the world.

We would like to thank all the contributing authors for lending their expertise to make the book truly unique. They have played a crucial role in the development of this book. Without their invaluable contributions this book wouldn't have been possible. They have made vital efforts to compile up to date information on the varied aspects of this subject to make this book a valuable addition to the collection of many professionals and students.

This book was conceptualized with the vision of imparting up-to-date information and advanced data in this field. To ensure the same, a matchless editorial board was set up. Every individual on the board went through rigorous rounds of assessment to prove their worth. After which they invested a large part of their time researching and compiling the most relevant data for our readers. Conferences and sessions were held from time to time between the editorial board and the contributing authors to present the data in the most comprehensible form. The editorial team has worked tirelessly to provide valuable and valid information to help people across the globe.

Every chapter published in this book has been scrutinized by our experts. Their significance has been extensively debated. The topics covered herein carry significant findings which will fuel the growth of the discipline. They may even be implemented as practical applications or may be referred to as a beginning point for another development. Chapters in this book were first published by Hindawi Publishing Corporation; hereby published with permission under the Creative Commons Attribution License or equivalent.

The editorial board has been involved in producing this book since its inception. They have spent rigorous hours researching and exploring the diverse topics which have resulted in the successful publishing of this book. They have passed on their knowledge of decades through this book. To expedite this challenging task, the publisher supported the team at every step. A small team of assistant editors was also appointed to further simplify the editing procedure and attain best results for the readers.

Our editorial team has been hand-picked from every corner of the world. Their multi-ethnicity adds dynamic inputs to the discussions which result in innovative outcomes. These outcomes are then further discussed with the researchers and contributors who give their valuable feedback and opinion regarding the same. The feedback is then collaborated with the researches and they are edited in a comprehensive manner to aid the understanding of the subject.

Apart from the editorial board, the designing team has also invested a significant amount of their time in understanding the subject and creating the most relevant covers. They scrutinized every image to scout for the most suitable representation of the subject and create an appropriate cover for the book.

The publishing team has been involved in this book since its early stages. They were actively engaged in every process, be it collecting the data, connecting with the contributors or procuring relevant information. The team has been an ardent support to the editorial, designing and production team. Their endless efforts to recruit the best for this project, has resulted in the accomplishment of this book. They are a veteran in the field of academics and their pool of knowledge is as vast as their experience in printing. Their expertise and guidance has proved useful at every step. Their uncompromising quality standards have made this book an exceptional effort. Their encouragement from time to time has been an inspiration for everyone.

The publisher and the editorial board hope that this book will prove to be a valuable piece of knowledge for researchers, students, practitioners and scholars across the globe.

List of Contributors

Yu-Yen Chang
Department of Physics and Graduate Institute of Astrophysics, National Taiwan University, Taipei 10617, Taiwan
Leung Center for Cosmology and Particle Astrophysics, National Taiwan University, Taipei 10617, Taiwan
Institute of Astronomy and Astrophysics, Academia Sinica, Taipei 10617, Taiwan

Wei Hao Wang
Institute of Astronomy and Astrophysics, Academia Sinica, Taipei 10617, Taiwan

Rikon Chao
Leung Center for Cosmology and Particle Astrophysics, National Taiwan University, Taipei 10617, Taiwan
Institute of Astronomy and Astrophysics, Academia Sinica, Taipei 10617, Taiwan
Department of Electrical Engineering, National Taiwan University, Taipei 10617, Taiwan

Pisin Chen
Department of Physics and Graduate Institute of Astrophysics, National Taiwan University, Taipei 10617, Taiwan
Leung Center for Cosmology and Particle Astrophysics, National Taiwan University, Taipei 10617, Taiwan
Institute of Astronomy and Astrophysics, Academia Sinica, Taipei 10617, Taiwan
Kavli Institute for Particle Astrophysics and Cosmology, SLAC National Accelerator Laboratory, Stanford University, Stanford, CA 94305, USA

Kei Kotake
Division of Theoretical Astronomy, National Astronomical Observatory of Japan, 2-21-1, Osawa, Mitaka, Tokyo 181-8588, Japan
Center for Computational Astrophysics, National Astronomical Observatory of Japan, Mitaka, Tokyo 181-8588, Japan

Tomoya Takiwaki
Center for Computational Astrophysics, National Astronomical Observatory of Japan, Mitaka, Tokyo 181-8588, Japan

Yudai Suwa
Yukawa Institute for Theoretical Physics, Kyoto University, Oiwake-cho, Kitashirakawa, Sakyo-ku, Kyoto 606-8502, Japan

Wakana Iwakami Nakano
Department of Aerospace Engineering, Tohoku University, 6-6-01 Aramaki-Aza-Aoba, Aoba-ku, Sendai 980-8579, Japan

Shio Kawagoe
Knowledge Dissemination Unit, Oshima Lab, Institute of Industrial Science, The University of Tokyo, 4-6-1 Komaba, Meguro-ku, Tokyo 153-8505, Japan

Youhei Masada
Department of Computational Science, Graduate School of System Informatics, Kobe University, Nada, Kobe 657-8501, Japan

Shinichiro Fujimoto
Kumamoto National College of Technology, 2659-2 Suya, Goshi, Kumamoto 861-1102, Japan

J.-P. Bruneton, M. Rinaldi and S. Schlogel
Namur Center for Complex Systems (naXys), University of Namur, 5000 Namur, Belgium

A. Kanfon
Faculte des Sciences et Techniques, Universite d'Abomey-Calavi, BP 526 Cotonou, Benin

A. Hees
Namur Center for Complex Systems (naXys), University of Namur, 5000 Namur, Belgium
Royal Observatory of Belgium, Avenue Circulaire 3, 1180 Brussels, Belgium
LNE-SYRTE, Observatoire de Paris, CNRS, UPMC, avenue de l'Observatoire 61, 75014 Paris, France

A. Fuzfa
Namur Center for Complex Systems (naXys), University of Namur, 5000 Namur, Belgium
Center for Cosmology, Particle Physics and Phenomenology (CP3), University of Louvain, 1348 Louvain-la-Neuve, Belgium

Marco Tucci
LAL, Universit´e Paris-Sud and CNRS/IN2P3, 91400 Orsay, France

Luigi Toffolatti
Departamento de Fisica, Universidad de Oviedo, C. Calvo Sotelo s/n, 33007 Oviedo, Spain
IFCA, Universidad de Cantabria, Avenida los Castros s/n, 39005 Santander, Spain

J. Bobin, J.-L. Starck and F. Sureau
Laboratoire AIM, IRFU, SEDI-SAP, Service d'Astrophysique, Orme des Merisiers, Bat 709, Piece 282, 91191 Gif-Sur-Yvette, France

J. Fadili
GREYC CNRS, ENSICAEN, Universite de Caen 6, Boulevard du Marechal Juin, 14050 Caen Cedex, France

N. Cappelluti
Osservatorio Astronomico di Bologna, INAF, Via Ranzani 1, 40127 Bologna, Italy
University of Maryland, Baltimore County, 1000 Hilltop Circle, Baltimore, MD 21250, USA

V. Allevato
Max Planck Institut fur Plasmaphysik and Excellence Cluster Universe, Boltzmannstrasse 2, 85748 Garching, Germany

A. Finoguenov
University of Maryland, Baltimore County, 1000 Hilltop Circle, Baltimore, MD 21250, USA
Max Planck Institute fur Extraterrestrische Physik, Giessenbachstrasse 1, 85748 Garching, Germany

Giuseppe Lodato
Dipartimento di Fisica, Universita degli Studi di Milano, Via Celoria 16, Milano, Italy

Jorge Perez Peraza and Victor Velasco
Instituto de Geofisica, Universidad Nacional Autonoma de Mexico, C.U., 04510 Coyoacan, DF, Mexico

Igor Ya. Libin
International Academy for Appraisal and Consulting (MAOK), Moscow, Russia

K. F. Yudakhin
IZMIRAN, Academy of Sciences of Russia, Troitsk, Moscow 142092, Russia

Tim Johannsen
Physics Department, University of Arizona, 1118 E. 4th Street, Tucson, AZ 85721, USA

John J. Quenby and Diana N. A. Shaul
Blackett Laboratory, Imperial College, London SW7 2BZ, UK

Tamitha Mulligan and J. Bernard Blake
Space Sciences Department, The Aerospace Corporation, Los Angeles, CA 90009, USA

Anca Constantin
Department of Physics and Astronomy, James Madison University, Harrisonburg, VA 22807, USA

Anil C. Seth
Department of Physics and Astronomy, University of Utah, Salt Lake City, UT 84112, USA

David Garofalo
Department of Physics, Columbia University, New York, NY 10027, USA

Thiem Hoang
Canadian Institute for Theoretical Astrophysics, University of Toronto, 60 St. George Street, Toronto, ON, Canada M5S 3H8

A. Lazarian
Department of Astronomy, University of Wisconsin-Madison, Madison, WI 53705, USA

CPSIA information can be obtained
at www.ICGtesting.com
Printed in the USA
LVOW06*1443111017

552032LV00004B/49/P

9 781632 395405